Colour Vision Deficiencies XIII

Documenta Ophthalmologica Proceedings Series

VOLUME 59

The titles published in this series are listed at the end of this volume.

Colour Vision Deficiencies XIII

Proceedings of the thirteenth Symposium
of the International Research Group on
Colour Vision Deficiencies, held in
Pau, France
July 27–30, 1995

Edited by
C.R. Cavonius

with associate editors:
A.J. Adams, J. Birch, S.J. Dain, J.D. Mollon,
J.D. Moreland, J. Pokorny, V. Smith,
H.G. Sperling, E. Zrenner

Kluwer Academic Publishers
Dordrecht / Boston / London

Library of Congress Cataloging-in-Publication Data is available.

ISBN 0-7923-4224-0

Published by Kluwer Academic Publishers BV,
PO Box 17, 3300 AA Dordrecht, The Netherlands

Kluwer Academic Publishers BV incorporates
the publishing programmes of
D. Reidel, Martinus Nijhoff, Dr W. Junk and MTP Press.

Sold and distributed in the United States and Canada
by Kluwer Academic Publishers, PO Box 358,
Accord Station, Hingham, MA 02018-0358, USA

In all other countries, sold and distributed
by Kluwer Academic Publishers Group, Distribution Center,
PO Box 322, 3300 AH Dordrecht, The Netherlands

Printed on acid-free paper

Printed and bound in Great Britain by Hartnolls Ltd., Bodmin, Cornwall.

Contents

PART TWO: STRUCTURE AND FUNCTION IN COLOUR VISION

PART THREE: ACQUIRED COLOUR VISION DEFICIENCIES

PART FIVE: PLATE TESTS, ARRANGEMENT TESTS, AND AUTOMATED TESTS

PART SIX: GENERAL CONSIDERATIONS IN TESTING COLOUR VISION

THE VERRIEST LECTURE

Preface

The 13th biennial Symposium of the International Research Group on Colour Vision Deficiencies was held from 27 July to 30 July, 1995, in the splendid *fin de siécle* Theatre Saint Louis, in Pau, France. A total of 80 papers and posters were presented during the four days, of which 56 have been selected for inclusion in this volume. Each has been reviewed by two of the Editors. Additionally, in numerous cases in which specialized knowledge was called for the Editors asked for external help; their assistance is acknowledged below.

While Pau now exists largely on agriculture and, more recently, oil extraction and refining, a century ago it was a winter resort that rivalled Cannes and Nice; indeed, Baedeker judged the air of Pau to be preferable for those of a nervous disposition. Pau and the surrounding Béarn district are rich in history and have close ties to the Basque region, both of which were exploited brilliantly by the local organizers, Doctors Jean and Véronique Lied, to introduce the participants to the gastronomy and traditions of the region.

The Committee of the IRGCVD owe a great debt of gratitude to the external reviewers, Oscar Estévez, Karl Gegenfurtner, Gunilla Haegerstrom-Portnoy, R.L. Hilz, John Krauskopf, Anne Kurtenbach, Barry B. Lee, Adam Reeves, Marilyn Schneck, L.T. Sharpe, Eugene Switkes, Wayne Verdon, and B.A. Wandell, whose advice helped materially to sharpen the logic and style of these papers. The value of their contribution is shown by the fact that many authors specifically thanked the reviewers for helping to improve their chapters.

Finally, it is a pleasure to thank Dagmar Winter, whose rapid and accurate processing of several thousand pages of manuscript and revisions made possible the timely publication of these Proceedings.

C.R. Cavonius

C.R. Cavonius (ed.), Colour Vision Deficiencies XIII, p. xi.

PART ONE

Intraindividual differences: genotypes and phenotypes

1. '...aus dreyerley Arten von Membranen oder Molekülen': George Palmer's legacy

J.D. MOLLON
(Cambridge, UK)

Abstract

It is 200 years since George Palmer suggested that colour vision depended on three classes of molecule or membrane. The present paper gives an introductory review of what is today known of the opsins, of the genes that encode them, and of the alterations to which they are subject. Several theories of anomalous trichromacy are briefly reviewed.

Introduction

We do not know the day on which George Palmer died, but he signed his last will on July 17 1795 and he was buried at Hendon parish church in the Middlesex countryside on August 8 1795[1]. So we can be sure that the 13th Symposium of the IRGCVD fell within a few days of the bicentennial of his death. And it is particularly appropriate that the Research Group should remember this singular man. He was born in Westminster in 1740. By trade he was a prominent dealer in glass, especially coloured glass, and by patrimony he was a member of the Drapers' Company (Mollon, 1993). In 1777 Palmer proposed that the retina contained three types of fibre, corresponding to three physical types of light. His physiological hypothesis, published 25 years before Thomas Young's Bakerian Lecture, is a recognizable version of the modern theory of trichromacy, although his physical theory – of three types of light – is mistaken. Perhaps the most remarkable version of his theory appears in a German popular science magazine of 1781, where the opinions are attributed to a certain 'Giros von Gentilly' and it is suggested that the retina is composed of three kinds of molecule or membrane (Voigt, 1781; Fig. 1). That article contains what is almost certainly the first suggestion that colour blindness arises from an inherited defect of receptor molecules (indeed this is probably the first suggestion that any hereditary defect is due to a molecular error). Palmer

---

[1]Public Record Office, London, Prob 11/1264, ff 372RH-374LH; Burial register, Parish of St Mary's, Hendon.

C.R. Cavonius (ed.), Colour Vision Deficiencies XIII, pp. 3–20.

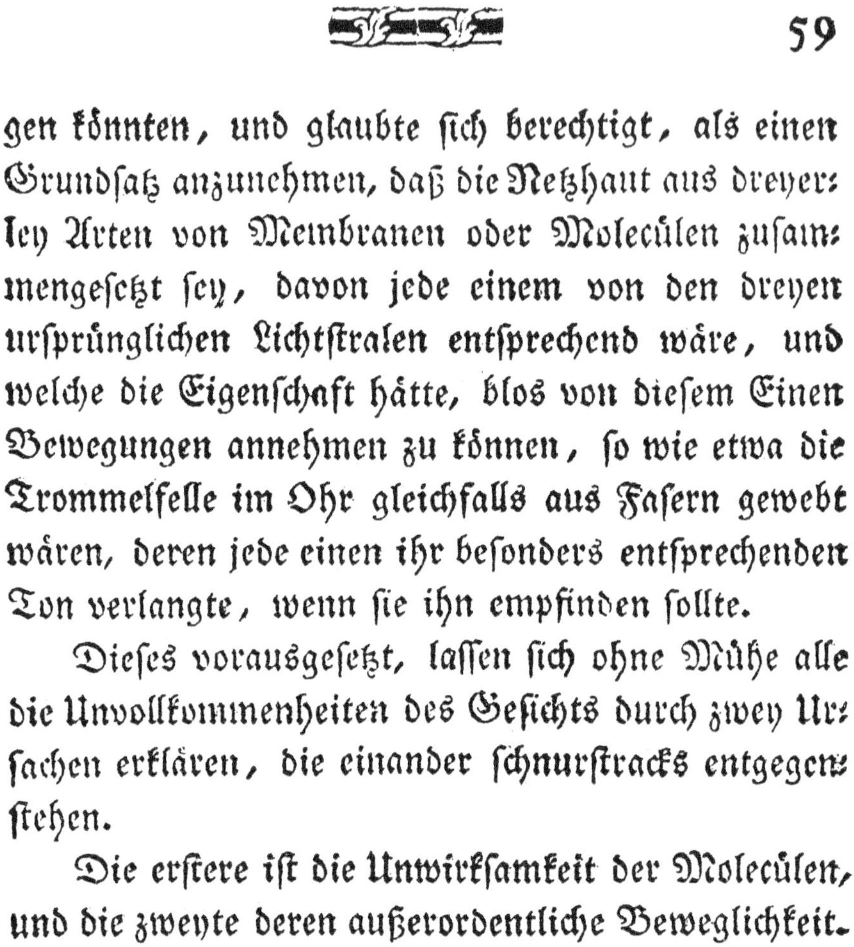

59

gen könnten, und glaubte sich berechtigt, als einen Grundsatz anzunehmen, daß die Netzhaut aus dreyerley Arten von Membranen oder Molecülen zusammengesetzt sey, davon jede einem von den dreyen ursprünglichen Lichtstralen entsprechend wäre, und welche die Eigenschaft hätte, blos von diesem Einen Bewegungen annehmen zu können, so wie etwa die Trommelfelle im Ohr gleichfalls aus Fasern gewebt wären, deren jede einen ihr besonders entsprechenden Ton verlangte, wenn sie ihn empfinden sollte.

Dieses vorausgesetzt, lassen sich ohne Mühe alle die Unvollkommenheiten des Gesichts durch zwey Ursachen erklären, die einander schnurstracks entgegenstehen.

Die erstere ist die Unwirksamkeit der Molecülen, und die zweyte deren außerordentliche Beweglichkeit.

Fig. 1. A passage from J. H. Voigt (1781) in which he describes the theory of 'Giros von Gentilly' (*alias* George Palmer, 1740–1795). This passage contains what may be the first explanation of an inherited defect in terms of molecular error.

proposed the existence of two alternative faults that can affect each of the three types of molecule: the first is that it can be inactive, the second is that it can be overactive, constitutively active to use the modern term. If only one of the three types of molecule is affected, then the defect is mild, but if two are affected then the condition is manifestly more severe. In a later monograph (Palmer, 1786), he further suggested that adaptation to saturated colours can transiently produce colour blindness in the normal eye, owing to selective fatigue of the receptors.

It was almost a century before Palmer's idea of light-sensitive molecules was systematically revived by Wilibald Kühne. The subsequent century has seen a slowly growing understanding of the photopigments, an understanding that accelerated with the analysis of the amino acid sequence of rhodopsin (Hargrave *et al.*, 1983) and the sequencing of the genes that code for both the rod and the cone pigments (Nathans and Hogness, 1984; Nathans *et al.*, 1986a, 1986b).

Palmer's suggestion that visual molecules may exhibit constitutive activity was confirmed in 1992, when it was shown that a particularly severe form of dominant retinitis pigmentosa arises from a mutation of rhodopsin that causes the molecule to signal continuously (Robinson *et al.*, 1992). The mutation (Lys-296-Glu) affects the site at which the chromophore normally binds to the protein.

Photopigments and their genes

The structure of opsins

To the left in Figure 2[2] is depicted the outer segment of a cone, with its lipid membranes multiply infolded and packed with molecules of photopigment. Photopigments consist of proteins, typically 348–364 amino acids long, bound to 11-*cis*-retinal, the aldehyde of vitamin A_1. The 11-*cis*-retinal acts as the 'chromophore': it gives the photopigment its colour, its selective spectral absorbance. The protein components of the pigments have come to be called **opsins**. Each consists of seven helices, which span the cell membrane and are linked by loops within and without the membrane (Hargrave and McDowell, 1992). The seven helices form a palisade that surrounds the 11-*cis*-retinal; the latter is bound to a lysine in the seventh helix (Fig. 2). We now know that the palisade is splayed, so that helices are closer together at the cytoplasmic surface, the side within the cell (Baldwin, 1993).

Figure 3 offers a plan view of a photopigment molecule, sketched on the basis of the analysis of Baldwin (1993). To visualize the three-dimensional arrangement of the helices, imagine taking hold of the left-most helix (1) in the cartoon of the molecule at the top, pulling it forwards from the plane of the paper, and bending it round to touch helix 7. If you then viewed the resulting structure from above (the intracellular surface), you would have the plan view shown in the lower part of the figure. The plan view indicates the approximate location of the chromophore, 11-*cis*-retinal, and the probable positions of some of the most important intramembrane amino acids that determine the difference in spectral sensitivity between long- and short-wave pigments.

---

[2]Figures 2, 3 and 4 will be found between pages 20 and 21 of the book.

The opsins are members of the superfamily of G-protein coupled receptors or **heptahelicals**. This family has more than 200 members and includes many of the receptors that mediate communication between our brain cells, and between hormones and neurons. The several forms of the serotonergic, dopaminergic, adrenergic and muscarinic acetylcholine receptor molecules, as well as the large subfamily of olfactory receptors, are all heptahelicals, distant cousins of the opsins (Watson and Arkinstall, 1994). In every case the heptahelical is embedded in the cell membrane, and initiates its signal within the cell by activating a G-protein. In the case of opsins, the G proteins are called **transducins** (Lerea *et al.*, 1989). But whereas photopigments are semi-permanently bound to 11-*cis*-retinal and separate from it after a photon has isomerized the chromophore to its all-*trans* form, other heptahelicals wait with an empty binding pocket and activate their G protein when they bind to a molecule of the right neurotransmitter or hormone or odorant.

The opsin genes

The genes that code for rhodopsin and for the short-wave pigment lie on chromosomes 3 and 7 respectively (Nathans *et al.*, 1986a; Fitzgibbon *et al.*, 1994). Several cases of congenital tritanopia have been found to be associated with point mutations of the gene that codes for the short-wave opsin (Weitz *et al.*, 1992a, 1992b). Thus tritanopia can be produced by a mutation that leads to the substitution of arginine for glycine at position 79 in the amino acid sequence of the opsin, or by mutations that give proline for serine at position 214 or serine for proline at position 264. In the case of the glycine-79-arginine substitution, all homozygotes, but only a minority of heterozygotes, appear to be affected, whereas the two other substitutions show high penetrance in the heterozygous condition (Nathans *et al.*, 1992).

As was predicted from the classical evidence for X-linkage of red–green colour deficiencies (Earle, 1845; Brunner, 1930; Kalmus, 1965), the genes for the long- (L) and middle-wave (M) cone pigments lie on the X-chromosome (Nathans *et al.*, 1986a). The following discussion will concentrate on these genes.

Figure 4 (top) illustrates how the L and M genes lie relatively close together on the q arm of the X-chromosome, although they are further apart than is suggested by traditional diagrams of this kind: the individual genes are separated by some 24 kilobases of non-coding DNA. Typically the X-chromosome carries more than one M gene, although the total number of L and M genes is controversial; Neitz *et al.* (1995) report subjects with as many as nine opsin genes. The complete cluster of L and M genes on the X-chromosome is often referred to as the **opsin gene array**.

It is thought that only one gene in the opsin array is expressed in any given cell. Lying 4 kilobases upstream of the first gene in the array is a locus control region (LCR*)*, which is thought to control which of the genes in the opsin array is expressed in a given cone cell. A man who lacks this LCR will be an S cone monochromat, that is to say, he will express none of the X-chromosome opsin genes (Nathans *et al.*, 1989; 1993).

How does the LCR exercise its choice in a given cone? Within the nucleus of a cell, the DNA of a given chromosome is not laid out linearly, and in the present case it is thought that the locus control region bends back to couple with one of the promoter regions that lie just upstream of each gene; this coupling makes the favoured gene available for expression. The idea has been advanced that the closer a gene is physically to the LCR the greater its chances of being expressed (Winderickx *et al.*, 1992a). This concept is important in the interpretation of colour deficiency.

Introns and exons

The structure of individual L and M genes is depicted in the lower part of Figure 4. The exons of the opsin genes, i.e. the regions that actually code for the amino acid sequence of the protein, are interrupted by 5 introns – long, non-coding stretches of DNA. In this feature the opsin genes differ from those for many of the heptahelicals, although the dopamine receptor genes are similarly interrupted by introns.

The intronic structure of the opsin genes must be remarkably stable, for introns 2–5 of the L and M genes are homologous in their positions to the introns of the rhodopsin gene, even though the rod and cone genes diverged long before the emergence of the mammals. Introns 2–5 are of equal length in the L and M genes, but in more than 99% of Caucasians and 98% of Japanese, intron 1 of the L gene is 1.9 kilobases longer than that of the M gene. This latter difference has proved useful to molecular biologists in the task of separating two such similar genes, but a curious finding is that 35% of Afro-Americans exhibit L and M genes that are identical in the length of intron 1 (Lund Jørgensen *et al.*, 1990).

The determination of spectral sensitivity

The M and L opsins are extremely similar, differing only by 15 amino acids (Nathans *et al.*, 1986a). This observation has been taken to suggest that the two pigments diverged relatively recently, in evolutionary terms – perhaps 30 million years ago. But are substitutions required at all 15 sites in order to change the spectral sensitivity from middle-wave to long-wave? Neitz *et al.* (1989) showed that much (at the time they suggested *all*) of the spectral difference depended on nucleotide differences in exons 4 and 5. Their inference was based on a protanope whose electroretinographic spectral sensitivity was similar to that of the normal M pigment and who appeared to have only a single X-chromosome opsin gene. The latter was a 'fusion gene': exons 2 and 3 resembled the L sequence of Nathans and colleagues, whereas exons 4 and 5 resembled the M sequence. Since exons 1 and 6 do not differ between the normal L and M genes, the implication was that exons 4 and 5 primarily controlled the spectral sensitivity of the resulting opsin.

Further detail has come from two sources. In many species of New World monkey the males carry only a single X-chromosome opsin gene but the gene is polymorphic, so that different conspecifics have different photopigments with peak sensitivities in the range 535–565 nm (Mollon *et al.*, 1984; Jacobs, 1993). By correlating genotypes and phenotypes, it was possible to identify the amino acids that are likely to account for most of the variation in spectral sensitivity (Williams *et al.*, 1992). Secondly, the pigments encoded by alternative genes have been measured *in vitro*. Merbs and Nathans (1992a, 1992b, 1993) constructed hybrid genes from cDNA clones for the normal M and L pigments, while Oprian and colleagues (Oprian *et al.*, 1991; Asenjo *et al.*, 1994) chemically synthesized artificial genes. In each case the genes were expressed in cultured mammalian kidney cells. The gene products – the artifical opsins – were then combined with 11-*cis*-retinal, the cell membranes were solubilized in detergent, and the spectral sensitivities of the resulting pigments were estimated from the difference in the absorption at each wavelength before and after bleaching.

The accumulated evidence suggests that the spectral difference between L and M pigments depends mainly on the seven amino acids indicated in Figure 2. In terms of size of spectral shift, the two most important sites are amino acids 277 and 285, which are determined by nucleotide substitutions in exon 5 of the gene. Particular interest has been excited by site 180, which is coded by exon 3 and is polymorphic in both L and M pigments. The substitution of serine for alanine at this site shifts the peak sensitivity (λ_{max}) by a few nanometres to longer wavelengths (Merbs and Nathans, 1992a); and Winderickx *et al.* (1992b) have shown that subjects whose L pigment carries the serine variant tend to require less red in a Rayleigh match than individuals who have alanine at site 180, although the two groups overlap and both groups lie within the normal range. The difference of a single nucleotide in our genomes may therefore mean that different individuals live in subtly different perceptual worlds.

In addition to sites 180, 277 and 285, the following amino acid sites have been reported to influence spectral sensitivity: 65 and 116 (encoded by exon 2), 230 and 233 (encoded by exon 4) and 309 (encoded by exon 5). Only Merbs and Nathans (1993) found a positive effect for site 65 and only Asenjo *et al.* (1994) found a positive effect for site 116; where a positive shift was found for these sites, it was no more than 1 nm.

Other than site 116, which is in the first extracellular loop of the molecule, all the critical sites lie in the transmembrane helices and the spectral shift requires the substitution of an hydroxyl-bearing amino acid for a non-polar amino acid. Individual substitutions are not additive in their effects on spectral sensitivity: the size of the shift may depend on the identity of the amino acids at other critical sites (Merbs and Nathans, 1993; Asenjo *et al.*, 1994). Owing to the number of critical sites and interactions betwen their effects, there appears to be an almost continuous distribution of possible photopigments in the range 530 to 565 nm – a fact that seems salient to the understanding of anomalous trichromacy.

What of the six or seven sites that differ between L and M cones but which do not affect spectral sensitivity (or affect it only minimally)? It is possible that substitutions which do affect spectral sensitivity require other structural adjustments if the molecule is to retain its optimal three-dimensional configuration (Williams *et al.*, 1992). Consider, for example, a gene with exons 5 and 6 drawn from the normal M gene and the remaining exons from the L gene. The opsin encoded by the gene might have a λ_{max} close to that of the normal M pigment but might be compromised in its function: it might transport less readily to the outer segment; it might be unstable in the membrane; it might be reduced in its quantum efficiency; or it might be impaired in its signalling within the cell. In their *in vitro* expression studies, Nathans and colleagues and Oprian and colleagues noted in passing that some of the expressed pigments were unstable or of reduced optical density, although it is difficult to judge whether these observations, in a very different system, are physiologically significant. It is certainly worth considering the possibility that there are forms of daltonism (or of cone dystrophy) that arise from opsins of impaired viability even though the corresponding gene exhibits no gross mutation and appears to code for a functional pigment.

Are the L and M cones labelled?

The midget bipolars and midget ganglion cells of the retina are thought to be the substrate for the phylogenetically recent subsystem of colour vision which compares the quantum catches of the long- and middle-wave cones (Mollon, 1989). In the foveal region, the centre input of a midget ganglion cell is necessarily cone specific, since this input is drawn from a single cone via a single midget bipolar; whether the surround input is cone specific (as suggested by Reid and Shapley, 1992) or is drawn promiscuously from both L and M cones (as suggested by Lennie *et al.*, 1991) is still debated.

The L and M cones of the catarrhine foveola appear to be randomly arranged (Mollon and Bowmaker, 1992). So if the surround input of the midget ganglion cell is specific to either L or M cones, there would seem to be, in principle, only two ways in which specificity could be achieved: either the L and M cones must carry a label that links them during development to their rightful bipolars, or the juvenile retina must undergo a learning process, perhaps reinforcing synapses that carry correlated inputs, and attenuating others, in a Hebbian way.

The same problem is faced, to a greater degree, by olfactory receptor neurons, which owe their stimulus selectivity to many different receptor molecules, all of which are heptahelicals and thus cousins of the opsins. How do such neurons make specific connections to the olfactory glomeruli? Recent evidence suggests that the heptahelical receptor molecules are present not only at the epithelial surface but also in the axons of the olfactory receptor neurones. Singer *et al.* (1995) have identified amino acids in the second extracellular loop of these heptahelicals that are correlated with (and perhaps act as labels for) amino acids thought to determine the specificity to a particular odorant; the latter amino acids lie in the transmembrane regions, within the binding pocket for the odorant. Might opsins carry an analogous label? Rhodopsin and the short-wave opsin do differ from each other and from the L and M opsins in the latter part of the second extracellular loop (Nathans *et al.*, 1986a), i.e. in the loop that carries the putative label of the olfactory receptors. However, the L and M opsins do not differ from each other in this loop. In their case, the only possible candidates for a label are site 116, in the first extracellular loop, and site 298, in the third. Of these sites, the latter is the more plausible candidate: the codon for site 116 does not differ for all L and M clones (Nathans *et al.*, 1986a) and it is separated by three long introns from the exon-5 codons that primarily determine spectral sensitivity, whereas the codon for site 298 differs for all clones and, being within exon 5, is very tightly linked to the critical codons. Chimpanzees, like man, differ at site 298, but arguing against this site being a label is the fact that gorillas and Old World monkeys have alanine at 298 in both sequences (Dulai *et al.*, 1994).

Ways in which the opsin genes may be compromised

Major deletions

One family is known where colour blindness is associated with a major deletion in an opsin gene (Reichel *et al.*, 1989). The 15-year-old propositus had near-normal acuity and a protan colour deficiency, while older males in the family showed a macular degeneration, with 20/200 vision and a reduced cone electroretinogram. Molecular analysis suggested a 6.5 kilobase deletion between exons 1 and 4 of the L opsin gene.

Point mutations

Some instances of X-linked colour deficiency are likely to arise from point mutations of the L or M opsin genes. One such mutation is the substitution of arginine for cysteine at position 203 in the amino acid sequence of the M gene. Site 203, encoded by exon 4, is in the second extracellular loop of the opsin molecule and the normal cysteine at the site is thought to form a disulphide bond with another cysteine in the first extracellular loop at position 126. Since these two cysteines are highly conserved in evolution (in opsins and in several other heptahelicals) and are thought to be critical for the three-dimensional structure of the opsin, the mutation presumably produces an ineffective molecule. Winderickx *et al.* (1992c) found the cys-203-arg substitution to be associated with extreme deuteranomaly in a man whose opsin array appeared to consist of one L gene and several M genes, the latter all exhibiting the same mutation. The reason why the subject was not a dichromat is a more general problem: some residual discrimination on the anomaloscope has been reported for several subjects with only a single gene in their opsin array. Winderickx and his colleagues and Nathans *et al.* (1993) found the cys-203-arg mutation occasionally associated with normal colour vision or with simple deuteranomaly, and in these cases the assumption is that the mutated gene lies in the downstream graveyard of the opsin array, distant from the LCR, and therefore not expressed. If the opsin array consists of a single gene exhibiting the cys-203-arg mutation, the result is S-cone monochromatism (Nathans *et al.*, 1993).

It is unlikely that the cys-203-arg substitution is the only opsin gene mutation circulating in man: there must be others that similarly produce a non-viable molecule and still others that impair the quantum efficiency or the stability of the molecule. Despite the large number of daltonians that have been examined by molecular biological techniques, in only a minority of cases have all exons of all their X-linked opsin genes been sequenced, or at least screened for point mutations. Moreover, mutations in the non-coding region immediately upstream of each gene could sometimes lead to failures of expression. If the gross deletion in the family of Reichel *et al.* (1989) produces a slow cone dystrophy, then there may be other daltonians who exhibit a very mild cone dystrophy – as postulated by Regan *et al.* (1994).

For mutations that produce achromatopsia, we should consider molecules that are common to the different cone types but are distinct from their rod counterparts. Thus a candidate gene for rod monochromatism is that for the alpha unit of cone transducin. This gene has been localized to chromosome 1 (Magovcevic *et al.*, 1995).

Unequal crossing-over

A major source of mischief is thought to be unequal crossing-over in the opsin gene array (Nathans *et al.*, 1986b). Corresponding chromosomes align themselves at meiosis, and it is at this stage that crossing-over occurs, stretches of DNA being exchanged between chromosomes. Owing to the juxtaposition and the homology of the L and M genes, they are quite likely to align themselves improperly at meiosis. If now crossing-over occurs and the breakpoint is between genes, one chromosome may lose a gene and the second may gain it. The simplest explanation of dichromacy is that the opsin array has been reduced to one gene by unequal crossing-over, and many examples do fit this hypothesis. The most celebrated of dichromats, John Dalton, appears to have had only a single L gene (Hunt *et al.*, 1995; Mollon *et al.*, 1996). If the breakpoint occurs within the mispaired genes, then hybrid genes will result. Since introns are longer than exons, the chances are that the breakpoint will be within an intron, and so each of the resulting genes will draw some exons from the L sequence and some from the M sequence. Many subjects, both normal and colour-deficient, appear to carry such genes. Hybrid genes, if expressed, will typically produce an opsin that has a λ_{max} intermediate between those of the normal L and M pigments. Some hybrid opsins may be compromised in their viability.

Although the L and M genes typically differ from each other in the size of intron 1, an interesting and provocative observation is that introns 2 and 4 are almost identical for the L and M genes and less divergent than the coding sequences in between (Shyue *et al.*, 1994): intron 2, which is 1987 base pairs long, differs only at six nucleotides (at least for clones derived from Jeremy Nathans), whereas intron 4, which is 1552 base pairs long, is identical for two of JN's clones (clone JN44 has a gap at position 32). This similarity of introns is curious, since these non-coding sequences might be thought to be free to diverge in the course of evolution – and indeed rapid divergence would have the positive advantage of minimizing the likelihood of unequal crossing-over (Mollon and Jordan, 1988). Either the homology of the introns has a function or it is an indication of how frequently unequal crossing-over occurs.

Theories of anomalous trichromacy

An agreed understanding of anomalous trichromacy has not come as readily as everyone expected in 1986. Any theory must explain how the colour matches of anomals come to be altered, but also faces the long-recognized problem of phenotypic variation: some anomalous trichromats are almost as impaired in their hue discrimination as are dichromats, whereas many are fair, and some have exquisite discrimination, achieving 'superior normal' scores on the Farnsworth-Munsell 100-hue test (Köllner, 1915; Nelson, 1938; McKeon and Wright, 1940; Pokorny *et al.*, 1979; Regan *et al.*, 1994). Moreover, on the Nagel anomaloscope, anomalous trichromats show little correlation between their mid-match points and their matching range (Willis and Farnsworth, 1952; Jameson *et al.*, 1982).

The following list of theories may be useful. It is intended only as a summary and for more detailed discussion the reader is referred to the papers cited below, to Pokorny *et al.* (1979), and to the many contributions to the present volume and its predecessors.

Partial reduction hypothesis

Most current theories of anomalous trichromacy suppose that at least one of the retinal photopigments has been altered in its spectral sensitivity, and this was certainly the assumption of König and Dieterici (1892). But a long-persisting view held that anomalous trichromacy was a partial reduction of normal colour vision: either the normal L or the normal M cones were present in reduced numbers or had reduced sensitivity. That this will not explain simple anomalous trichromacy has often been pointed out (von Kries, 1899; de Vries 1948a; Jameson and Hurvich, 1956; Rushton, 1972) Suppose there were only one L cone left on each side of a foveal matching field: to make a Rayleigh match, the subject must still choose the red–green ratio that produces the same quantum catch in one L cone as the yellow light produces in the single L cone on the other side.

However, there is one phenotype to which the partial reduction hypothesis might well apply, and that is extreme anomaly. By definition, the extreme anomal has a matching range that includes the normal match as well as either the protanomalous or the deuteranomalous match (Pokorny *et al.*, 1979). From the results of Rayleigh matches alone we cannot, therefore, exclude the interesting possibility that some extreme anomals represent partial reductions of normal vision and that their residual colour discrimination depends upon a population of either L or M cones that is reduced in numbers or is in some other way handicapped. If such phenotypes were found, it would be inappropriate to continue to class them with anomalous trichromats. Extreme anomals are a neglected phenotype and it would be valuable to have more information on their colour matching functions for both large and small fields.

The WDW heresy

Wright (1946, pp 353–4) suggested that the 'red' and 'green' cones might be normal in deuteranomaly, but that the 'green' channel might draw its input from both red and green cones; the spectral sensitivity of the channel is thus altered, being shifted to long wavelengths. As an explanation of simple deuteranomaly, this hypothesis faces the same problem as the partial reduction hypothesis (de Vries, 1948a). When an observer makes a colour match in a bipartite field, we believe that he sets the triplet of quantum catches in the three types of cone on the right to be the same as the triplet on the left. The two sides of the field must then match for him whether or not the subsequent neural wiring is normal or abnormal. To accept Wright's position we should have to abandon the precious dogma that colour matches are determined only by the spectral sensitivities of the cones.

Most (but not all) of the remaining hypotheses below are versions of the 'single-pigment shift model' of anomalous trichromacy (Pokorny and Smith, 1977): such theories suppose that the colour vision of the anomalous trichromat depends on two cone types with normal sensitivity and a third cone type with altered spectral sensitivity.

The Schouten hypothesis

De Vries (1948b) attributes to Schouten the idea that there is a single anomalous pigment, intermediate in spectral position between the normal L and M pigments and common to both protanomaly and deuteranomaly. In support of this idea, de Vries cited a dichromat who accepted the matches of both protanomals and deuteranomals. The implication is that the three phenotypes share two pigments, the short-wave pigment and an 'anomalous pigment', and differ only in whether they retain one or other (or neither) of the normal L and M pigments. In other words, de Vries' dichromat is a reduction form of both protanomaly and deuteranomaly. For further discussion of the Schouten hypothesis, see MacLeod and Hayhoe (1974) and Pokorny *et al.* (1975).

The protanolabe/deutanolabe hypothesis

Rushton and his collaborators favoured the view that there were different anomalous pigments in protanomaly and deuteranomaly. They even gave them names: **protanolabe** and **deutanolabe** (Rushton *et al.*, 1973). This position is taken by DeMarco *et al.* (1992), but it does not marry well with contemporary genetics. It would have made sense in the days when there were held to be separate 'protan' and 'deutan' loci: within that framework it is possible to imagine that only certain alterations are possible to the gene at each locus (Pokorny *et al.*, 1979). If, however, we allow that the X-chromosome opsin genes lie in an undifferentiated array and that all that distinguishes them is the spectral sensitivity of the photopigment for which they code, then there is no reason to expect that in protanomaly a normal M gene is always expressed in conjuction with a certain hybrid gene and in deuteranomaly a normal L gene is invariably expressed in conjunction with a different hybrid.

The continuing influence of the two-loci hypothesis has led to an unfortunate contradiction in terminology. DeMarco *et al.* (1992) denote protanolabe by L′ and deutanolabe by M′, reserving L and M for the normal long-wave and middle-wave pigments. Yet their L′ is closer in λ_{max} to the normal middle-wave pigment than to the normal long-wave pigment, while their M′ is closer to the normal long-wave pigment than to the normal middle-wave pigment. Other recent authors have used L and L′ to denote alternative long-wave pigments (e.g. two opsins that differ only at site 180) and M and M′ to denote alternative middle-wave pigments. Although the opsin array may yet have secrets, and the two-loci hypothesis could arise in a new form, current evidence argues against the usage of DeMarco and colleagues.

The Alpern hypothesis

The late Mathew Alpern and his collaborators proposed that there were several normal variants of erythrolabe, the long-wave pigment, and several normal forms of chlorolabe, the middle-wave pigment (Alpern and Moeller, 1977). In short, there was a 'cluster' of erythrolabes with different peak sensitivities and similarly a 'cluster' of chlorolabes. This proposal led to the interesting hypothesis that there are no distinct 'anomalous' pigments. Rather, the protanomal draws two pigments from the chlorolabe cluster, and the further apart their spectral positions, the better we might expect to be his chromatic discrimination to be, *ceteris paribus*. Similarly, the deuteranope draws two pigments from the erythrolabe cluster. Some early evidence for more than one kind of long-wave pigment came from Dartnall *et al.* (1983) who, by means of microspectrophotometry, measured cones from two colour-normal patients with different Π_5 sensitivity; we know now, of course, that the site 180 polymorphism does generate alternative forms of the long-wave pigment.

The colour matches of carriers of anomaly might be seen as supporting Alpern's view that there are no truly anomalous pigments: the mothers of simple anomalous boys seldom make Rayleigh matches that lie outside the normal range (Jordan and Mollon, 1993). Owing to the phenomenon of X-chromosome inactivation (Lyon, 1972), a subset of cones in the heterozygous retina should express the abnormal X-chromosome which is inherited by an average of half of their sons. If this X-chromosome carries a gene for a true anomalous pigment (i.e. one that is abnormal in its spectral sensitivity) and if this gene is expressed in the anomalous son, then we might expect it also to be expressed in some cones of the mother's retina, and thus to affect her colour matches.

The reduced Alpern hypothesis

A minimalist version of Alpern's hypothesis was advanced by Neitz and Neitz (1992). They suggested that not only are there no anomalous cones but there are only two forms of the normal L cone and two forms of the normal M cones. Protanomaly depends on a comparison of the two possible types of L cone and deuteranomaly on a comparison of the two types of M cone. This hypothesis derived from the idea that only three sites control the spectral sensitivity of the L and M pigments, sites 180, 277 and 285 in the amino acid sequence (Neitz *et al.*, 1991). An implicit assumption is that the latter two sites are so close together that they never become unyoked by unequal crossing-over: they are always tyrosine and threonine (long-wave) or phenylalanine and alanine (middle-wave). Thus, the only source of variation is site 180, which is polymorphic for both L and M genes. A protanomalous observer is one who lacks an effective L gene (through complete deletion or local mutation) but is fortunate enough to have and to express the genes for the two alternative M photopigments. A deuteranomal lacks an M gene, but is fortunate enough to have and to express the genes for the two different L pigments.

Theories that invoke variations in optical density

A photopigment will be changed in its spectral sensitivity if its optical density is reduced from the relatively high value thought to obtain in the normal cone outer segment (Knowles and Dartnall, 1977). When the pigment is present at high density, there is self-screening at wavelengths close to the λ_{max}: as light passes along the cylinder of the outer segment, the available light reaching a given molecule of pigment will be disproportionately attenuated at wavelengths near the λ_{max} but will be less attenuated at the skirts of the absorbance spectrum. Consequently the shape of the pigment's sensitivity curve is broader at high optical densities. We have considered above the possibility that an altered genetic sequence might produce an opsin of virtually unaltered λ_{max} but altered optical density.

Baker (1966) explored the hypothesis that protanomalous vision depends on a normal M pigment in normal density and a normal L pigment in reduced concentration. His experimental test was to measure Rayleigh matches in normals after a far-red bleach: in fact, his subjects did not generate protanomalous matches even during the earliest stages of recovery, when their erythrolable (L-pigment) should have been dilute. Ruddock and Naghshineh (1974) examined the possibility that normal L and M pigments are mixed in one class of cone of the anomalous retina and that optical density is high enough for self-screening, and screening of one pigment by the other, to modify the spectral sensitivity of the components. He and Shevell (1995; see also this volume) showed by computation that variations in optical density can substantially affect the Rayleigh match midpoint and range if the two underlying pigments of anomalous vision are only a few nanometers apart in nominal λ_{max}. These effects arise chiefly from the change in slope of the long-wave limb when self-screening is changed. He and Shevell pointed out that some chromatic discrimination could be achieved if two pigments with the same nominal λ_{max} but different optical densities were present.

An account based on current data

The work of Merbs and Nathans (1992b; 1993) and of Asenjo *et al.* (1994) shows that there is potentially an almost continuous range of photopigments with λ_{max} between, say, 532 nm and 563 nm. Thus in the series of pigments measured *in vitro* by Asenjo and colleagues the largest gap is 4 nm (between 534 and 538 nm). It is true that experimental error might reduce the size of any gap in the series and it is also true that some opsins may not be viable *in vivo.* On the other hand, there are many combinations of substitutions that have not been tried *in vitro* and which might prove to fill in even the small gaps remaining.

The genomic evidence suggests a single array of opsin genes with a single locus control region: there are not, or so it seems, separate L and M loci in the classical sense. In both normal and colour-deficient subjects a variety of genes may be present in the array, some genes having sequences close to the recognized M and L sequences, some having 'hybrid' sequences. Only a subset (perhaps only two) of the genes are thought to be expressed. If we take the *in vitro* and the genomic evidence at its face value, it ought to be possible to find phenotypes with virtually any combination of pigments in the range 530 to 565 nm. Rayleigh match midpoints would depend on the spectral positions of the expressed pigments and chromatic discrimination upon their **spectral proximity**, the distance between their absorbance spectra. Some potential pigments are likely to be reduced in optical density; this would additionally affect matches and discrimination. However, the fact remains that the pool of X-chromosome opsin genes in the human population is dominated by genes that code for λ_{max} values close to 530 and 560 nm. Moreover, the two most potent amino acid sites, 277 and 285, are tightly linked and there has been no report of their being unyoked in a natural human gene (although viable pigments of this kind do occur in platyrrhine monkeys). Although a continuous range of pigments is available in principle there are, therefore, likely to be strong biases in the frequency distribution within this range: at one end there is likely to be a cluster of L and near-L pigments, at the other a cluster of M and near-M pigments. The typical deuteranomal, we may suppose, draws his two expressed pigments from the first cluster, the typical protanomal expresses two pigments from the second. In both cases, hue discrimination depends primarily upon the spectral proximity of the pigments expressed and secondarily on factors such as optical density. Pigments falling midway between M and L probably exist (they could arise, for example, from the unyoking of sites 277 and 285) but they are likely to be confined to rare phenotypes, phenotypes that would correspond to anomals of the Schouten type. Whether an account of this kind can accommodate the discontinuous distribution of anomals and normals at the anomaloscope (Schmidt, 1953) remains to be seen.

Acknowledgements

I thank C.R. Cavonius, D. Hunt and B.C. Regan for comments on the text.

References

Alpern, M. and Moeller, J. (1977). The red and green cone visual pigments of deuteranomalous trichromacy. J. Physiol. (Lond.), 266: 647–675.

Asenjo, A.B., Rim, J. and Oprian, D.D. (1994). Molecular determinants of human red/green color discrimination. Neuron, 12: 1131–1138.

Baker, H.D. (1966). Single-variable anomaloscope matches during recovery from artificial red blindness. J. Opt. Soc. Am., 56: 686–689.

Baldwin, J.M. (1993). The probable arrangement of the helices in G protein-coupled receptors. EMBO J., 12: 1693–1703.

Brunner, W. (1930). Über den Verebungsmodus der verschiedenen Typen der angeborenen Rotgrünblindheit. Graefes Arch. Ophthalmol., 124: 1–52.

Dartnall, H.J.A., Bowmaker, J.K. and Mollon, J.D. (1983). Human visual pigments: microspectrophotometric results from the eyes of seven persons. Proc. Roy. Soc. Lond., 220: 115–130.

de Vries, Hl. (1948a). The luminosity curve of the eye as determined by measurements with the flickerphotometer. Physica, 14: 319–348.

de Vries, Hl. (1948b). The fundamental response curves of normal and abnormal dichromatic and trichromatic eyes. Physica, 14: 367–380.

DeMarco, P., Pokorny, J. and Smith, V.C. (1992). Full-spectrum cone sensitivity functions for X-chromosome-linked anomalous trichromats. J. Opt. Soc. Am., 9: 1465–1476.

Dulai, K.S., Bowmaker, J.K., Mollon, J.D. and Hunt, D.M. (1994). Sequence divergence, polymorphism and evolution of the middle-wave and long-wave visual pigment genes of great apes and Old World monkeys. Vision Res. 34: 2483–2491.

Earle, P. (1845) On the inability to distinguish colors. Am. J. Med. Sci, n.s. 9: 346–354.

Fitzgibbon, J., Appukuttan, B., Gayther, S., Wells, D., Delhanty, J. and Hunt, D.M. (1994). Localisation of the human blue cone pigment gene to chromosome band 7q31.3-32. Hum. Genet., 93: 79–80.

Hargrave, P.A. and McDowell, J.H. (1992). Rhodopsin and phototransduction. Int. Rev. Cytol., 137: 49–97.

Hargrave, P.A., McDowell, J.H., Curtis, D.R. *et al.* (1983). The structure of bovine rhodopsin. Biophys. Struct. Mech. 9: 235–244.

He, J.C. and Shevell, S.K. (1995). Variation in color matching and discrimination among deuteranomalous trichromats: Theoretical implications of small differences in photopigments. Vision Res. 35: 2579–2588.

Hunt, D.M., Dulai, K.S., Bowmaker, J.K. and Mollon, J.D. (1995). The chemistry of John Dalton's color blindness. Science, 267: 984–988.

Jacobs, G.H. (1993). Mammalian colour vision. Biol. Rev., 67: 413–471.

Jameson, D. and Hurvich, L.M. (1956). Theoretical analysis of anomalous trichromatic color vision. J. Opt. Soc. Am., 46: 1075–1089.

Jameson, D., Hurvich, L.M. and Varner, D. (1982). Discrimination mechanisms in color deficient systems. In: Verriest, G. (ed), Colour Vision Deficiences VI: 295–301. W. Junk, The Hague.

Jordan, G. and Mollon, J.D. (1993). A study of women heterozygous for colour deficiencies. Vision Res. 33: 1495–1508.

Kalmus, H. (1965). Diagnosis and Genetics of Defective Colour Vision. Pergamon, Oxford.

Knowles, A. and Dartnall, H.J.A. (1977). The photobiology of vision. In: Davson, H. (ed.), The Eye, vol 2B: 1-689. Academic Press, New York.

Köllner, H. (1915). Über Übergänge zwischen normalen Farbensinn und angeborene Rotgrünblindheit und über die Möglichkeit ihrer quantitativen Bestimmung. Arch. Augenheilk., 78: 302–335.

König, A. and Dieterici, C. (1892). Die Grundempfindungen in normalen und anomalen Farbensystemen und ihre Intensitätsverteilung im Spektrum. Z. Psychol., 4: 241–347.

Kries, J. von (1899). Ueber die anomalen trichromatischen Farbensysteme. Z. Psychol., 19: 63–69.

Lennie, P., Haake, P.W. and Williams, D.R. (1991). The design of chromatically opponent receptive fields. In: Landy, M.S. and Movshon, J.A. (eds.), Computational Models of Visual Processing: 71–82. MIT Press, Cambridge, MA.

Lerea, C.L., Bunt-Milam, A.H. and Hurley, J.B. (1989). A transducin is present in blue-, green-, and red-sensitive cones photoreceptors in the human retina. Neuron, 3: 367–376.

Lund Jørgensen, A., Deeb, S.S. and Motulsky, A.G. (1990). Molecular genetics of X chromosome-linked color vision among populations of African and Japanese ancestry: high frequency of a shortened red pigment gene among Afro-Americans. Proc. Natl. Acad. Sci. USA, 87: 6512–6516.

Lyon, M.F. (1972). X-chromosome inactivation and developmental patterns in mammals. Biol. Rev., 47: 1–35.

McKeon, K.M. and Wright, W.D. (1940). The characteristics of protanomalous vision. Proc. Phys. Soc., 52: 464–479

MacLeod, D.I.A. and Hayhoe, M.M. (1974). Three pigments in normal and anomalous color vision. J. Opt. Soc. Am., 64: 92–96

Magovcevic, I., Weremowicz, S., Morton, C.C., Fong, S.L., Berson, E.L. and Dryja, T.P. (1995). Mapping of the human cone transducin alpha-subunit (GNAT2) gene to 1p13 and negative mutation analysis in patients with Stargardt disease. Genomics, 25: 288–290.

Merbs, S.L. and Nathans, J. (1992a). Absorption spectra of the human cone pigments. Nature, 356: 433.

Merbs, S.L. and Nathans, J. (1992b). Absorption spectra of the hybrid pigments responsible for anomalous color vision. Science, 258: 464–466.

Merbs, S.L. and Nathans, J. (1993). Role of hydroxyl-bearing amino acids in differentially tuning the absorption spectra of the human red and green cone pigments. Photochem. Photobiol., 58: 706–710.

Mollon, J.D. (1989). "Tho' she kneel'd in that Place where they grew...". J. Exp. Biol., 146: 21–38.

Mollon, J. D. (1993). George Palmer. In: Nicholls, C. S. (ed), The Dictionary of National Biography: 509–510. Oxford University Press, Oxford.

Mollon, J.D. and Bowmaker, J.K. (1992). The spatial arrangement of cones in the primate fovea. Nature, 360: 677–679.

Mollon, J.D. and Jordan, G. (1988). Eine evolutionäre Interpretation des menschlichen Farbensehens. Die Farbe, 35/36: 139–170.

Mollon, J.D., Bowmaker, J.K. and Jacobs, G.H. (1984). Variations of colour vision in a New World primate can be explained by polymorphism of retinal photopigments. Proc. Roy. Soc., Lond., B, 222: 373–399.

Mollon, J.D., Dulai, K. and Hunt, D.M. (1996). Dalton's colour blindness: an essay in molecular biography. In: Dickinson, C., Murray, I. and Carden, D. (eds) John Dalton's Colour Vision Legacy. Taylor and Francis, London, in press.

Nathans, J. and Hogness, D.S. (1984). Isolation and nucleotide sequence of the gene encoding human rhodopsin. Proc. Natl. Acad. Sci. USA, 81: 4851–4855.

Nathans, J., Thomas, D. and Hogness, D.S. (1986a). Molecular genetics of human color vision: The genes encoding blue, green and red pigments. Science, 232: 193–202.

Nathans, J., Piantanida, T.P., Eddy, R.L., Shows, T.B. and Hogness, D.S. (1986b). Molecular genetics of inherited variation in human color vision. Science, 232: 203–210.

Nathans, J., Davenport, I.H., Maumenee, R.A. *et al.* (1989). Molecular genetics of human blue cone monochromacy. Science, 245: 831–838.

Nathans, J., Sung, C.-H., Weitz, C.J., Davenport, C.M., Merbs, S.L. and Wang, Y. (1992). Visual pigments and inherited variation in human vision. In: Corey, D.P. and Roper, S.D. (eds.), Sensory Transduction: 110–131. Rockefeller University Press, New York.

Nathans, J., Maumenee, I.H., Zrenner, E., *et al.* (1993). Genetic heterogeneity among blue-cone monochromats. Am. J. Hum. Genet., 53: 987–1000.

Neitz, J. and Neitz, M. (1992). Do people with anomalous color vision have anomalous cones? Invest. Ophthalmol. Vis. Sci., 33: 754.

Neitz, J., Neitz, M. and Jacobs, G.H. (1989). Analysis of fusion gene and encoded photopigment of colour-blind humans. Nature, 342: 679–682.

Neitz, M., Neitz, J. and Jacobs, G.H. (1991). Spectral tuning of the pigments underlying red-green color vision. Science, 252: 971–974.

Neitz, M., Neitz, J. and Grishok, A. (1995). Polymorphism in the number of genes encoding long-wavelength-sensitive cone pigments among males with normal color vision. Vision Res. 35: 2395–2407.

Nelson, J.H. (1938). Anomalous trichromatism and its relation to normal trichromatism. Proc. Phys. Soc., 50: 661–697.

Oprian, D.D., Asenjo, A.B., Lee, N. and Pelletier, S.L. (1991). Design, chemical synthesis and expression of genes for the three human color vision pigments. Biochemistry, 30: 11367–11372.

Palmer, G. (1777). Theory of Colours and Vision. S. Leacroft, London.

Palmer, G. (1786). Théorie de la Lumière, applicable aux arts, et principalement à la peinture. Hardouin et Gattey, Paris.

Pokorny, J. and Smith, V.C. (1977). Evaluation of single-pigment shift model of anomalous trichromacy. J. Opt. Soc. Am., 67: 1196–1209.

Pokorny, J., Moreland, J.D. and Smith, V.C. (1975). Pigments in anomalous trichromats. J. Opt. Soc. Am., 65: 1522–1524.

Pokorny, J., Smith, V.C., Verriest, G. and Pinckers, A.J.L.G. (1979). Congenital and Acquired Color Vision Defects. Grune and Stratton, New York.

Regan, B.C., Reffin, J.P. and Mollon, J.D. (1994). Luminance noise and the rapid determination of discrimination ellipses in colour deficiency. Vision Res. 34: 1279–1299.

Reichel, E., Bruce, A.M., Sandberg, M.A. and Berson, E.L. (1989). An electroretinographic and molecular genetic study of X-linked cone degeneration. Am. J. Ophthal., 108: 540–547.

Reid, R.C. and Shapley, R.M. (1992). Spatial structure of cone inputs to receptive fields in primate lateral geniculate nucleus. Nature, 356: 716–718.

Robinson, P.R., Cohen, G.B., Zhukovsky, E.A. and Oprian, D.D. (1992). Constitutively active mutants of rhodopsin. Neuron, 9: 719–725.

Ruddock, K.H. and Naghshineh, S. (1974). Mechanisms of red–green anomalous trichromacy. In: Verriest, G. (ed), Colour Vision Deficiencies II: 210–214. Karger, Basel.

Rushton, W.A.H. (1972). Pigments and signals in colour vision. J. Physiol. (Lond.), 220: 1–31P.

Rushton, W.A.H., Powell, D.P. and White, K.D. (1973). Pigments in anomalous trichromats. Vision Res. 13: 2017–2031.

Schmidt, I. (1953). Some problems related to testing colour vision with the Nagel anomaloscope. J. Opt. Soc. Amer., 45: 514–522.

Shyue, S.-K., Li, L., Chang, B.H.-J. and Li, W.-H. (1994). Intronic gene conversion in the evolution of human X-linked color vision genes. Mol. Biol. Evol., 1: 548–551.

Singer, M.S., Shepherd, G.M. and Greer, C.A. (1995). Olfactory receptors guide axons. Nature, 377: 19–20.

Voigt, J.H. (1781). Das herrn Giros von Gentilly Muthmassungen über die Gesichtsfehler bey Untersuchung der Farben. Magazin für das Neueste aus der Physik und Naturgeschichte (Gotha), 1: 57–61.

Watson, S. and Arkinstall, S. (1994). The G-protein Linked Receptor Facts Book. Academic, London.

Weitz, C.J., Miyake, Y., Shinzato, K. *et al.* (1992a). Human tritanopia associated with two amino acid substitutions in the blue-sensitive opsin. Ann. Hum. Genet., 50: 498–507.

Weitz, C.J., Went, L.N. and Nathans, J. (1992b). Human tritanopia associated with a third amino acid substitution in the blue-sensitive visual pigment. Ann. Hum. Genet., 51: 444–446.

Williams, A.J., Hunt, D.M., Bowmaker, J.K. and Mollon, J.D. (1992). The polymorphic photopigments of the marmoset: spectral tuning and genetic basis. EMBO J., 11: 2039–2045.

Willis, M.P. and Farnsworth, D. (1952). Comparative evaluation of anomaloscopes. Report No. 190, Medical Research Laboratory, Bureau of Medicine and Surgery, US Navy.

Winderickx, J., Battisti, L., Motulsky, A.G. and Deeb, S.S. (1992a). Selective expression of human X chromosome-linked green opsin genes. Proc. Natl. Acad. Sci. USA, 89: 9710–9714.

Winderickx, J., Lindsay, D.T., Sanocki, E., Teller, D.Y., Motulsky, A.G. and Deeb, S.S. (1992b). Polymorphism in red photopigment underlies variation in colour matching. Nature, 356: 431–433.

Winderickx, J., Sanocki, E., Lindsey, D.T., Teller, D.Y., Motulsky, A.R. and Deeb, S.S. (1992c). Defective colour vision associated with a missense mutation in the human green visual pigment gene. Nature Genet., 1: 251–255.

Wright, W.D. (1946). Researches on Normal and Defective Colour Vision. Henry Kimpton, London.

Department of Experimental Psychology
University of Cambridge
Downing Street
Cambridge, CB2 3EB
UK

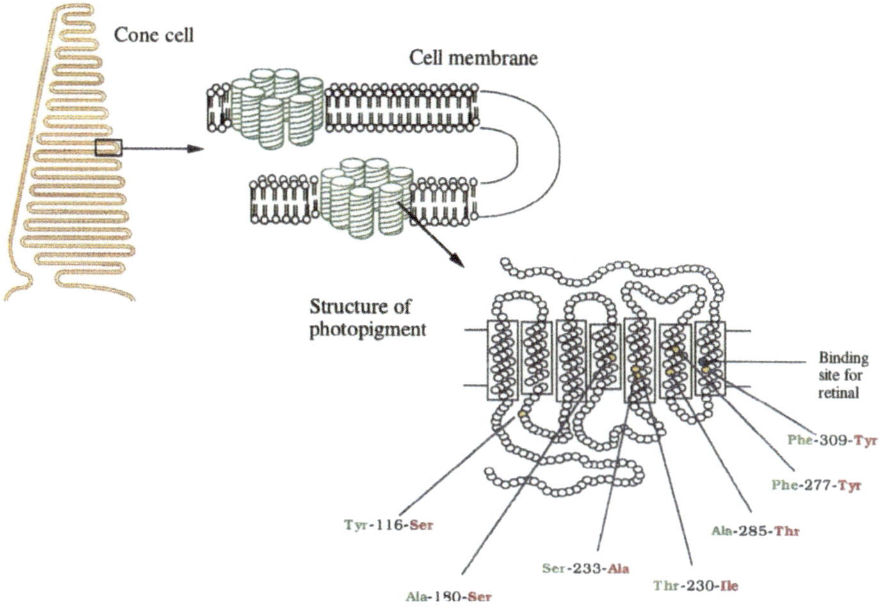

Chapter 1 Fig. 2. The enfolded membrane of the cone outer segment (upper left) is packed with photopigment molecules. Each of these consists of seven helices which span the membrane of the cell and are linked by loops outside the membrane. The seven helices form a palisade surrounding the chromophore, 11-*cis*-retinal. Represented at the bottom right is the sequence of amino acids that make up this heptahelical molecule. Highlighted in yellow are seven amino acid sites that are thought to shift the spectral sensitivity of the opsin from long-wave to middle-wave. The codes below represent the alternative amino acids at these positions: in each case the amino acid indicated in green is the alternative that shifts the peak sensitivity of the molecule to shorter wavelengths, that in red the one that shifts sensitivity to longer wavelengths.

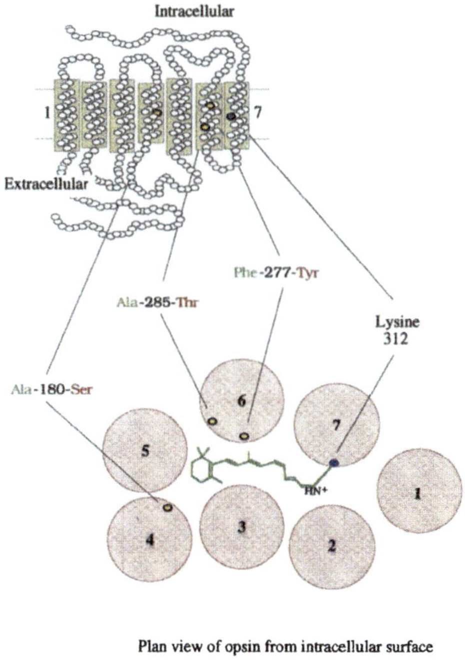

Chapter 1 Fig. 3. A plan view of the opsin molecule showing the approximate positions of the amino acids that have a major influence on the peak sensitivity of the pigment.

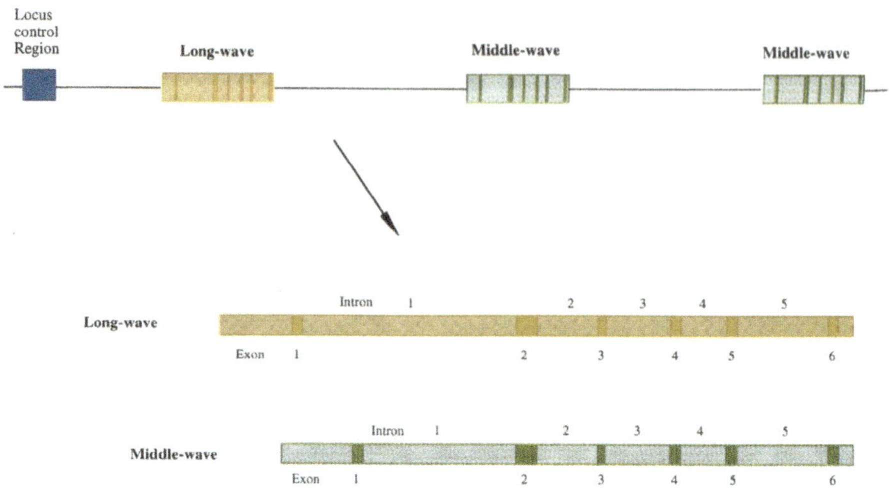

Chapter 1 Fig. 4. Above: the arrangement of L and M genes on the X chromosome. Below: the arrangement of introns and exons for each gene.

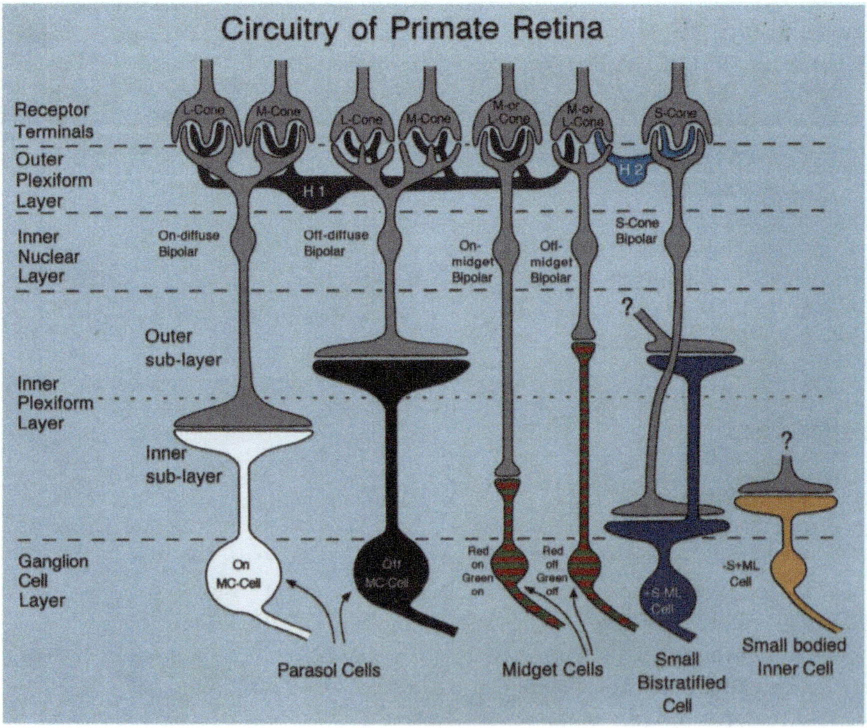

Chapter 10 Fig. 2. A sketch of hypothetical circuitry in primate retina. On- and off-centre cell types (each of which form separate mosaics) are connected to the cones through specific bipolar cells. Red and green on-centre cells share the same mosaic, as do red and green off-centre cells, and they are shown in red and green stripes. The identification of the –S+ML, yellow on, blue off cell is still tentative. Latest results identifying horizontal cell types are also included.

2. Levels of expression of the red, green and green–red hybrid pigment genes in the human retina

T. YAMAGUCHI, A.G. MOTULSKY and S.S. DEEB
(Seattle, WA, USA)

Abstract

Correlations between the genotype at the X-chromosome-linked red/green pigment genes and the colour vision phenotype will be aided by considering the relative levels of expression of these genes into mRNA and protein in the retina. The red and green pigment genes exist in head-to-tail arrays: one red pigment gene is found 5′ upstream of one or more green pigment genes. In addition to normal red and green pigment genes, 5′green–red3′ hybrid genes, typically associated with deuteranomaly, may also be found in subjects with normal colour vision. A locus control region (LCR), located 5′ upstream of the red pigment gene, is shown to be required for expression of the pigment genes in cone photoreceptors. We hypothesize that the expression of any pigment gene in an array is inversely proportional to its distance from this LCR. Green–red hybrid genes that occupy proximal positions in the array are more likely to be expressed and cause deuteranomaly, whereas those that occupy distal positions are less likely to be expressed, and do not result in defective colour vision.

The ratios of expression of the red and green pigment genes in human retinas of donors who have only normal red and green pigment genes shows a wide distribution with a mean of 4:1.

Introduction

The fundamental spectral sensitivity curves at the cornea of the human red, green and blue cones were first estimated psychophysically from colour matching data derived from dichromatic observers (König and Dieterici, 1893). Since then, a number of investigators have refined this approach (e.g., Smith and Pokorny, 1975; Wyszecki and Stiles, 1982). Cone-sensitivity functions estimated by colour matching, performed on observers with normal colour vision employing transient chromatic adaptation (Stockman *et al.*, 1993), were shown to agree broadly with those derived from dichromats.

Direct physical methods have also been used to determine the fundamental spectral sensitivity curves of the visual pigments. These include retinal reflectance densitometry (Mitchell and Rushton, 1971), microspectrophotometry of

C.R. Cavonius (ed.), Colour Vision Deficiencies XIII, pp. 21–31.

individual human cones (Dartnall *et al.*, 1983), electroretinography (Neitz *et al.*, 1989, 1991), and more recently, measurements of absorption spectra of recombinant pigments expressed in tissue culture (Merbs and Nathans, 1992a, 1993; Asenjo *et al.*, 1994).

The red and green photopigments differ in a maximum of 15 from a total of 364 amino acid residues (Nathans *et al.*, 1986a). The presence of hydroxyl-bearing vs. nonpolar residues constitutes the difference at 8 amino acid residues at positions 65, 116, 180, 230, 233, 277, 285 and 309. All but residue 116 are located in the transmembrane pocket and therefore are likely to influence the absorption spectrum. Comparing differences in amino acid sequence with spectral characteristics of pigments of the fish *Astyanax fasciatus*, Yokoyama and Yokoyama (1990) suggested that amino acid residues at positions 180, 277 and 285 play a major role in determining the difference in absorption maxima (30 nm) between the human red and green visual pigments. Based on a comparative study of the sequence and spectral characteristics of the colour vision pigments of Old and New World monkeys, Neitz *et al.* (1991) confirmed the importance of these three residues in spectral tuning. Evidence for the contribution of residue 233 was later derived from comparison of primate pigments (Williams *et al.*, 1992; Ibbotson *et al.*, 1992).

Expression of recombinant human red and green pigments allowed direct, and more detailed, assessment of the contribution of a number of amino acid residues to spectral tuning. Residues at positions 277 and 285 were shown to contribute 7 nm and 14 nm, respectively, to the 30 nm difference between the absorption maxima of these pigments (Merbs and Nathans, 1992a, 1993; Asenjo *et al.*, 1994). The recombinant human red pigment variants bearing either Ser or Ala at position 180, which occur in the general Caucasian population at a frequency of 60% and 40%, respectively (Winderickx *et al.*, 1992a), were shown to differ in their wavelength of maximal absorption (λ_{max}) by approximately 5 nm (Ser-180 = 557 nm; Ala-180 = 552 nm; Merbs and Nathans, 1992b). In addition, substitutions at positions 116, 230, 233, and 309 produced small spectral shifts (Merbs and Nathans, 1993, Asenjo *et al.*, 1994). No change was observed upon substitution at position 65 (Asenjo *et al.*, 1994). The spectral characteristics of recombinant human photopigments are generally quite similar to those determined by psychophysical and microspectrophotometric methods.

In conclusion, differences at only 7 amino acid positions (116, 180, 230, 233, 277, 285, and 309) are responsible for the difference in spectral characteristics between the human red and green visual pigments (Fig. 1). Residues at positions 180, 277, and 285 contribute the majority of the difference in absorption.

The 5′red–green3′ and 5′green–red3′ hybrid genes were first discovered in groups of individuals with protan and deutan colour vision defects, respectively (Nathans *et al.*, 1986b). The structure of hybrid genes observed among colour-deficient individuals and the absorption maxima of the encoded chimeric pigments are shown in Figure 2. The absorption maxima of recombinant chimeric pigments expression in cultured cells reconstituted with the chromo-

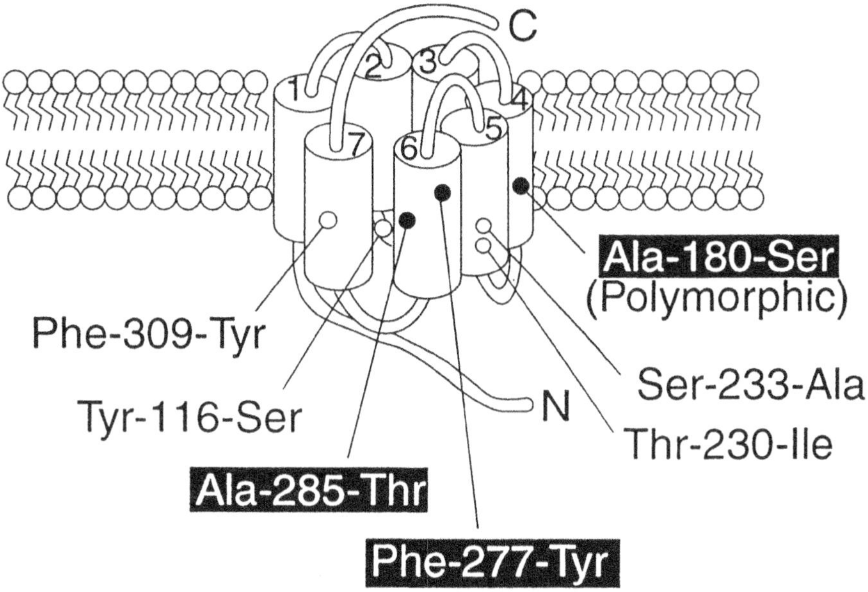

Fig. 1A. Diagram of a photopigment molecule embedded within the photoreceptor outer segment membrane. The cylinders represent the 7-a helices that form a bundle, which holds the chromophore 11-*cis*-retinal. Shown are the 7 amino acid positions that are responsible for the difference in spectral sensitivity between the red and green pigments (adapted from Mollon, 1993). Amino acids to the left of residue numbers are found in the green pigment and those to the right are found in the red pigment. Amino acid residues at positions 180, 277, and 285 contribute the majority of the spectral difference.

phore 11-*cis*-retinal (Merbs and Nathans, 1992a; Asenjo *et al.*, 1994) were intermediate between those of the normal red and green. Subsequently, we observed that a small fraction (4–8% of unselected Caucasian males) with normal colour vision had, in addition to normal red and green pigment genes, 5′green–red3′ hybrid genes (Drummond-Borg *et al.*, 1989; Deeb *et al.*, 1992; Fig. 3). This is in contrast to the finding by Neitz and Neitz (1995) of such hybrid genes in approximately 44% of males with normal colour vision. The finding that such arrays are associated with both normal and deuteranomalous colour vision (Fig. 3) raises the question of what causes deuteranomaly.

We hypothesized that not all opsin genes in an array are expressed in the retina in a sufficient number of cones to influence colour vision. We suggested that in deuteranomalous individuals the red and green–red hybrid, but not the normal green pigment, genes were expressed, whereas in normal individuals the normal red and green, but not hybrid, genes are expressed (Deeb *et al.*, 1992;

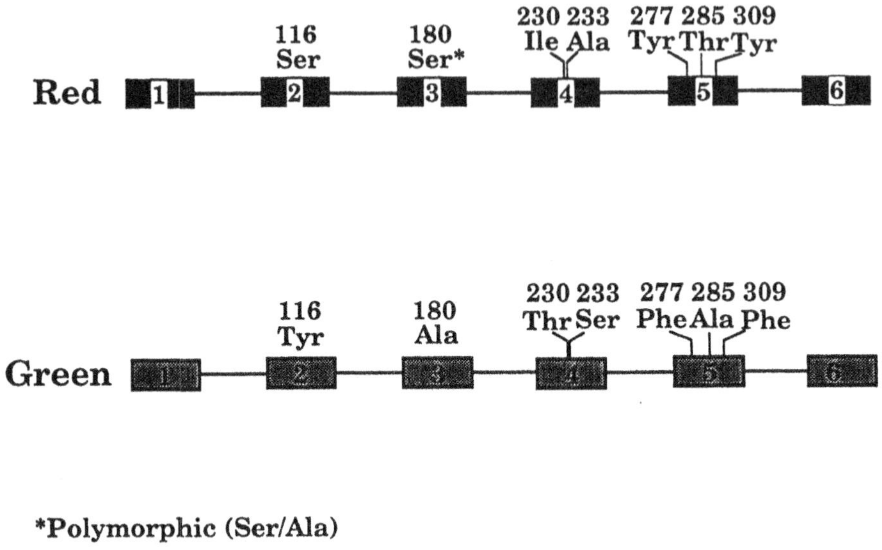

Fig. 1B. The intron/exon structure of the genes encoding the red and green pigments indicating the 7 amino acid residues (positions given above) responsible for the difference in absorption maxima between the red and green pigments. Residues at positions 277 and 285 (Tyr and Thr in red vs. Phe and Ala in green) in exon 5 account for the majority of the difference in absorption maximum between the red and green pigments. Thus, exon 5 plays a major role in determining whether a pigment is red-like or green-like. Rectangles (with numbers) represent exons and connecting lines represent introns (diagram not to scale).

Winderickx *et al.*, 1992b). Furthermore, we proposed that the level of expression of the genes of a single array varies as a function of distance from the locus control region (LCR), located approximately 3.6 kb 5′ upstream of the red pigment gene, and shown to be required for expression of all genes of the array (Nathans *et al.*, 1993; Wang *et al.*, 1992). We have addressed directly the question of relative extent of expression of the red, green and hybrid pigment genes by analysing retinal mRNA encoding the red, green and hybrid pigments.

Subjects and methods

Postmortem human retinal tissue and blood samples from male donors were collected between 4 and 8 h after death. These specimens were obtained through the Lions Eye Bank of the University of Washington, Seattle. The preparation of genomic DNA and RNA as well as their use as templates for analysis of the red and green exonic sequences by amplification with the polymerase chain reaction followed by single strand conformation polymorphism analysis (SSCP) has been

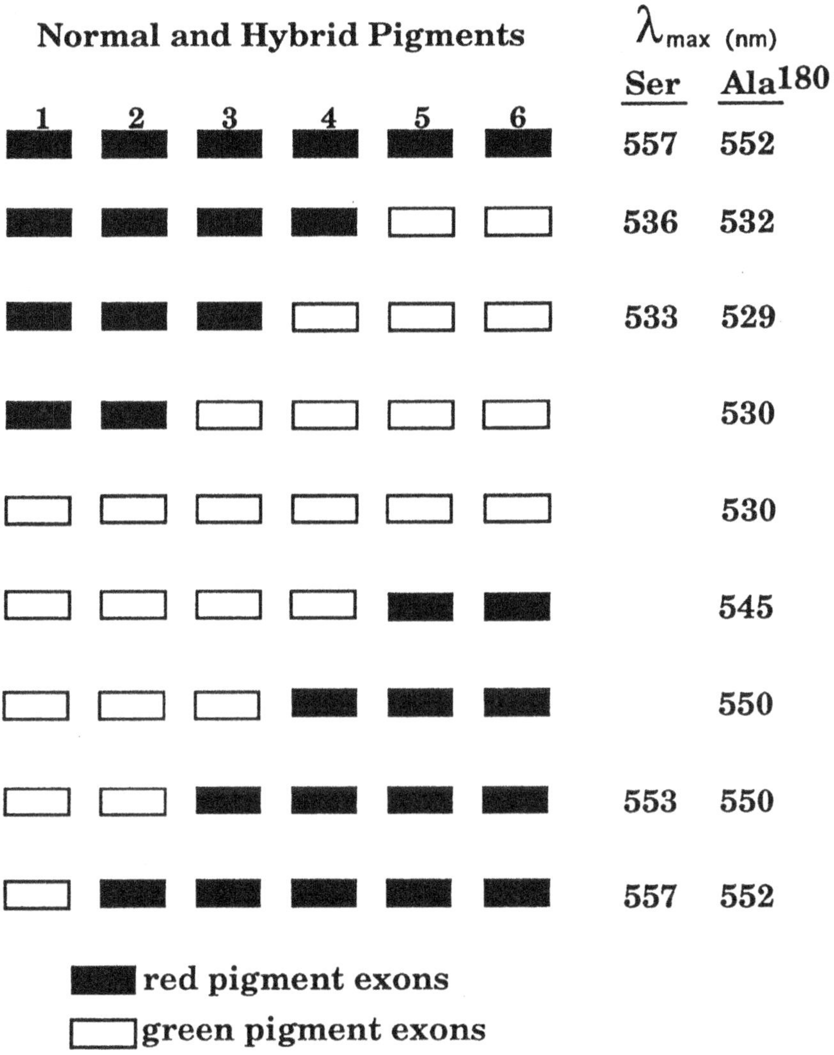

Fig. 2. Absorption maxima of red/green hybrid pigments. Solid and open rectangles represent exons (numbers on top) of the red and green pigment genes, respectively. The absorption maxima of the encoded pigments are according to Merbs and Nathans (1992a,b). The two columns on the right indicate absorption maxima (λ_{max}) of the pigment with either Ser or Ala at the polymorphic position 180. The pigments in the first and fifth rows represent normal red and green pigments; other rows represent red–green and green–red hybrid pigments.

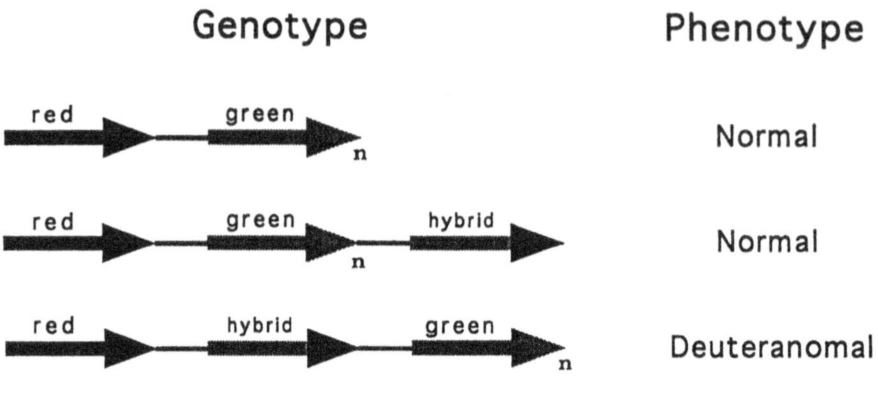

Fig. 3. X-chromosome visual pigment gene arrays found in males with normal and deuteranomalous colour vision. Solid arrows represent red pigment gene sequences and cross-hatched arrows represent green pigment gene sequences. In hybrid pigment genes, the first 4 exons are derived from the green pigment gene and the last two exons are derived from the red pigment gene. 'n' indicates that there can be one or more copies of the green opsin gene in an array. Whether the hybrid gene occupies a proximal or distal position with respect to the red pigment gene is unknown. However, we propose that when the hybrid gene occupies a proximal position, it is preferentially expressed in the retina. This would result in deuteranomaly since the number of photoreceptors with hybrid pigments will be much greater than that with normal green pigment (see Fig. 5).

described in detail (Winderickx *et al.*, 1993). Band intensities on SSCP gels were quantified by phosphorimaging. The proximal 230 bp of the promoters of the red and green opsin genes were analysed for sequence variants by amplification and SSCP analysis.

Results

Ratio of red to green mRNA in human retinae

Total cellular RNA was prepared from whole retinae of 27 male donors whose genomic DNA contained one normal red and one or more normal green pigment genes but no 5′green–red3′ hybrid genes. The RNA was analysed for the ratio of red to green exon 4 mRNA sequences by reverse transcription, followed by quantitative amplification and separation of red- from green-specific exon 4 sequences by SSCP analysis.

The results (Fig. 4) indicate a wide variation in this ratio. In 26 of 27 cases it is clear that the red pigment gene, which occupies the proximal position with respect to the locus control region (LCR) in any given array, is represented in whole retinae at higher levels (on average 4-fold) than the green pigment gene.

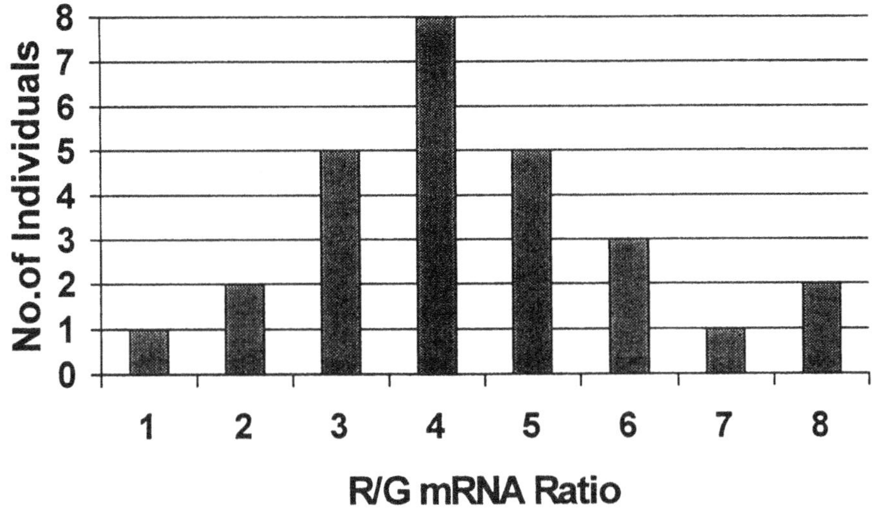

Fig. 4. Frequency distribution of the ratio of red to green pigment mRNA in whole retinae. Total RNA was extracted from retinae of human male donors and analysed for the concentration of red relative to green exon 4 encoded sequences.

Ratio of green to hybrid mRNA transcripts in human retinae

Another set of six postmortem retinae were of unknown colour vision status, the male donors of which had arrays comprised of one red, one 5′green–red3′ hybrid and one or more green pigment genes. Exons 1–4 of the hybrid genes in these individuals were derived from the green pigment gene and exons 5 and 6 from the red pigment gene. Thus these hybrid genes encoded a red-like hybrid pigment. We found that the great majority of mRNA transcripts in the retinae of these donors were encoded by the normal red and green pigment genes. The hybrid genes were expressed, if at all, at extremely low levels ($<0.5\%$ of that of the red pigment gene) and we therefore infer that these donors probably had normal colour vision. The average ratios of expression of these genes (exon 4 in this particular experiment), as estimated by mRNA, was 200(red) : 50(green) : 1(hybrid). The details of these studies will be published elsewhere.

The lack of expression of the green–red hybrid genes in the retinae studied is unlikely to be due to the presence of inactivating mutations, since the sequences of the proximal 230 bp of the promoter, as well as of all coding sequences, were normal.

Discussion

The ratio of red to green mRNA transcripts observed in extracts of whole retinae varies widely (a range of 1 to 8) among individuals, with a modal value of 4. This ratio may, to a large extent, reflect the ratio of red to green cones in the normal retina.

A number of investigators, using psychophysical techniques, have observed wide individual variation in the ratio of red to green cones, but with a modal value of about 2 (de Vries, 1947; Rushton and Baker, 1964; Vimal *et al.*, 1989; Wesner *et al.*, 1991). A narrower range of values (1.46–2.36) was obtained by Cicerone and Nerger (1989). All these studies used small and centrally located test fields and the observed ratios were therefore representative of the foveolar region. In addition, microspectrophotometric measurements on 7 eyes of individuals with normal colour vision gave a mean red:green cone ratio in the foveolar region of 1.39 (Dartnall *et al.*, 1983).

One possible explanation for the discrepancy between our results on cone ratios derived from mRNA analysis and the foveal ratio of about 2 is that the ratio of red to green cones is higher in the peripheral retina than it is in the fovea. Since only about 10% of all cones are in the fovea, the inferred ratios we obtained from whole retinal mRNA represent largely extrafoveal cones. There is evidence to suggest that the ratio of red to green cones is higher in the peripheral than in the central retina (see Hagstrom *et al.*, this volume). However, estimates of red:green cone ratios (average of 2, range 0.67–9.0) obtained on 16 subjects with normal colour vision by flicker photometric electroretinography (ERG) were similar for both small and very large test fields (Jacobs and Neitz, 1993), suggesting that the red:green ratio does not change significantly across the retina.

Correlation between mRNA ratios and cone ratios may be influenced by variation in the level of expression of the red and green pigment genes in individual cones. A difference in expression level may, therefore, not reflect precisely a difference in the ratio of cones.

The results that indicate selective gene expression in the retina are consistent with our earlier findings in which differential expression of green pigment genes in the retina was observed (Winderickx *et al.*, 1992b). Analysis in 10 male subjects who had two or more green pigment genes in their genomic DNA showed clearly that whenever the two alleles of exon 5 were present in genomic DNA only one was represented in retinal mRNA in sufficient quantities to be detected by SSCP analysis.

The model shown in Figure 5 was proposed (Winderickx *et al.*, 1992b) to explain such selective expression in the retina. In this model, the LCR is postulated to regulate expression of the opsin genes of the array in a position-dependent manner. Thus, if the LCR forms a stable transcriptionally active complex with the red opsin gene promoter, a red-sensitive cone will result. If the LCR forms a stable complex with the proximal green pigment gene promoter instead, a green-sensitive cone is formed. Transcriptionally active complexes

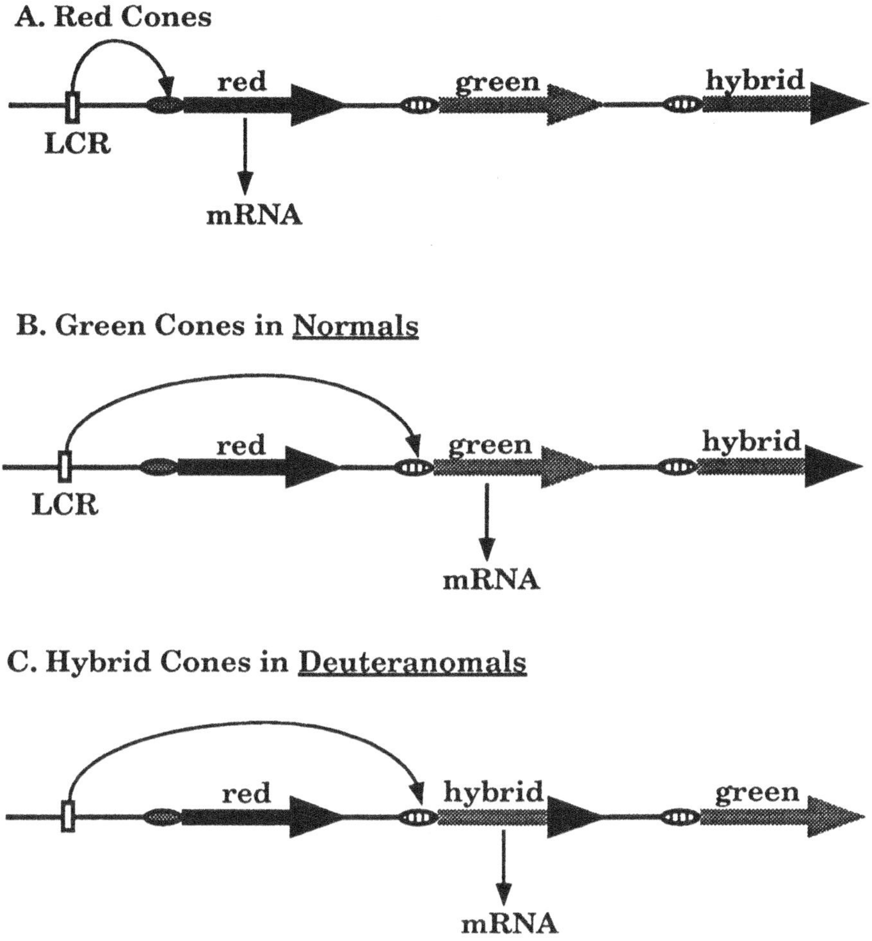

Fig. 5. Model for differential expression in the X-linked red/green pigment gene complex. Red cones result from the exclusive expression of the red pigment gene as a result of stable coupling (mediated by DNA binding proteins) of the locus control region (LCR) to the red pigment gene promoter. Green cones are formed if the LCR preferentially and permanently couples to the proximal green pigment gene promoter and turns on its expression. Distal green pigment gene promoters (including those that belong to green–red hybrid genes) are activated with very low probability (adapted from Winderickx *et al.*, 1992b).

between the LCR and promoters of pigment genes located downstream of the two proximal pigment genes are much less favoured. We suggest, therefore, that only the two most proximal pigment genes of an array are expressed in sufficient cones to influence colour vision as measured anomaloscopically. Based on this model, if a male subject has a normal red, normal green and a hybrid gene, deuteranomaly will result if the green–red hybrid gene occupies the proximal position and is expressed preferentially over the normal green opsin gene.

Acknowledgement

This work was supported by National Institutes of Health Grant number EY08395. We thank the Lions Eye Bank at the University of Washington for supplying the donated eyes.

References

Asenjo, A.B., Rim, J. and Oprian, D.D. (1994). Molecular determinants of human red/green colour discrimination. Neuron 12: 1131–1138.

Cicerone, C.M. and Nerger, J.L. (1989). The relative numbers of long-wavelength-sensitive and middle-wavelength-sensitive cones in the human fovea centralis. Vision Res. 19: 115–128.

Dartnall, H.J.A., Bowmaker, J.K. and Mollon, J.D. (1983). Human visual pigments: microspectrophotometric results from the eyes of seven persons. Proc. Roy. Soc. Lond., B 220: 115–130.

Deeb, S.S., Lindsey, D.T., Hibiya, Y., Sanocki, E., Winderickx, J., Teller, D.Y. and Motulsky, A.G. (1992). Genotype-phenotype relationships in human red/green colour vision defects: Molecular and psychophysical studies. Am. J. Hum. Genet. 51: 687–700.

de Vries, H.L. (1947). The heredity of the relative numbers of red and green receptors in the human eye. Genetica 24: 199–212.

Drummond-Borg, M., Deeb, S.S. and Motulsky, A.G. (1989). Molecular patterns of X chromosome-linked colour vision genes among 134 men of European ancestry. Proc. Natl. Acad. Sci. USA 86: 983–987.

Ibbotson, R.E., Hunt, D.M., Bowmaker, J.K. and Mollon, J.D. (1992). Sequence divergence and copy number of the middle- and long-wave photopigments in Old World monkeys. Proc. Roy. Soc. Lond. B 247: 145–154.

Jacobs, G.H. and Neitz, J. (1993). Electrophysiological estimates of individual variation in the L/M cone ratio. In: Drum B. (ed.) Colour Vision Deficiencies XI. Doc. Ophthal. Proc. Ser. 56: 107–112. Kluwer, Dordrecht.

König, A. and Dieterici, C. (1893). Die Grundemfindungen in normalen und anomalen Farbensystemen und ihre Intensitätsverteilung im Spektrum. Z. Psychol. Physiol. Sinnesorg. 4: 241–347.

Merbs, S.L. and Nathans, J. (1992a). Absorption spectra of the hybrid pigments responsible for anomalous colour vision. Science 258: 464–466.

Merbs, S.L., and Nathans, J. (1992b). Absorption spectra of human cone pigments. Nature 356: 433–435.

Merbs, S.L. and Nathans, J. (1993). Role of hydroxyl-bearing amino acids in differentially tuning the absorption spectra of the human red and green cone pigments. Photochem. Photobiol. 58: 869–881.

Mitchell, D.E. and Rushton, W.A. (1971). Visual pigments in dichromats. Vision Res. 11: 1033–1043.

Mollon, J.D. (1993). Mixing genes and mixing colours. Current. Biol. 3: 82–85.

Nathans, J., Thomas, D., and Hogness, D.S. (1986a). Molecular genetics of human colour vision: the genes encoding blue, green and red pigments. Science 232: 193–202.

Nathans, J., Piantanida, T.P., Eddy, R.L., Shows, T.B. and Hogness, D.S. (1986b). Molecular genetics of inherited variation in human colour vision. Science 232: 203–210.

Nathans, J., Maumenee, I.H., Zrenner, E. *et al.* (1993). Genetic heterogeneity among blue-cone monochromats. Am. J. Hum. Genet. 53: 987–1000.

Neitz, M. and Neitz, J. (1995). Numbers and ratios of visual pigment genes for normal red–green colour vision. Science 267: 1013–1016.

Neitz, J., Neitz, M. and Jacobs, G.H. (1989). Analysis of fusion gene and encoded photopigment of colour-blind humans. Nature 342: 679–682.

Neitz, M. and Neitz, J. and Jacobs, G.H. (1991). Spectral tuning of pigments underlying red–green colour vision. Science 252: 971–974.

Rushton, W.A.H. and Baker, H.D. (1964). Red/green sensitivity in normal vision. Vision Res. 4: 75–85.

Smith, V.C. and Pokorny, J. (1975). Spectral sensitivity of the cone photopigments between 400 and 500 nm. Vision Res. 15: 161–171.

Stockman, A., MacLeod, D.I.A. and Johnson, N.E. (1993). Spectral sensitivities of the human cones. J. Opt. Soc. Am. A 10: 2491–2521.

Vimal, R.L.P., Pokorny, J., Smith, V.C. and Shevell, S.K. (1989). Foveal cone thresholds. Vision Res. 29: 61–78.

Wang, Y., Macke, J.P., Merbs, S.L. *et al.* (1992). A locus control region adjacent to the human red and green visual pigment genes. Neuron 9: 429–440.

Wesner, M.F., Pokorny, J., Shevell, S.K. and Smith, V.C. (1991). Foveal cone detection statistics in colour normals and dichromats. Vision Res. 31: 1021–1037.

Williams, A.J., Hunt, D.M., Bowmaker, J.K. and Mollon, J.D. (1992). The polymorphic pigments of the marmoset: spectral tuning and genetic basis. EMBO J. 11: 2039–2045.

Winderickx, J., Lindsey, D.T., Sanocki, E., Teller, D.Y., Motulsky, A.G. and Deeb, S.S. (1992a). A Ser/Ala polymorphism in the red photopigment underlies variation in colour matching among colour-normal individuals. Nature 356: 431–433.

Winderickx, J., Battisti, L., Motulsky, A.G. and Deeb, S.S. (1992b). Selective expression of the human X-linked green pigment genes. Proc. Natl. Acad. Sci. USA 89: 9710–9714.

Winderickx, J., Battisti, L., Hibiya, Y., Motulsky, A.G. and Deeb, S.S. (1993). Haplotype diversity in the human red and green opsin genes: evidence for frequent sequence exchange in exon 3. Hum. Mol. Genet. 2: 1413–1421.

Wyszecki, G. and Stiles, W.S. (1982). Colour Science, 2nd edn.: 604–633. Wiley, New York.

Yokoyama, S. and Yokoyama, R. (1990). Convergent evolution of the red- and green-like visual pigment genes in fish, Astyanax fasciatus, and human. Proc. Natl. Acad. Sci. USA 87: 9315–9318.

Samir Deeb
Genetics Department
Box 357360
University of Washington
Seattle, WA 98195

3. Variety of photopigment genes underlying red–green colour vision

M. NEITZ and J. NEITZ
(Milwaukee, WI, USA)

Abstract

To understand the relationship between colour vision genotype and phenotype one needs experimental approaches that allow accurate determination of both. An important aspect of colour vision genotype is the number and ratio of genes that encode the long- and middle-wave-sensitive cone pigments. It has become standard practice to estimate these by quantitative Southern analysis. This method has many inherent problems that limit its usefulness as a quantitative measure, and its validity in estimating the number and ratio of X-linked pigment genes has never been demonstrated. Recently, a new method to determine accurately and reliably the number of long- and middle-wave genes on the X-chromosome was developed. Genotypes determined by the new method have been confirmed by independent experiments including direct sequence analysis of the genes. These results show that men with normal colour vision often have more than one long-wave gene and that the number of X-linked visual pigment genes in individuals is much greater than was previously suspected. The issue of which of the X-linked visual pigment genes are expressed in the retina is critical to understanding colour vision genotype/phenotype relationships, particularly in light of the new understanding of genotype. Since expression of any one pigment gene, sometimes in a very small number of cones, can have a significant impact on colour vision, it is important to examine levels of pigment gene expression using the most sensitive methods available. We have demonstrated that expression of as many as three or four different X-linked pigment genes in individual male retinas at visually relevant levels is common.

Colour vision genotypes

Classically, all men with normal colour vision were supposed to have the same genotype. It was believed that the X-chromosome carries a single gene encoding the long-wave pigment and a single gene encoding the middle-wave pigment. The molecular genetic analysis of genotypes underlying normal red–green colour vision was reported for the first time in 1986 (Nathans *et al.*, 1986a,b). In those experiments, the number of long- and middle-wave genes on the X-

C.R. Cavonius (ed.), Colour Vision Deficiencies XIII, pp. 33–43.

chromosomes of 25 men with normal colour vision were deduced from the results of quantitative Southern analysis. The Southern data showed very clearly that there is variability in the number of visual pigment genes on the X-chromosome. Nathans *et al.* (1986b) proposed that men with normal colour vision have a single long-wave pigment gene and usually one, two or three middle-wave pigment genes. Re-examination of the results of quantitative Southern analysis (Nathans *et al.*, 1986a, 1986b) indicates that the data more strongly support an alternative model in which there is variability in the number of long-wave genes as well as in the number of middle-wave genes, and that the number of genes per array is greater and more variable than was first suggested (for a review, see Neitz and Neitz, 1994). In the original molecular genetic study, four restriction fragment length polymorphisms were used to identify long-versus middle-wave genes. Only one of these was used to deduce the structure of the arrays in 25 men with normal colour vision (Fig. 1A). Quantitative Southern analysis was used to measure the ratio of fragment A_g/A_r and the genotype for each individual was deduced from one such measurement. The number of visual pigment genes was calculated from the A_g/A_r fragment ratio by assuming that fragment A_r and the long-wave gene are always present in single copy number, but the validity of this measurement as an indicator of gene number was never demonstrated. In order to evaluate the validity of this estimate, two questions need to be answered. First, is fragment A_r really always present in single copy? Second, do the A fragments contain sequences that functionally distinguish long- from middle-wave pigments?

The answer to the first question comes from an examination of the recombination events that occur in the visual pigment gene cluster. The visual pigment genes are prone to homologous, unequal crossing-over because the genes are arranged in a head-to-tail tandem array, and the long- and middle-wave genes share more than 98% nucleotide sequence identity (Nathans *et al.*, 1986b). Fragment A_r is larger than A_g because there is an insert in the first intron of some long-wave genes (Fig. 1A). About one-third of the long-wave gene is 5′ to this insert. The probability of a recombination event occurring is directly related to the length of the DNA, so about 1 in 3 homologous unequal recombination events will occur upstream of the insert. As illustrated in Figure 1B, crossovers in the first third of the gene will duplicate fragment A_r and the long-wave gene coding sequence. The second long-wave gene will have exon 1 and upstream sequences from the parent middle-wave gene. The long- and middle-wave genes appear to have the identical exon 1 sequence (Nathans *et al.*, 1986b). The DNA immediately upstream of each gene does not contain sequences that produce a functional difference between long- and middle-wave pigments. Rather, these sequences distinguish the first gene in the array from all others (Neitz and Neitz, 1995b). Thus, it is clear that the relative ratio of fragments A_g/A_r does not provide information from which the number or ratio of long- to middle-wave genes can be calculated.

The answer to the question of whether the A fragments contain sequences that functionally distinguish long- from middle-wave genes comes from spectral

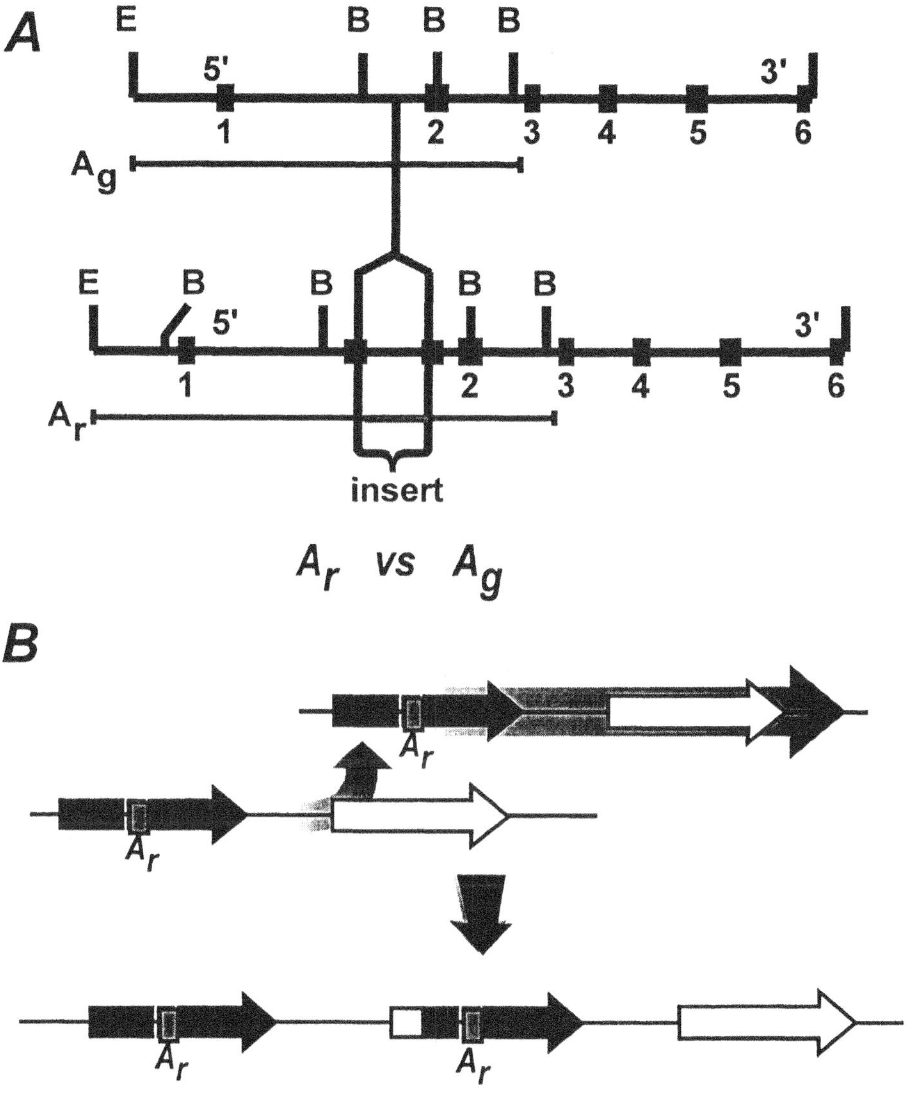

Fig. 1. **(A)** Restriction map of long- and middle-wave genes. The numbered boxes indicate the exons. Restriction sites are indicated as: E = *Eco*RI, B = *Bam*HI. **(B)** Unequal homologous recombination in the first third of the genes duplicates fragment A_r and the long-wave gene. Black arrows indicate long-wave genes, white arrows indicate middle-wave genes. Stippled boxes, labelled A_r, indicate the insert in intron 1 that makes fragment A_r longer than A_g. The duplicated long-wave gene has 5′ sequences derived from the parent middle-wave gene indicated in white.

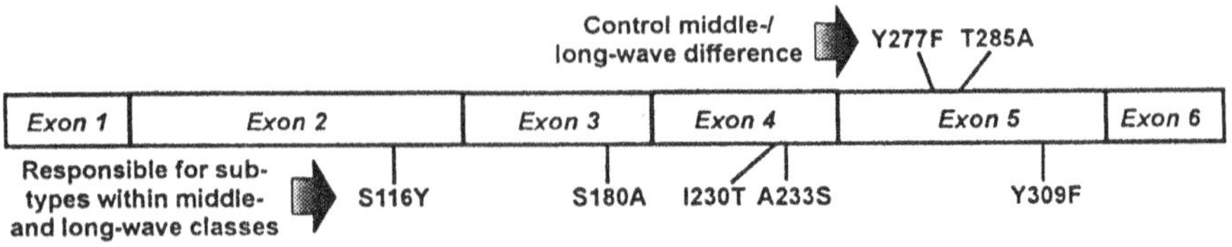

Fig. 2. Amino acid substitutions involved in spectral tuning. Shown are the seven amino acid substitutions required to shift the spectral peak from that of the shortest middle-wave pigment to that of the longest long-wave pigment (Asenjo *et al.*, 1994).

tuning studies, which are summarized in Figure 2. Substitutions at seven amino acid positions are required to shift the spectral peak from that of the shortest middle-wave pigment to that of the longest long-wave pigment (Asenjo *et al.*, 1994). Exon 5 of each gene encodes two amino acid differences that produce the majority of the spectral difference between long- and middle-wave pigments (Chan *et al.*, 1992; Merbs and Nathans, 1992b, 1993; Neitz *et al.*, 1991). Other spectrally active amino acid positions encoded by exons 2–5 of the genes produce relatively small spectral shifts and can be thought of as producing subtypes of long- and middle-wave pigments (Asenjo *et al.*, 1994; Merbs and Nathans, 1992b, 1993). Restriction fragments A_g and A_r contain exons 1 and 2 and thus do not contain sequences that functionally distinguish long- from middle-wave genes.

Recently, Neitz and Neitz (1995b) developed a new method to quantify the number of long-wave genes and the number of middle-wave genes on the X-chromosomes of individual males. The results of the analysis of the X-linked pigment gene arrays in 24 Caucasian men with normal colour vision using the new method are summarized in Table 1 and Figure 3. On average, the number of visual pigment genes per X-chromosome is much greater than previously suspected. Remarkably, more than half of the men in this small sample have multiple long-wave genes. This finding was independently confirmed by sequence analysis of the visual pigment genes in each subject (Neitz *et al.*, 1995). Nine men had multiple long-wave genes that differed in nucleotide sequence, thereby confirming the presence of more than one long-wave gene. Five of those men had genes encoding at least two spectrally different long-wave pigments. Another five men had genes encoding at least two spectral subtypes of middle-wave pigment (Table 1). Thus, some individuals have the genetic potential for expressing more than three spectrally distinct cone pigments in their retinas.

Table 1. Estimated numbers of long- and middle-wave genes.

Subject no.	No. of X-linked pigment genes Mean ± SEM	Ratio of middle-/long-wave genes Mean ± SEM	No. long-wave genes	No. middle-wave genes
Subjects with more than one long-wave gene				
07	9.0 ± 1.3	3.8 ± 0.3	2	7
09	7.7 ± 1.2	3.2 ± 0.4	2	6
23	6.5 ± 1.0	0.5 ± 0.0	4*	2
14	6.7 ± 0.7	2.1 ± 0.1	2*	5
25	6.2 ± 0.6	2.1 ± 0.1	2*	4
06	6.0 ± 0.6	2.5 ± 0.1	2*	4*
22	5.6 ± 1.0	2.0 ± 0.2	2	4
20	4.8 ± 0.4	1.9 ± 0.1	2	3*
04	4.5 ± 0.8	1.0 ± 0.1	2	2
26	5.7 ± 0.4	0.5 ± 0.1	4*	2
08	3.3 ± 0.2	0.9 ± 0.0	2	2*
19	3.3 ± 0.3	0.9 ± 0.1	2	2*
Subjects with only one long-wave gene				
02	4.4 ± 0.3	2.0 ± 0.1	1	3
05	4.0 ± 0.3	2.0 ± 0.0	1	3
03	3.6 ± 0.4	1.9 ± 0.0	1	2
11	3.5 ± 0.1	2.1 ± 0.1	1	2
24	3.4 ± 0.4	1.8 ± 0.1	1	2
15	3.3 ± 0.5	1.9 ± 0.1	1	2*
12	2.9 ± 0.3	1.2 ± 0.1	1	2
27	2.7 ± 0.4	1.8 ± 0.1	1	2
21	2.2 ± 0.1	1.0 ± 0.0	1	1
10	2.2 ± 0.1	0.9 ± 0.1	1	1
13	2.2 ± 0.1	0.9 ± 0.1	1	1
28	2.1 ± 0.0	1.0 ± 0.0	1	1

*Indicates individuals who have genes encoding at least two spectrally distinct long- or middle-wave pigments.

Hybrid genes

As indicated in Figure 3, many men with normal colour vision have 5′middle-3′long-wave hybrid genes. Hybrid genes were originally identified in men with defective red–green colour vision (Nathans *et al.*, 1986a). From this, it was suggested that the hybrid genes encode the anomalous pigments that underlie anomalous trichromacy. More recent studies show that hybrid genes represent normal variants of the human X-linked pigment genes and that they are a characteristic feature of normal colour vision (Fig. 3). For example, we reported three men with normal colour vision, each of whom had a hybrid gene as their only gene that could potentially encode a long-wave pigment (Neitz and Neitz,

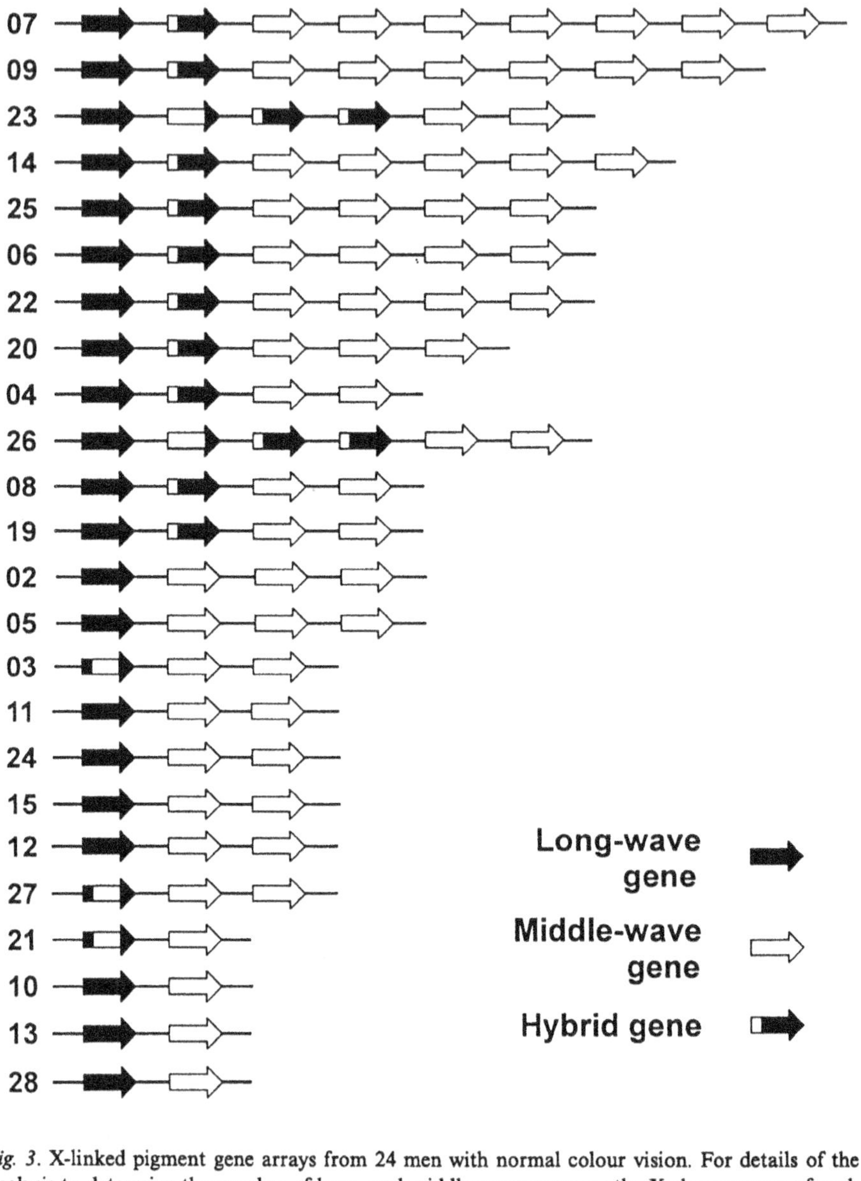

Fig. 3. X-linked pigment gene arrays from 24 men with normal colour vision. For details of the analysis to determine the number of long- and middle-wave genes on the X-chromosome of each subject see Neitz and Neitz (1995b). In arrays with more than one long-wave gene, the additional long-wave genes are indicated as hybrid genes, and they contain varying amounts of sequence from the parent middle-wave gene (Neitz *et al.*, 1995).

1995b; Neitz *et al.*, 1995). The arrays for these three men are shown in Figure 3 (subjects 03, 21 and 27). The hybrid long-wave gene of each subject is reported to have middle-wave exon 4 and long-wave exon 5 (Neitz *et al.*, 1995). Since each of these subjects has only one long-wave gene, and each has normal colour vision, we can be confident that the hybrid long-wave gene is expressed. Clearly, expression of the hybrid gene need not cause colour anomaly.

Visual pigment gene expression

The issue of which of the X-linked pigment genes are expressed is critical to understanding the relationship between colour vision phenotype and genotype. How many of the X-linked visual pigment genes on the X-chromosome are expressed in the eyes of individual males?

Two studies have been reported in which the expression of specific genes was investigated. In a first set of experiments, Winderickx *et al.* (1992) identified male eye donors who each had multiple middle-wave genes differing at a specific restriction site in exon 5. In RT-PCR analysis of retinal mRNA from each eye donor, only middle-wave pigment cDNAs lacking the restriction site were detected. We have replicated this experiment (unpublished) and we found that middle-wave genes with this site are not expressed at detectable levels in the retina when the donor also has a middle-wave gene that lacks the restriction site. In a second study, Yamaguchi *et al.* (1995) looked specifically for expression of a 5′middle-3′long-wave hybrid gene in male eye donors who expressed normal long-wave and middle-wave pigment genes. Expression of the hybrid gene was detected only at low levels ($< 1\%$) or not at all. Yamaguchi *et al.* suggested that although more than three of the X-linked pigment genes can be expressed, only two are expressed at a level that contributes significantly to colour vision.

We have addressed the issue of whether more than just two of the X-linked pigment genes are expressed by characterizing the visual pigment genes expressed in donor eyes from Caucasian men. We examined the mRNA expressed in the retinas from four male eye donors for which previous analyses had suggested the expression of more than one long-wave gene or more than one middle-wave gene (Neitz *et al.*, 1992). To detect expression of multiple subtypes of long- or middle-wave genes, we used the restriction digestion assay previously developed to detect the presence of more than one subtype of long- or middle-wave genes on the X-chromosome (Neitz *et al.*, 1995). The experimental design is illustrated in Figure 4A. mRNA was isolated from retinas of each donor and long- and middle-wave genes were selectively, and separately, amplified using the polymerase chain reaction (PCR). A second round of PCR was done to obtain exon 3 from long- or middle-wave genes. Exon 3 is highly polymorphic among the X-linked pigment genes. Three exon 3 polymorphisms produce restriction site differences that can be used to distinguish between subtypes within the long-wave class of genes and within the middle-wave class of genes. These are *Bst*YI, which distinguishes between subtypes that differ at nucleotide

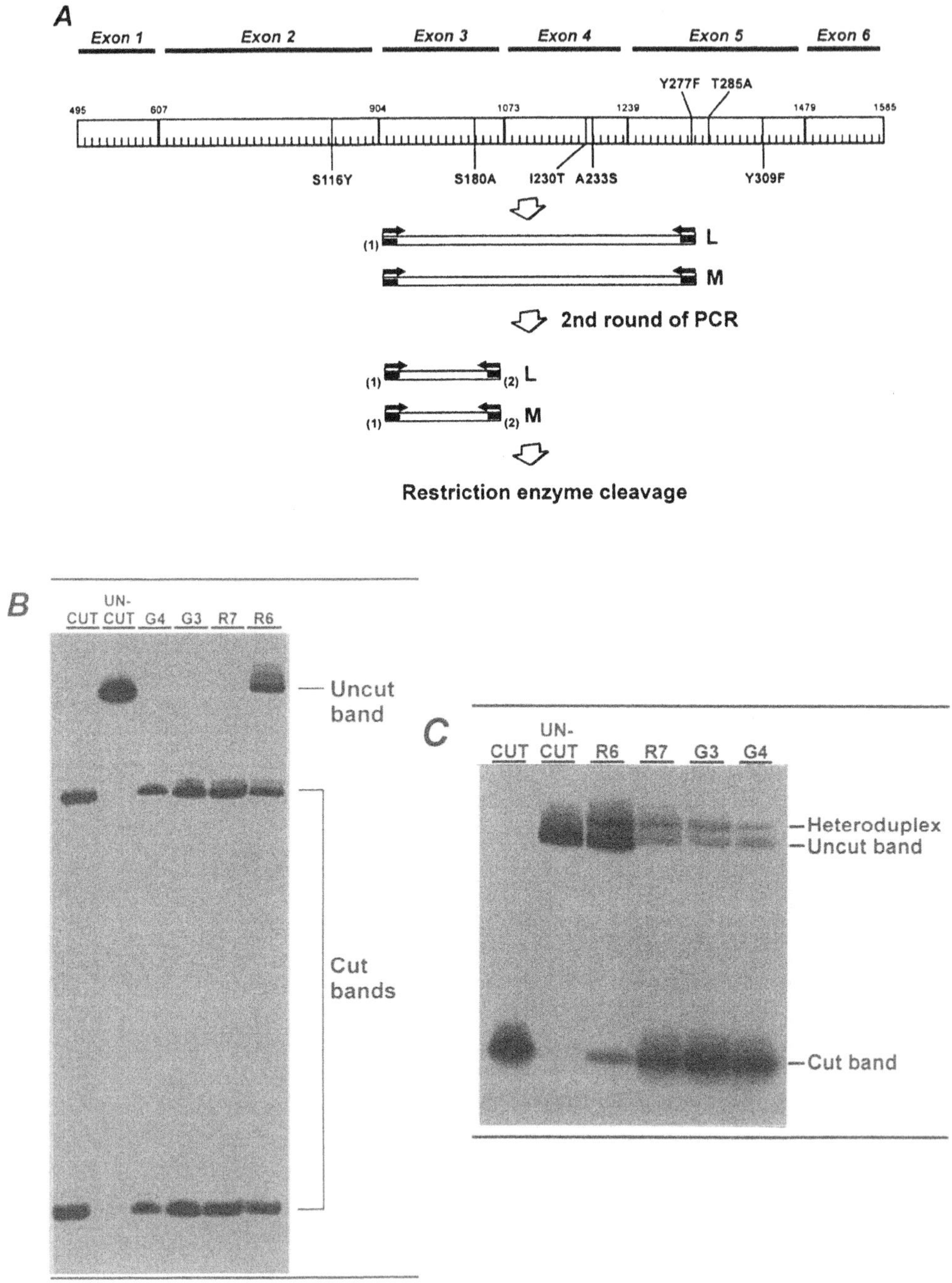
A
Exon 1
Exon 2
Exon 3
Exon 4
Exon 5
Exon 6
Y277F T285A
495 607 904 1073 1239 1479 1585
S116Y
S180A
I230T A233S
Y309F
(1) L
M
2nd round of PCR
(1) (2) L
(1) (2) M
Restriction enzyme cleavage
B
CUT UN-CUT G4 G3 R7 R6
Uncut band
Cut bands
C
CUT UN-CUT R6 R7 G3 G4
Heteroduplex
Uncut band
Cut band

position 1026; *Fok*I, which distinguishes subtypes that differ at position 951; and *Bso*FI, which distinguishes subtypes differing at position 1032. The PCR products, which are either exon 3 fragments from long-wave genes or exon 3 fragments from middle-wave genes, are radioactively end-labelled and digested with a specific restriction enzyme that cuts some subtypes but not others. If, for example, all of the expressed long-wave genes are the same at a particular polymorphic nucleotide position, then they would be the same with regard to the presence or absence of the restriction site that cuts at that position. In our assay either only cut or only uncut exon 3 fragments would be observed. However, if two subtypes that differ at the specific nucleotide position are expressed, one with the restriction site and one without it, then a mixture of both cut and uncut exon 3 would be observed.

Previous analyses suggested that two eye donors (3 and 4) each expressed more than one middle-wave gene and that another two eye donors (6 and 7) each expressed at least two long-wave genes (Neitz *et al.*, 1992). These four eye donors are presumed to have had normal colour vision since both long- and middle-wave cDNAs could be amplified from each eye. The restriction digestion assay was used to examine visual pigment gene expression in the eyes from these four donors. The results are shown in Figures 4B and C.

Fig. 4. (**A**) Strategy for investigating expression of multiple long- and multiple middle-wave genes in individual retinas. Retinal mRNA was isolated using the QuickPrep mRNA purification kit from Pharmacia. Reverse transcription and PCR amplification of cDNA was carried out using the Perkin Elmer RT-PCR kit. Primers that specifically amplify long- or middle-wave genes have been described previously (Neitz *et al.*, 1995) and were used to amplify a genomic DNA template. In the figure, the long- and middle-wave specific primers are aligned to exon 5 to indicate that they correspond to sequences in exon 5 that functionally distinguish between long- and middle-wave genes. These primers differ by nucleotide differences that encode one of two amino acids responsible for producing the majority of the spectral difference between long- and middle-wave pigments. The long-wave specific primer sequence which corresponds to sequences in exon 5 of long-wave genes is 5′GCAGTACGCAAAGATCATCACC 3′; the middle-wave specific primer sequence which corresponds to sequences in exon 5 of middle-wave genes is 5′AAGCAGAATGC-CAGGACCATC 3′. Primer 1 is specific to exon 3 sequences shared by long- and middle-wave genes; its sequence is 5′GGATCACAGGTCTCTGGTCTC 3′. To amplify long- or middle-wave specific gene fragments, the thermal cycling parameters were one cycle at 94°C for 5 minutes, followed by 30 cycles of 94°C for 1 minute, 61°C for 1 minute, 72°C for 1 minute and 30 seconds, followed by 1 cycle at 72°C for 10 minutes. Amplified long- or middle-wave gene fragments were used, separately, in a second round of amplification to amplify exon 3. The sequence of primer 2 is 5′CTGCTCCAACCAAAGATGG 3′, and it corresponds to sequences shared by long- and middle-wave genes. The thermal cycling parameters were as above except that the duration of the 72°C extension step was reduced to 1 minute for the 30 cycles of amplification. (**B**) Results of restriction digestion assay using enzyme *Bst*YI. Cut and uncut controls are indicated. The cut control shows that the enzyme is working with 100% efficiency. The uncut control shows that there is no extraneous nuclease activity. Lanes G3 and G4 contain middle-wave cDNA from two eye donors. Lanes R6 and R7 contain long-wave cDNAs from two eye donors. (**C**) Results of restriction digest assay using enzyme *Fok*I. Cut and uncut controls are indicated and show that the enzyme is working with perfect efficiency. The heteroduplex band contains heteroduplexes of exon 3 DNA from the different subtypes of long- or middle-wave genes.

Exon 3 amplified from middle-wave genes expressed in the eyes from donors 3 and 4 yielded only uncut fragments when incubated with *Bst*YI (Fig. 4B) or with *Bso*FI. However, when exon 3 from middle-wave genes expressed in these two eye donors was incubated with *Fok*I a mixture of cut and uncut fragments was observed. This indicates that eye donors 3 and 4 each express at least two subtypes of middle-wave genes, and that these subtypes differ at nucleotide position 951. This results in an amino acid difference at position 153 in the opsins. Spectral tuning studies suggest that this amino acid difference does not produce a shift in spectral peak (Asenjo *et al.*, 1994).

Exon 3 amplified from long-wave genes expressed in eyes from donor 7 yielded only uncut fragments when digested with *Bst*YI (Fig. 4A). However, a mixture of cut and uncut fragments was observed following incubation with *Fok*I (Fig. 4B) and *Bso*FI. Thus, eye donor 7 expresses at least two different long-wave genes.

Exon 3 amplified from long-wave genes expressed in the eyes of donor 6 yielded a mixture of cut and uncut fragments with all three enzymes (Fig. 4A,B). Quantitation of the ratio of the signal of the cut and uncut bands by Phosphor-Image analysis indicates that the two subtypes detected by *Bst*YI were expressed in the ratio of 2 cut:1 uncut. However, the ratio of uncut:cut subtypes detected by *Fok*I was about 10:1. This suggests that eye donor 6 expresses at least three different subtypes of long-wave genes.

Restriction enzyme *Bso*F I cleaves at nucleotide position 1032 when the nucleotide G is present but not when the nucleotide T is present. The polymorphism at nucleotide 1032 encodes the difference between the amino acids serine and alanine at position 180 of the opsins. Long-wave pigments with serine180 differ in spectral peak from long-wave pigments with alanine180 by about 5–7 nm (Asenjo *et al.*, 1994; Chan *et al.*, 1992; Merbs and Nathans, 1992a; Neitz and Neitz, 1995a; Neitz *et al.*, 1991). The results of the restriction analysis of the long-wave genes expressed in eyes from donors 6 and 7 were confirmed by direct sequence analysis. Thus, eye donors 6 and 7 each express at least two spectrally-different long-wave pigments.

Do the visual pigment genes beyond three expressed in individual eyes significantly contribute to colour vision?

Expression of multiple spectral subtypes of long- and middle-wave pigments may contribute to variations in normal colour vision (Neitz and Jacobs, 1986, 1989; Neitz *et al.*, 1993), but the effect is likely to be quite subtle. However, in individuals with colour vision defects in which all functional long-wave or all functional middle-wave genes have been deleted, the expression of more than one type of long- or middle-wave pigment contributes significantly to colour vision. That is, these individuals can use the spectral difference between subtypes of long- or subtypes of middle-wave pigments to derive enough colour information to allow them the pleasure of anomalous trichromacy and they are rescued from dichromacy.

References

Asenjo, A.B., Rim, J. and Oprian, D.D. (1994). Molecular determinants of human red/green colour discrimination. Neuron 12: 1131–1138.
Chan, T., Lee, M. and Sakmar, T. P. (1992). Introduction of hydroxyl-bearing amino acids causes bathochromic spectral shifts in rhodopsin: amino acid substitutions responsible for red–green colour pigment spectral tuning. J. Biol. Chem. 267: 9478–9480.
Merbs, S.L. and Nathans, J. (1992a). Absorption spectra of human cone pigments. Nature 356: 433–435.
Merbs, S.L. and Nathans, J. (1992b). Absorption spectra of the hybrid pigments responsible for anomalous colour vision. Science 258: 464–466.
Merbs, S.L. and Nathans, J. (1993). Role of hydroxyl-bearing amino acids in differentially tuning the absorption spectra of the human red and green cone pigments. Photochem. Photobiol. 58: 706–710.
Nathans, J., Piantanida, T.P., Eddy, R.L., Shows, T.B. and Hogness, D.S. (1986a). Molecular genetics of inherited variation in human colour vision. Science 232: 203–210.
Nathans, J., Thomas, D. and Hogness, D.S. (1986b). Molecular genetics of human colour vision: the genes encoding blue, green, and red pigments. Science 232, 193–202.
Neitz, J. and Jacobs, G.H. (1986). Polymorphism of the long-wavelength cone in normal human colour vision. Nature 323: 623–625.
Neitz, J. and Jacobs, G.H. (1989). Polymorphism of cone pigments in colour normals: evidence from colour matching. In: Verriest, G. and Drum, B. (eds.), Colour Vision Deficiencies IX: 27–34, Kluwer, Dordrecht.
Neitz, J. and Neitz, M. (1994). Colour vision defects. In: Wright, A.S. and Jay, B. (eds.), Molecular Genetics of Inherited Eye Disorders: 217–258. Harwood Academic Publishers, Chur, Switzerland.
Neitz, J., Neitz, M. and Jacobs, G.H. (1992). More than three cone types in normal colour vision. Adv. Colour Vision 4: 14–16.
Neitz, J., Neitz, M. and Jacobs, G.H. (1993). More than three different cone pigments among people with normal colour vision. Vision Res. 33: 117–122.
Neitz, M. and Neitz, J. (1995a). Genetic basis of photopigment variations in human dichromats. Vision Res. 35: 2095–2103.
Neitz, M. and Neitz, J. (1995b). Numbers and ratios of visual pigment genes for normal red–green colour vision. Science 267: 1013–1016.
Neitz, M., Neitz, J. and Grishok, A. (1995). Polymorphism in the number of genes encoding long-wavelength sensitive cone pigments among males with normal colour vision. Vision Res. 35: 2395–2407.
Neitz, M., Neitz, J. and Jacobs, G. H. (1991). Spectral tuning of pigments underlying red–green colour vision. Science 252: 971–974.
Winderickx, J., Battisti, L., Motulsky, AG. and Deeb, S.S. (1992). Selective expression of human X-chromosome-linked green opsin genes. Proc. Natl. Acad. Sci. USA 89: 9710–9714.
Yamaguchi, T., Motulsky, A.G. and Deeb, S.S. (1995). Expression of human cone pigment genes in the retina. Invest. Ophthalmol. Vis. Sci., Suppl. 36: S1054.

Department of Ophthalmology and Department of Cellular Biology and Anatomy
Medical College of Wisconsin
925 North 87th Street
Milwaukee, WI 53226-4812, USA

4. Phenotypes of anomalous trichromacy

S.K. SHEVELL and J.C. HE
(Chicago, IL, USA)

Abstract

This paper considers the substantial variation among anomalous trichromats, even among those within a discrete subtype such as simple deuteranomalous trichromacy. First, theoretical calculations show that relatively small individual differences in wavelengths of peak sensitivity of pigments and/or in optical densities can substantially affect Rayleigh matching, which is the procedure typically used to diagnose different types of anomalous trichromacy. Second, an ensemble of colour matches is introduced as a technique to assess individual differences in peak wavelength and density. Finally, results are reported from simple deuteranomals, two of whom are shown to have pigments with different peak wavelengths and two of whom are shown to have pigments with different optical densities. We argue that the standard classifications of anomalous trichromacy are too broad to relate phenotypes to genotypes.

Introduction

Diversity is a characteristic of human trichromatic vision. There is great variation in colour matching and discrimination among observers. The larger differences are recognized as specific subtypes of abnormal vision, such as simple deuteranomaly or extreme protanomaly, but even within a particular subtype there remain substantial individual differences. Advances in molecular genetics have revived interest in the physiological mechanisms that cause these individual differences in colour vision (phenotypes) because the mechanisms are fundamental to the relation between genotype and phenotype.

What phenotypes?

A *phenotype* is an observable characteristic of an individual, including a behaviour. The term will be used here to refer to a particular classification of a visual system that can be determined by some kind of measurement. Simple deuteranomalous trichromacy, for example, is a phenotype characterized by a particular range of measurements on a standard test of colour matching (that is, a particular behaviour). *Genotype* refers to characteristics of an individual's

C.R. Cavonius (ed.), Colour Vision Deficiencies XIII, pp. 45–58.

genes. The empirical relation between a genetic code and an expressed phenotype is important but considers only a part of the scientific question. The other part of the question concerns physiological mechanisms affected by the gene and which contribute to the phenotype. Physiological mechanisms are fundamental to understanding the link between genotype and phenotype.

From the beginning, only 10 years ago (Nathans *et al.*, 1986a,b), molecular geneticists have sought the relationship between the genotypes and the phenotypes of abnormal colour vision. Most work has focused on the genotypes, with the phenotypes specified no more finely than the classical distinctions of Franceschetti (1928). Substantial individual differences are well known, however, even among anomalous trichromats within a single Franceschetti subtype (Nelson, 1938; Wright, 1946; Willis and Farnsworth, 1952; Hurvich, 1972; Regan *et al.*, 1994). If these individual differences result from differences in genes coding for visual pigments, then there is a many-to-one relation between genotypes and each phenotype specified at the Franceschetti level. Alternatively, if other inherited or acquired processes can substantially affect abnormal colour vision, such as the post-receptoral differences posited by Jameson and Hurvich (1956), then there would be a one-to-many relation between cone-pigment genotype and colour-vision phenotype. A many-to-many relation would result from both of the above possibilities operating simultaneously (or many other possibilities). In general, it is simplistic to assume there is a one-to-one relation between cone-pigment genes and the classical types of abnormal colour vision.

One thesis of this paper is that the classical categories of anomaly are too broad to relate genotypes to phenotypes. Instead, the relation between cone-pigment genes and an individual's phenotypic colour vision requires an understanding of the physiological mechanisms mediating differences in abnormal colour vision. Conceptually, it might be possible to discover the relation between genotypes and phenotypes without considering the physiological elements mediating colour vision, if the relation were simple (say, one-to-one with Franceschetti classes). As a practical matter, however, simplicity seems not to hold. For example, in our collaboration with Drs. M. and J. Neitz at the Medical College of Wisconsin, we found a normal observer and a simple deuteranomalous observer who are undifferentiated with respect to their cone-pigment genes (one-to-many); and also two simple protanomalous observers with very different pigment genes (many-to-one). We propose, therefore, that phenotypes should be more finely distinguished than classical subtypes, and that these finer distinctions depend, in part, on properties of an individual's expressed visual pigments.

Properties of photopigments

Two important physiological causes of individual differences in colour matching are the wavelength of peak sensitivity of photopigments (λ_{max}) and the optical

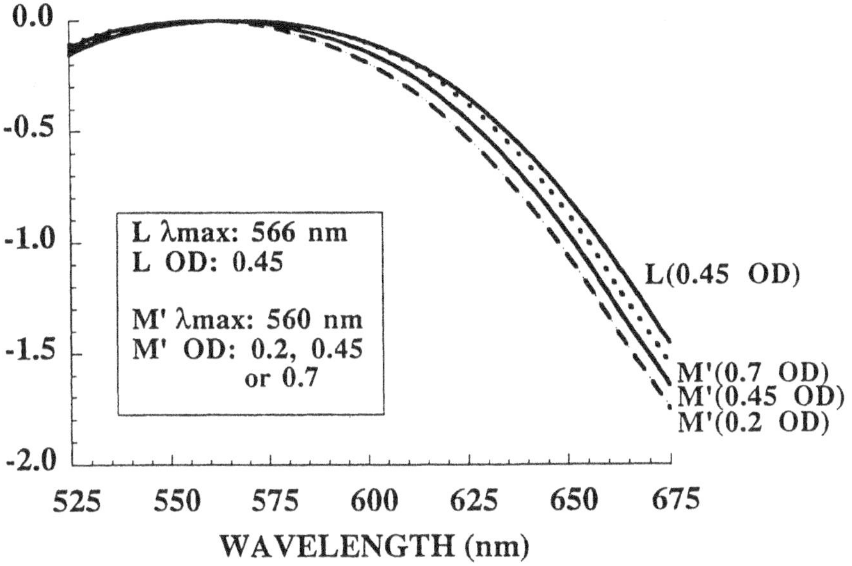

Fig. 1. Spectral sensitivities of hypothetical deuteranomalous photopigments: StdDA L and M′ (solid lines), M′ with 0.7 optical density (dotted line), M′ with 0.2 optical density (dashed line).

densities of the photopigments. Most genetic work has assumed a polymorphism of only the λ_{max} of pigments. Optical density, however, also can have a strong effect on colour matching and discrimination, and thus upon the phenotype, when the two pigments in the red-green range have similar λ_{max}, as in anomalous trichromacy (He and Shevell, 1995).

The wavelength λ_{max} specifies the location of the peak of the cone's spectral sensitivity function. Varying λ_{max} shifts the spectral sensitivity curve to the left or right but does not alter its shape (technically, shape invariance depends on units of log wavenumber on the abscissa; Bowmaker *et al.*, 1980). Optical density, on the other hand, alters only the shape of the sensitivity curve, which is broader at higher than at lower optical density. The spectral sensitivities of four hypothetical photopigments of deuteranomaly are shown in Figure 1. The solid curves, labeled L and M′ (0.45), have the peak wavelengths and optical density of the 'standard' deuteranomalous (StdDA) observer specified by DeMarco *et al.* (1992): 566 and 560 nm, respectively (energy based at the cornea), and optical density 0.45. The separation between these two curves, at least theoretically, is the basis of deuteranomalous chromatic discrimination from 545 to 700 nm.

The dashed and dotted lines in Figure 1 are plots of M′ sensitivity ($\lambda_{max} = 560$ nm) but with optical density of 0.20 or 0.70, respectively. At longer wavelengths, the sensitivity functions with the same λ_{max} but different densities are about as distinct as the 'standard' deuteranomalous L and M′ sensitivities with different λ_{max} but equal density. In general, a modest difference in only λ_{max} or a difference in only optical density can cause a similar difference in relative sensitivity at long wavelengths. Therefore, variation in optical density or variation in λ_{max} could cause similar individual differences in anomalous trichromacy. Of course, these two mechanisms need not be mutually exclusive.

The effect of λ_{max} and optical density of pigments on the Rayleigh match

The Rayleigh match is the definitive measurement to diagnose and classify deuteranomalous and protanomalous trichromatic colour vision. This colour match in modern instruments uses a split circular field of about 2° visual angle. One hemifield is typically an admixture of 545+670 nm, and the other hemifield is 589 nm. The proportion of 670 nm in the admixture field is set on each trial by the test administrator. The observer controls the radiance of the 589 nm light, and reports whether it can be set to match the admixture field. Repeating this procedure at many admixture proportions yields the range of proportions perceived by the observer to match 589 nm. This range is the measure from the Rayleigh match.

For any observer, the quantal match between the 545+670 nm admixture and the 589 nm field can be calculated if pre-receptoral spectrally selective transmissivity and photopigment spectral sensitivities are known. Estimating the match range requires an additional assumption about chromatic discrimination. We assume here the model of Boynton and Kambe (1980), although the exact model of discrimination is not critical (for a fuller description see He and Shevell, 1995). For deuteranomalous and normal observers, the proportion of 670 nm light in the admixture is expressed in the usual 'deutan mode' of anomaloscopic calibration (Mitchell and Rushton, 1971; Pokorny *et al.*, 1973); for protanomals, the admixture proportion is given in 'protan mode'.

Consider the effect of variation in photopigment λ_{max} on the Rayleigh matches of anomalous trichromats. Figure 2 shows calculations of the deuteranomalous match midpoint (dashed line) and range (solid lines) when the λ_{max} of the M′ pigment or L pigment is shifted by ± 3 nm from the StdDA values (top left and right panels, respectively). Optical density is StdDA: 0.45 for both pigments. Variation in λ_{max} has only a small effect on the deuteranomalous match, even when the two pigments are separated by as little as 3 nm. Shifts of λ_{max} alter primarily the match range. Shifting the λ_{max} of the protanomal's M or L′ cone has an even smaller effect on the Rayleigh match (bottom left and right panels). An important implication of these results is that genotypes that cause modest individual differences in pigment λ_{max} may not cause observable differences in Rayleigh matches.

The effect of optical density on the Rayleigh match is complex, and depends on the difference in λ_{max} between the two pigments. Figure 3 shows the effect of

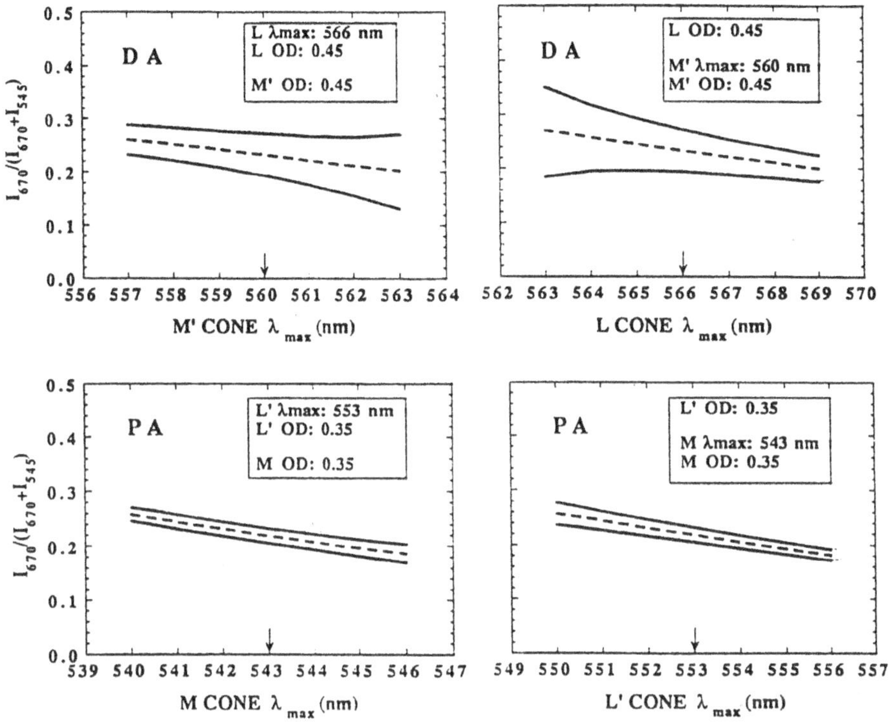

Fig. 2. Rayleigh quantal match (dashed line) and range (solid lines) as a function of the λ_{max} of one anomalous pigment, with optical density and the other pigment's λ_{max} held fixed at 'standard' values (see text). Arrow on each horizontal axis indicates the 'standard' value. Upper left, DA M′ varied; upper right, DA L varied; lower left, PA M varied; lower right, PA L′ varied.

varying optical density over a wide range, assuming that the densities of an observer's two photopigments are the same. For economy, only protanomalous results are shown (similar calculations for deuteranomals are given in He and Shevell, 1995). The separation between the two cones' λ_{max} is different in each panel: 10 (the 'standard' protanomalous [StdPA] value of DeMarco *et al.*, 1992), 7, 4 or 2 nm. With a given λ_{max} separation, varying density equally for both photopigments affects primarily the match midpoint, as for normal observers (Pokorny and Smith, 1976). Varying λ_{max} separation affects primarily the match range, which can be moderately large when the separation is only 2 nm.

When the density of one pigment is held fixed, varying the density of the other pigment can affect both midpoint and range, particularly when the separation between pigments is only a few nanometers. Figure 4 shows for protanomals the Rayleigh midpoint and range when only the density of L′ is varied (M density fixed at 0.35), with λ_{max} separation of 10, 7, 4 or 2 nm (each separation in a

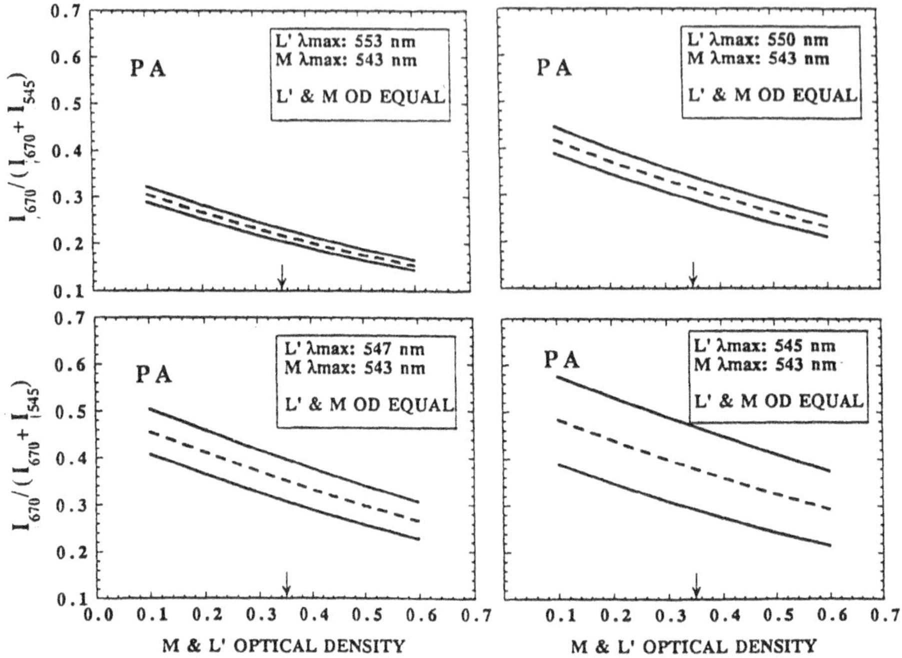

Fig. 3. Protanomalous Rayleigh quantal match (dashed line) and range (solid lines) as a function of optical density held equal for both pigments. Arrow on each horizontal axis indicates the 'standard' value. The separation between the two pigments' λ_{max} values is different in each panel: upper left, 10 nm; upper right, 7 nm; lower left, 4 nm; lower right, 2 nm.

different panel). The match range can be quite wide when λ_{max} separation is only a few nanometers.

The above theoretical results show that variation in optical density could strongly affect anomalous colour matches, particularly when combined with individual differences in λ_{max}. We now turn to seeking empirical evidence of individual differences in λ_{max} and optical density among anomalous trichomats.

Measuring individual differences in λ_{max} and optical density

It is not trivial to determine by psychophysical measurement whether two anomals differ from each other in λ_{max} or in optical density (or in both). Most psychophysical procedures used to assess individual differences in colour vision are designed to eliminate effects of pre-receptoral, spectrally selective filtering (von Kries, 1924; Wright, 1929). Residual individual differences, after taking account of pre-receptoral filtering, are interpreted as differences in receptors,

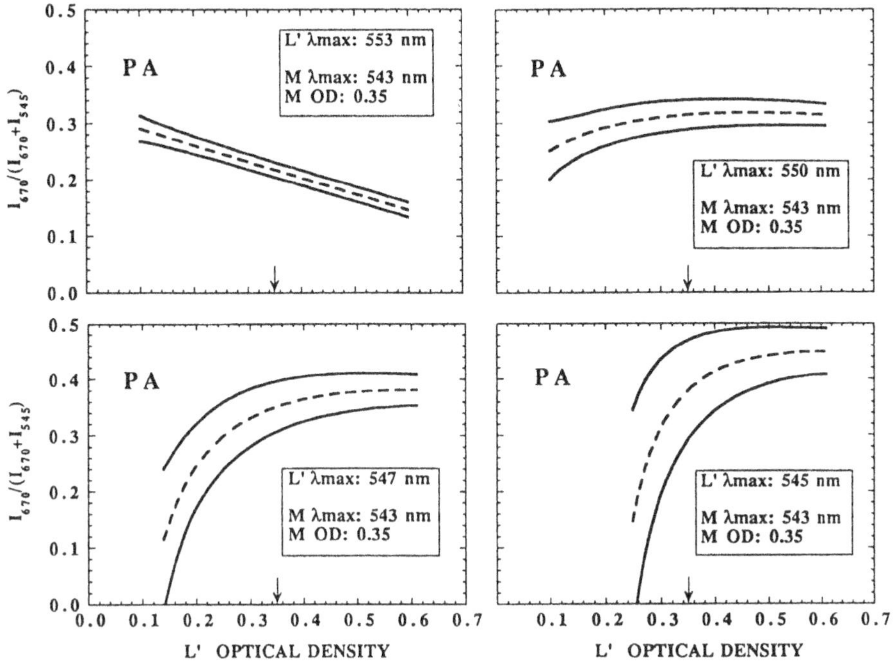

Fig. 4. As Fig. 3 but the optical density of only the L′ pigment is varied (optical density of M at StdPA value of 0.35).

which may reflect either differences in λ_{max} or in density (or in both). For example, the colour matches of two deuteranomalous trichromats were found to be different even after taking account of pre-receptoral filtering. This has been interpreted as evidence of individual differences in the λ_{max} of deuteranomalous pigments (Alpern and Moeller, 1977). The measured differences might, however, reflect individual differences in only optical density.

The Rayleigh match cannot distinguish individual differences in λ_{max} from differences in density. Calculations like those in Figures 2–4 show that the Rayleigh quantal-match values that would result from all pairs of deuteranomalous M′ and L pigments with λ_{max} within ± 6 nm of the StdDA values span from 0.16 to 0.37 (with optical density fixed at 0.45, and of course λ_{max} of M′ less than L). Quantal matches that would occur with fixed StdDA λ_{max} values and any combination of M′ and L optical densities from 0.2 to 0.7 are between 0.00 and 0.36. Therefore, varying only density results in a larger range of quantal-match values than varying only λ_{max}. Clearly, moderate differences in λ_{max} between two observers of a given anomalous subtype cannot be determined definitively using the Rayleigh match alone.

An ensemble of several different colour matches, however, has the potential to distinguish differences in λ_{max} from differences in optical density. The overall design is to use some measures that are more sensitive to λ_{max}, and other measures that are more sensitive to optical density. All measures are chosen to be insensitive to pre-receptoral filtering. While very extreme differences in only optical density might account for a single unusual measurement, extreme values of density can be eliminated as a possibility if other measurements are inconsistent with the extreme values needed to account for the first measurement. Thus no single colour match is definitive but a set of several matches from each observer can assess individual differences in λ_{max} or in optical density.

An ensemble of colour matches

In addition to the Rayleigh match, described above, three other colour matches are useful for detecting individual differences in λ_{max} and/or optical density. These are the Rayleigh match with bleached photopigment, the extended Rayleigh equation (von Kries, 1924) and the dual Rayleigh match (He and Shevell, 1994). These colour matches, as a set, can be used to infer properties of photopigments.

The *Rayleigh match with bleaching* is similar to the typical Rayleigh match except that it is performed with a substantial proportion of photopigment bleached to the transparent state. A reduction of normal optical density in both pigments, which can be achieved by pre-exposure to a bleaching light above about 20 000 td, increases the proportion of 670 nm light in the admixture required for a match. We use a Rayleigh-match measurement with a 40 000–100 000 td bleaching light (~65–85% bleach) in comparison to a Rayleigh-match measurement at the typical light level of about 60 td (virtually no bleach). A large difference between these two measurements is expected if the initial density is substantial; in this case, an 85% reduction is a loss of a large *magnitude* of density. A small or negligible difference, on the other hand, is expected with low initial density because 85% reduction of a small initial density is a small change in magnitude that has little effect on colour matching. In order to eliminate any influence of pre-receptoral transmissivity, the quantitative index of bleaching is based on the admixture ratio, I_{670}/I_{545}, at each level of bleaching light: $(I_{670}/I_{545})^{100\,000}$ divided by $(I_{670}/I_{545})^{60}$.

The *extended Rayleigh match* follows a concept introduced by von Kries. It is designed to eliminate any effect of pre-receptoral factors on individual differences in colour matching. If a standard observer is defined with pigments of given λ_{max} and optical density, then the quantal match can be calculated for the admixture 545+670 to match any test wavelength above 550 nm. If an individual subject differs from the standard observer only because of pre-receptoral filtering, then the ratio of the individual's matches to the standard theoretical matches will be the same at all test wavelengths. The index that is independent of pre-receptoral spectrally selective filtering is $(I_{670}/I_{545})^{Obs,\lambda}/(I_{670}/I_{545})^{Std,\lambda}$, where λ indicates the test wavelength.

Although the extended Rayleigh match is designed to eliminate individual

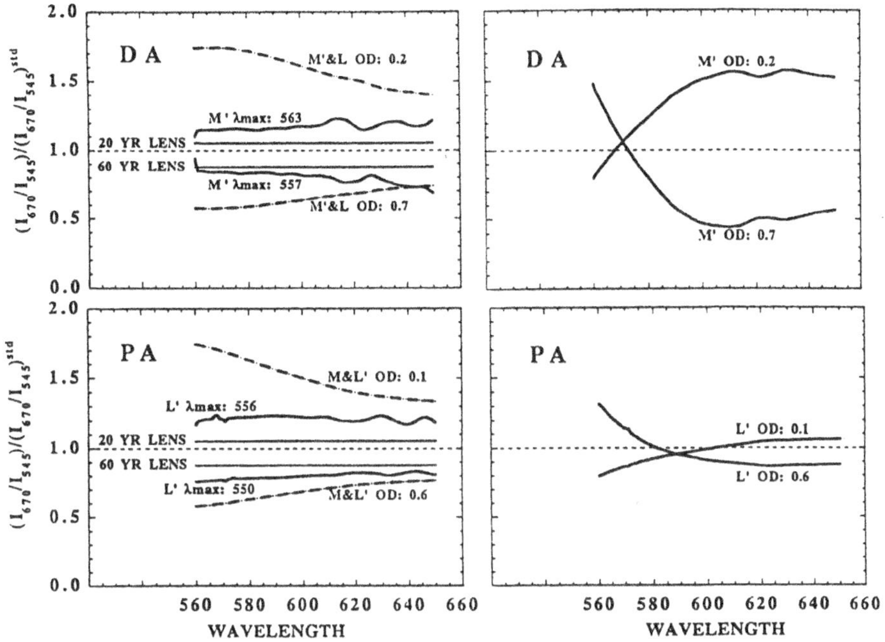

Fig. 5. Extended Rayleigh ratio as a function of test wavelength, for deuteranomals (top panels) and protanomals (bottom panels). Left panels show values for the following differences from the 'standard' anomalous observer: 20-year-old lens, 60-year-old lens, ±3 nm shift of M′ or L′ λ_{max}, or low or high optical density held equal for both pigments. Right panels show values for low or high optical density of only one pigment.

differences in pre-receptoral factors, quantitative modelling shows that the procedure attenuates also individual differences in λ_{max}. This is shown in Figure 5, where the extended Rayleigh-match ratio, $(I_{670}/I_{545})^{Obs,\lambda}/(I_{670}/I_{545})^{Std,\lambda}$ (ordinate), is plotted as a function of the wavelength in the test field (abscissa). Solid horizontal lines show the extended Rayleigh-match ratios for deuteranomals (top left panel) and protanomals (bottom left panel) who differ in only lens transmissivity (20 or 60 years of age). The ratio is constant at all test wavelengths, as understood by von Kries. A modest difference between observers in pigment λ_{max}, on the other hand, does not result in a perfectly horizontal line. The solid lines labelled with a λ_{max} wavelength show the effect of modestly lower or higher λ_{max} of the deuteranomalous M′ cone (top left panel) or the protanomalous L′ cone (bottom left panel). Note that the deviations from horizontal due to differences in λ_{max} are modest. The location of these solid lines relative to the dotted reference line at 1.0 shows that λ_{max} affects the extended Rayleigh measurement, but not in a way that depends strongly on test wavelength. Nagy *et al.* (1985) have made a similar observation.

Individual differences in optical density could, however, cause large deviations from the horizontal. The dashed lines in the left panels show the effect of varying density equally in both pigments. The lines in the right panels show the effect of varying optical density of only one pigment, with the other pigment's density held fixed at the standard value (DeMarco *et al.*, 1992). The extended Rayleigh ratio is more sensitive to plausible variation in optical density than in λ_{max}. Extended Rayleigh matches, in fact, may be used to infer constraints on individual differences in optical density. Note that previous studies that inferred individual differences in λ_{max} using the extended Rayleigh procedure (or derivatives) may instead have detected individual differences in optical density.

The *dual Rayleigh ratio* uses the typical Rayleigh match together with a second match in which the long-wave light in the admixture is changed from 670 to 620 nm (while the test wavelength remains at 589 nm). The ratios from the admixture field for these two matches, $(I_{620}/I^*_{545})/(I_{670}/I_{545})$, is an index that is affected very little by pre-receptoral factors (He and Shevell, 1994). The dual Rayleigh ratio is very sensitive to individual differences in λ_{max} but it also varies with optical density. It cannot by itself detect individual differences in λ_{max} among anomalous trichromats of a given subtype. An inference about differences in λ_{max} depends on the dual Rayleigh ratio together with other measures that constrain the possible range of individual differences in optical density.

Case examples from deuteranomaly

Two deuteranomals with pigments that differ in λ_{max}

The ensemble of matches reveals that two simple deuteranomals we have studied, observers BT and DB, have pigments with different λ_{max}. The dual Rayleigh ratios for these two observers are plotted as points in the left panel of Figure 6. The shaded region represents the mean ± 1 S.E.M based on three replications of the matches. Also plotted in this panel are theoretical values for the dual Rayleigh ratio, assuming individual differences in only optical density. These results, taken in isolation, cannot exclude the possibility that these observers might have pigments with the same λ_{max} values but different optical densities. The measured value for BT is consistent with many plausible densities; the measurement from DB would occur only if the M′ density were above 0.30. This M′ density, however, is inconsistent with this observer's extended Rayleigh measurement (right panel). His extended Rayleigh value would require a very low M′ density (less than 0.10). These two measurements together show that BT and DB cannot have 'standard' DeMarco *et al.* λ_{max} values, with individual differences in only optical density.

A possible alternative is that these observers have the same pigment λ_{max} values but that these differ from those of the StdDA. This is evaluated by comparing the complete ensemble of matches to theoretical values for all

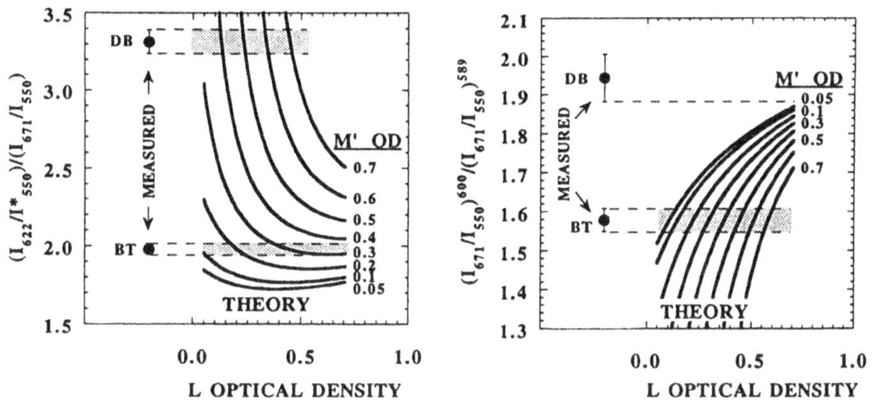

Fig. 6. The left panel (right panel) shows measurements of the dual Rayleigh ratio (extended Rayleigh ratio) for deuteranomalous observers BT and DB (solid circles, with shaded areas covering ± 1 S.E.M.). Solid lines in each panel are theoretical values for the measurement based on StdDA λ_{max} wavelengths and the various optical densities shown.

combinations of λ_{max} (M′ range 555–575 nm, L range 561–576 nm) and optical density (range 0.05–0.70 for M′ and L). As a worst-case analysis, the possible values for a measurement from an observer are calculated using the extremes of the observer's match ranges: the error-free value for a measurement of ratio A/B, where A is a midpoint $[1/2(A_L+A_H)]$ of a match with low and high endpoints A_L and A_H and similarly B is a midpoint of a match with endpoints B_L and B_H, is taken as within the interval from A_L/B_H to A_H/B_L. If a theoretical value falls within this interval, then that combination of λ_{max} and optical density values is taken as plausible for that measurement for that observer. Results show that observers BT and DB have no values of λ_{max} in common that can account for their complete sets of measurements. In general, λ_{max} is higher for DB, who requires L-cone λ_{max} of at least 572 nm and M′-cone λ_{max} of at least 566 nm (except for four specific combinations: L λ_{max} = 570 nm and M′ λ_{max} of 569 nm, or L λ_{max} of 571 nm and M′ λ_{max} of 568, 569 or 570 nm). Observer BT, on the other hand, requires L λ_{max} of 569 nm or lower unless M′ λ_{max} is at or below 561 nm. Therefore, these two observers have photopigments with different wavelengths of peak sensitivity.

Two deuteranomals with pigments that differ in optical density

The ensemble of matches shows also that the pigments of two other observers, JH and PM, do not have the same optical densities. Results from the Rayleigh match with bleaching (Fig. 7) indicate that PM has low M′ density, while the plausible range of L density is wide (though L density cannot be extremely high).

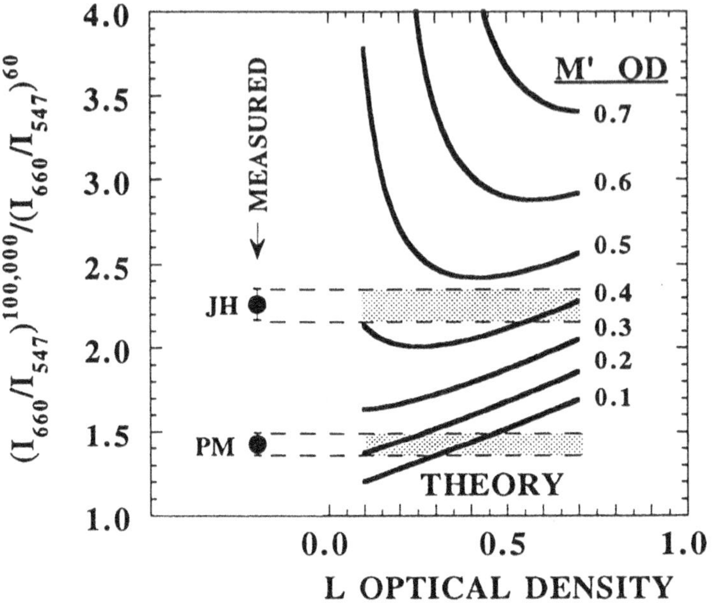

Fig. 7. Measurements of the Rayleigh match with bleaching for deuteranomalous observers JH and PM (solid circles, with shaded areas covering ± 1 S.E.M.). Solid lines are theoretical values for the measurement based on StdDA λ_{max} wavelengths and the various optical densities shown.

The bleaching measurement for JH, on the other hand, shows M′ density near 0.4 while L density is not constrained. If one assumes these two observers have the StdDA λ_{max} values, then one could infer from Fig. 7 that they differ from each other in optical density of the M′ pigment. This is an unjustified assumption, however, so again the worst-case analysis is used to find those combinations of λ_{max} and optical density for M′ and L that are consistent with their ensembles of measurements. This analysis reveals that observer PM has low density for both pigments: M′ density is constrained to less than 0.15, and L density to between 0.15–0.30. Optical density for observer JH is substantially higher. The minimum plausible L density is above 0.45. If L density is 0.55 or less then M′ density can range from 0.20–0.45, while if L density is higher (up to 0.70) then M′ density is little constrained (0.10–0.55).

It is worth noting that these two observers have nearly identical Rayleigh matches on the Nagel anomaloscope (18–20 and 19–21 for PM and JH, respectively), and have similar values for the extended Rayleigh ratio (means 1.80 and 1.74, respectively) and the dual Rayleigh ratio (means 2.42 and 2.64, respectively). Detecting their difference in phenotype depends on completing all of the matches. Further, lower optical density for PM, inferred from the bleach, allows an inference about the λ_{max} of his photopigments in the light of how his other measurements are affected by density.

Acknowledgment

Supported in part by PHS grant EY-04802. Publication partially supported by Research to Prevent Blindness.

References

Alpern, M. and Moeller, J. (1977). The red and green cone visual pigments of deuteranomalous trichromacy. J. Physiol. (Lond.) 266: 647–675.

Bowmaker, J.K., Dartnall, H.J.A. and Mollon, J.D. (1980). Microspectrophotometric demonstration of four classes of photoreceptor in the Old World primate, *Macaca fascicularis*. J. Physiol. (Lond.) 298: 131–143.

Boynton, R.M. and Kambe, N. (1980). Chromatic difference steps of moderate size measured along theoretically critical axes. Colour Res. Appl. 5: 13–23.

DeMarco, P., Pokorny, J. and Smith, V.C. (1992). Full-spectrum cone sensitivity functions for X-chromosome-linked anomalous trichromats. J. Opt. Soc. Am. A 9: 1465–1476.

Franceschetti, A. (1928). Die Bedeutung der Einstellungsbreite am Anomaloskop für die Diagnose der einzelnen Typen der Farbensinnstörungen, nebst Bemerkungen über ihren Vererbungsmodus. Schweiz. Med. Wschr. 58: 1273–1279.

He, J.C. and Shevell, S.K. (1994). Individual differences in cone photopigments of normal trichromats measured by dual Rayleigh-type color matches. Vision Res. 34: 367–376.

He, J.C. and Shevell, S.K. (1995) Variation in color matching and discrimination among deuteranomalous trichromats: theoretical implications of small differences in photopigments. Vision Res. 35: 2579–2588.

Hurvich, L.M. (1972). Color vision deficiencies. In: Jameson, D. and Hurvich, L.M. (eds.), Handbook of Sensory Physiology, Vol. VII/4: 582–624. Springer, Berlin.

Jameson, D. and Hurvich, L.M. (1956). Theoretical analysis of anomalous trichromatic color vision. J. Opt. Soc. Am. 46: 1075–1089.

Mitchell, D.E. and Rushton, W.A.H. (1971). Visual pigments in dichromats. Vision Res. 11: 1033–1043.

Nagy, A.L., Purl, K.F. and Houston, J.S. (1985). Cone mechanisms underlying the color discrimination of deutan color deficients. Vision Res. 25: 661–669.

Nathans, J., Thomas, D. and Hogness, D.S. (1986a). Molecular genetics of human color vision: the genes encoding blue, green, and red pigments. Science 232: 193–202.

Nathans, J., Piantanida, T.P., Eddy, R.L., Shows, T.B. and Hogness, D.S. (1986b). Molecular genetics of inherited variation in human color vision. Science 232: 203–210.

Nelson, J.H. (1938). Anomalous trichromatism and its relation to normal trichromatism. Proc. Phys. Soc. (Lond.) 50: 661–697.

Pokorny, J., and Smith, V.C. (1976). Effect of field size on red-green color mixture equations. J. Opt. Soc. Am. 66: 705–708.

Pokorny, J., Smith, V.C. and Katz, I. (1973). Derivation of the photopigment absorption spectra in anomalous trichromats. J. Opt. Soc. Am. 63: 232–237.

Regan, B.C., Reffin, J.P. and Mollon, J.D. (1994). Luminance noise and the rapid determination of discrimination ellipses in colour deficiency. Vision Res. 34: 1279–1299.

von Kries, J. (1924). Normal and anomalous color systems. In: von Helmholtz, H., Treatise on Physiological Optics (3rd ed), Vol. II (Southall, J.P.C., Trans.): 395–425. Optical Society of America, Rochester, NY.

Willis, M.P. and Farnsworth, D. (1952). Comparative evaluation of anomaloscopes. Med. Res. Lab. Rep. No 190, Bur. Med. Surg. U. S. Navy Dept., Washington D. C.

Wright, W.D. (1929). A re-determination of the trichromatic coefficients of spectral colors. Trans. Opt. Soc. 30: 141–164.

Wright, W.D. (1946). Researches on Normal and Defective Colour Vision. Henry Kimpton, London.

Departments of Psychology, and Ophthalmology & Visual Science
University of Chicago
939 East Fifty-seventh Street
Chicago, IL 60637, USA

5. Ratio of M/L pigment gene expression decreases with retinal eccentricity

S.A. HAGSTROM, J. NEITZ and M. NEITZ

(Milwaukee, WI, USA)

Abstract

Evidence has accumulated to indicate that, on average, there are about 1.5–2.0 long-wave cones in the human fovea for each middle-wave cone. Much less is known about how the ratio of these cone populations might change with retinal eccentricity. We have examined how the ratio of middle- to long-wave cone pigment mRNA changes with eccentricity in individual human retinas. Retinas were examined from seven male eye donors. Patches of retina, 6 mm in diameter were removed using a trephine at three different eccentricities. mRNA was reverse transcribed, photopigment cDNAs were amplified, and the relative amounts of middle- and long-wave pigment mRNA were determined for each patch. For 6 mm patches centred on the fovea, the average ratio of middle- to long-wave mRNA was about 2:3. This value is similar to the average ratio of middle- to long-wave cones previously estimated for the human fovea. There were significant changes in the cone ratios with increasing eccentricities ($p = 0.0004$). For patches of retina taken from the most peripheral locations (centred 12 mm eccentric from the fovea), the average middle- to long-wave mRNA ratio decreased to about 1:3. These results can be explained by a decrease in the ratio of middle- to long-wave cone populations with increasing eccentricity from central retina to periphery.

Introduction

Knowing about the relative numbers of the different cone types and their ratios in the retina is important for understanding the biological mechanisms of human colour vision. Psychophysical experiments (Cicerone and Nerger, 1989; Pokorny *et al.*, 1991; Nerger and Cicerone, 1992) and microspectrophotometry data on single cone cells (Dartnall *et al.*, 1983) suggest that long-wave cones outnumber middle-wave cones by a factor of 1.5–2.0. Reports also suggest that there is considerable variability in this ratio across individuals (Rushton and Baker, 1964; Vimal *et al.*, 1989; Wesner *et al.*, 1991; Pokorny *et al.*, 1991). These individual differences in cone populations have long been offered to explain the large individual differences in relative sensitivity to red and green flickering

C.R. Cavonius (ed.), Colour Vision Deficiencies XIII, pp. 59–65.

lights in people with normal colour vision (e.g. de Vries, 1947). Much less is known about this ratio in the peripheral retina and how it might change with retinal eccentricity (however, see Cicerone *et al.*, 1994). In the present study, the relationship between middle- to long-wave cone pigment ratio and eccentricity in individual human retinas is addressed by examining the relative abundance of middle- to long-wave mRNA at different retinal eccentricities.

Methods

Tissue and nucleic acid isolation

Human eyes were obtained through the Wisconsin Lions Eye Bank. One eye from each of seven male donors ranging from 39 to 77 years of age was used in these experiments. Each eye was enucleated within 3 h of death, placed on ice, and the retina dissected out immediately. A trephine was used to punch 13 patches (6 mm in diameter) of retina from each eye as shown in Figure 1. A central patch was centred approximately on the fovea. The six patches surrounding the central one were combined to provide a mid-peripheral retinal sample. The six outermost patches were combined to provide a far-peripheral retinal sample. Nucleic acid (DNA and RNA) was isolated from each of the three retinal tissue samples from each eye by incubating the tissue in 200 μl of STE (0.1 M NaCl, 10 mM Tris–HCl pH 8.0, 1 mM EDTA pH 8.0) 1% SDS and 0.9 mg proteinase K for 8–12 h at 55°C. Samples were extracted first with a 1:1 mixture of phenol/chloroform, then with chloroform. Nucleic acids were ethanol precipitated and resuspended in 100 μl 10 mM Tris–HCl, 1 mM EDTA pH 8.0.

Determination of the ratio of middle- to long-wave cone pigment cDNAs expressed in retinal tissue

From the nucleic acid preparations for each tissue sample, mRNA was reverse transcribed into cDNA and the visual pigment cDNA was amplified using the RT-PCR kit from Perkin Elmer. A segment of the long- and middle-wave pigment cDNAs containing the 3′ end of exon 4 and the 5′ end of exon 5 was amplified. The upstream primer sequence was 5′ ATCCGAGCGGTGG-CAAAGC 3′, and the downstream primer sequence was 5′ TGGCAGCAG-CAAAGCATGC 3′. Thermal cycling parameters were: 1 cycle at 94°C for 5 min, followed by 40 cycles of 94°C for 45 sec, 57°C for 45 sec, and 72°C for 45 sec, and 1 cycle of 72°C for 10 min. Polymerase chain reaction (PCR) products were gel purified as previously described (Neitz and Neitz, 1995; Neitz *et al.*, 1995a). An aliquot of the amplified cDNA was used in a second round of amplification as described above.

Visual pigment cDNAs were analysed using a restriction endonuclease assay. The PCR product was purified by ultrafiltration, end-labelled with ^{32}P and

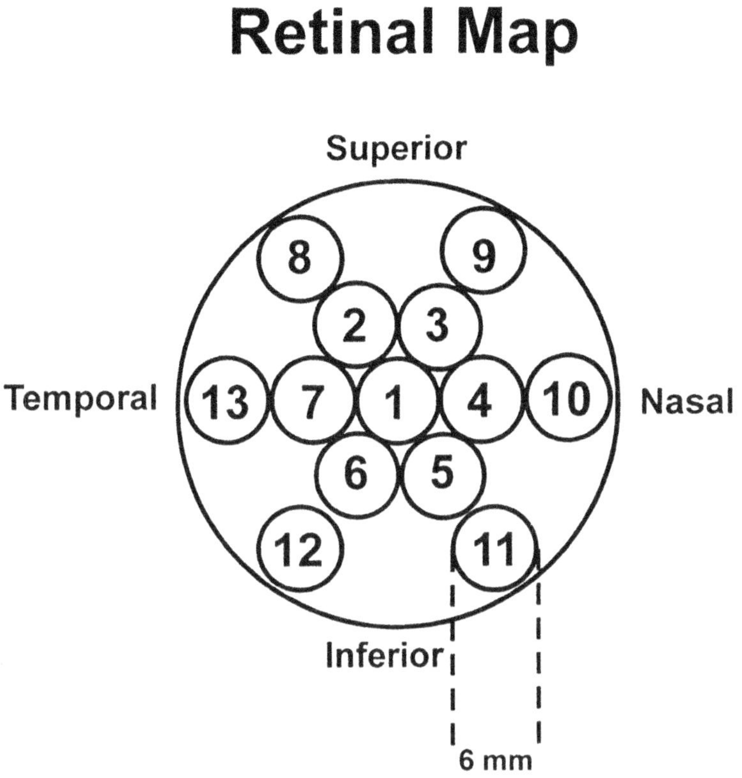

Fig. 1. Location of the patches of retina. A trephine was used to punch 13 patches (6 mm diameter) of retina from each eye. The central retina, patch 1, is centred on the fovea. Patches 2–7 were combined to produce the mid-peripheral sample. Patches 8–13 were combined to produce the far-peripheral sample.

digested with *Rsa*I as described previously (Neitz and Neitz, 1995). Digestion products were electrophoretically separated on 8% polyacrylamide. The gels were fixed, dried onto filter paper, and PhosphorImaged. Measurements were repeated on each sample. Independent replications (2–4/sample) were made on a different aliquot of the nucleic acid preparation.

Results

We examined the ratio of middle- to long-wave cDNAs at three retinal eccentricities from seven human eyes. Three tissue samples from each eye were: a 6 mm patch centred approximately on the fovea, an eccentric ring of tissue with a radius of approximately 6 mm, and a ring with a radius of approximately 12 mm.

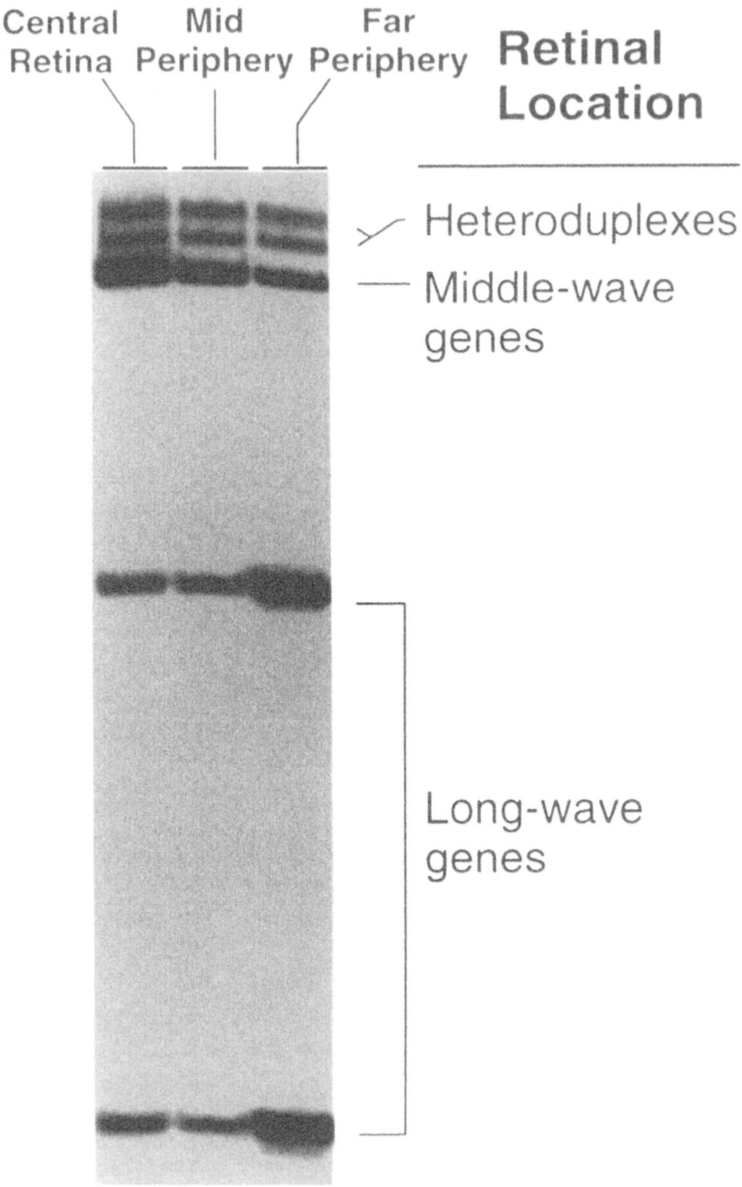

Fig. 2. Analysis of the ratio of middle- to long-wave cDNAs. Five bands are observed – two heteroduplex bands and three homoduplex bands. This segregation of DNA into homoduplex and heteroduplex pairing is a by-product of the PCR process. The heteroduplexes each contain one strand from a long-wave gene and the other strand from a middle-wave gene. One homoduplex corresponds to uncut middle-wave cDNA fragments and the other two are *RsaI* cleavage products of long-wave cDNA fragments. The middle- to long-wave cDNA ratio was determined for each of the regions of retina as previously described (Neitz and Neitz, 1995).

Table 1. Ratio of middle- to long-wave cDNAs for three retinal regions.

Subject	Age	Eye	No. of experiments	M/L cDNA (Mean ± SEM) Central	Mid-periphery	Far-periphery
84	39	O.D.	2	0.89 ± 0.06	0.48 ± 0.04	0.30 ± 0.03
87	41	O.D.	4	0.58 ± 0.09	0.48 ± 0.10	0.40 ± 0.02
28	45	O.S.	2	1.17 ± 0.06	0.98 ± 0.02	0.21 ± 0.01
97	48	O.D.	3	0.54 ± 0.13	0.48 ± 0.07	0.31 ± 0.03
23	55	O.D.	2	0.50 ± 0.07	0.69 ± 0.01	0.18 ± 0.02
56	59	O.D.	3	0.61 ± 0.14	0.48 ± 0.03	0.31 ± 0.05
69	77	O.D.	4	0.67 ± 0.05	0.47 ± 0.02	0.31 ± 0.02
			Mean	0.71	0.58	0.29

Amino acid substitutions encoded by exon 5 of the pigment genes produce the majority of the spectral difference between long- and middle-wave pigments (Asenjo *et al.*, 1994; Chan *et al.*, 1992; Merbs and Nathans, 1993; Neitz and Neitz, 1994; Neitz *et al.*, 1989, 1991, 1995b). Nucleotide differences in exon 5 generate a *Rsa*I restriction site in long-wave genes that is absent in middle-wave genes (Nathans *et al.*, 1986). We have taken advantage of this polymorphism to distinguish long- from middle-wave cDNAs (Neitz and Neitz, 1995). Figure 2 shows an example of the results from one eye for three eccentricities. Two cut bands represent long-wave cDNA fragments produced by *Rsa*I cleavage, and one uncut band represents middle-wave cDNA fragments. Two heteroduplex bands are present when both long- and middle-wave genes are expressed. From measurements of the radioactivity in the five bands, the middle- to long-wave cDNA ratio was determined for each retinal region (Neitz and Neitz, 1995). Half the radioactive counts in the heteroduplexes (H/2) are added to the radioactive counts in the uncut band (U), and the other half are added to the sum of the counts in the two cut bands (C). The final quotient (U+H/2)/(C+H/2) yields an estimate of the ratio of middle- to long-wave cDNAs in the sample.

Table 1 shows the results from seven eyes. On average, the middle- to long-wave cDNA ratio decreases gradually across the retina. The decrease from the patch centred on the fovea to the most eccentric is about two-fold. This change in ratio with eccentricity is gradual but highly statistically significant ($r^2 = 0.49$; $p = 0.0004$). There also appears to be more variability in the middle- to long-wave cDNA ratio in the central retina than in the most peripheral samples. The changes in middle- to long-wave cDNA ratio with eccentricity are plotted in Figure 3. Each data point is the mean of the middle- to long-wave cDNA ratio for experiments on one eye at the plotted retinal region.

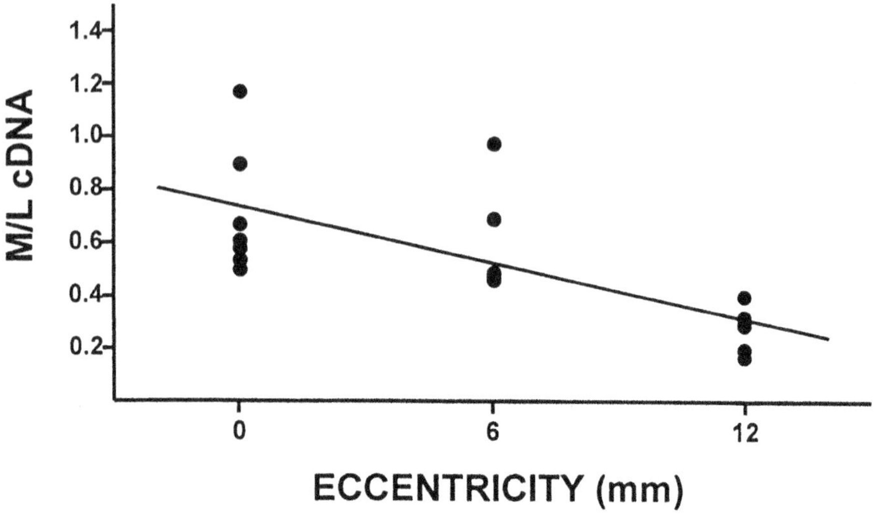

Fig. 3. Change in the middle- to long-wave cDNA ratio with increasing eccentricity. Each data point is the mean middle- to long-wave cDNA ratio for experiments on one eye at the plotted retinal region.

Discussion

Our results indicate that the middle- to long-wave cDNA ratio decreases with increasing retinal eccentricity by a factor of about 2. There were also considerable individual differences, similar to those reported earlier for estimates of foveal middle- to long-wave cone ratios. Interestingly, there is a suggestion in our results that the individual differences are greater for the central retina than for the most peripheral samples. We would suggest that the variability in the middle- to long-wave cDNA ratio that is measured in these experiments, both across individuals and across the retina, reflects differences in the ratio of middle- to long-wave cones.

Acknowledgements

This work was supported by NEI Grant EY09620 and a Research to Prevent Blindness Career Development Award to M. Neitz.

References

Asenjo, A.B., Rim, J. and Oprian, D.D. (1994). Molecular determinants of human red/green color discrimination. Neuron, 12: 1131–1138.

Chan, T., Lee, M. and Sakmar, T.P. (1992). Introduction of hydroxyl-bearing amino acids causes bathochromic spectral shifts in rhodopsin: amino acid substitutions responsible for red–green color pigment spectral tuning. J. Biol. Chem. 267: 9478–9480.

Cicerone, C.M. and Nerger, J.L. (1989). The relative numbers of long-wavelength-sensitive to middle-wavelength-sensitive cones in the human fovea centralis. Vision Res. 29: 115–128.

Cicerone, C.M., Gowdy, P.D. and Otake, S. (1994). Composition and arrangement of the cone mosaic in the living human eye. Invest. Ophthal. Vision Sci. (Suppl). 35: 1571.

Dartnall, H.J.A., Bowmaker, J.K. and Mollen, J.D. (1983). Human visual pigments: microspectrophotometric results from the eye of seven persons. Proc. Roy. Soc. Lond. B 220: 115–130.

de Vries, H.L. (1947). The heredity of the relative numbers of red and green receptors in the human eye. Genetica, 24: 199–212.

Merbs, S.L. and Nathans, J. (1993). Role of hydroxyl-bearing amino acids in differentially tuning the absorption spectra of the human red and green cone pigments. Photochem. Photobiol. 58: 706–710.

Nathans, J., Thomas, D. and Hogness, D.S. (1986). Molecular genetics of human color vision: the genes encoding blue, green, and red pigments. Science, 232: 193–202.

Neitz, J. and Neitz, M. (1994). Color vision defects. In: Wright, A.S. and Jay, B. (eds.), Molecular Genetics of Inherited Eye Disorders. Harwood Academic, Chur, Switzerland.

Neitz, M. and Neitz, J. (1995). Numbers and ratios of visual pigment genes for normal red-green color vision. Science, 267: 1013–1016.

Neitz, M., Neitz, J. and Jacobs, G.H. (1989). Analysis of fusion gene and encoded photopigment of colour-blind humans. Nature, 342: 679–682.

Neitz, M., Neitz, J. and Jacobs, G.H. (1991). Spectral tuning of pigments underlying red–green color vision. Science, 252: 971–974.

Neitz, M., Neitz, J. and Grishok, A. (1995a). Polymorphism in the number of genes encoding long-wavelength sensitive cone pigments among males with normal color vision. Vision Res. 35: 2395–2407.

Neitz, M., Neitz, J. and Jacobs, G.H. (1995b). Genetic basis of photopigment variations in human dichromats. Vision Res. 35: 2095–2103.

Nerger, J.L. and Cicerone, C.M. (1992). The ration of L cones to M cones in the human parafoveal retina. Vision Res. 32: 879–888.

Pokorny, J., Smith, V.C. and Wesner, M.F. (1991). Variability in cone populations and implications. In: Valberg, A. and Lee, B.B. (eds.), From Pigments to Perception: 23–34. Plenum, New York.

Rushton, W.A.H. and Baker, H.D. (1964). Red/green sensitivity in normal vision. Vision Res. 4: 75–85.

Vimal, R.L.P., Pokorny, J., Smith, V.C. and Shevell, S.K. (1989). Foveal cone thresholds. Vision Res. 29: 61–78.

Wesner, M., Pokorny, J., Shevell, S. and Smith, V. (1991). Foveal cone detection statistics in color-normals and dichromats. Vision Res. 31: 1021–1037.

S.A. Hagstrom, J. Neitz* and M. Neitz
Department of Cellular Biology and Anatomy and Department of Ophthalmology
Medical College of Wisconsin
8701 Watertown Plank Road
Milwaukee, WI 53226, USA
* To whom correspondence should be sent

6. Unique hues in heterozygotes for protan and deutan deficiencies

G. JORDAN and J.D. MOLLON
(Cambridge, UK)

Abstract

We measured the wavelengths of unique blue, green, and yellow in a population of normal women and carriers of different forms of red–green deficiencies. In the case of unique blue and green none of the groups of heterozygotes differed significantly from normals. In the case of unique yellow, at a retinal illuminance of 800 td, carriers of protanomaly made settings at significantly shorter wavelengths than any other group. Carriers of deuteranomaly, deuteranopia and protanopia did not differ from normal individuals. At 20 td none of the groups of heterozygotes differed from normal. Our results offer no support for the hypothesis of Cicerone (1987) that the wavelength seen as unique yellow is determined primarily by the relative numbers of L to M cones. Nor do we have any evidence for the report of Donders (1884) that the settings of unique yellow vary with Rayleigh matches.

Introduction

In the circle of hues there are four colours, the *Urfarben* of Hering (1878), that appear phenomenologically unmixed: unique blue, green, yellow, and red. For example, a unique yellow contains neither greenness nor redness. Hering further noted that the four unique hues are organized into two opponent pairs and that we never experience mixtures of the two components of one pair. Thus we never see a reddish-green or a bluish-yellow.

What are the physiological or ecological determinants of the unique hues? Why do observers differ in the wavelengths they choose for unique yellow, green and blue? There is little agreement on the answers to these questions or indeed on the domain in which the answers are to be sought. Historically, the unique hues have been taken to correspond to the null-points of Hering's putative opponent mechanisms: unique red and green are seen when the yellow–blue process is in equilibrium, and the red–green mechanism is polarized; unique blue and yellow are seen when the red–green (R–G) mechanism is in equilibrium and the blue–yellow (B–Y) system is polarized (Hering, 1878; Hurvich and Jameson, 1957).

C.R. Cavonius (ed.), Colour Vision Deficiencies XIII, pp. 67–76.

In the case of unique yellow, two current hypotheses relate individual judgments of unique yellow to properties of the cones. The first is the *cone ratio hypothesis*. Cicerone (1987) suggested that the variability in unique yellow derives from individual variability in the relative numbers of M and L cones. Within the tradition of opponent colour theory, unique yellow is seen when

$$[N_L \cdot k_L \cdot L(\lambda)] - [N_M \cdot k_M \cdot M(\lambda)] = 0 \qquad (1)$$

where N_L and N_M are the relative numbers of L and M cones, k_L and k_M are neural weighting factors, and $L(\lambda)$ and $M(\lambda)$ are the spectral sensitivities of the L and M cones at wavelength λ. Cicerone proposed that the numerosity of each class of cone is the main factor that determines the overall weighting of those cones at the input to the opponent channel. The greater the proportion of L cones, the shorter the wavelength of unique yellow.

The second hypothesis is the *cone sensitivity hypothesis*. Donders (1884) reported a strong correlation between unique yellow and Rayleigh matches: the more red a subject required in the match the shorter the wavelength of unique yellow. Westphal (1910) termed this relationship 'Donders' rule'. If this law holds, it suggests that the wavelength of unique yellow depends on the spectral positions of the M and L photopigments, since the latter variations are now recognized to be major determinants of Rayleigh matches (Winderickx *et al.*, 1992; Neitz and Neitz, 1994). In an analysis of anomalous trichromacy, Pokorny and Smith (1977) postulated that unique yellow corresponds to the wavelength that produces the same quantum catch in the long- and middle-wave cones as does white light. Their hypothesis could today be extended to relate the variations in unique yellow in normals to the known polymorphisms of visual pigments. An interesting question arises as to what white is the reference white. One possibility is that it is the average illuminant of the subject's recent environment (Mollon, 1982).

In the case of the other unique hues, theories of the relationship between receptor variations and phenomenological equilibria are less well developed. This is perhaps because the short-wave cones, as well as the long- and middle-wave cones, are thought to determine unique red, green and blue.

Heterozygotes for red–green deficiencies offer an interesting approach to receptoral theories of the variation in unique hues. A random process of X-chromosome inactivation determines which of the two X-chromosomes is expressed in any individual cell of a woman's body (Lyon, 1972). Thus the retinal mosaic of a heterozygote is thought to contain a subset of cones that express the abnormal chromosome that her colour-deficient son inherits. This assumption is consistent with the finding that heterozygotes exhibit altered luminous–efficiency functions (Schmidt, 1934; Jordan and Mollon, this volume) and with the finding that very small (1 min) probes reveal colour-deficient patches in the parafoveas of heterozygotes (Born *et al.*, 1976).

A test of the cone-ratio hypothesis is offered by carriers for dichromacy. Carriers of protanopia are expected to have a reduced number of long-wave cones and carriers of deuteranopia a reduced number of middle-wave cones. According to the cone ratio hypothesis these variations in cone numerosities should alter the weightings of L- and M-cone inputs to opponent channels and thus lead to clear differences in unique yellow and perhaps in the other unique hues.

According to the pigment-shift model of anomalous trichromacy (Pokorny and Smith, 1977), the anomalous retina contains cones that differ in sensitivity from the normal L and M cones. If these anomalous pigments do exist, then carriers of anomalous trichromacy offer a test of what we have above called the cone sensitivity hypothesis; a subset of their cones should express the shifted photopigment that their anomalous sons inherit, and their unique hues should be altered.

Method

The subjects were 14 normal controls and 37 carriers (22 deutan and 15 protan) whose genotypes were inferred from their sons' phenotypes. Sons and mothers were tested on a battery of colour vision tests including the Nagel anomaloscope and the Farnsworth-Munsell 100-hue test.

We measured unique green, yellow and blue for each subject. Since unique-red is extra-spectral (Dimmick and Hubbard, 1939) we did not attempt its determination with our present apparatus. The stimulus was a 1° circular field that was flashed on for 1 sec on a dark background. The interstimulus interval depended upon the spectral separation of successive stimuli but had a minimum value of 4 sec when the stimuli were 1–2 nm apart. Before the subject started, the concept of each unique hue was explained to her, and the corresponding spectral range was shown in Maxwellian view to familiarize her with the stimuli. Our Maxwellian-view optical system incorporates a computer-controlled monochromator with an integral stepping motor. Thus it was possible to determine the unique hues by a staircase method. The subject indicated by push buttons the direction in which a given stimulus departed from a unique hue, e.g. 'too red' or 'too green' in the case of unique yellow. First a two-staircase procedure with starting values 50 nm apart was used to familiarize the subjects with the task and to obtain an initial estimate of the unique hue. Subsequently, four staircases, centred on the subject's initial mean setting, were randomly interleaved. The last six reversals of each independent staircase were used to estimate the wavelength of the unique hue and the values plotted in the figures are the means of the four estimates.

In our first experiment, the luminance of the stimuli varied with the output of the monochromator at different wavelengths. For unique blue the value was approximately 20 td, for unique green approximately 100 td, and for unique yellow approximately 800 td. In a second experiment, we repeated the measurements for unique yellow, now holding the troland value at 20 td: under computer control, adjustments of a neutral density wedge were yoked to

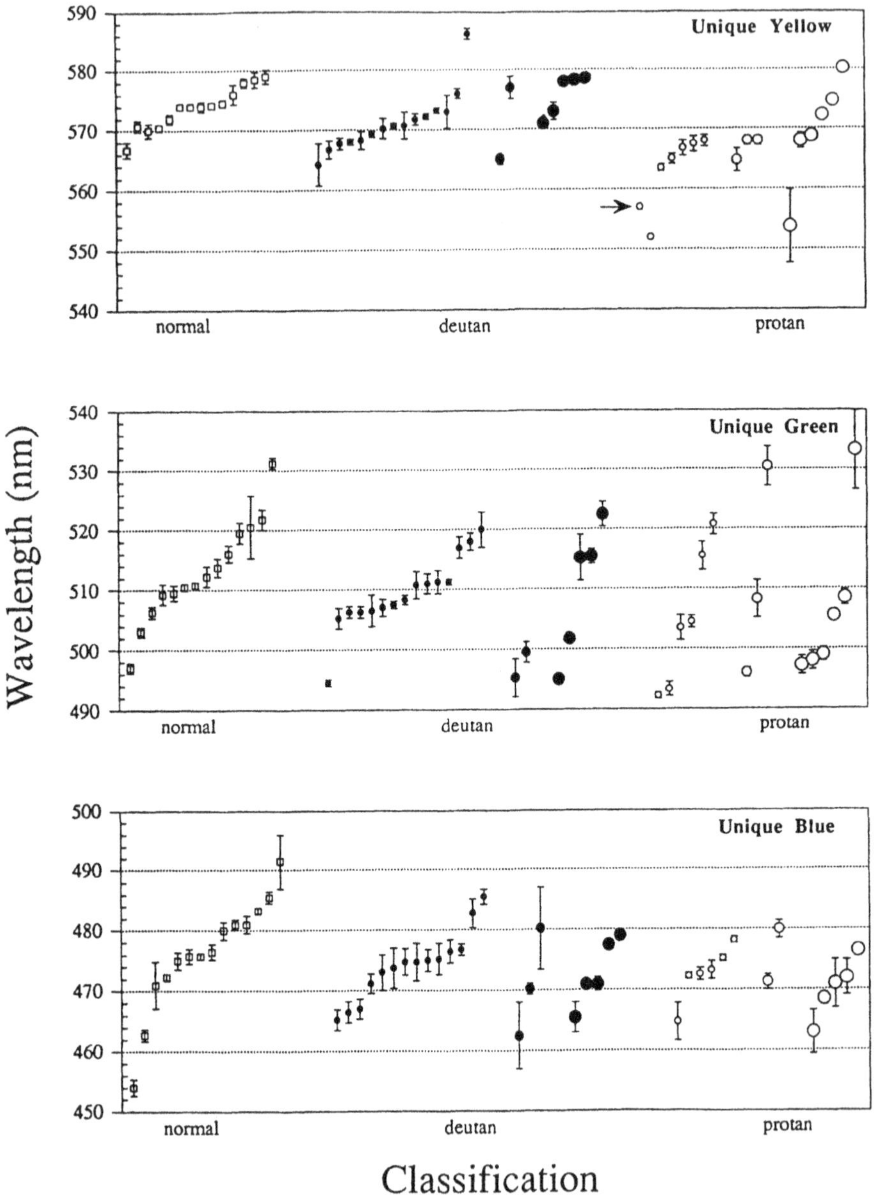

Fig. 1. The wavelengths seen as unique yellow (top panel), green (middle), and blue (bottom panel) for normals (open squares), deutan carriers (filled circles), and protan carriers (open circles). Gradation in symbol size represents the degree of deficiency of the abnormal gene: small symbols represent carriers for anomalous trichromacy, large symbols carriers for dichromacy, and intermediate symbols carriers for extreme anomalies. The error bars are standard deviations. The carriers for protanomaly set the wavelength of unique yellow significantly shorter than any other group.

variations in wavelength, so as to hold constant the monochromator output.

In addition, Rayleigh matches were measured with the Nagel anomaloscope (Schmidt and Haensch) and relative sensitivity to long and middle wavelengths was measured with the OSCAR test (see Jordan and Mollon, this volume).

Results

Figure 1 shows a summary of all unique hue settings for the first experiment: unique yellow at the top; unique green in the middle; and unique blue at the bottom. The ordinate shows wavelength in nm. Open squares represent the settings of the normal controls, filled circles represent the deutan carriers and open circles represent the protan carriers. Within both groups of carriers the symbols are graded in size: the smallest symbols indicate the carriers of simple anomaly, the intermediate ones the carriers of extreme anomaly and the largest symbols the carriers of dichromacy.

There are no significant differences between carriers and normal controls in the settings of unique green or of unique blue. In the case of unique yellow, a one-way analysis of variance for normal, protan and deutan groups showed a significant effect of group ($F = 6.09$, $df = 2$, $p = 0.004$). This effect is due to the carriers of protanomaly, who set the wavelength of unique yellow very significantly shorter than the normal group ($p < 0.001$). Note that one carrier of protanomaly sets the wavelength of unique yellow at a remarkably short value of 552 nm. When retested 6 months later, with the output of the monochromator held constant at a similar troland value, she still made a setting of 557 nm. Her second setting is represented by the arrow in Figure 1. (This subject recalls the female observer of Akita *et al.* (1982), who made a Rayleigh match within the normal range but chose a wavelength of 536 nm for her unique yellow. That subject might also be a protanomalous carrier.) Carriers of protanomaly were the only divergent group: neither the dichromat carriers nor the carriers of extreme anomaly nor the carriers of simple deuteranomaly differed from the normal women.

Figure 2 shows the results for the second experiment on unique yellow, where the stimulus was held constant at the lower value of 20 td. There is now no significant difference in the wavelength judged to be unique yellow between any group of carrier and the normal controls. The carriers of protanomaly have shifted towards longer wavelengths. Despite the absence of differences in settings, the subjective comments of some subjects in this experiment may be worth mentioning. Some carriers of dichromacy reported an almost colourless sensation at wavelengths near their final setting. Two carriers of deuteranomaly (whose settings lie at the short-wave end of the distribution) said that the hue they ended up with was neither red nor green and yet not pure yellow: one did not know what to call the colour, the other said that, if forced, she would say a 'bluish-yellow' (a forbidden colour in the Hering scheme).

Figure 3 shows the relationship between settings on the OSCAR test and

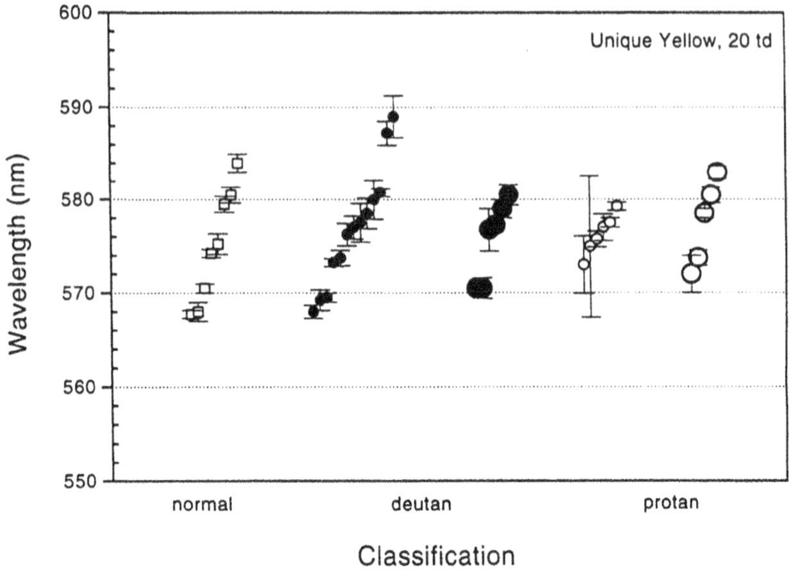

Fig. 2. The unique yellow settings at a retinal illuminance of 20 td for normals and carriers. There is no significant difference in the settings between any group. Symbols as in Figure 1

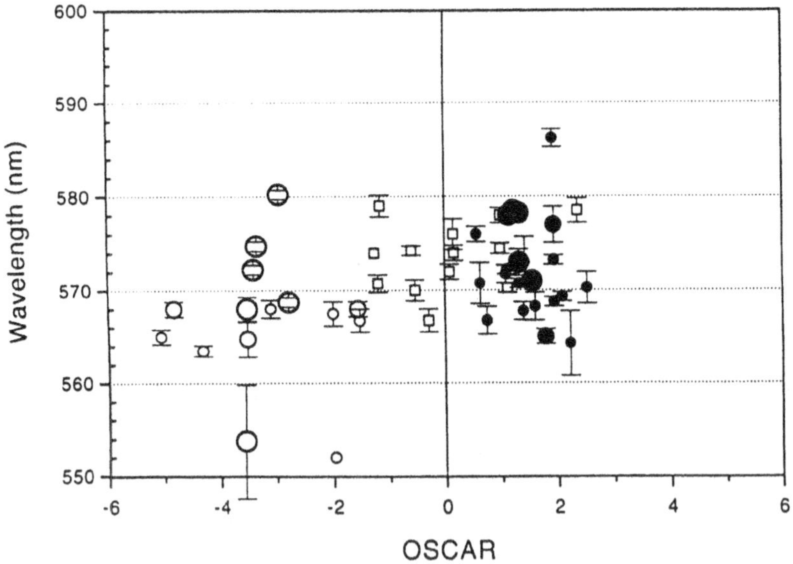

Fig. 3. Relationship between the wavelengths seen as unique yellow and the subjects' OSCAR settings. Symbols are used as before. Zero on the abscissa represents the mean OSCAR setting of normals. As expected carriers for protan deficiencies make settings on the left-hand side of zero, indicating a lowered sensitivity for long wavelengths (Schmidt's sign). Carriers for deutan deficiency make settings on the right-hand side of zero, indicating a lowered sensitivity for middle wavelengths; but there is considerable overlap with the normals. Symbols as in Figure 1

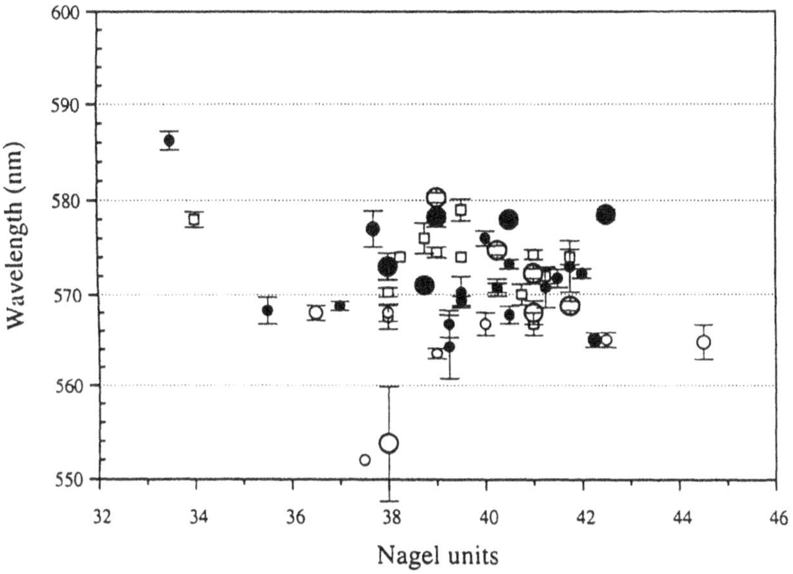

Fig. 4. The relationship between the wavelengths seen as unique yellow and the subjects' Rayleigh matches. Rayleigh matches are expressed in raw Nagel units. There is no significant correlation between the two variables.

estimates of unique yellow in the first experiment. As we show elsewhere (Jordan and Mollon, this volume), protan carriers tended to make settings on the left-hand side, indicating a reduced sensitivity to long wavelengths (Schmidt's sign), whereas deutan carriers tended to make settings on the right-hand side, indicating a reduced sensitivity to middle wavelengths. Normal individuals made settings in the middle range. However, there is only a modest, although significant, correlation between unique yellow and OSCAR settings ($R = 0.38$, $p = 0.005$). Similar analyses showed no significant relationship between the other unique hues and OSCAR settings.

Figure 4 shows the relationship between the subjects' Rayleigh matches and their settings of unique yellow in the first experiment. The abscissa is in Nagel units: subjects on the right need more red in the mixture to match the standard yellow; the ordinate shows the wavelength seen as unique yellow. There was no overall correlation between Rayleigh matches and the wavelength seen as unique yellow ($R = -0.13$, $p = 0.34$) and there are no obvious trends within any of the individual groups of carrier or the normal controls. Similar analyses showed no relationship between Nagel matches and the other two unique hues that we tested. Our result for unique yellow is concordant with that of Hailwood and Roaf (1937).

Discussion

Cone ratio hypothesis

Our results offer no support for the hypothesis that the relative numbers of L and M cones determine the wavelengths of unique hues in general and unique yellow in particular. If the cone-ratio hypothesis of Cicerone were correct, then we should expect carriers of protanopia (who are thought to have reduced numbers of L cones) to set unique yellow at longer wavelengths than normal controls, whereas carriers of deuteranopia (who are thought to have fewer M cones) should make settings at shorter wavelengths than normals. In fact, the two kinds of carrier of dichromacy did not differ from normal individuals in our study, even though they make clearly different settings on the OSCAR test, a flicker-photometric measure of the type often taken to reflect cone ratios (de Vries, 1948).

If Alpern's account of anomalous trichromacy is correct – if, that is, the residual discrimination of anomals depends on two types of normal M cone or two types of normal L cone (Alpern and Moeller, 1977) – then our argument can be extended to include the carriers of anomalous trichromacy. For the retina of the carrier should contain only normal cones, but in abnormal proportions. Here again there is no support for Cicerone's proposal. In fact, at 800 td the carriers of protanomaly differed from normals in the direction opposite to that predicted from the cone-ratio hypothesis.

One counter-argument, put to us by J. Pokorny, would be that the fovea of the carrier contains relatively coarse patches of normal and defective retina and that settings of unique yellow (unlike flicker-photometric settings) are achieved only by the colour-normal patches. Against this argument may be cited the report of Grützner *et al.* (1976), who probed the retina with brief, tiny stimuli and found that dichromat carriers, though abnormal in the parafovea, were able to correctly identify the colours of 1 min targets for all eccentricities less than 2°: the implication is that cones expressing alternative X-chromosomes are well intermingled in the carrier's foveola and are not segregated in large patches. A second counter-argument, put to us by a referee, requires one to assume the replacement hypothesis for dichromacy, i.e. to assume that those cones that should have become L cones became M cones in the protanope and in the corresponding carrier. If these extra M cones contributed equally to the two expressions of equation 1, then unique yellow might be left unchanged in the carrier. This hypothesis requires the additional hypothesis that the extra M cones are neither labelled as M cones nor as L cones.

Cone-sensitivity hypothesis

In our population as a whole, we found no relationship between unique hues and the subjects' Rayleigh matches. In particular, we did not confirm the inverse relationship between unique yellow and Rayleigh match that was reported by

Donders (1884). (There is good reason to suppose that the result of Donders would be found for a sample of observers that included protanomalous and deuteranomalous trichromats as well as normal individuals, since anomalous individuals differ in the expected way in their settings of unique yellow (Westphal, 1910; Hurvich and Jameson, 1964; Pokorny and Smith, 1977).) Our result offers no support to hypotheses that relate individual differences in unique hues to variations in the spectral positions of the receptors. Our results for carriers of anomaly bear upon this issue in a more complicated way. In their actual settings of unique yellow, carriers of deuteranomaly did not differ from normals in either experiment: this result implies either that unique yellow is not affected by the presence of cones with abnormal spectral sensitivities, or that Alpern's hypothesis is correct and the retina of the deuteranomalous carrier contains only normal cones. In contrast, carriers of protanomaly, under the high-intensity conditions of the first experiment, differed systematically from normals, the shift in their settings being in the same direction as the shift seen in actual protanomalous observers (Hurvich and Jameson, 1964). This implies either that there are cones with abnormal spectral sensitivity in the retinae of heterozygotes for protanomaly or that their colour judgements are influenced both by a normal L/M channel and a channel that differences two types of M cones, the channel on which according to the Alpern hypothesis their sons rely. This conclusion must remain tentative, since the protanomalous carriers, like other carriers, did not differ from the normal individuals for the 20 td stimuli.

Acknowledgements

Supported by MRC grant G9113630N and by a Wellcome Trust Fellowship to GJ.

References

Akita, M., Ejima, Y. and Takahashi, S. (1982). Differences in unique-yellow loci between individuals. Color Res. Appl. 7: 168–172.

Alpern, M. and Moeller, J. (1977). The red and green cone visual pigments of deuteranomalous trichromacy. J. Physiol. (Lond.) 266: 647–675.

Cicerone, C.M. (1987). Constraints placed on color vision models by the relative numbers of different cone classes in human fovea centralis. Die Farbe 34: 59–66.

Born, G., Grützner, P. and Hemminger, H. (1976). Evidenz für eine Mosaikstruktur der Netzhaut bei Konduktorinnen für Dichromasie. Hum. Genet. 32: 189–196.

de Vries, H. (1948). The hereditary of the relative numbers of red and green receptors in the human eye. Genetica 24: 199–212.

Dimmick, F.L. and Hubbard, M.R. (1939). The spectral components of psychologically unique red. Am. J. Psych. 52: 348–353.

Donders, F.C. (1884). Farbengleichungen. Arch. Anat. Physiol.: 518–552.

Grützner, P., Born, G. and Hemminger, H. (1976). Coloured stimuli within the central visual field of carriers of dichromatism. Mod. Probl. Ophthal. 17: 147–150.

Hailwood, J.G. and Roaf, H.E. (1937). The sensation of yellow and anomalous trichromatism. J. Physiol. (Lond.) 91: 36–47.
Hering, E. (1878). Zur Lehre vom Lichtsinne. Sechs Mittheil. an die k. Akad. Wissensch. Wien, 2nd ed. Carl Gerold's Sohn, Wien.
Hurvich, L.M. and Jameson, D. (1957). An opponent-process theory of color vision. Psychol. Rev. 64: 384–404.
Hurvich, L.M. and Jameson, D. (1964). Does anomalous color vision imply color weakness? Psychon. Sci. 1: 11–12.
Lyon, M.F. (1972). X-chromosome inactivation and developmental patterns in mammals. Biol. Rev. 47: 1–35.
Mollon, J.D. (1982). Color vision. Annu. Rev. Psychol. 33: 41–85.
Neitz, J. and Neitz, M. (1994). Color vision defects. In: Wright, A.F. and Jay, B. (eds.), Molecular Genetics of Inherited Eye Disorders: 217–257. Harwood Academic, Reading.
Pokorny, J. and Smith, V.C. (1977). Evaluation of single-pigment shift model of anomalous trichromacy. J. Opt. Soc. Am. 67: 1196–1209.
Schmidt, I. (1934). Über manifeste Heterozygotie bei Konduktorinnen für Farbensinnstörungen. Klin. Mbl. Augenheilk. 92: 456–467.
Westphal, H. (1910). Unmittelbare Bestimmungen der Urfarben. Z. Sinnesphysiol. 44: 182–230.
Winderickx, J., Lindsey, D.T., Sanocki, E., Teller, D.Y., Motulsky, A.G. and Deeb, S.S. (1992). Polymorphism in the red photopigment underlies variation in color matching. Nature 356: 431–433.

Department of Experimental Psychology
University of Cambridge
Downing St.
Cambridge, CB2 3EB
UK

7. Sensitivity and spectral tuning of the red–green chromatic pathway in heterozygous carriers of congenital colour vision defect

W.H. SWANSON and M. FIEDELMAN
(Dallas, TX, USA)

Abstract

We measured sensitivity and spectral tuning of the red–green chromatic pathway in 20 women heterozygous for congenital colour vision deficiencies, and 13 age-matched normal individuals. Spectral increment thresholds from 510 to 660 nm were measured in 30 nm steps, for a 4.8 test superimposed on a contiguous 4000 K, 3.7 log td white pedestal. Stimulus intensity was varied with temporal 2AFC staircases, and maximum likelihood estimation was used to derive thresholds. There was no difference between normal individuals and carriers either for clinical colour vision tests or for spectral increment thresholds. The carriers appear to have normal sensitivity and spectral tuning for the red–green chromatic system, despite heterogeneous expression of cone photopigment genes. This is consistent with the assumption that the primary effect of heterogeneity in the cone mosaic on sensitivity of the midget ganglion cell mosaic is to reduce the number of ganglion cells mediating detection.

Introduction

Psychophysical models of normal colour vision typically assume that there are exactly three classes of cone photopigments, distinguished as L, M, and S (for long-, middle- and short-wavelength-sensitive). However, recent genetic studies indicate that normal trichromatic males often have genes for more than three types of cone photopigments, and may express more than three photopigment genes (Hagstrom *et al.*, 1995). The heterogeneity in peak wavelengths of cones providing input to the midget ganglion cell mosaic (which mediates red–green chromatic sensitivity; see Dacey, 1994) may, therefore, cause variations in chromatic sensitivity among cells in that mosaic. In this study, we investigated the effects of cone inhomogeneity on chromatic sensitivity.

Evidence for extreme forms of inhomogeneity in the cone mosaic has been reported in heterozygous carriers of X-linked congenital colour vision defects, for whom X-chromosome inactivation early in embryological development is

C.R. Cavonius (ed.), Colour Vision Deficiencies XIII, pp. 77–86.

thought to result in cone mosaics that contain both normal and colour-defective patches (Lyon, 1972; Nagy *et al.*, 1981; Swanson, 1991; Jordan and Mollon, 1993). This random X-chromosome inactivation can result in regions that have four different photopigments in separate cones (in carriers of anomalous trichromacy), or only two photopigments (in carriers of dichromacy).

The effects of inhomogeneity in the cone mosaic upon chromatic sensitivity should be more extreme in heterozygous carriers of colour vision defects than in males with normal colour vision expressing more than three photopigments. Indeed, carriers have been found to perform abnormally for colour naming of small spots in the periphery (Grützner *et al.*, 1976), and on the Ishihara pseudoisochromatic plates (Jordan and Mollon, 1993). However, it is not clear whether these defects are due to a general loss in chromatic sensitivity, or whether these tests are particularly sensitive to the effects of cone inhomogeneity on responses of cells in the midget ganglion cell mosaic. The colour naming experiment would have stimulated small numbers of cells, and the Ishihara requires comparisons across elements in the mosaic. We have previously presented evidence that cone inhomogeneity in carriers produces inhomogeneity in spectral sensitivity of the ganglion cells mediating high-frequency flicker sensitivity (Swanson, 1991). If cone inhomogeneity also produces inhomogeneities in the spectral sensitivities of the midget ganglion cells, then the colour naming and Ishihara abnormalities may reflect this inhomogeneity, rather than an overall reduction in chromatic sensitivity.

In the most comprehensive study of carriers to date, Jordan and Mollon (1993) found normal chromatic sensitivity when measured with the FM 100-hue test and large-field Rayleigh match width, yet abnormal chromatic sensitivity when measured with the small-field Rayleigh match width. It is difficult to compare results from these three measures of chromatic sensitivity, since they do not provide methods for controlling or evaluating the subject's criteria (Swanson, 1993). In the current study we measured chromatic sensitivity in obligate heterozygotes using spectral increment thresholds gathered with criterion-free methods. In order to obtain good isolation of the chromatic response, we used monochromatic increments on a contiguous white pedestal, with Gaussian modulation to avoid temporal transients. In order to stimulate a large number of midget ganglion cells, we used a 5° circular target.

Methods

Observers

Twenty obligate carriers of congenital colour vision deficiency, age 32–47 years, were identified by diagnosis of one or more sons: five carriers each for protanopia, deuteranopia, deuteranomalous trichromacy and protanomalous trichromacy. The sons were tested with 2° Rayleigh matches (Swanson, 1993), flicker photometry, the Ishihara plates, the Farnsworth D-15, Adams desaturated D-15 and Lanthony New Colour Test (Swanson, 1991; Swanson and Everett, 1992; Stager *et al.*, 1993) and were required to have consistent performance across tests. Carriers were required to have letter acuity of 20/20 or better and to pass the Regan Contrast Sensitivity Charts. Carriers were also tested with the Ishihara plates, anomaloscopy, the Farnsworth D-15, the Adams desaturated D-15, the Standard Pseudoisochromatic Plates II (SPPII) and the Farnsworth-Munsell 100-hue test.

We did not want to include women who lacked one of the the normal photopigments, and excluded one carrier of protanopia who appeared to be a protanomalous trichromat. Her Rayleigh match had a midpoint of 56 Nagel units, a range of 26 Nagel units, and protan brightness settings. She had four protan crossovers on the D-15, and correctly read only two digits on the Ishihara (she read four 'transformation' digits). Her spectral increment data from 510–636 nm were fitted well by the middle-wavelength cone fundamental with no evidence of chromatic opponency.

Thirteen clinically normal volunteers were selected from a larger set to approximately match the ages of the carriers (Table 1). They were required to have letter acuity of 20/20 or better, to pass the Regan Contrast Sensitivity Charts and Ishihara plates, to have normal Rayleigh matches, to pass the D-15, desaturated D-15, and SPPII tests and to have FM 100-hue scores no more than 1 S.D. worse than the age-matched mean normal value of Verriest *et al.* (1982).

Table 1. Mean, standard deviation (S.D.), Student's *t*-test, and probability of difference in means, for age, clinical test results and parameters to fit of data. Results shown for heterozygous carriers of congenital colour vision defects and age-matched normals.

	Carriers		Controls		Statistics	
	Mean	S.D.	Mean	S.D.	t	*p*
Age (years)	39.7	3.9	39.6	6.3	0.06	0.95
Results of standard clinical tests of colour vision						
FM 100-hue (square root of error score)	5.9	1.9	6.5	3.1	0.55	0.59
Match width (Nagel units)	1.35	0.62	1.04	0.76	1.29	0.21
Match midpoint (Nagel units)	38.6	2.3	39.5	4.0	0.80	0.43
Parameters for fits to spectral increments						
Chromatic sensitivity (log quanta^{-1} sec deg^{2})	–7.69	0.26	–7.58	0.22	1.11	0.27
Ratio of L and M weights (log[L/M])	–0.08	0.09	–0.09	0.04	0.50	0.62
Amount of isolation (log sensitivity difference)	1.05	0.24	1.16	0.21	1.44	0.16

Apparatus

A three-channel Maxwellian-view optical system with a 2 mm artificial pupil and a tungsten-halogen light source was used to present spectral increments on a contiguous white pedestal. A Wratten 80A daylight filter was used to provide a colour temperature of 4000 K for the 3.7 log td white pedestal and the 2.2 log td surround. Triple-cavity interference filters (10 nm bandwidth) were used to provide monochromatic increments from 510 to 660 nm in 30 nm steps. The increments were presented by modulating a calibrated liquid crystal variable attenuator (Meadowlark Optics, Longmont, CO), driven by a 12-bit digital-to-analogue converter and DMA-8 timing board (National Instruments, Austin, TX); this provides 3 log units of attenuation, and was used to present a Gaussian increment with a 100 msec time constant (stimulus duration 800 msec, with 70% of the energy of the increment in the central 200 msec). A computer-controlled filter wheel provided a 4 log unit range of attenuation for the increments, and the variable attenuator was used to provide attenuations of 0.0 to 1.0 log unit in the increment.

Procedure

A two-alternative forced-choice method was used. The observer initiated each trial by pushing a button. A single trial was comprised of two intervals (before and after a beep), and the observer was required to indicate which of the intervals contained the increment. Staircases continued until 14 reversals were obtained, and maximum-likelihood estimation was used to determine thresholds. If the mean-of-reversals and maximum-likelihood threshold estimates differed by 0.15 log unit or more, the staircase was considered too noisy for reliable threshold estimation (Swanson and Birch, 1992); in these cases (less than 10% of total), a second staircase was run and trials from the two staircases were combined for a new maximum-likelihood estimate.

Analysis

Data were analysed with a three-parameter model which is similar to (but simpler than) those of Thornton and Pugh (1983) and Miyahara (1993). We found this model to give good descriptions of the data for a wide range of normals and patients, with a minimum number of parameters. The model assumes that, for both L and M cones, cone response is normalized to the mean quantal catch; a white adapting field was used to provide relatively neutral adaptation. It is assumed that thresholds near 570 nm are mediated by a mechanism with the spectral sensitivity of V_λ for 510–660 nm; this could reflect activity in either an S-(L+M) mechanism or in a luminance mechanism. The Smith and Pokorny (1975) L and M cone fundamentals were used to produce an |L-M| response; like Miyahara, but unlike Thornton and Pugh, both L-M and M-L had the same weight (which reduces the number of free parameters). The three parameters are: W_C (sensitivity of the red–green chromatic system), a (weight of the L cone input to the chromatic system), and W_L (sensitivity of the system with V_λ spectral sensitivity). Sensitivity is computed as

$$[R_C^2 + R_L^2]^{0.5}, \text{ where}$$
$$R_C = W_C\,[a \times \mathrm{LWS}(\lambda) - (1-a) \times \mathrm{MWS}(\lambda)],\ 0 \leqslant a \leqslant 1, \text{ and}$$
$$R_L = W_L \times V_\lambda \quad (1)$$

and LWS is the L-cone quantal catch and MWS the M-cone quantal catch. The amount of isolation of the chromatic response is log (W_C/W_L).

Results

The carriers performed relatively well on the Ishihara: 14 (70%) read all digits correctly (compared with 85% of controls), four (20%) missed one digit (compared with 15% of controls), and two (10%) missed two or three digits (no normal individuals missed more than one digit). The non-parametric Mann–Whitney test showed no difference between carriers and normals ($U = 151$, $p > 0.20$). All 20 carriers missed exactly one digit on the SPPII (all but one normal individual also missed this digit, and two normal individuals each missed two digits). All carriers had perfect scores on the D-15, and made only two errors (minor tritan reversals) on the desaturated D-15 (as did two controls). The carriers had normal performance for the FM 100-hue and for 2° Rayleigh matches, and normal values for the three parameters of the model (Table 1). There was no difference in sensitivity or shapes of the mean functions for carriers and for age-matched normal controls (Fig. 1, left). There was a tendency for carriers of dichromacy to have higher sensitivity than carriers of anomalous trichromacy (Fig. 1, right); the mean difference in sensitivity of the two groups was 0.23 log unit ($t = 2.15$, $p = 0.045$). There was also a trend for carriers of anomalous trichromacy to have lower sensitivity than the age-matched controls ($t = 2.09$, $p = 0.048$).

Parameters derived by fitting each observer's data with the three-parameter model are shown in Figure 2; in both panels the vertical axis is chromatic sensitivity. For each scatterplot, a bivariate Gaussian ellipse is shown which includes 95% of the distribution for normal individuals. One carrier had reduced sensitivity and less isolation than normal, and another had an abnormal ratio of L/M cone weights; both were carriers of protanomaly.

Discussion

We found no group differences between carriers and age-matched controls in performance on clinical colour vision tests, chromatic sensitivity, spectral tuning, or amount of isolation of the red–green chromatic pathway. Carriers of

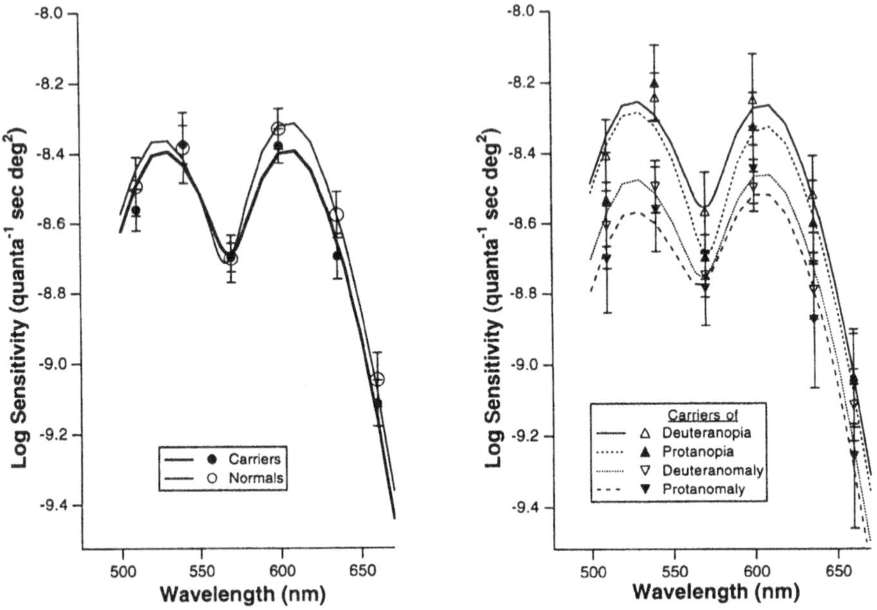

Fig. 1. Mean increment thresholds shown as symbols, best fit of model shown as curves; error bars $\pm$S.E.M. Left: means for normal controls and heterozygous carriers of congenital colour defects. Right: means for 4 subgroups of heterozygous carriers of congenital colour defects.

anomalous trichromacy tended to have poorer chromatic sensitivity than carriers of dichromacy or normal controls.

Jordan and Mollon (1993) also measured chromatic sensitivity in obligate carriers of congenital colour vision defects. Like us, they found normal FM 100-hue scores and normal Rayleigh match midpoints for the carriers, and no significant difference in match midpoints or match widths between carriers of dichromacy and carriers of anomalous trichromacy. Unlike us, Jordan and Mollon found an increased number of errors on the Ishihara and slightly enlarged Rayleigh match widths for the carriers. These differences may be due either to differences in populations tested or to procedural differences. Jordan and Mollon used the tenth edition of the Ishihara, illuminated by a Macbeth Easel lamp; we used the 1984 24-plate edition, illuminated with a GE Chromline F40/75 fluorescent lamp (colour rendering index 92 at 7500 K). This might have resulted in slight differences in the stimuli. For instance, Jordan and Mollon found that 11 of 30 heterozygotes misread transformation plate 9 (74 transforms to 21), while none of our 20 heterozygotes misread our corresponding plate 7 (same transformation). Jordan and Mollon used manual anomaloscopy, making

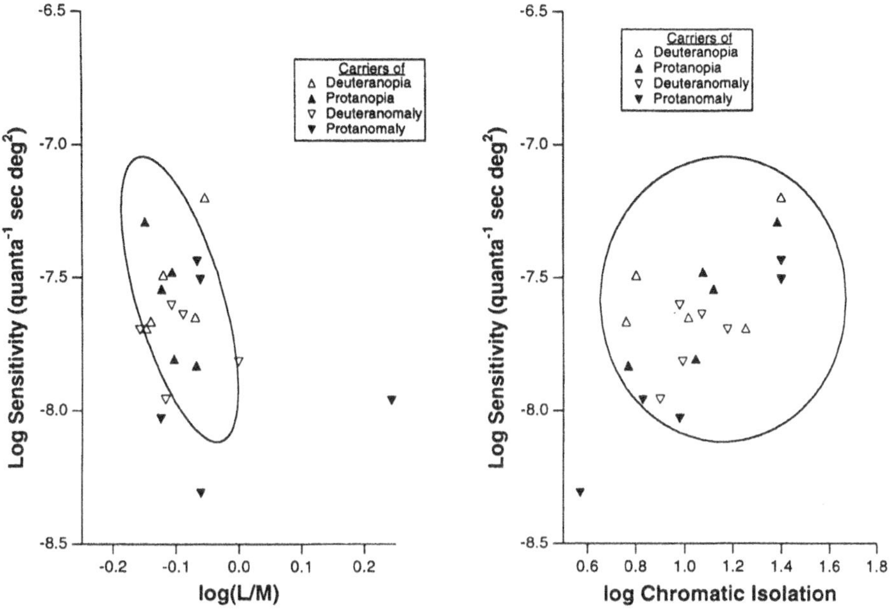

Fig. 2. Parameters for fits to data of individual heterozygous carriers. Ellipses include 95% of the distribution for normals. Left: sensitivity of the chromatic mechanism vs. ratio of L and M cone weights. Right: sensitivity of the chromatic mechanism vs. amount of chromatic isolation.

a few matches to determine match endpoints; we used automated anomaloscopy, with 40–80 trials, and computed match endpoints as 50% points on psychometric functions. We believe that automated anomaloscopy is less affected than manual anomaloscopy by operator criteria or by factors which cause shallow psychometric functions (Swanson, 1993).

Our results for spectral increment thresholds contrast with our earlier study of heterochromatic modulation photometry, which found abnormal shapes of functions for carriers of protan defects, with altered spectral sensitivity and less clearly-defined isoluminant points than for normals and carriers of deutan defects (Swanson, 1991). We suggest that differences in the receptive field properties of the two classes of retinal ganglion cells that mediate the detection of the stimuli used in these two studies account for differences in the effects of heterogeneity in the cone mosaic upon psychophysical spectral tuning functions.

Flicker photometry (Lee *et al.*, 1988) appears to be mediated by the parasol ganglion cells, whose receptive fields sum responses of many cones, without selectivity for L or M cones (Dacey, 1994). Variations in weights of L and M cone inputs to parasol cells, with no variation in overall sensitivity, would be expected from heterogeneity of the cone mosaic. Therefore heterochromatic modulation threshold functions will presumably measure the envelope of responses of parasol cells with a greater range of L and M cone weights than seen in the normal population (and hence a less well-defined isoluminant point).

The ganglion cell mosaic mediating red–green colour vision is presumably composed of the midget ganglion cells (Merigan, 1989), which comprise 90% of foveal retinal ganglion cells (Dacey, 1994). The receptive field centre of a foveal midget ganglion cell receives input from a single cone, and it is currently uncertain whether the receptive field surrounds of midget ganglion cells make specific or non-specific connections with cones (Lennie *et al.*, 1991; Shapley and Reid, 1992). Here we consider several ways that heterogeneity of the cone mosaic could affect sensitivity of midget retinal ganglion cells.

For carriers of dichromacy, heterogeneity will cause variations in the numbers of L and M cones. For spectral increment thresholds on white pedestals, the weight of L and M cone inputs to the red–green chromatic response in normals appears to be determined by their adaptation states (Thornton and Pugh, 1983), and not upon their relative numbers. If this holds for carriers of dichromacy, then chromatic sensitivity of individual cells should not be affected by changes in cone numbers, regardless of whether surrounds are cone-specific. But if this does not hold, then if surrounds are not cone-specific the ratio of L/M cone inputs should be shifted to higher values for carriers of deuteranopia, and lower values for carriers of protanopia. We found no evidence of differences in L/M cone weights for carriers of deuteranopia and protanopia (Fig. 2, left panel).

A second possibility for midget ganglion cells in carriers of dichromacy is suggested by the hypothesis that in protanopes L cones are replaced by M cones, and the centres and surrounds of midget ganglion cells are no longer opponent (Cicerone *et al.*, 1987). For carriers of protanopia, this would result in ganglion cells that have greater than normal sensitivity at middle wavelengths (where L and M cones have similar sensitivities, and chromatic opponency reduces midget ganglion cell sensitivity below that of centre or surround alone). These cells would then mediate detection near 570 nm (where sensitivity of normal cells is lowest), decreasing the amount of chromatic isolation. We found no evidence of such a decrease in carriers of protanopia (Fig. 2, right panel). Note that Schwartz (1994) found evidence against non-opponent replacement in dichromats.

For carriers of dichromacy these considerations concerning the midget ganglion cell mosaic predict that spectral tuning would be normal in most heterozygous carriers, and chromatic sensitivity losses should be due only to reduced numbers of ganglion cells mediating detection. To estimate the effect of reduced numbers of ganglion cells, increment thresholds at 660 nm were gathered from one observer on contiguous white pedestals spanning a 2.3 log unit range of stimulus areas (diameters from 0.4° to 5.7°). Foveal threshold vs. log area data were described well by a line having a slope of –0.25, consistent with probability summation of responses of midget ganglion cells. Therefore a reduction by 50% in the number of midget ganglion cells mediating detection would only be expected to cause an 0.075 log unit decrease in sensitivity, a very small difference that would require a large sample size to be detectable.

For carriers of anomalous trichromacy, X-chromosome inactivation may result in four distinct photopigments. Consider the example of a carrier of protanomaly, whose pigments would be S, M, L, and L′, where L′ is the protanomalous pigment, with a peak wavelength shifted close to that of the M pigment. If a foveal midget ganglion cell had centre input from a cone containing the L′ pigment, then the ganglion cell would have reduced chromatic sensitivity whether or not the surround received mixed cone inputs (since an M cone surround or a mixed M, L and L′ cone surround would have spectral sensitivity closer to an L′ cone centre than to an L cone centre). If a foveal midget ganglion cell had centre input from a cone containing the M pigment, then if the surround received both L and L′ cone inputs the ganglion cell would also have reduced chromatic sensitivity. Therefore, whether or not surrounds receive cone-specific input, heterozygosity for anomalous trichromacy should result in some cells in the midget ganglion cell mosaic having reduced chromatic sensitivity as compared to ganglion cells lacking cone inputs from anomalous cones. The percentage of ganglion cells with reduced chromatic sensitivity will depend on the extent to which anomalous cones are intermingled with normal cones; if this intermingling is extensive, then the majority of foveal midget ganglion cells could be affected.

In conclusion, we found no differences in mean chromatic sensitivity between a group of heterozygous carriers of X-linked congenital colour defects and a group of age-matched normals. There was a tendency for carriers of anomalous trichromacy to have poorer chromatic sensitivity than carriers of dichromacy or normal individuals. This is consistent with the assumption that variations in numbers of L and M cones (as in carriers of dichromacy) will have little effect on chromatic sensitivity of individual midget ganglion cells, whereas increased numbers of photopigments may result in some midget ganglion cells having reduced sensitivity. Heterogeneity of cone photopigments would only affect the midget ganglion cell mosaic by reducing the number of ganglion cells mediating detection. Unless this reduction is extreme, chromatic thresholds are unlikely to be abnormal. Since heterogeneity in the cone mosaics of these carriers should be greater than for normal males who express more than three photopigments, these data indicate that existence of more than three cone photopigments in normals should have little effect on chromatic sensitivity.

Acknowledgements

Supported in part by the United States Department of Health and Human Services, NIH EY 07716, and by the Home Interiors & Gifts Fund of the Communities Foundation of Texas. Joost Felius made helpful comments on several versions of this manuscript.

References

Cicerone, C.M., Nagy, A.L. and Nerger, J.L. (1987). Equilibrium hue judgements of dichromats. Vision Res. 27: 983–991.

Dacey, D.M. (1994). Physiology, morphology and spatial densities of identified ganglion cell types in primate retina. In: Goode, J. and Morgan, M. (eds.). Higher-order Processing in the Visual System: 12–34. Wiley, Chichester.

Grüntzner, P., Born, G. and Hemminger, H.J. (1976). Coloured stimuli within the central visual field of carriers of dichromatism. Mod. Prob. Ophthalmol. 17: 147–150.

Hagstrom, S.A., Neitz, J. and Neitz, M. (1995). Four different cone types in a human eye identified from mRNA in individual photoreceptors. Invest. Ophthalmol. Vis. Sci. 36: S1054.

Jordan, G. and Mollon, J.D. (1993). A study of women heterozygous for colour deficiencies. Vision Res. 33: 1495–1508.

Lee, B.B., Martin, P.R. and Valberg, A. (1988). The physiological basis of heterochromatic flicker photometry demonstrated in the ganglion cells of the macaque retina. J. Physiol. (Lond.) 404: 323–347.

Lennie, P., Haake, P.W. and Williams, D.R. (1991). The design of chromatically opponent receptive fields. In: Landy, M.S. and Movshon, J.A. (eds.) Computational Models of Visual Processing: 71–82. MIT Press, Cambridge.

Lyon, M.F. (1972). X-chromosome inactivation and developmental patterns in mammals. Biol. Rev. 47: 1–35.

Merigan, W.H. (1989). Chromatic and achromatic vision of macaques: role of the P pathway. J. Neurosci. 9: 776–783.

Miyahara, E., Pokorny, J. and Smith, V.C. (1996). Increment threshold and purity discrimination spectral sensitivities of X-chromosome-linked color-defective observers. Vision Res. 36: 1597–1613.

Nagy, A.L., MacLeod, D.I.A., Heyneman, N.E. and Eisner, A. (1981). Four cone pigments in women heterozygous for colour deficiency. J. Opt. Soc. Am. 71: 719–722.

Schwartz, S.H. (1994). Spectral sensitivity of dichromats: role of postreceptoral processes. Vision Res. 34: 2983–2990.

Shapley, R.M. and Reid, R.C. (1992). Spatial structure of cone inputs to receptive fields in primate lateral geniculate nucleus. Nature 356: 716–718.

Smith, V.C. and Pokorny, J. (1975). Spectral sensitivity of the foveal cone photopigments between 400 and 500 nm. Vision Res. 15: 161–171.

Stager, D.R., Everett, M. and Swanson, W.H. (1993). The importance of controlled illumination in colour vision testing in a pediatric ophthalmology clinical practice. J. Ped. Ophthalmol. Strab. 30: 39–42.

Swanson, W.H. (1991). Heterochromatic modulation photometry in heterozygous carriers of congenital colour defects. In: Drum, B., Moreland, J.D. and Serra, A. (eds.) Colour Vision Deficiencies X, Doc. Ophthalmol. Proc. Ser. 54: 457–471. Kluwer, Dordrecht.

Swanson, W.H. (1993). Analysis of Rayleigh match data with psychometric function. J. Opt. Soc. Am. A 10: 1807–1817.

Swanson, W.H. and Birch, E.E. (1992). Extracting thresholds from noisy psychophysical data. Percept. Psychophys. 51: 409–422.

Swanson, W.H. and Everett, M. (1992). Colour vision screening of young children. J. Ped. Ophthalmol. Strab. 29: 49–54.

Thornton, J.E. and Pugh, E.N.J. (1983). Red/green colour opponency at detection threshold. Science 219: 191–193.

Verriest, G., Laethem, J. v. and Uvijls, A. (1982). A new assessment of the normal ranges of the Farnsworth-Munsell 100-hue test scores. Am. J. Ophthalmol. 93: 635–642.

Retina Foundation of the Southwest
9900 North Central Expressway, Suite 400
Dallas, TX 75231, USA

8. Rod and cone inputs to parvo- and magnocellular cells in the dichromatic common marmoset (*Callithrix jacchus*)

J. KREMERS, S. WEISS, E. ZRENNER and J. MAURER

(Tübingen, Germany)

Abstract

We measured the responses of lateral geniculate cells of dichromatic common marmosets (*Callithrix jacchus*) to 500 msec flashes of narrow bandwidth chromatic lights upon a white background. The background intensity was either 1, 10 or 40 cd/m^2. The flash intensities varied between 0 and 40 cd/m^2. A response was defined as maximal response in a 50 msec time window after correction for the mean activity. From the response as a function of the Michelson contrast, the contrast gains of cells were obtained for each wavelength, from which their spectral sensitivities could be inferred. We found that up to relatively high luminances the cells received rod input, and that the rod input was never opponent to the cone input.

Introduction

The marmoset is a New World primate which has only one gene on the X-chromosome that codes for a cone pigment in the middle- and long-wavelength spectral region. As a result all males and homozygotic females are dichromats, whereas heterozygotic females are trichromats (Mollon *et al.*, 1984; Jacobs and Neitz, 1985). Recent data indicate that cells in the marmoset lateral geniculate nucleus (LGN) can receive relatively strong rod input (Yeh *et al.*, 1995). The object of the present study was to quantify the strength of rod and cone inputs to LGN cells. We therefore measured spectral sensitivities of LGN cells in dichromatic animals to narrow bandwidth chromatic flashes upon achromatic backgrounds of different intensities. We found that cells had significant rod input up to relatively high luminances. Furthermore, rod and cone inputs were always non-opponent.

C.R. Cavonius (ed.), Colour Vision Deficiencies XIII, pp. 87–97.

Methods

Animal preparation

Six adult dichromatic common marmosets (*Callithrix jacchus*, 300–400 g, three males and three females) were initially anaesthetized by an intramuscular injection of 15–30 mg/kg ketamine hydrochloride (Ketanest) and 3.5 mg/kg xylazine hydrochloride with 0.15 mg/kg methyl-4-hydroxybenzoat (0.15 ml/kg Rompun 2% solution). Additional doses of ketamine hydrochloride were administered if necessary. After tracheotomy the animals were ventilated through a tracheal canula with a 70%/30% N_2O/O_2 mixture containing 0.2–0.8% enflurane (Ethrane; 0.4–0.8% during surgery; 0.2–0.4% during measurements). Eye movements were suppressed by intravenous administration of gallamin triethiodide (Flaxedil; 5 mg/kg/h) dissolved in Ringer, together with glucose and Solu Decortin (0.6 ml/h). A rectal probe connected to a warming pad was used to maintain a rectal temperature of 37.2°C. EEG and ECG were continuously monitored to check the depth of anaesthesia.

Pupils were dilated with atropine sulphate (1%) and neosynephrine (5%). Contact lenses (radius of curvature: 3.5 mm; diameter: 5 mm) with appropriate correction (measured with a slit skiascope) protected the eye against drying. The stimuli were presented through a 2 mm diameter artificial pupil. Due to the smaller eye of the marmoset we calculated that the retinal illuminance was about 4.9 times greater than for humans.

A craniotomy was performed and tungsten-in-glass microelectrodes were lowered into the LGN for extracellular recordings. The stereotaxic location of the craniotomy was based on an atlas of the marmoset brain (Stephan *et al.*, 1980). Occasionally, small lesions were made to check the recording site after histological processing of the brains. At the end of the experiments the animals were euthanized with an overdose of sodium pentobarbital (Nembutal) and blood samples were taken for a genetic analysis of the X-chromosomes. The purpose of the genetical analysis was to determine which of three possible present cone types in the long- and middle-wavelength range was present, using the sequences described by Williams *et al.* (1992) and Hunt *et al.* (1993). After perfusion, the brain and eyes were removed for histology.

Visual stimulation

The visual stimuli were generated with a two-pathway optical bench. The first path was used for the white background, the second for the chromatic flashes. For the different flash colours a filter wheel with 15 narrow bandwidth interference filters (width at half maximum varied between 20 and 26 nm) was placed in the second channel. Up to 15 different flash intensities were provided by a circular neutral density wedge. In addition, the spontaneous activity of the cells was measured in the absence of a stimulus. Both the colour wheel and the neutral density wedge were moved by stepping motors (Zebotronics with

ELBAG drives), which were controlled by the host computer. The chromatic light was flashed in 500 msec on, 500 msec off cycles using a shutter (Compur electronics). The two light paths were combined with a beam splitter and back-projected onto a screen. The projections were circular with a diameter of 5 cm and were otherwise spatially uniform. The screen was placed either 57 or 114 cm from the animal.

The spectral output of the two pathways was measured as luminances and as radiances with a SpectraScan spectroradiometer (Photo Research). The luminances were recalibrated at regular intervals using an UDT luminance detector and an International Light IL1700 Photometer.

Recording procedure

The spikes were converted into TTL pulses using a window discriminator. The time of occurrence of each spike was stored with a resolution of 0.5 msec by a CED 1401 on-line computer. Each stimulus condition was measured for 6 sec. These recordings were send to the host computer and were stored on hard disk and magnetic optical disk, and PSTHs were created by the host computer. The CED 1401 also controlled the shutters, thus minimizing errors due to delays between shutter actions and spike recording. Complete measurement of all flash colours and all intensities took about 0.5–1.5 h.

Results

We recorded responses of 14 parvocellular (P) cells, 6 magnocellular (M) cells and three cells with S cone input in six dichromatic animals (three males and five females). Figure 1 shows some original responses of a P cell to flashes of three different wavelengths, each at three different intensities against a white background of 1 cd/m^2. This cell responds best to 500 nm flashes. With the exception of the three cells with S cone input we never found any indication of colour opponent inputs to any of the cells.

A peak detecting algorithm was applied to the cell responses. For each stimulus, we searched for the maximal response within a 50 msec time window at the stimulus phase in which an excitatory response was elicited (flash-on for on-cells and flash-off for off-cells). From this we subtracted the mean activity of the cell in the last 100 msec before onset of the excitatory stimulus. The rationale for this approach was that a peak detector with a time window of about 50 msec applied to macaque retinal ganglion cell responses gave thresholds that closely resembled human psychophysical thresholds (Lee *et al.*, 1993, 1995; Kremers *et al.*, 1993). We did not fix the 50 msec time window to a certain position relative to the stimulus (as was done in the work cited) but 'scanned' through the whole response. This introduced more noise but we felt it was a more realistic model for a cortical detector. The responses were plotted as function of the radiant Michelson contrast (*C*) of the stimulus ($C = E_{flash}/[E_{flash} + 2 \, . \, E_{bckgr}]$ where E_{flash}

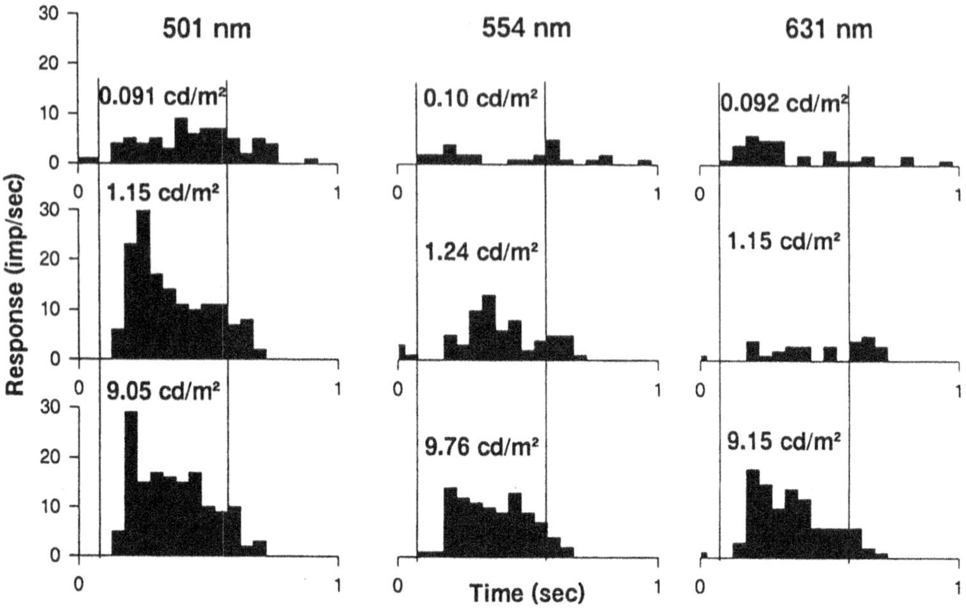

Fig. 1. Responses of one parvocellular cell to three different wavelengths, each at three different intensities (as indicated) upon a 1 cd/m^2 background. The time of flash onset (at $t = 100$ msec) is indicated by the first vertical line. Flash offset was at $t = 600$ msec (second vertical line). The bandwidth is 50 msec.

and E_{bckgr} are the radiances of flashed light and of background). A Naka–Rushton function was fitted through the data. These functions are special Michaelis–Menten functions. We found satisfactory fits with a function having the form: $R - R_{spont} = R_{max} \cdot C/[C{+}b]$ where C is the Michelson radiant contrast, b the contrast at half-maximal response, R_{max} the maximal response and R_{spont} the spontaneous response of the peak detector at the absence of a stimulus. This equation is identical to that used by Lee *et al.* (1990), except that our correction for spontaneous activity was different for each measurement, since we found that the spontaneous activity was variable both between cells and within a cell at different times. We also used the radiant Weber fraction as a measure for stimulus intensity. With 10 and 40 cd/m^2 backgrounds, the results were very similar. With a 1 cd/m^2 background, the Weber fraction gave good fits for all data points, whereas with Michelson contrast good fits were obtained near threshold. Although the Weber fraction is generally better for describing the stimulus intensity offlashes, we nevertheless used the Michelson contrast data, since we were interested in threshold responses, for which the state of adaptation is determined primarily by the background and not by the stimulus.

The cell's spectral sensitivity was defined as contrast gain (the slope of the Naka–Rushton function when contrast is zero) as a function of the peak

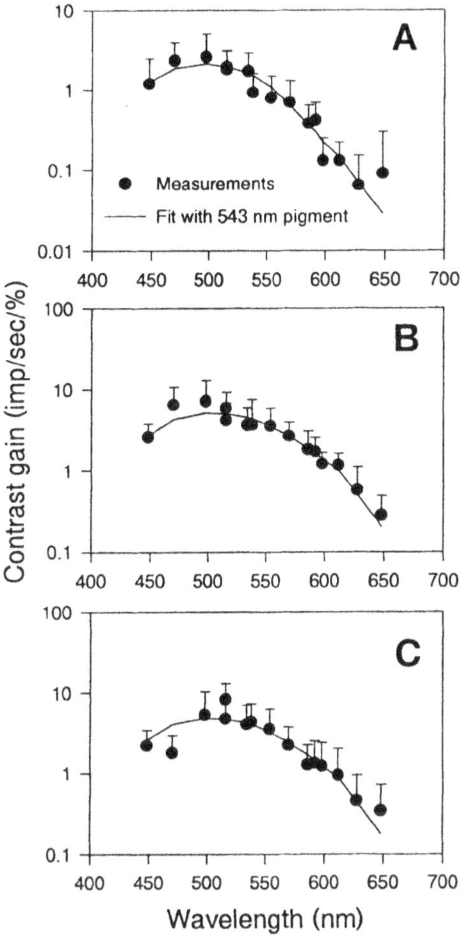

Fig. 2. Mean spectral sensitivity of P and M cells of one animal putatively having the 543 nm pigment (circles). The data are therefore fitted with an addition of the absorption of the rods and the 543 nm photoreceptors (lines). Wavelengths are the peak wavelengths of the chromatic stimuli. A: P cells with a 1 cd/m^2 background ($n = 6$); B: M cells with 1 cd/m^2 background ($n = 3$); C: P cells with 10 cd/m^2 background ($n = 7$). All the cells are very sensitive to 500 nm flashes. The decrease in sensitivity at the long wavelength limb is larger for P cells with a 1 cd/m^2 background (A) than for the other two conditions.

wavelength of the chromatic flashes. Figure 2 shows the mean spectral sensitivity of P cells with 1 and 10 cd/m^2 backgrounds and of M cells with a 1 cd/m^2 background in a female animal for which the genetic analysis identified the 543 nm photopigment. All three spectral sensitivities have a maximum around 500 nm, which is an indication of strong rod input to the cells. The

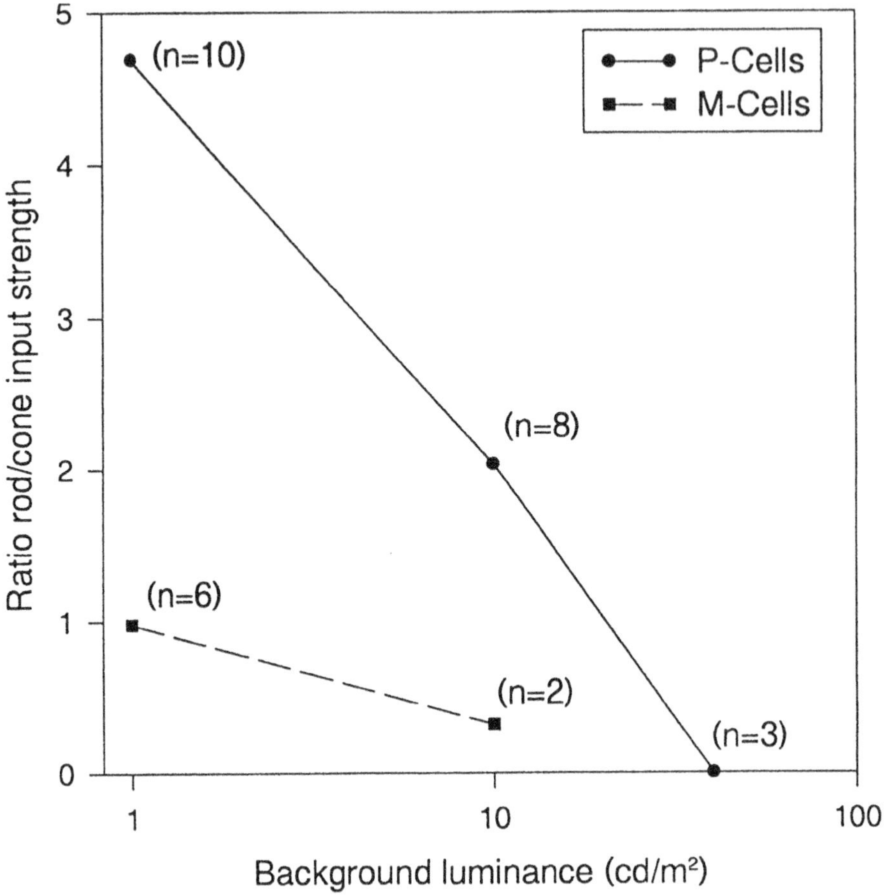

Fig. 3. Ratio of rod to cone input strength for P and M cells at different background intensities. There is a clear decrease of rod to cone ratio with increasing background intensity. The number of measured cells for each condition is indicated between brackets.

spectral sensitivities of the P cells at 1 cd/m^2 are narrower than those for the other conditions, indicating that these cells receive stronger rod input. The sensitivities were fitted with an arithmetic sum of the rod and cone absorptances to the flash colours. Although we used narrow band interference filters, we did not use the absorption spectra of the rods and cones directly; instead we calculated the photoreceptor absorptance to the stimuli by multiplying the spectral output of the coloured lights with the absorption spectra of the cones and rods after correction for lens absorption (data kindly provided by J. Bowmaker). Since most cells' receptive fields were peripheral we did not correct for absorption in the macula pigment. The photoreceptor absorption spectra were obtained from Dartnall's standard spectrum for frog rhodopsin, assuming maximal absorption at 500 nm for the rods, 423 nm for the S cones and 543, 556 and 563 nm for the L-M cones (Knowles and Dartnall, 1977). The template was fitted with an empirical mathematical function ($D = a + b \times \exp[1-(f-c)/d-exp\{-(f-c)/d\}]$, where D is the optical density, f is wavenumber and a, b, c and d are free parameters), which gave satisfactory fits. We recalculated the measurements using a template fitted through microspectrophotometric data (also provided by J. Bowmaker) and found only minor differences in cone and rod sensitivities. Optical density was converted to absorptance assuming an axial absorbance of 0.3 (Tovée *et al.*, 1992). The absorbance curves were transformed from quantal units into radiant units by dividing the absorptance by the wavelength. The data were fitted with all three different cone absorbances, although only the fits of the pigment that was present according to the genetical analysis are shown. Due to the large rod inputs, the other pigments also gave good fits, but the goodness of fit was normally better when using the genetically determined pigment.

The ratio of rod to cone input strengths is shown in Figure 3 as a function of the background luminance. The ratio clearly decreases with increasing background luminance. Surprisingly, M cells received relatively smaller rod signals than P cells at all background levels.

We wondered whether this difference between P and M cells was caused by actual differences in input strengths, or by different dynamics between the responses of P and M cells to the flashes. We therefore determined the time of occurrence of the peak response for each condition that elicited a response larger than 50 impulses/sec. Figure 4A shows the combined results for eight cells (three on-centre and two off-centre P cells; one on-centre and two off-centre M cells) in one animal using a 1 cd/m^2 background. Clearly, M cells have shorter latencies than P cells to the excitatory stimulus. The latency of the peak response decreases with increasing response strength. Figure 4B shows the mean latency after the excitatory stimulus in different cells and for different background conditions. The response occurred earlier in M cells. Rod signals might be slower than cone signals and therefore might not be so obvious in the faster peak response of M cells. Alternatively, a smaller rod input strength might itself cause a faster response. Response time decreased with increasing background intensity, indicating that a shift from a rod-dominated to a cone-dominated response does influence the response time. We conclude that M and P cells receive large rod inputs up to relatively high retinal illuminances. The rod input to P cells seems to be stronger. The rod-to-cone input strength is related to the latency of the peak response.

The cells with S cone input were excited by short-wavelength flashes, and inhibited by long-wavelength flashes. The spectral sensitivities of four measurements on three different cells are shown in Figure 5. These cells barely responded to 500 nm flashes.

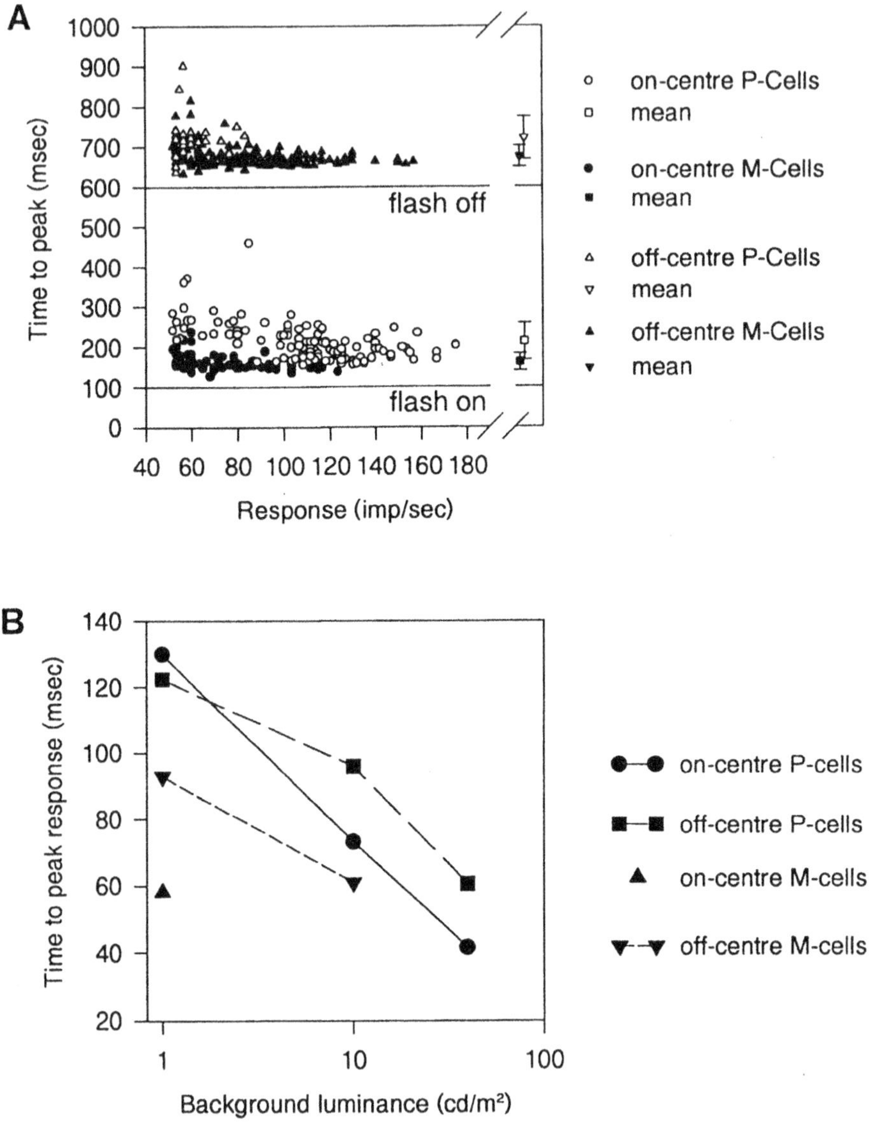

Fig. 4. (A): The time to peak response of M and P cells as a function of response amplitude. The peak response tends to occur earlier when the response is larger. Furthermore, M cells respond earlier than P cells. On-centre cells have peak responses after flash-onset (at 100 msec), whereas off-centre cells are excited by flash-offsets (at 600 msec). Responses below 50 imp/sec were too noisy to give reliable estimates of the response time. (B): The mean time to peak response after the excitatory stimulus for P and M cells as a function of background intensity. The time to peak response clearly decreases with increasing background intensity and is smaller in M cells.

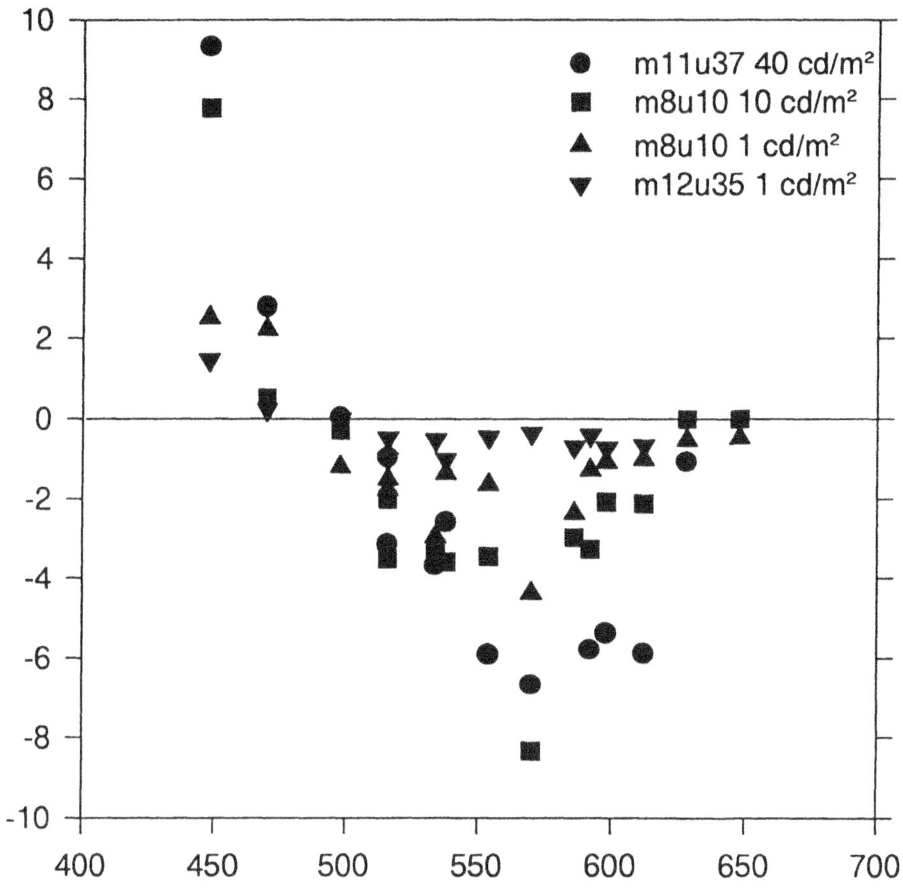

Fig. 5. Spectral sensitivities of cells with (excitatory) S cone input at different background levels. Negative contrast gains indicate an inhibition to flash-onset and an excitation by flash-offset. Positive contrast gains indicate that the flash-onset was excitatory. The neutral point was always close to 500 nm.

Discussion

The spectral sensitivity of dichromatic marmoset LGN cells to chromatic flashes on a white background revealed some interesting results. We found that both parvocellular cells with L/M cone input and magnocellular cells received rod input up to relatively high luminances. Even at 10 cd/m^2 background (resulting in a retinal illuminance equivalent to about 150 td in the human eye) the cells received rod input. Moreover, since the flashes were presented upon a background, the mean luminance was higher than the background luminance. Indications of a large rod input to marmoset LGN cells have also been found in other studies (Yeh *et al.*, 1995; Weiss *et al.*, 1995). We found that parvocellular cells received more rod input than magnocellular cells. This may be due to different dynamics of P and M cell responses to the flashes, rather than to actual differences in input strength. M cells responded faster, presumably thereby favouring cone responses, whereas the slower P cell responses might reveal more rod input. Alternatively, the difference may be caused by the small cell sample. With another paradigm, in which the ratio of counterphase red–green sinusoidal modulation contrast was varied, we found that M and P cells had similar rod/cone amplitude ratios (Weiss *et al.*, 1995).

Macaque retinal ganglion cells show less strong rod input (B.B. Lee, personal communication), despite the fact that macaques have a lower cone density and higher rod density (Troilo *et al.*, 1993; Goodchild *et al.*, 1995). It is unclear whether the large rod input is specific for marmosets, or typical of dichromats in general.

We further found that in both P and M cells the rods had a non-opponent input, i.e. the rod signal always had the same sign as the L/M cone. This means that the loss of chromatic sensitivity in dichromats cannot be compensated by opponency of the rod system.

Several aspects made it difficult to determine the cone pigment on the basis of the spectral sensitivities. The strong rod input obscured the spectral absorption of the cones, especially at low retinal illuminances. Furthermore, due to cortical inhibition, the cells often responded weakly to the flash stimuli, especially at high retinal illuminances. We therefore have only a limited number of reliable measurements at high retinal illuminances. We fitted the data of each animal with all three possible cone absorption spectra. Best fits were usually obtained using the pigment that could be expected to be present from the genetic analysis, although the differences in goodness of fit were relatively small.

In cells with S cone input the neutral wavelength (the wavelength eliciting no response) is very close to 500 nm. Wiesel and Hubel (1966) and Zrenner (1983) obtained identical results for macaque LGN and retinal ganglion cells with S cone input (Wiesel and Hubel's type II cells). We fitted the data with the combination of absorbances of rods and the two cone types present, but due to the larger number of free parameters, the results were less well constrained by the data, making conclusions concerning photoreceptor input difficult.

Acknowledgements

The authors wish to thank Eva Burkhardt for technical assistance, and Bernd Wissinger for assistance with the genetical analysis. We also thank Jim Bowmaker for providing the microspectrophotometric and lens absorption data, and Anne Kurtenbach for comments on the manuscript. Supported by DFG Grant Zr 1/9-2 TP 3.1.

References

Goodchild, A.K., Ghosh, K.K. and Martin P.R. A comparison of photoreceptor spatial density and ganglion cell morphology in the retina of human, macaque monkey, cat, and the marmoset *Callithrix jacchus*. J. Comp. Neurol. (in press).

Hunt, D.M., Williams, A.J., Bowmaker, J.K. and Mollon, J.D. (1993). Structure and evolution of the polymorphic photopigment gene of the marmoset. Vision Res. 33: 147–154.

Jacobs, G.H. and Neitz, J. (1985). Colour vision in squirrel monkeys: sex-related differences suggest the mode of inheritance. Vision Res. 25: 141–144.

Knowles, A. and Dartnall, H.J.A. (1977). The characterization of visual pigments by absorption spectroscopy. In: Davson, H. (ed.) The Eye Vol. 2B. The Photobiology of Vision: 53–101. Academic, New York.

Kremers, J., Lee, B.B., Pokorny, J. and Smith, V.C. (1993). Responses of macaque ganglion cells and human observers to compound periodic waveforms. Vision Res. 33: 1997–2011.

Lee, B.B., Pokorny, J., Smith, V.C., Martin, P.R. and Valberg, A. (1990). Luminance and chromatic modulation sensitivity of macaque ganglion cells and human observers. J. Opt. Soc. Am. A. 7: 2223–2236.

Lee, B.B., Wehrhahn, C., Westheimer, G. and Kremers, J. (1993). Macaque ganglion cell responses to stimuli that elicit hyperacuity in man: detection of small displacements. J. Neurosc. 13: 1001–1009.

Lee, B.B., Wehrhahn, C., Westheimer, G. and Kremers, J. (1995). The spatial precision of macaque ganglion cell responses in relation to vernier acuity in human observers. Vision Res. 35: 2743–2758.

Mollon, J.D., Bowmaker, J.K. and Jacobs, G.H. (1984). Variations in colour vision in a new world primate can be explained by polymorphism of retinal photopigments. Proc. Roy. Soc. Lond. B. 222: 373–399.

Stephan, H., Baron, G. and Schwerdtfeger, W.K. (1980). The Brain of the Common Marmoset (*Callithrix jacchus*). A Stereotaxic Atlas. Springer, Berlin.

Tovée, M.J., Bowmaker, J.K. and Mollon, J.D. (1992). The relationship between cone pigments and behavioral sensitivity in a new world monkey (*Callithrix jacchus jacchus*). Vision Res. 32: 867–878.

Troilo, D., Howland, H.C. and Judge, S.J. (1993). Visual optics and retinal cone topography in the common marmoset (*Callithrix jacchus*). Vision Res. 33: 1301–1310.

Weiss, S., Kremers, J. and Zrenner, E. (1995). Cone and rod input of parvocellular geniculate cells in the common marmoset (*Callithrix jacchus*). Invest. Ophthalmol. Vis. Sci. Suppl.: 691.

Wiesel, T.N. and Hubel, D.H. (1966). Spatial and chromatic interactions in the lateral geniculate body of the rhesus monkey. J. Neurophysiol. 29: 1115–1156.

Williams, A.J., Hunt, D.M., Bowmaker, J.K. and Mollon, J.D. (1992). The polymorphic photopigments of the marmoset: spectral tuning and genetic basis. EMBO J. 11: 2039–2045.

Yeh, T., Lee, B.B., Kremers, J. *et al.* (1995). Visual responses in the lateral geniculate nucleus of dichromatic and trichromatic marmosets (*Callithrix jacchus*). J. Neurosci. 15: 7892–7904.

Zrenner, E. (1983). Neurophysiological Aspects of Color Vision in Primates. Comparative Studies on Simian Retinal Ganglion Cells and the Human Visual System. Springer, Berlin.

Department of Experimental Ophthalmology
University Eye Hospital
Röntgenweg 11
D-72076 Tübingen, Germany

9. Results of clinical colour vision tests of 'Pigmentfarbenanomale'

S. TANABE and K. HUKAMI
(Nagoya, Japan)

Abstract

One hundred and twelve patients who were classified as red–green defective on the basis of pseudoisochromatic plates showed almost normal results on the anomaloscope. These so-called 'Pigmentfarbenanomale' (Heinsius 1959) were in a group of 3785 congenital red–green defective patients. Their performance on conventional tests was: pseudoisochromatic plates: mild or moderate anomaly, mostly deutan; Farnsworth's Panel D-15 Test: pass; lantern test: in most cases 4/9 or more errors; anomaloscope: AQ 0.6–1.3. These findings, except for those of the anomaloscope, are identical with those of ordinary deuteranomaly. We established pedigrees for five families with Pigmentfarbenanomalie. Of these, two also had members who were deuteranomalous. None of the cases fits a protan pattern.

Introduction

It is taken for granted that congenital red–green colour vision defects are best determined by anomaloscopic examination. Therefore, so-called Pigmentfarbenanomalie, which yields a normal Rayleigh match on the anomaloscope, has long been of interest (Engelbrecht, 1944; Heinsius, 1959). Available reports concerning Pigmentfarbenanomalie have not fully explained the discrepancy between the test results of pseudoisochromatic plates and the anomaloscope.

We have now studied 112 cases of Pigmentfarbenanomalie whose test results are presented below, although a satisfactory explanation of the condition has not been obtained.

Subjects and methods

We reviewed 112 cases of Pigmentfarbenanomalie, which were among 3785 patients with congenital red–green defects referred to our special clinic for colour vision defects during the past 20 years. None of the subjects had ophthalmologic abnormalities other than colour vision defects, and all had

C.R. Cavonius (ed.), Colour Vision Deficiencies XIII, pp. 99–104.

corrected vision of 1.0 or better and normal visual fields. Colour vision tests included the Ishihara, Tokyo Medical College Colour Vision Test (TMC), AO-HRR plates, Standard Pseudoisochromatic Plates Part 1, Panel D-15, Ichikawa's lantern (which resembles the Farnsworth lantern) and the Schmidt-Haensch Nagel Model 1 anomaloscope. The results were compared with those of 100 randomly sampled deuteranomals.

Results

With the pseudoisochromatic plates, responses typical of anomalous trichromats were obtained in all but three very mild cases. These three subjects made few or no errors with some pseudoisochromatic plates, but were identified as colour defective by other tests. The results of the diagnostic plates are shown in Table 1. On the basis of all three plates together, the type of defect was diagnosed as deutan in all 112 cases. The degree of severity is shown on Table 2.

Table 1. Protan/deutan classification with SPP-1, TMC, and AO-HRR.

		Deutan	Unclassified	Protan
SPP-1	Pig.	96%	4%	0%
	DA	97%	0%	3%
TMC	Pig.	75%	21%	4%
	DA	80%	18%	2%
AO-HRR	Pig.	73%	23%	3%
	DA	76%	20%	4%

Table 2. Severity determined with TMC and AO-HRR

		I	II	III
TMC	Pig.	24%	43%	33%
	DA	18%	53%	29%
		Mild	Medium	Strong
AO-HRR	Pig.	47%	49%	3%
	DA	44%	53%	2%

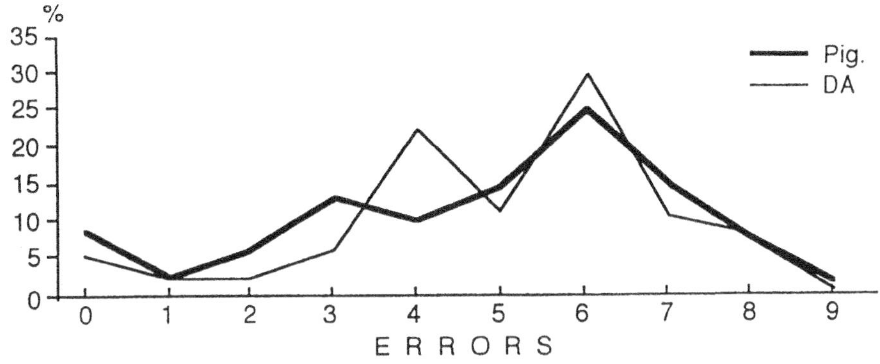

Fig. 1. Error scores of lantern test. Comparison between Pigmentfarbenanomalie (Pig.) and deuteranomaly (DA).

These test results were almost identical with ordinary deuteranomaly. All but one subject passed the D-15 test; this subject had a borderline score. The error scores of the lantern test are shown in Figure 1. No significant differences were found compared with ordinary deuteranomaly.

Figure 2 shows matching ranges on the anomaloscope. The luminance settings of all the subjects showed a deutan pattern. Asterisked cases are of the same family. Cases marked # have deuteranomaly in the family, whose matching ranges are shown on the same line. The five pedigrees are shown in Figure 3.

Discussion

It has recently been established by molecular genetic studies that congenital colour vision defects arise from genetic anomalies that result in abnormal visual pigments (Nathans *et al.*, 1986). Furthermore, various gene mutations and subsequent abnormal amino acid sequences and their spectral properties have been investigated (Merbs and Nathans, 1992; Neitz *et al.*, 1989; Asenjo *et al.*, 1994). The common congenital colour vision defects can now be clearly identified by molecular analysis. However, clinical manifestations, such as degree of defects or ability to discriminate colours, are not always clear. The inconsistent features of Pigmentfarbenanomalie is one of these unsolved problems.

To account for the normal Rayleigh match in Pigmentfarbenanomalie, hypotheses of abnormal visual pigment, which incidentally include a normal matching point (Alpern, 1981; Yamade, 1990), or an abnormality of neural mechanism (Hurvich and Jameson, 1962) with normal visual pigments, have

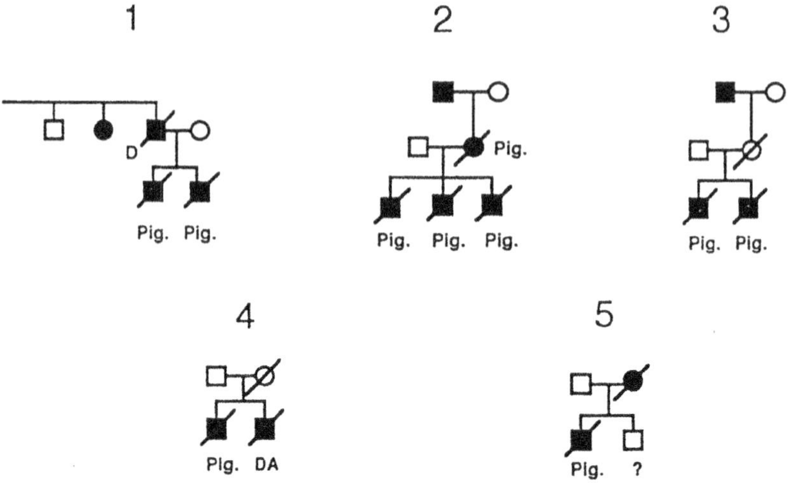

Fig. 2. In pedigrees 1, 2, and 3, all the members with colour vision defect are Pigmentfarbenanomalie (Pig). In pedigree 1, deuteranope (father) is only coincidence. In pedigrees 4 and 5, both Pigmentfarbenanomalie and deuteranomaly (DA) occurred.
■: colour-defective male; □ : normal male;
●: colour-defective female; ○: phenotypically normal female.

been proposed. However, these do not explain why the same mechanism performs normally on the anomaloscope but shows typical colour confusions on pseudoisochromatic plates.

Pigmentfarbenanomalie seems to be an atypical form of red–green colour vision defect, most commonly resembling deuteranomaly, and does not belong to the category of normal colour vision termed 'neuteranope' (Hurvich and Jameson, 1962) or 'Farbenasthenopie' (Grützner, 1971) because all of our subjects made deuteranopic colour matches, i.e. matches under Umstimmung between the mixture field and the yellow field of the anomaloscope, and deutan-type errors on pseudoisochromatic plates. Families 4 and 5, which include both Pigmentfarbenanomalie and deuteranomaly, fit this pattern. On the other hand, families 1, 2, and 3, whose anomaloscope settings are almost normal, suggest that there may be some distinct difference between Pigmentfarbenanomalie and ordinary deuteranomaly.

The genotype-phenotype relationship is still confusing, as shown in the reports that some genotypes may be expressed either as a dichromat or as an anomalous trichromat (Deeb *et al.*, 1992); that the degree of defect is not determined by the fusion point of hybrid gene (Neitz *et al.*, 1989; Deeb *et al.*, 1992); and that there are colour-normal subjects who have hybrid genes (Deeb *et al.*, 1992). Nevertheless, there must logically be some correspondence between genotype and phenotype. A comparison of clinical diagnoses of Pigmentfarbenanomalie and molecular genetic analysis is needed.

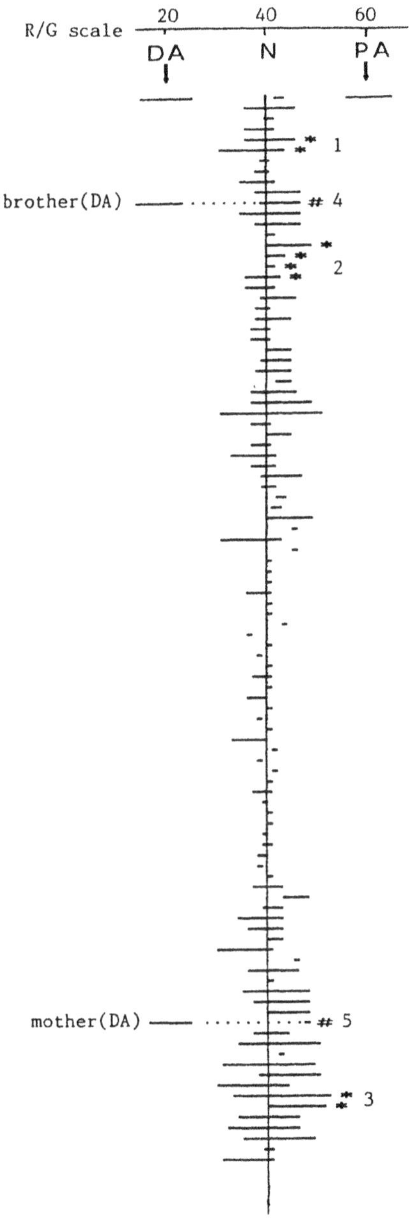

Fig. 3. Matching ranges of Pigmentfarbenanomales. Arrows show the average protanomalous Rayleigh match and deuteranomalous Rayleigh match respectively.
*: members of the same family;
#: members of a family that includes deuternomals, whose matching ranges are shown to the left.
Numbers refer to the pedigrees in Fig. 2.

References

Alpern, M. (1981). The colour vision of the colour blind. Jpn. J. Ophthalmol. 25: 1–17.

Asenjo, A.B., Rim, J. and Oprian, D.D. (1994). Molecular determinants of human R/G colour discrimination. Neuron 12: 1131–1138.

Deeb, S.S., Lindsey, D.T., Hibiya, Y. *et al.* (1992). Genotype-phenotype relationship in human red/green colour vision defects: Molecular and psychological studies. Am. J. Hum. Genet. 51: 687–700.

Engelbrecht, K. (1944). Grenzen und Grenzfälle der Farbenempfindung. Klin. Monatsb. Augenheilk. 110: 630–637.

Grützner, P. (1971). Über Farbenanomalie, Farbenamblyopie, und Farbenasthenopie. Klin. Monatsb. Augenheilk. 158: 89–96.

Heinsius, E. (1959). Über die verschiedenen Formen der Trichromasie, sowie über die Grenze zwischen Farbentüchtigkeit und Farbenuntüchtigkeit. Klin. Monatsb. Augenheilk. 135: 95–107.

Hurvich, L.M. and Jameson, D. (1962). Colour theory and abnormal red–green vision. Doc. Ophthalmol. 16: 409–442.

Merbs, S.J. and Nathans, J. (1992). Absorption spectra of the hybrid pigments responsible for anomalous colour vision. Science 258: 464–466.

Nathans, J., Piantanida, T.P., Eddy, R.L., Shows, T.B. and Hogness, D.S. (1986). Molecular genetics of inherited variation in human colour vision. Science 232: 203–210.

Neitz, J., Neitz, M. and Jacobs, G.H. (1989). Analysis of fusion gene and encoded photopigment of colour-blind humans. Nature 342: 679–682.

Yamade, S. (1990). Cases of colour anomaly displaying normal Rayleigh matches (pigment colour defect). Folia Ophthalmol. Jpn. 41: 452–457.

Japanese Red Cross First Nagoya Hospital
3-35, Michishita-cho, Nakamura-ku
Nagoya 453, Japan

PART TWO

Structure and function in colour vision

10. Structure and function in primate retina

B.B. LEE and D.M. DACEY
(Göttingen, Germany and Seattle, WA, USA)

Abstract

Recent results obtained using both *in vivo* and *in vitro* recording techniques have suggested that primate retinal circuitry is organized in a precise and elegant manner, and that luminance and chromatic detection channels defined psychophysically have firm anatomical and physiological substrates at a retinal level. These results are reviewed, and it is suggested that parallel pathways beginning in the retina are best defined in terms of the anatomical mosaics of ganglion cell types in which they originate.

Introduction

The concept of separable chromatic and achromatic channels in vision is almost as venerable as the trichromatic theory. The latter now has firm electrophysiological and genetic foundations, but the electrophysiological and anatomical basis for separable channels is not yet so well established. However, recent evidence points to a retinal organization in which chromatic and luminance signals are carefully segregated and are derived from very specific sets of anatomical connections. In this brief review, we describe some of the recent developments which lead to this conclusion. A more detailed overview of recent anatomical and physiological results may be found elsewhere (Lee, 1996).

Experiments involving detection of chromatic and luminance modulation have yielded evidence for separable chromatic and achromatic mechanisms (e.g. Kelly and van Norren, 1977; Swanson *et al.*, 1987). It seems reasonable to seek substrates for such mechanisms early and peripherally in the visual system. These mechanisms differ from the red–green, blue–yellow and achromatic processes as formulated in classical opponent process theory (Hering, 1878; Hurvich and Jameson, 1957), which may have a physiological substrate more centrally, at a cortical level (e.g. DeValois and DeValois, 1993). Here we are concerned with the first type of mechanism and with whether one can find physiological and anatomical substrates in the retina.

In all Old World simians, including humans, a number of parallel systems begin within the retina, relay through the lateral geniculate nucleus (LGN), and remain at least partially segregated deep into the cortex. Chromatically

C.R. Cavonius (ed.), Colour Vision Deficiencies XIII, pp. 107–117.

opponent cells with antagonistic input from the middle- (M) and long-wavelength (L) cones, and also cells with input from short wavelength (S) cones opposed to the other two types, project through the parvocellular (PC) layers of the LGN. Cells with receptive fields of on- or off-centre type, receiving synergistic input from M and L cones to centre and surround, project through the magnocellular (MC) layers of the LGN.

The relationship between these cell types and systems and psychophysical channels and mechanisms has been controversial (e.g. Gouras and Zrenner, 1979), but recent physiological evidence suggests strongly that the magnocellular system forms the substrate of a luminance channel that is responsible for performance on tasks such as heterochromatic flicker photometry (HFP; Lee *et al.*, 1988), while the substrate of mechanisms responsible for the detection of chromatic change is formed by chromatically-opponent cells (Cole *et al.*, 1993; Lee *et al.*, 1993a). A great deal of indirect anatomical evidence exists as to cell types and connectivity within the retina (Wässle and Boycott, 1991). For example, the parasol (Polyak, 1941) morphological type projects to the magnocellular layers of the LGN.

It has recently become possible to maintain primate retinae *in vitro* and this permits a more direct correlation of physiology and anatomy. Visual responses may be recorded intracellularly from a selected cell and its physiological type can be identified, after which the cell is stained and its morphology reconstructed (Dacey and Lee, 1994b). This method provides a powerful means of correlating physiological and anatomical cell types and determining whether each forms its own mosaic across the retina.

The magnocellular pathway and parasol cells

Cells of the magnocellular pathway make up 5–10% of the retinal ganglion cells (Perry *et al.*, 1984), even in the foveal region (Grünert *et al.*, 1993). The evidence that this pathway provides a substrate for heterochromatic flicker photometry is strong, since many features of HFP find a direct substrate in MC cells (Lee *et al.*, 1988; Smith *et al.*, 1992). Figure 1 illustrates the responses of MC cells in this context. At the left of the figure are drawn axes as first used in physiological experiments by Derrington *et al.* (1984). Responses to three directions of modulation are shown: along the luminance axis (+Lum to –Lum), along a constant S cone axis, where M and L cones are modulated in counterphase (+L-M to +M-L), and along the tritan S cone axis, where only the S cone is modulated (+S to –S). In each panel, the upper histogram shows the extracellular response *in vivo* of an MC retinal ganglion cell from parafoveal retina to two cycles of modulation. The middle trace shows an intracellular recording from a parasol ganglion cell from peripheral retina. The modulation of the membrane potential can be seen, and impulses are superimposed upon it. The lowest histogram shows a spike histogram from the intracellular recording. Both recordings yield very similar patterns of responses. There is a strong response to luminance modulation, no response to S cone modulation, and a

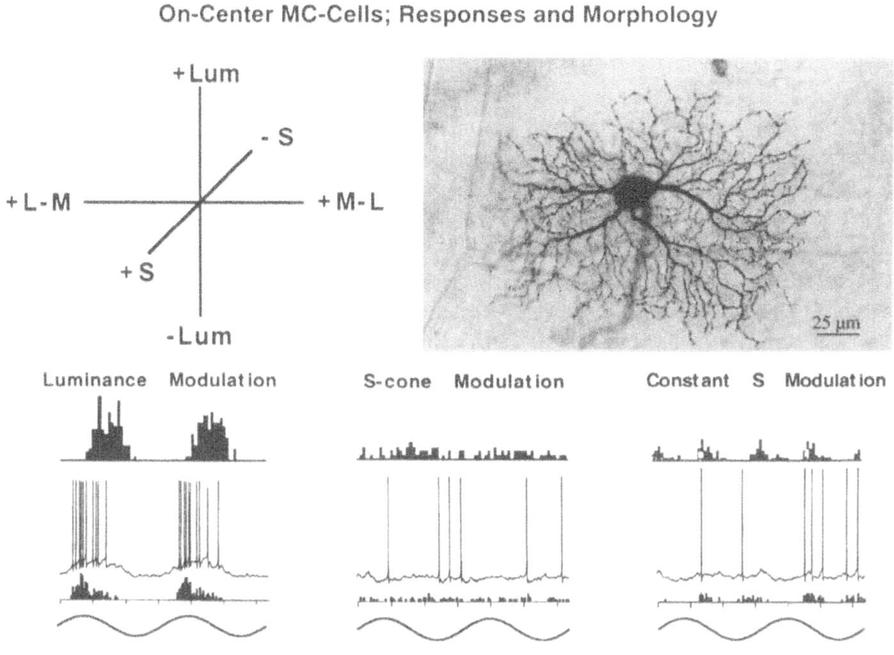

Fig. 1. Responses of on-centre parasol, MC cells to luminance, constant S cone and S cone modulation. Three panels show electrophysiological responses to each modulation type. In each, the upper histogram shows the response of a parafoveal ganglion cell *in vivo* (2 cycles, 10 Hz, 50% stimulus contrast, peak response ca. 160 imp/sec). The middle and lower panels show *in vitro* data from a cell from peripheral retina obtained under similar stimulus conditions. The middle panel shows a single response trace. Membrane potential was ca. –60 mV and spike amplitude 65 mV. The lowest histogram shows the averaged spike response from the intracellular recording. Inset is the cell morphology revealed by neurobiotin staining. It is that of an inner stratifying parasol cell. Stimuli were produced with LEDs of dominant wavelengths of 470 nm, 554 nm and 638 nm as light sources. LED outputs were set to the same mean luminance, to give a total retinal illumination of ca. 1000 td. See Yeh *et al.* (1995) for further details.

small response at twice the stimulus frequency to constant S cone modulation. This seems to be a result of a non-linearity in summation of the M and L cone signals (Lee *et al.*, 1989a).

After intracellular staining with neurobiotin, the morphology shown in the inset photomicrograph could be reconstructed. It is typical of a parasol cell, with a circular, densely branching dendritic tree. This neurone was an on-centre cell, and the dendritic tree ramified in the vitreal half of the inner plexiform layer (IPL). Off-centre cells had dendritic trees ramifying in the scleral half of the IPL, so these cells conform to the usual pattern found in mammalian retina (Famiglietti and Kolb, 1976).

Figure 2* shows a hypothetical wiring diagram for the most frequent ganglion cell types encountered. On the left are sketched on- and off-centre parasol/ MC cells. These receive input from invaginating and flat diffuse bipolar cells respectively (Boycott and Wässle, 1991). All appear to contact all cones within their receptive fields, and several bipolars converge onto the dendritic tree of the ganglion cell. This convergence determines centre diameter, which corresponds closely to the dendritic tree size.

The psychophysical mechanism underlying performance on HFP is also thought to be responsible for the detection of luminance modulation. There is now considerable evidence consistent with the hypothesis that the magnocellular pathway also provides the physiological signal for this performance (Lee *et al.*, 1989b, 1990, 1993a; Kremers *et al.*, 1992, 1993) The construction of an achromatic channel for spatial vision is more controversial. The high density of M/L cone opponent cells might suggest an important role in fine spatial vision, but it is more difficult to construct an achromatic signal from the outputs of these cells (e.g. Lennie and D'Zmura, 1988) than appears at first sight (Valberg *et al.*, 1992; Lee, 1996). For fine spatial tasks with achromatic patterns the signals in the PC pathway seem too weak to provide a useful signal (Lee *et al.*, 1993b, 1995). The magnocellular pathway may provide a much more significant input to an achromatic channel for fine spatial vision than currently supposed.

The parvocellular pathway and midget ganglion cells

If psychophysical thresholds are plotted in a cone space, the threshold contours often consist of combinations of straight line segments (Stromeyer *et al.*, 1985; Cole *et al.*, 1993). Each of these segments may be viewed as a region within which a single chromatic or luminance detection mechanism determines the threshold (Poirson *et al.*, 1991). Cell 'thresholds' plotted in a similar space contain straight segments which can provide a substrate for the psychophysical data (Lee *et al.*, 1993a).

Figure 3 shows recordings from cells of the PC pathway with M/L cone input, in the same format as in Figure 1. The upper histograms show *in vivo* responses of a red-on cell from the parafovea. There is a weak response to luminance modulation, a strong response to modulation along the constant S cone axis, and no response to S cone modulation. Anatomically, M/L cone opponent cells are thought to correspond to midget ganglion cells. The intracellular recordings are from an inner green-on midget ganglion cell from peripheral retina. The cell shown displays a chromatic response to red–green, constant S cone stimulation, but no response to S cone modulation. However, there is also a relatively large luminance response.

---

*Figure 2 can be found opposite page 21 of the book.

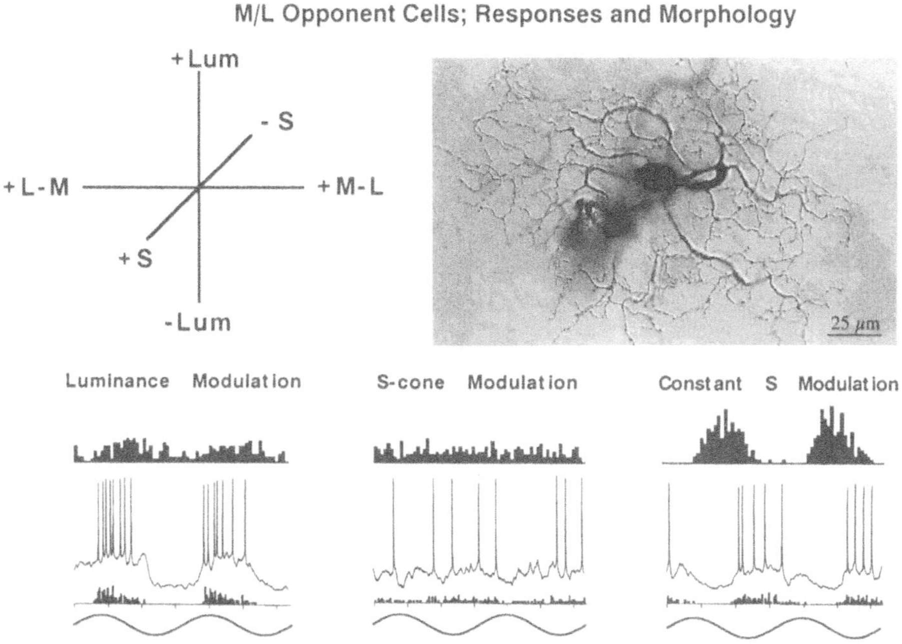

Fig. 3. Responses of two on-centre cells of the midget PC pathway to the same stimuli as employed in Figure 1. The *in vivo* recording was obtained from a red-on, +L-M cell, and the *in vitro* recording from a +M-L green-on cell. Both give an excitatory response to the luminance modulation, but the response to the constant S modulation is approximately phase-inverted. The inset photomicrograph shows peripheral midget morphology.

In the central retina, a midget ganglion cell connects to a single midget bipolar, and its dendritic tree is very compact. In the far peripheral retinà, midget cells no longer show this morphology but make contact with several bipolars. This can be inferred from the reconstruction of the cell's morphology. Although the morphology is quite distinct from that of the parasol cell of Figure 1, the dendritic tree covers the terminal fields of a number of midget bipolars. If the tree makes non-selective contact with the bipolar terminals, then chromatic opponency will be lost. We have found peripheral midgets with little obvious M/L cone opponency. This is extremely rare for PC cells from central retina (Padmos and van Norren, 1975; Derrington *et al.*, 1984). The relationship between opponency, eccentricity and morphology awaits more detailed study.

Midget ganglion cells are drawn in the midsection of Figure 2. There are two anatomically distinguishable types ramifying in the inner and outer sublaminae of the IPL. It is generally assumed that the inner midgets correspond to red- and green-on cells and the outer midgets to red- and green-off cells; this mixture of functional type with one anatomical form is stressed in Figure 2. There is only one mosaic of inner and one of outer midgets, so that two different physiological

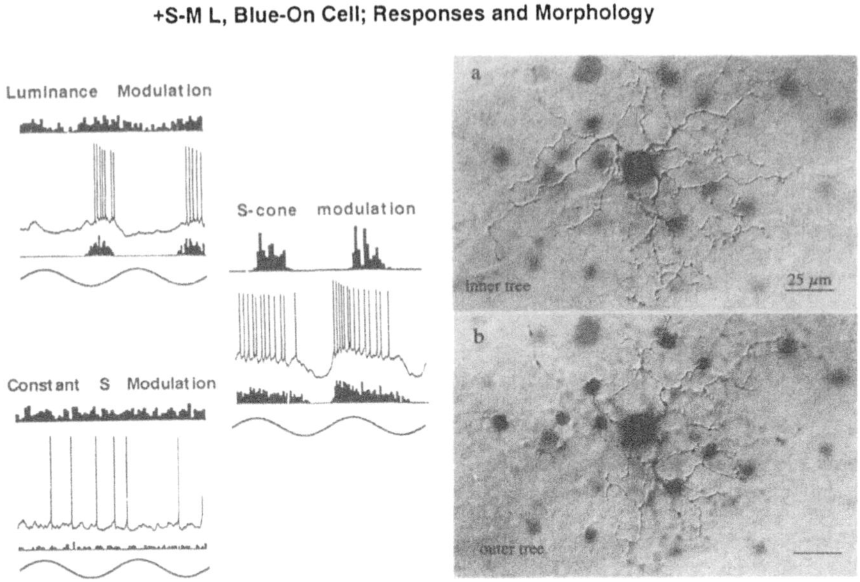

Fig. 4. Response of a +S-ML to the same stimuli as employed as in Figure 1. The cells give a vigorous response to the S cone stimulus, and no response to the constant S stimulus. The inset photomicrographs show the inner and outer dendritic trees of the small bistratified ganglion cell.

cell types are included within a single retinal cell mosaic (Dacey, 1993b). This is most unusual, and is the only known example other than the M and L cones themselves.

Midget ganglion cells have been drawn in Figure 2 as receiving input from midget bipolars, of which there are invaginating and flat types. It has recently been shown that midget bipolars extend far into the retinal periphery, presumably providing a cone-specific path into the IPL (Boycott and Wässle, 1991; Milam *et al.*, 1993; Wässle *et al.*, 1994).

Cells with S cone input

The S cone pathway is phylogenetically older than the M/L cone opponent system, so it is somewhat surprising that these two opponent systems have so long been subsumed into the midget cell system. Two physiological types of ganglion cell with S cone input have so far been identified. In both, excitatory or inhibitory S cone input is opposed by some combination of the other cone types. It is now known that these retinal ganglion cells have a specific morphology and a circuitry separate from the midget system. Confirming an earlier anatomical suggestion (Dacey, 1993a), the small bistratified ganglion cell corresponds to the blue-on, excitatory S cone physiological type (Dacey and Lee, 1994a). Responses of these ganglion cells are shown in Figure 4. As in earlier figures, the upper histograms are spike-density tracings from a blue-on cell in parafoveal retina. There is a vigorous response to S cone modulation, a weak response to luminance modulation, and no response to red–green modulation along a constant S cone axis. The intracellular recording from a peripheral small bistratified cell shows a very similar pattern of responses.

The inset morphology is distinct from that shown in Figures 1 and 2. Focusing through the retina reveals two strata of dendrites, one close to the cell body in the inner sublamina of the IPL and another in the outer, off sublamina. Again, dendritic tree size closely corresponded to receptive field diameter.

The right-hand section of Figure 2 shows a hypothetical circuitry for this cell type. The small bistratified cell has two dendritic layers, one in the inner sublamina of the IPL near the termination of the S cone bipolars and another in the outer sublamina, with unknown bipolar input. It is likely that these two dendritic inputs help generate opponency, if the S cone bipolar provides excitatory input and one of the flat diffuse bipolar types provides an 'off' M/L cone input. The anatomical identity of the blue-off S cone cell is still unclear, but a tentative identification with an inner ramifying cell with a small cell body and large dendritic tree has been made on the basis of a very limited sample.

It is possible that S cone cells (and others) might project through a distinct distribution of relay neurones within the geniculate to the cortex (Hendry, 1995). S cones exist in almost all mammals and it seems quite reasonable for the pathway carrying their signals to be anatomically distinguishable beyond the retina. A distinct retinal circuitry for S cone cells confirms that the assumption that the retino-cortical projection consists only of MC and PC pathways is an oversimplification. Each ganglion cell matrix defines a cell type, carrying distinct retinal information.

Other ganglion cell types

The commonly encountered physiological and anatomical types sketched in Figure 2 probably make up 80–90% of the ganglion cell population. Many of the remaining cells probably project to midbrain structures (Schiller and Malpeli, 1977). Their physiology and anatomy is as yet poorly understood. Of more interest are additional cell classes that project to the geniculate. It has been proposed that a separate class of PC cell codes for color (Rodieck, 1991): these are physiologically Type II cells with co-extensive opponent mechanisms (Wiesel and Hubel, 1966), with the small bistratified ganglion cell as an anatomical substrate. However the small bistratified cell corresponds exclusively to the excitatory S cone cell. For M/L cone opponent cells, careful quantitative work has failed to establish the Type II cell as a distinct class (Derrington and Lennie, 1984; Reid and Shapley, 1992). Despite the lack of any evidence for the Type II colour cell hypothesis, the question of what other ganglion cell types might project through the geniculate into striate cortex remains of substantial interest (Hendry and Yoshioka, 1994).

Other cellular elements in the retina

Recording from the primate retina *in vivo* permits registration of only ganglion cell activity, but the *in vitro* approach does not have this limitation. It thus offers the possibility of investigating how ganglion cell responses are derived from interactions at the horizontal, bipolar, and amacrine cell levels. As a first step, we have recently analysed the network of horizontal cells (Dacey *et al.*, 1996).

The number of different horizontal cell types and their connectivities has been controversial. Two types of horizontal cell were first described, HI and HII, each making indiscriminate contact with all cones within their dendritic fields (Boycott *et al.*, 1987; Wässle *et al.*, 1989). However, on the basis of anatomical evidence, HI cells were later divided into two types, HI and HIII, and it was proposed that HII cells made specific contact with S cones (Kolb *et al.*, 1994).

Intracellular recording from horizontal cells has only been able to identify two types, HI and HII. HI cells receive synergistic, hyperpolarizing input from M and L cones, as suggested by earlier work (Dacheux and Raviola, 1990). No S cone response could be elicited, and dendritic trees appeared to avoid S cone pedicles. On the other hand, dendrites of HII cells converge on S cone pedicles and HII cells show a strong, hyperpolarizing, S cone response. They also make scattered contacts with other cone pedicles, and hyperpolarize to both M and L cone stimulation, so no cone opponency is present. These horizontal cells and their connectivities have also been drawn in Figure 2.

These new findings suggest that cone opponency in the primate retina is not generated in the outer plexiform layer in any obvious manner through the horizontal cell network. It is likely that HII cells in primates (and analogous horizontal cells in other mammals) are concerned specifically with the S cone pathway. Since MC and M/L cone-opponent PC cells show such a striking lack of S cone input (Figs. 1, 3), HI cells presumably invaginate the cone pedicles to form triads with bipolar cells belonging to both these ganglion cell types.

Horizontal cells are often thought to contribute to the surrounds of ganglion cells and it is plausible that HI horizontal cells contribute to the surrounds of MC cells. The contribution of HI cells to M/L cone-opponent PC cell fields is less clear. Although the cone specificity of the surrounds of these cells is a topic of recent debate (Lennie *et al.*, 1991), physiological evidence has been strongly in favour of cone-specific surrounds (Reid and Shapley, 1992; Smith *et al.*, 1992; Kremers *et al.*, 1995). However, until a model for the generation of cone opponency through circuitry in the inner plexiform layer is available, this controversy will remain unresolved.

Conclusions

The results summarized here suggest that the primate retina is organized in a simple, elegant, and very specific manner. Separate sets of bipolar and ganglion cells process cone signals to provide a summed M and L cone signal in the magnocellular pathway, and two chromatically opponent signals, one M/L cone-opponent and the other involving the S cone, each carried by different ganglion cell types. The luminance and chromatic detection mechanisms seen in psychophysical experiments now have an anatomical as well as a physiological basis.

Some combination of signals in these parallel pathways may occur at a cortical level (Merigan and Maunsell, 1993). This is not inconsistent with separable psychophysical mechanisms for detection, if it is assumed that thresholds are primarily determined by the signal-to-noise ratio of peripheral neurones. If suprathreshold stimuli are used, it would be surprising if cortical mechanisms did not attempt to utilize useful signals no matter which afferent path they take. However, it seems likely that any combination of signals occurs in a very specific manner, rather than some form of indiscriminate pooling.

It is uncertain how this retinal circuitry may be modified in individuals with congenital colour vision deficiency. No advanced Old World simian except man has been shown to exhibit such inherited defects. New World simians are genetically different from those of the Old World in the inheritance of pigment genes, and all males are dichromats. How far these animals may serve as an anatomical and physiological model of human dichromacy remains to be investigated.

Acknowledgements

We would like to thank Jan Kremers, Joel Pokorny, Vivianne Smith and Iris Yeh for their experimental assistance and for permission to quote unpublished data. This work was partially supported by NIH grant EY06678.

References

Boycott, B.B. and Wässle, H. (1991). Morphological classification of bipolar cells of the primate retina. Eur. J. Neurosci. 3: 1069–1088.

Boycott, B.B., Hopkins, J.M. and Sperling, H.G. (1987). Cone connections of the horizontal cells of the rhesus monkey's retina. Proc. Roy. Soc. Lond. B 229: 345–379.

Cole, G.R., Hine, T. and McIlhagga, W. (1993). Detection mechanisms in L-, M- and S-cone contrast space. J. Opt. Soc. Am. A 10: 38–51.

Dacey, D.M. (1993a). Morphology of a small field bistratified ganglion cell type in the macaque and human retina: is it the blue-ON cell? Vis. Neurosci. 10: 1081–1098 .

Dacey, D.M. (1993b). The mosaic of midget ganglion cells in the human ret ina. J. Neurosci. 13: 5334–5355.

Dacey, D.M. and Lee , B.B. (1994a). The blue-ON opponent pathway in primate retina originates from a distinct bistratified ganglion cell type. Nature 367: 731–735.

Dacey, D.M. and Lee, B.B. (1994b). Physiology of identified ganglion cell types in an in vitro preparation of macaque retina. Invest. Ophthalmol. Vis. Sci. 35: 2001.

Dacey, D.M., Lee, B.B., Stafford, D.M., Smith, V.C. and Pokorny, J. (1996). Horizontal cells of the primate retina: Cone specificity without cone opponency. Science 271: 656–658.

Dacheux, R.F. and Raviola, E. (1990). Physiology of HI horizontal cells in the primate retina. Proc. Roy. Soc. Lond. B 239: 213–230.

De Valois, R.L. and De Valois, K.K. (1993). A multi-stage color model. Vision Res. 33: 1053–1065.

Derrington, A.M. and Lennie, P. (1984). Spatial and temporal contrast sensitivities of neurons in lateral geniculate nucleus of macaque. J. Physiol. (Lond.) 357: 219–240.

Derrington, A.M., Krauskopf, J. and Lennie, P. (1984). Chromatic mechanisms in lateral geniculate nucleus of macaque. J. Physiol. (Lond.) 357: 241–265.

Famiglietti, E.V. and Kolb, H. (1976). Structural basis for ON- and OFF-center responses in retinal ganglion cells. Science 194: 193–195.

Gouras, P. and Zrenner, E. (1979). Enhancement of luminance flicker by color-opponent mechanisms. Science 205: 587–589.

Grünert, U., Greferath, U., Boycott, B.B. and Wässle, H. (1993). Parasol (Pa) ganglion cells of the primate fovea: Immunocytochemical staining with antibodies against $GABA_A$-receptors. Vision Res. 33: 1–14.

Hendry, S.H. (1995). Striate and extrastriate output of a third geniculate channel in macaques. Invest. Ophthalmol. Vis. Sci. 36: S292.

Hendry, S.H.C. and Yoshioka, T. (1994). A neurochemically distinct third channel in the macaque dorsal lateral geniculate nucleus. Science 264: 575–578.

Hering, E. (1878). Zur Lehre vom Lichtsinne. Carl Gerold's Sohn, Wien.

Hurvich, L.M. and Jameson, D. (1957). An opponent-process theory of color vision. Psychol. Rev. 6: 384–404.

Kelly, D.H. and van Norren, D. (1977). Two-band model of heterochromatic flicker. J. Opt. Soc. Am. 67: 1081–1091.

Kolb, H., Fernandez, E., Schouten, J., Ahnelt, P., Linberg, K.A. and Fisher, S.K. (1994). Are there three types of horizontal cell in the human retina? J. Comp. Neurol. 343: 370–386.

Kremers, J., Lee, B.B. and Kaiser, P.K. (1992). Sensitivity of macaque retinal ganglion cells and human observers to combined luminance and chromatic modulation. J. Opt. Soc. Am. A 9: 1477–1485.

Kremers, J., Lee, B.B., Pokorny, J. and Smith, V.C. (1993). Responses of macaque ganglion cells and human observers to compound periodic waveforms. Vision Res. 33: 1997–2011.

Kremers, J., Lee, B.B. and Yeh, T. (1995). Receptive field dimensions of monkey retinal ganglion cells. In: Drum, B. (ed.), Color Vision Deficiencies XII, Doc. Ophth. Proc. Ser. 399–406. Kluwer, Dordrecht.

Lee, B.B. (1996). Receptive fields in primate retina. Vision Res. 36: 631–644.

Lee, B.B., Martin, P.R. and Valberg, A. (1988). The physiological basis of heterochromatic flicker photometry demonstrated in the ganglion cells of the macaque retina. J. Physiol. (Lond.) 404: 323–347.

Lee, B.B., Martin, P.R. and Valberg, A. (1989a). Nonlinear summation of M- and L-cone inputs to phasic retinal ganglion cells of the macaque. J. Neurosci. 9: 1433–1442.

Lee, B.B., Martin, P.R. and Valberg, A. (1989b). Sensitivity of macaque retinal ganglion cells to chromatic and luminance flicker. J. Physiol. (Lond.) 414: 223–243.

Lee, B.B., Pokorny, J., Smith, V.C., Martin, P.R. and Valberg, A. (1990). Luminance and chromatic modulation sensitivity of macaque ganglion cells and human observers. J. Opt. Soc. Am. A 7: 2223–2236.

Lee, B.B., Martin, P.R., Valberg, A. and Kremers, J. (1993a). Physiological mechanisms underlying psychophysical sensitivity to combined luminance and chromatic modulation. J. Opt. Soc. Am. A 10: 1403–1412.

Lee, B.B., Wehrhahn, C., Westheimer, G. and Kremers, J. (1993b). Macaque ganglion cell responses to stimuli that elicit hyperacuity in man: Detection of small displacements. J. Neurosci. 13: 1001–1009.

Lee, B.B., Wehrhahn, C., Westheimer, G. and Kremers, J. (1995). The spatial precision of macaque ganglion cell responses in relation to Vernier acuity of human observers. Vision Res. 35: 2743–2758.

Lennie, P. and D'Zmura, M. (1988). Mechanisms of color vision. CRC Crit. Rev. Neurobiol. 3: 333–400.
Lennie, P., Haake, P.W. and Williams, D.R. (1991). The design of chromatically opponent receptive fields. In: Landy, M.S. and Movshon, J.A. (eds.), Computational Models of Visual Processing, 71–82. MIT Press, Cambridge, MA, USA.
Merigan, W.H. and Maunsell, J.H.R. (1993). How parallel are the primate visual pathways? Annu. Rev. Neurosci. 16: 369–402.
Milam, A.H., Dacey, D.M. and Dizhoor, A.M. (1993). Recoverin immunoreactivity in mammalian cone bipolar cells. Vis. Neurosci. 10: 1–12.
Padmos, P. and van Norren, D. (1975). Cone systems interaction in single neurons of the lateral geniculate nucleus of the macaque. Vision Res. 15: 617–619.
Perry, V.H., Oehler, R. and Cowey, A. (1984). Retinal ganglion cells that project to the dorsal lateral geniculate nucleus in the macaque monkey. Neuroscience 12: 1110–1123.
Poirson, A.B., Wandell, B.A., Varner, D.C. and Brainard, D.H. (1991). Surface characterizations of color thresholds. J. Opt. Soc. Am. A 7: 783–789.
Polyak, S.L. (1941). The Retina. University of Chicago Press, Chicago.
Reid, R.C. and Shapley, R.M. (1992). Spatial structure of cone inputs to receptive fields in primate lateral geniculate nucleus. Nature 356: 716–718.
Rodieck, R.W. (1991). Which cells code for color? In: Valberg, A. and Lee, B.B. (eds.), From Pigments to Perception, 83–93. Plenum, New York.
Schiller, P.H. and Malpeli, J.G. (1977). Properties and tectal projections of monkey retinal ganglion cells. J. Neurophysiol. 40: 428–445.
Smith, V.C., Lee, B.B., Pokorny, J., Martin, P.R. and Valberg, A. (1992). Responses of macaque ganglion cells to the relative phase of heterochromatically modulated lights. J. Physiol. (Lond.) 458: 191–221.
Stromeyer, C.F. III, Cole, G.R. and Kronauer, R.E. (1985). Second-site adaptation in the red–green chromatic pathways. Vision Res. 25: 219–237.
Swanson, W.H., Ueno, T., Smith, V.C. and Pokorny, J. (1987). Temporal modulation sensitivity and pulse detection thresholds for chromatic and luminance perturbations. J. Opt. Soc. Am. A 4: 1992–2005.
Valberg, A., Lee, B.B., Kaiser, P.K. and Kremers, J. (1992). Responses of macaque ganglion cells to movement of chromatic borders. J. Physiol. (Lond.) 458: 579–602.
Wässle, H. and Boycott, B.B. (1991). Functional architecture of the mammalian retina. Physiol. Rev. 71: 447–480.
Wässle, H., Boycott, B.B. and Röhrenbeck, J. (1989). Horizontal cells in the monkey retina: cone connections and dentritic network. Eur. J. Neurosci. 1: 421–435.
Wässle, H., Grünert, U., Martin, P.R. and Boycott, B.B. (1994). Immunocytochemical characterization and spatial distribution of midget bipolar cells in the macaque monkey retina. Vision Res. 34: 561–579.
Wiesel, T. and Hubel, D.H. (1966). Spatial and chromatic interactions in the lateral geniculate body of the rhesus monkey. J. Neurophysiol. 29: 1115–1156.
Yeh, T., Lee, B.B. and Kremers, J. (1995). The temporal response of ganglion cells of the macaque retina to cone-specific modulation. J. Opt. Soc. Am. A 12: 456–464.

B.B. Lee
Department of Neurobiology
Max Planck Institute for Biophysical Chemistry
D-37077 Göttingen, Germany

D.M. Dacey
Biological Structure
University of Washington
Seattle, WA 98195, USA

11. Four cone subtypes in a primitive mammal, the South American opossum – immunocytochemical characterization

P.K. AHNELT, P. RÖHLICH and J.N. HOKOÇ

(Vienna, Austria; Budapest, Hungary; Rio de Janeiro, Brazil)

Abstract

We studied the outer retina of the marsupial South American opossum (*Didelphis marsupialis aurita*) by immunolabelling the retina with cone-specific peanut agglutinin (PNA) and with antibodies (COS-1 and OS-2) specific against mammalian photopigments. A small cone population was found ($< 3000/mm^2$) in which four subtypes were distinguishable. Two cone types containing a colourless oil droplet (double and single) and another single cone subtype were labelled by COS-1 suggesting they all carry the same or similar medium/long wavelength photopigments. A small proportion ($< 10\%$) of single cones (without oil droplets) was stained by OS-2 and can be assumed to be shortwave-sensitive.

Despite this surprising complexity the two single cone subtypes may be homologous to the dual set found in most modern mammals. The role of the peculiar group of oil droplet-containing cones remains unclear, although their topography may indicate a role for increased mesopic sensitivity.

Introduction

As has been shown by immunohistochemistry (Szél *et al.*, 1988, 1992; Wikler and Rakic, 1990) and morphology (Ahnelt *et al.*, 1987) most mammalian species have a small cone population containing a short-wavelength-sensitive pigment as well as a dominant population with one or two long-wavelength-sensitive pigments. In contrast, the cones of sauropsids (reptiles and birds) show a variety of morphologies (e.g. double cones), photopigments and oil droplets (Bowmaker, 1977; Cserháti *et al.*, 1989; Ohtsuka and Kawamata, 1990; Kolb and Lipetz, 1991; Kawata *et al.*, 1992). There exists a gap with respect to evolutionary transitions: it is unclear how the design of modern mammalian retinas can be related to the common reptilian ancestors.

By studying primitive mammals we may detect and understand intrinsic phylogenetic trends and constraints of mammalian retinal organization. The

C.R. Cavonius (ed.), Colour Vision Deficiencies XIII, pp. 119–125.

small group of American marsupials includes the opossums (*Monodelphis, Didelphis*) which are supposed to perhaps have the most conservative habitus (Marshall *et al.*, 1990). Previous studies have investigated different aspects of their visual system (Walls, 1939; Hokoç and Oswaldo-Cruz, 1979; Hokoç *et al.*, 1992) including some ability for colour discrimination (Friedman, 1967). Two of the present authors have reported recently the presence of three types of horizontal cells in the retina of the South American opossum *Didelphis marsupialis aurita* (Hokoç *et al.*, 1993), suggesting rather elaborate neural connections in the outer plexiform layer.

In a number of placental mammals, including primates, monoclonal antibody COS-1 (Szél *et al.*, 1988; Röhlich and Szél, 1993) labels cones sensitive to middle-to-long wavelengths (M/L), whereas antibody OS-2 binds selectively to short-wavelength-sensitive (S) cones (Szél *et al.*, 1988). We have now used these antibodies to characterize opossum cone types and identify possible relationships with cones in modern mammals.

Methods

Whole eye cups or retinas were obtained from adult South American opossums (*D. marsupialis aurita*) fixed in 4% paraformaldehyde in 0.1 M phosphate buffer (pH 7,4). For peanut agglutinin (PNA) labelling (Blanks and Johnson, 1984), retinas were preincubated in 0.5 mg/ml BSA and incubated in 200 μg/ml PNA or BSA/PBS for 48 h, then incubated with ABC. Reaction sites were visualized by DAB.

For photopigment immunocytochemistry, isolated retinas were washed with 0.2% Triton X-100 and treated with 1% bovine serum albumin. Retinas were then incubated in the first antibody, either COS-1 or OS-2 (for long-wavelength- or short wavelength-sensitive photopigments, respectively) for 24 h (Szél *et al.*, 1988; Röhlich and Szél, 1993). The second antibody (biotinylated anti-mouse IgG, Vector) then bound to the primary antibody and was located histochemically with the avidin–biotinylated peroxidase complex (Vector Elite) and and DAB-reaction (Sigma) for peroxidase activity. The labelled whole mounts were coverslipped with glycerol or embedded in Epon for semithin sectioning.

For counting and mapping of whole mounts, digitized samples were obtained under differential interference contrast (DIC) and analysed using the software packages NIH-Image, Delta-Graph and CANVAS for Macintosh computers and the 'Spatial Patterns' program (Fernández *et al.*, 1993).

Results

Rods and cones – general morphology and topography

Rods were dominant throughout the retina (Fig.1), reaching 400 000/mm^2 in

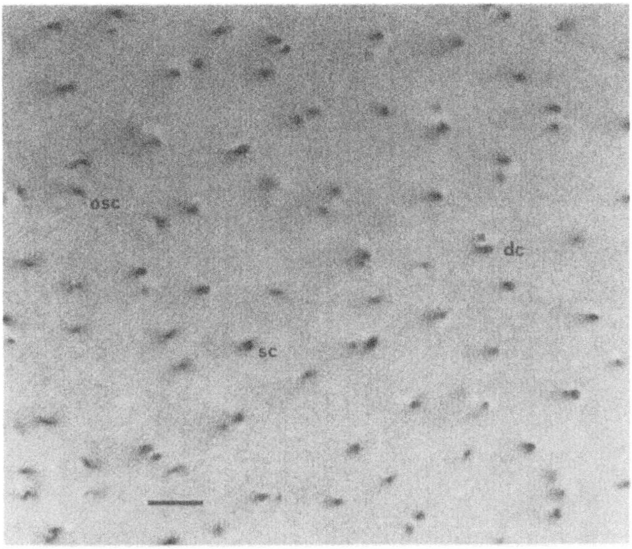

(a)

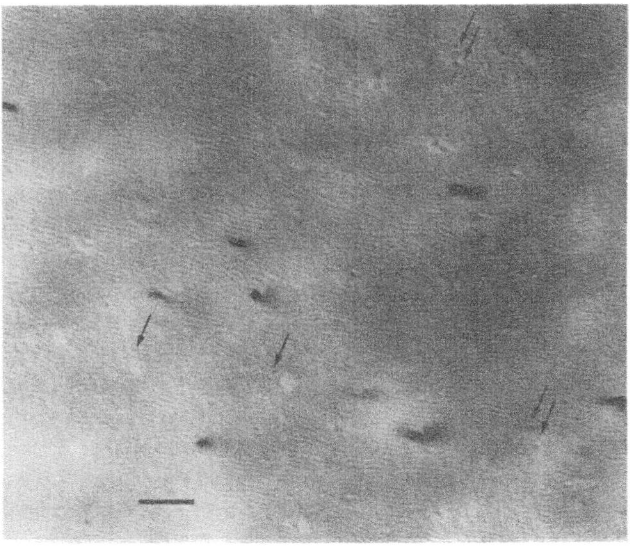

(b)

Fig. 1. Complementary antibody labeling of cone photopigments in inferior retina. Differential interference contrast. Bar: 20 μm. (**a**) COS-1. Labelled outer segments are oriented obliquely and belong to various cone subtypes: single cones without oil droplet (sc), single cones with oil droplet (osc) and double cones (dc). (**b**) OS-2. Only outer segments of a subpopulation of single cones without oil droplet are stained. Arrows: unstained outer segments of single cones and double cones with oil droplets.

the upper temporal region. Cone outer and inner segments were very 'rod-like'. While some of the cones could be identified by a clear oil droplet in the inner segments they could not always be distinguished unequivocally in whole mounts.

For quantitative evaluation of total cone densities the extracellullar matrix surrounding cone outer segments was labelled by binding to PNA (Blanks and Johnson, 1984). In the South American opossum PNA labelled the outer segments of all three morphological cone types (double cones are considered as one unit in counting). The peak density of PNA-labelled cones coincided with that of ganglion cells (Hokoç and Oswaldo-Cruz, 1979) in the upper temporal (tapetal) area, about 2–3 mm from the optic disc, corresponding to the area centralis. From this circular zone with about 2500 cones/mm^2, the density gradually declined to 1300–1500 cones/mm^2 in the periphery. An eccentric expansion of isodensity contours was notable towards the lower hemisphere.

Immunocytochemistry

Immunocytochemical labelling with antibodies directed against cone photopigment allowed further characterization of spectral cone subtype morphology and also revealed surprising differences with respect to topography. The overall density of COS-1-labelled cones (Fig. 1a) followed the pattern for PNA labelling, albeit at slightly lower values. Maxima were reached in the temporal area centralis (2400 cones/mm^2) and relatively high densities (2000) were maintained towards the lower hemisphere. This distribution is the accumulative result from the three COS-1-labelled subpopulations: (1) Single cones without oil droplets were the dominant subtype of the COS-1-labelled population. COS-1-positive single cones formed the major component of the area centralis (2300 cells/mm^2) and had regular intercone distances. (2) Double cones were strongly concentrated in the inferior retinal half (600/mm^2; 25% of COS-1-labelled cones). This distribution coincided with the non-tapetal region of the retinal pigment epithelial cells, and they formed the main component of the ventrad extensions of isodensity lines. None were found in the area centralis. (3). The density of single cones containing an oil droplet was much lower (200/mm^2) but their maximum again was found in the lower hemisphere. OS-2 antibody (Fig. 1b) labelled a small population (220/mm^2) of single cones. The mosaic of OS-2-labelled cones showed considerable local variation and contributed about 10% of the total population throughout the retina.

Discussion

Two cone systems seem to coexist in the opossum retina. The two types of single cones without oil droplets (Fig. 2, right column) may provide the basis for a photopic system capable of both optimizing spatial acuity and (dichromatic) colour discrimination (Jacobs, 1993). In addition the opossum retina contains a

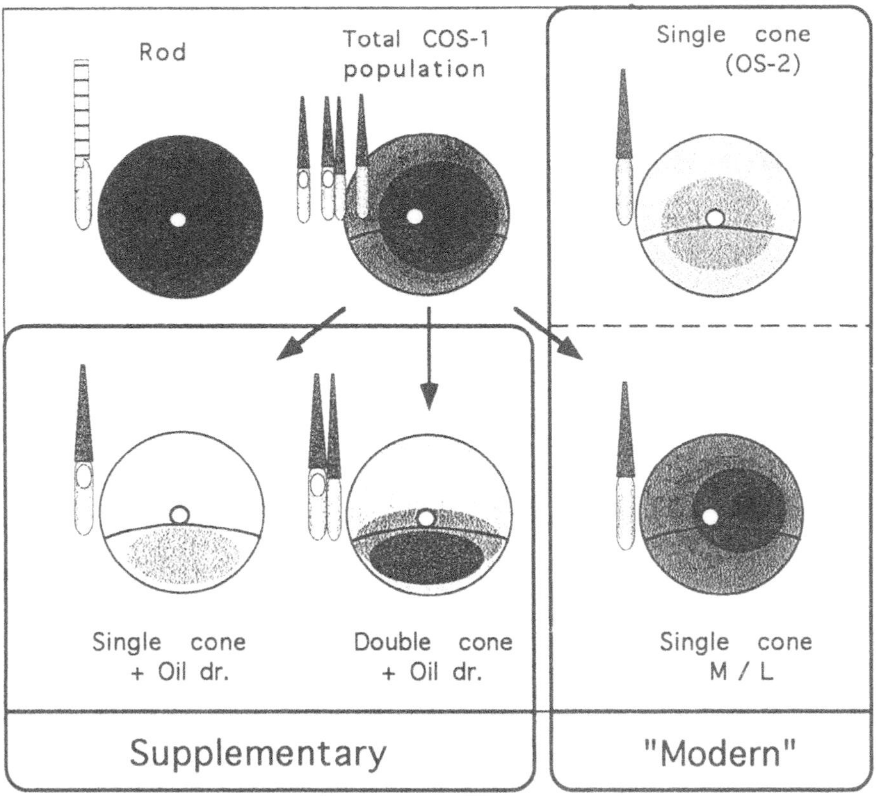

Fig. 2. Opossum cone subtypes – classification and distribution. By comparison with placental mammals the opossum cones may be interpreted as consisting of a set of 'modern' single cones accompanied by a 'supplementary' set containing oil droplets. Curved line: border of superior tapetal area, circle: papilla. Maximum grey levels represent ca. 3000 cones and 300 000 rods/mm^2 respectively. Upper series: rod density (left) is about two orders of magnitudes greater than that of cones. Overall density of COS-1-labelled cones (centre) matches closely the distribution of ganglion cells and PNA-labelled cones. COS-2-labelled cones (right) show low uniform density (150–300/mm^2). Lower series: three cone subtypes labelled by COS-1. The density of single cones with oil droplet (left) is approaching zero in the upper hemisphere and is somewhat higher in the inferior areas (200/mm^2). COS-1-labelled double cones containing an oil droplet (centre). This type of cone is concentrated in the inferior part of the retina (max. 600/mm^2). COS-1-labelled single cone (right). Relatively high density with maximum in the area centralis (2400/mm^2)

'supplementary' set of double cones and single cones. These oil droplet-associated cone types deviate from the 'modern' system not only by their morphology (and possibly by their wiring) but also with respect to their distribution. They are concentrated in the (non-tapetal) inferior retina and are missing entirely in the area centralis. Studying their relation to ganglion cell topography (Hokoç and Oswaldo-Cruz, 1979) suggests a higher ratio of convergence for the oil droplet cones. This may extend the opossum's cone-based vision towards more mesopic intensities.

Mammalian evolution included a long period (ca. 90 million years) during which the lifestyle was predominantly nocturnal. This period, when early mammals coexisted with the (mostly photopic) dinosaurs, has led to the differentiation and optimization of a rod-based scotopic visual system. During this phase only two cone types remained to serve the photopic retinal functions in the ancestors of placental mammals. The marsupial line appears to have continued to use some 'reptilian' oil-droplet cones. Whether they express specific connectivity patterns, merely support, or qualitatively expand the classically dimorphic mammalian cone system remains to be studied.

References

Ahnelt, P.K., Kolb, H. and Pflug, R. (1987). Identification of a subtype of cone photoreceptor, likely to be blue sensitive, in the human retina. J. Comp. Neurol. 255: 18–34.

Blanks, J.C. and Johnson, L.V. (1984). Specific binding of peanut lectin to a class of retinal photoreceptor cells. Invest. Ophthalmol. Vis. Sci. 5: 546–557.

Bowmaker, J.K. (1977). The visual pigments, oil droplets and spectral sensitivity of the pigeon. Vision Res. 17: 1129–1138.

Cserháti, P., Szél, A. and Röhlich, P. (1989). Four cone types characterized by anti-visual pigment antibodies in the pigeon retina. Invest. Ophthalmol. Vis. Sci. 30: 74–81.

Fernández, E., Cuenca, N. and De Juan, J. (1993). A compiled BASIC program for analysis of spatial point patterns, application to retinal studies. J. Neurosci. Methods 50: 1–15.

Friedman, H. (1967). Colour vision in the Virginia opossum. Nature 213: 835–936.

Hokoç, J.N. and Oswaldo-Cruz, E. (1979). A regional specialization in the opossum's retina: quantitative analysis of the ganglion cell layer. J. Comp. Neurol. 183: 385–396.

Hokoç, J.N., de Oliveira, M.M.M. and Ahnelt, P. (1993). Three types of horizontal cells in a primitive mammal, the opossum (*Didelphis marsupialis aurita*): a Golgi-LM study. Invest. Ophthalmol. Vis. Sci. 34 (Suppl.): 1152.

Jacobs, G.H. (1993). The distribution and nature of colour vision among the mammals. Biol. Rev. 68: 413–471.

Kawata, A., Oishi, T., Fukada, Y., Shichida, Y. and Yoshizawa, T. (1992). Photoreceptor cell types in the retina of various vertebrate species: immunocytochemistry with antibodies against rhodopsin and iodopsin. Photochem. Photobiol. 56: 1157–1166.

Kolb, H. and Lipetz, L.E. (1991). The anatomical basis for colour vision in the vertebrate retina. In: Cronly-Dillon, J. (ed.), Vision and Visual Dysfunction VI, Gouras, P. (ed.). The Perception of Colour: 128–145, Macmillan, London.

Marshall, L.G., Case, J.A. and Woodburne, M.O. (1990). Phylogenetic relationships of the families of marsupials. Curr. Mammal. 2: 433–505.

Ohtsuka, T. and Kawamata, K. (1990). Telodendrial contact of HRP-filled photoreceptors in the turtle retina: pathways of photoreceptor coupling. J. Comp. Neurol. 292: 599–613.

Röhlich, P. and Szél, A. (1993). Binding sites of photoreceptor-specific antibodies COS-1, OS-2 and AO. Curr. Eye Res. 12: 935–944

Szél, A., Diamantstein, T. and Röhlich, P. (1988). Identification of the blue sensitive cones in the mammalian retina by anti-visual pigment antibody. J. Comp. Neurol. 273: 593–602

Szél, A., Röhlich, P. Caffe, A.R., Juliusson, B., Aguirre, G. and van Veen, T. (1992). Unique topographic separation of two spectral classes of cones in the mouse retina. J. Comp. Neurol. 325: 327–342
Walls, G.H. (1939). Notes on the retinae of two opossum genera. J. Morphol. 64: 67–87.
Wikler, K.C. and Rakic, P. (1990). Distribution of photoreceptor subtypes in the retina of diurnal and nocturnal primates. J. Neurosci. 10: 3390–3401.
Young, H.M. and Pettigrew, J.D. (1991). Cone photoreceptors lacking oil droplets in the retina of the echidna, *Tachyglossus aculeatus* (Monotremata). Vis. Neurosci. 6: 409–420.

P.K. Ahnelt
Department of General and Comparative Physiology
University of Vienna
Vienna, Austria

12. Physical and sensory analysis of successive scotopic colour contrast

S.L. BUCK and M. AYERS
(Seattle, WA, USA)

Abstract

Two approaches sought to identify which portions of the colour pathways are influenced by rod signals in successive scotopic colour contrast (SC), the hues induced by a coloured adapting field into a successively presented, purely scotopic test field. We determined (1) the dependence of SC on the degree of adaptation of isolated single cone types and (2) the comparison of SC to successive photopic colour contrast (PC) with adapting fields that spanned the spectrum. Both scotopic and photopic test fields, which appear colourless when presented alone, took on characteristic hues after adapting field presentation. SC hues were described as violet, blue, green, yellow or orange and were hue-matched with 400–590 nm comparison lights. The SC hues were influenced by adaptation of each of the three cone types and showed consistent differences from the PC hues induced by the same adapting fields. We conclude that rod signals influence (1) pathways carrying signals from each type of cone and (2) two channels that code for hue. However, rod signals must have differential access to, or effectiveness in, the various colour pathways.

Introduction

Stabell and Stabell (1971, 1972) first showed that a coloured adapting field can induce a hue into a successively presented, purely scotopic, test field. These scotopic colour contrast hues were initially described as identical to the photopic contrast colours induced by the same adapting fields but were later recognized as systematically shifted (Buck, 1995; Buck and Brandt, 1993, 1995; Stabell and Stabell, 1994). An implication of these shifts is that scotopic colour contrast involves reduced or altered colour pathways. We sought to identify which portions of the colour pathways are influenced by rod signals in successive scotopic colour contrast by determining (1) the dependence of scotopic colour contrast on the degree of adaptation of isolated single cone types and (2) the comparison of scotopic colour contrast to photopic colour contrast obtained with adapting fields that spanned the spectrum. A preliminary report has been presented (Buck *et al.*, 1994).

C.R. Cavonius (ed.), Colour Vision Deficiencies XIII, pp. 127–134.

Methods

Three colour-normal observers served in all conditions. The procedure used to induce and measure successive colour contrast was as follows. Ten-second presentations of an 8° diameter adapting field alternated with 10 sec presentations of an intermittent 2° diameter test field. The test field alternated between 1 sec on and 1 sec off during the 10 sec test. Both stimuli were presented to the observer's right eye and centred about 5° to the right of fixation. The adapting fields were formed by interference filters spaced at 10 nm intervals from 400–700 nm, except for 460, 610 and 690 nm. Adapting fields were presented at light levels 1–3 log units above an observer's absolute cone detection level, depending on the condition.

In order to specify the hue of the contrast colour, observers adjusted the wavelength of a 1° foveal monochromatic comparison spot in the left eye to match the hue percept induced into the test field seen by the right eye. The optics prevented observers from seeing stimuli with both eyes at once. Observers had to make eye movements to alternate between views of the test and comparison stimuli. Observers easily learned to make hue matches, even in the presence of differences of saturation and brightness between test and comparison fields.

For scotopic colour contrast, the test eye was dark adapted for 30 min and illuminance of the scotopically detected, 500 nm test field was kept at least 0.3 log unit below absolute cone detection threshold. For photopic colour contrast, the test eye was strongly light-adapted to desensitize rods, and the test field was a photopically detected broadband light that was adjusted by each observer to a perceptually neutral, 'unique' white (no hue percept).

The S, M, and L cone fundamentals of DeMarco *et al.* (1992) were used to calculate adapting field conditions that allowed variation of adaptation of each cone type in turn, while keeping a constant stimulation of the other two cone types. For S cone variation, pairs of adapting fields were selected that were isoluminant 'tritan pairs' (wavelengths, in nm, of 410 and 540, 420 and 530, 430 and 510, 430 and 520, or 440 and 500). The two members of a tritan pair differentially stimulated S cones by 1.46–2.74 log units (depending on the specific pair) while keeping M and L cone stimulation effectively constant (maximum 0.02 log change of L/M cone ratio). Isoluminance matches of the two members of a pair were determined by 20 Hz flicker photometry.

For M and L cone variation, the adapting fields were 550, 600, 650, and 700 nm lights adjusted in relative irradiance to differentially stimulate either M or L cones, while keeping constant the stimulation of the other of these two cone types. Under both sets of conditions, S cone stimulation was kept effectively constant and near zero by choosing adapting field wavelengths from the Rayleigh region. These conditions provided up to 1.24 log units differential stimulation of either M or L cones.

Results and discussion

Figure 1 shows the wavelength of the comparison spot that matched the hue of the contrast colour induced into the test field, as a function of the wavelength of the adapting field. For the data shown here, adapting fields were 2 log units suprathreshold for cones and were physically identical for both scotopic and photopic measurements. For this and all subsequent conditions, the results for all three observers were qualitatively similar, and only data from only one observer are shown for each condition. Compared with photopic contrast hues, the scotopic contrast hues were more blue and/or less green after long-wavelength adaptation and were more blue and/or less red after adaptation just below 500 nm. The difference between scotopic and photopic colour contrast shows that rod signals pass through an altered or reduced set of colour pathways.

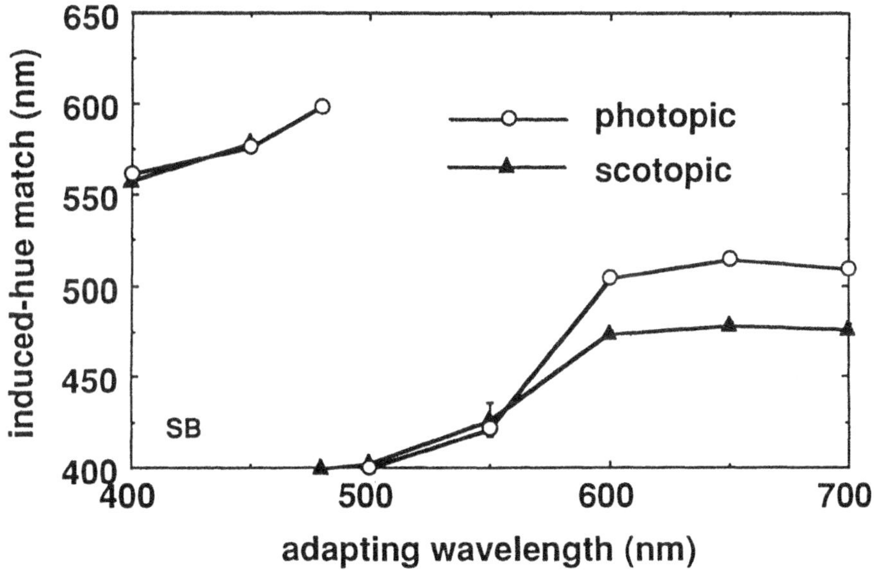

Fig. 1. Monochromatic lights matched to the hue of scotopic (filled symbols) and photopic (open symbols) successive contrast colours as a function of adapting field wavelength. The differences between scotopic and photopic colour contrast show that rod signals pass through an altered or reduced set of colour pathways. In all figures, error bars represent ± 1 S.E. of the mean of at least three sessions and are smaller than the symbol height if not visible.

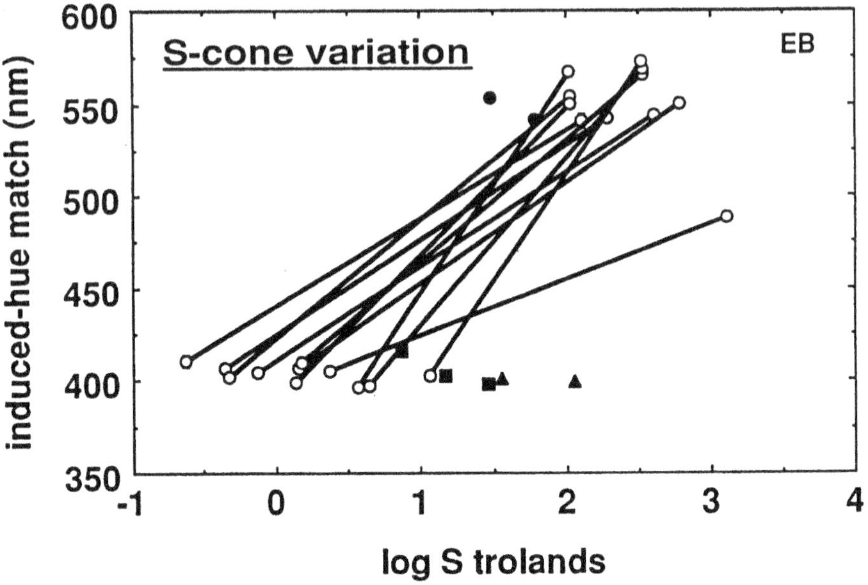

Fig. 2. Large variation of scotopic colour contrast hue as a function of excitation of S cones by the adapting field. For each line, excitation of M and L cones was held constant for the two adapting fields shown by the endpoints: only the excitation of S cones changed. The unconnected points represent different illuminances of unpaired 540 nm (squares), 500 nm (triangles) and 420 nm (circles) adapting fields.

In an attempt to identify the alteration/reduction, we examined the change of scotopic colour contrast hue with variation of the adaptation of each cone type in isolation. Figure 2 shows that the hue of the scotopic contrast colour changed markedly with changing adaptation of S cones. The end points of each line represent a tritan pair of adapting fields. For each pair, the excitation of M and L cones was the same for both members: only the excitation of S cones changed. All five wavelength combinations were presented at two light levels (1.5 and 2.0 log above cone threshold for the shorter wavelength of each pair). The 410/540 nm combination was also presented at 2.5 log above cone threshold. The steep slopes for all tritan pairs show the large effect of varying S cone adaptation for every wavelength combination and light level. The scotopic contrast hues fell into two groups: the 410–440 nm members of each pair of adapting fields yielded longer wavelength scotopic contrast hues, while the 500–540 nm members yielded shorter wavelength scotopic contrast hues. The unconnected points are discussed below.

Figure 3 similarly shows large variations of scotopic colour contrast hue with variation of adaptation of just M cones or of just L cones. The scotopic contrast

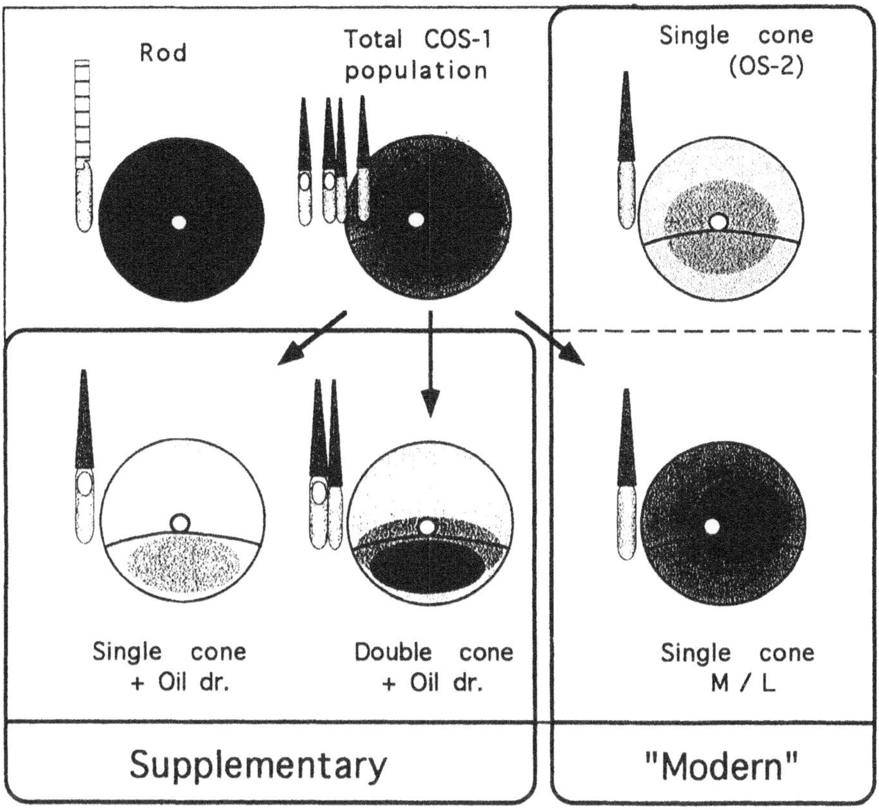

Fig. 2. Opossum cone subtypes – classification and distribution. By comparison with placental mammals the opossum cones may be interpreted as consisting of a set of 'modern' single cones accompanied by a 'supplementary' set containing oil droplets. Curved line: border of superior tapetal area, circle: papilla. Maximum grey levels represent ca. 3000 cones and 300 000 rods/mm^2 respectively. Upper series: rod density (left) is about two orders of magnitudes greater than that of cones. Overall density of COS-1-labelled cones (centre) matches closely the distribution of ganglion cells and PNA-labelled cones. COS-2-labelled cones (right) show low uniform density (150–300/mm^2). Lower series: three cone subtypes labelled by COS-1. The density of single cones with oil droplet (left) is approaching zero in the upper hemisphere and is somewhat higher in the inferior areas (200/mm^2). COS-1-labelled double cones containing an oil droplet (centre). This type of cone is concentrated in the inferior part of the retina (max. 600/mm^2). COS-1-labelled single cone (right). Relatively high density with maximum in the area centralis (2400/mm^2)

'supplementary' set of double cones and single cones. These oil droplet-associated cone types deviate from the 'modern' system not only by their morphology (and possibly by their wiring) but also with respect to their distribution. They are concentrated in the (non-tapetal) inferior retina and are missing entirely in the area centralis. Studying their relation to ganglion cell

hue changed in opposite directions with increased adaptation of M cones compared with L cones.

The preceding results show that the influence of rod signals on the colour pathways (the scotopic contrast colour) is affected by adaptation of each of the three cone types. The differences between scotopic and photopic colour contrast cannot, therefore, be explained by the lack of interaction of rods with particular cone types. The explanation must lie with a differential interaction of rod signals with portions of the subsequent colour pathways. Two additional observations support this view.

Figure 4 shows that scotopic contrast colours depend on the wavelength of the adapting field, and not the absolute excitation of any one cone type, which varied in this study by up to 1.5 log units for adapting fields of the same wavelength. The 'constant L' and 'constant M' functions are taken from the data shown in Figure 3 (the M and L cone variation functions, respectively). For the 'constant L+M' condition, the illuminance of the adapting fields was adjusted so that the sum of L cone and M cone stimulation was constant. The three functions overlap so well that they cannot be resolved. Thus, the scotopic contrast colour depends on the relative excitation (adaptation) of the three cone types, which is always constant for a given wavelength of adapting field.

Another demonstration that scotopic colour contrast hues depend on the relative adaptation across the cone types can be seen in Figure 2. The unconnected points are non-paired adapting fields composed of 420, 500 or 540 nm lights presented at the extremes of our range of light levels. They allow us to see that, near the mid-range of S cone adaptation, the same degree of S cone adaptation (i.e. the same log S troland value) is associated with very different scotopic contrast hues. What differs is the wavelength of the adapting lights and, therefore, the ratio of excitation of S cones relative to M and L cones.

What is the nature of the post-receptoral pathways in which rod signals act in colour vision? Adaptation of M and L cones had opposite directions of effect. Adaptation of S cones had the same direction of effect as adaptation of L cones. At first this seems to suggest that scotopic colour contrast arises only from a r/g-type opponent channel, in which M cones oppose both L and S cones.

However, we conclude instead that rod signals are active in two chromatic channels, the comparison of whose output determines hue. Figure 5 (left) shows that the asymptotically rising data pattern shown in Figure 4 cannot be modelled by a r/g channel alone, which predicts a dramatically non-monotonic change in contrast colour between 550 and 700 nm. Instead, a ratio of the outputs of two chromatic channels (Fig. 5 right) more closely models both scotopic and photopic contrast results. (Note other similarities to the spectral data of Figure 1, such as the discontinuity near 500 nm.) All functions shown in Figure 5 were constructed from the cone fundamentals of DeMarco *et al.* (1992). Ignoring weighting coefficients, the r/g channel was modelled as L+S−M, and the b/y channel was modelled by S−(M+L). Equivalent results would have been obtained using the perceptual opponent response functions of Hurvich and Jameson (Hurvich, 1981).

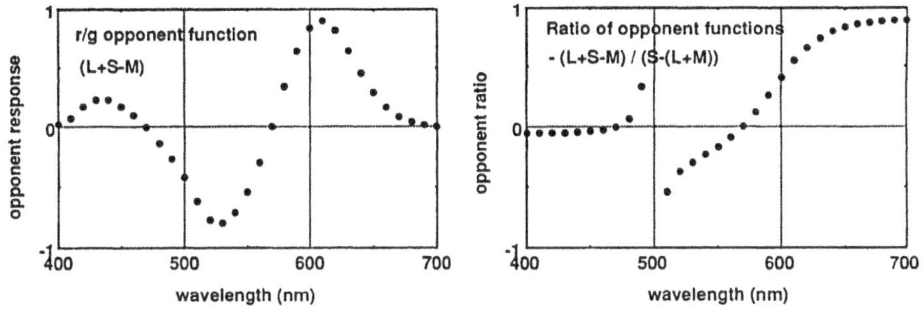

Fig. 5. Two potential post-receptor models of the pattern of change of scotopic colour contrast hues. *Left:* An r/g opponent function for which L and S cone inputs are opposed by M cone inputs. *Right:* A ratio of r/g and b/y opponent functions. The right function captures features of the data shown in Figures 1 and 4, supporting the idea that scotopic colour contrast hues are determined by the presence of rod signals in two post-receptor channels whose comparison determines hue.

The chromatic ratio model of colour contrast reflects two widely suggested ideas: first, that contrast colours result from adaptation and 'rebound' in opponent channels and second, that hue is coded by outputs of the opponent channels. Cone input weights were chosen to provide spectral unique hues at 470, 500 and 570 nm. Taking the ratio of the r/g to b/y response functions yields the familiar hyperbolic-like relation of adapting and induced spectral hues. The applicability of the model to both scotopic and photopic colour contrast supports the idea that both depend on a comparison of the outputs of two opponent channels, each of which receives inputs from all three cone types.

In summary, the influence of rod signals on the colour pathways, at least as revealed by the scotopic contrast colour hue, is affected by adaptation of each of the three cone types. Rod signals are active in two chromatic channels, the comparison of whose output determines hue. We cannot explain the differences between scotopic and photopic colour contrast by the interaction of rods with isolated cone types. The explanation must lie with a differential interaction of rod signals with portions of the subsequent colour pathways. The present results show that, under some conditions, rod signals appear to enhance blue and may also reduce red and green. A perceptual analysis (e.g. Buck, 1995) may be needed to investigate further the interaction of rod signals at this level.

Acknowledgements

This work was supported by USPHS/NIH grants EY03221 and EY07031.

References

Buck, S.L. (1995). Influence of rod signals on the perceptual opponent colour channels: evidence from successive scotopic contrast. Invest. Ophthalmol. Vis. Sci. Suppl. 36: 960.

Buck, S.L. and Brandt, J. (1995). The range of scotopic contrast colours matches that associated with S-cone activation. In: Drum, B. (ed.). Colour Vision Deficiencies XII, Doc. Ophthalmol. Proc. Ser. 57: 309–316, Kluwer, Dordrecht.

Buck, S.L. and Brandt, J. (1993). Scotopic contrast colours show preferential involvement of rod signals with S-cone colour pathways. Invest. Ophthalmol. Vis. Sci. Suppl. 34: 764.

Buck, S.L., Ayers, M. and Brandt, J. (1994). Successive and simultaneous scotopic contrast compared. Invest. Ophthalmol. Vis. Sci. Suppl. 35: 1657.

DeMarco, P., Pokorny, J. and Smith, V. (1992). Full-spectrum cone sensitivity functions for X-chromosome-linked anomalous trichromats. J. Opt. Soc. Am. A 9: 1465–1476.

Hurvich, L.M. (1981). Colour Vision. Sinauer, Sunderland, MA.

Stabell, U. and Stabell, B. (1971). Chromatic rod vision – III Wavelength of pre-stimulation varied. Scand. J. Psychol. 12: 282–288.

Stabell, B. and Stabell, U. (1972). Chromatic rod vision – II Duration of pre-stimulation varied. Scand. J. Psychol. 13: 136–140.

Stabell, U. and Stabell, B. (1994). Mechanisms of chromatic rod vision in scotopic illumination. Vision Res. 34: 1019–1027.

University of Washington
Department of Psychology
Box 351525
Seattle, WA 98195-1525, USA

13. Visual colour processing is completed in the retina

H.G. SPERLING
(Houston, TX, USA)

Abstract

Analysis of mammalian visual pathways is limited by difficulty in recording from single cells earlier than the third-order ganglion cells, but the electroretinogram (ERG) has separable components for each of the three afferent orders. The flash ERG a-wave is generated by the photoreceptors (Tomita and Yanagida, 1981; Heynen and van Norren, 1985), the b-wave by bipolar cells (Heynen and van Norren, 1985; Dick *et al.*, 1985) and the pattern ERG by the ganglion cells (Fiorentini *et al.*, 1981; Maffei and Fiorentini, 1981). We have obtained action spectra from each of these components and behavioural thresholds from the final common path. At the photoreceptor stage the data are fit by the sum of middle- and long-wave cone response. At the bipolar stage some inhibitory interaction between the red and green responses begins. At the ganglion stage there is greatly enhanced red–green inhibition and amplification of the blue response. There is no significant change between the ganglion cells and the final common path, so it may be concluded that colour processing is complete in the retina.

Introduction

The most recent of a series of experiments in my laboratory on the spectral sensitivity of the primate visual system has used the pattern ERG (PERG) (Sperling and Mishra, 1995). This study used an alternating pattern stimulus, time-base averaging the voltage response triggered at each alternation and recording from the cornea. Our goal was to obtain evidence on the details of the spectral sensitivity function at the ganglion cell – third afferent level – for comparison with our evidence from photoreceptor and bipolar levels obtained from a- and b-wave criteria (Sperling and Mills, 1987; Mills and Sperling, 1990). Other studies had shown that the PERG can be separated from the b-wave response since it is absent in diseases such as dominant optic atrophy, which affects the ganglion cells (Fiorentini *et al.*, 1981) and the response also disappears when the ganglion cells degenerate following severance of the optic nerve (Fiorentini *et al.*, 1981; Marrone *et al.*, 1994). In both cases the b-wave is

C.R. Cavonius (ed.), Colour Vision Deficiencies XIII, pp. 135–139.

unaffected. Experiments in which the recording electrode was advanced through the retina (Baker *et al.*, 1988; Seiving and Steinberg, 1985) have localized the b-wave response to the distal retina, the region of photoreceptors and bipolar cells, while pattern-generated potentials have been localized to the proximal retina, the region of amacrine and ganglion cells. Taken together, this is strong evidence that the PERG is generated by ganglion cell responses. The present study addressed where in the afferent chain red/green colour opponency and amplification of blue response occur, since these are the major differences between the first order, a-wave criterion, and the final common path, spectral sensitivity functions obtained by psychophysical studies of the whole organism (Sperling and Harwerth, 1971; Sperling and Mills, 1987). To the extent that our model, which links colour discrimination to spectral sensitivity, is correct, these two changes are taken to characterize colour processing.

The method used to obtain the a-wave and b-wave spectral sensitivity functions (Fig. 1c, d) is fully described in Mills and Sperling (1990). An intensity series of briefly flashed spectral lights was run, producing a voltage–intensity function, which was fit to a hyperbolic tangent function for each wavelength. Spectral sensitivity values were interpolated for fixed voltage criteria, 10 μV for the b-wave and 5 μV for the a-wave functions. In all cases the spectral flashes were 12.5° in diameter against a 30 000 td equal energy white background. They were flashed with a 70 msec on, 330 msec off duty cycle.

The psychophysical data (Figure 1a) were obtained by the method of limits with a foveated 2° test light against a 3000 td, 17° equal energy white background, 17° in diameter. The threshold criterion was the first of two successive misses starting above threshold for each wavelength and descending in fixed steps of 0.1 log transmission units.

The pattern ERG was recorded using a corneal thread electrode placed under the lower eyelid, and the potential difference was recorded between the cornea and the upper cheek bone, with a ground electrode on the contralateral ear. The pattern was focused on the retina by a triple-path Maxwellian view optical system which produces intense spectral patterns. We used a 17° checkerboard pattern with 1 cpd checks, viewed against a 72°, 4000 td, equal energy white background. Adjacent spectral and white checks alternated 180° out of phase at 8 Hz. The (equal energy) white checks were held constant in illuminance at 12 000 td. The spectral checks (at 24 narrow-band wavelengths through the visible spectrum) were varied in small steps from dimmer to brighter than the white checks, through a point of equal luminance where, if there were no chromatic differences, the pattern would have disappeared and the ERG response would have been expected to reach zero. As the spectral checks were varied in luminance, however, the voltage of the ERG went through a minimum but did not reach zero (measured peak-to-peak in the sine-wave like response wave form). The spectral sensitivity functions were derived as the reciprocal number of quanta at these minima for each wavelength.

The photoreceptor level (Figure 1d) shows neither red–green opponency nor amplification of blue response, since it is well fit by the simple scalar sum of

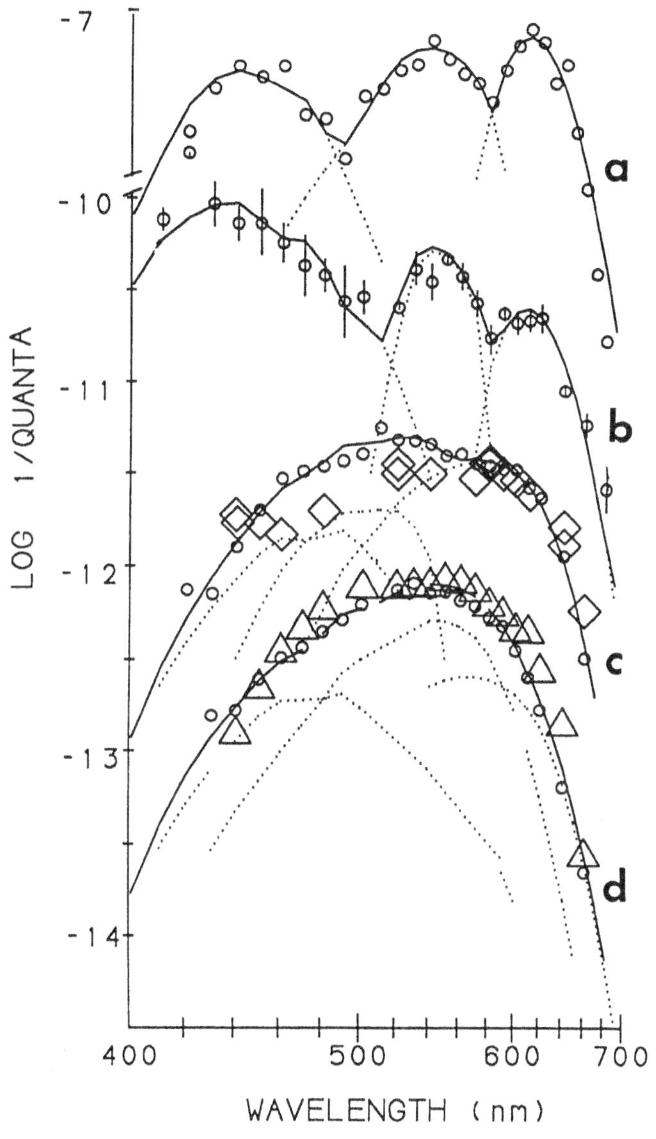

Fig. 1. Series of spectral sensitivity functions which convey the major points at the beginning and completion of colour processing at different afferent stages of the visual system. (a) Spectral sensitivity to whole-organism psychophysical increment threshold criteria (on one of the human eyes that entered into the mean PERG function in b). (b) Mean normal PERG spectral sensitivity from a number of determinations at each wavelength on two colour-normal humans averaged with the same determinations from two rhesus monkeys. Vertical bars denote 1 S.E.M. (c) Spectral sensitivity to b-wave criteria. Diamonds represent data points obtained by identical procedures to the PERG points, but with the pattern removed. (d) Spectral sensitivity to a-wave criteria. Triangles represent data obtained by identical procedures to the b-wave data but with GABA transmission blocked by bicuculline.

middle- and long-wavelength-sensitive cone response, with some intrusion of rods; the best fitting (solid) curve was computed as $SS = K_1R + K_2G + K_3$ rods, where R, G (and B, below) are the cone sensitivity functions proposed by Smith and Pokorny (1975). At the bipolar level (Fig. 1c) there is the beginning of red–green opponency revealed by the shoulder in the yellow. This function is best fit (solid curve) by $SS = (K_1R–K_2G) + (K_3G–K_4R) + K_5$ rods.

The PERG results (Fig. 1b), shows deepening of the shoulder in the yellow to a distinct notch and a very large peak in the blue region. The solid curve fitted to the data (open circles) is $SS = (K_1R–K_2G) + (K_3G–K_4R) + K_5B$, but with much greater values of K_2 and K_4 than those in Figure 1c. The final common path behavioural data (Figure 1a) is fitted by the same three terms, but with the significant difference that the best fit is obtained by combining the terms by probability summation, rather than scalar summation. Of considerable import is the fact that the values of K_2 and K_4, which control the subtractive terms, modelling red–green inhibition, are nearly the same magnitude to best fit the pattern ERG data as to best fit the final common path behavioural data.

We varied the experimental procedure by which we obtained the b-wave and pattern ERG functions in a controlled way and were able to completely reverse the critical aspects of the results. Opponency first appears at the bipolar level, as revealed by the b-wave criterion (Fig. 1c). We completely removed this colour opponency in macaques by injecting bicuculline into the vitreous (Mills and Sperling, 1990), resulting in a function that is identical with the a-wave function. These data are shown as the triangles superimposed on the a-wave data in Figure 1d. The PERG spectral sensitivity showed greatly increased red–green opponency and amplification of the blue response. This was reversed by performing the experiment with the checkerboard grating pattern removed. The resulting stimulus was a homogenous 17° circular field alternating between spectral and white. The same operations of determining the minima were performed and the reciprocal quanta at the minima are plotted as the diamond-shaped symbols superimposed on the b-wave data of Figure 1c. They show no notch in the yellow and much lower sensitivity in the blue. It seems clear that they are very similar to the b-wave function. The difference found in the 500 nm region may be attributed to the intrusion of rods from outside the focal area in the b-wave data (open circles), which the 8 Hz alternation minimizes.

It follows, as concluded in the non-colour studies cited above (Fiorentini *et al.*, 1981; Maffei and Fiorentini, 1981) that introducing luminance alternation around sharply focused borders evokes the ganglion cell (PERG) response, where red–green opponency is enhanced and blue sensitivity is amplified. Removing the pattern defaults the response to the bipolar level, probably because colour opponency at the ganglion cell level is spatially organized and at the bipolar level is not. Similarly, blocking GABA with bicuculline changes bipolar cell response back to photoreceptor level colour processing by blocking the colour opponent feedback pathway. Thus, these controlled experimental manipulations define three distinct stages of spectral information processing and allow us to move from one to another of them and back. These data fully associate the functional stage of colour processing with the major anatomical levels of the retina and the data indicate that red–green opponency and blue amplification have been completed in the retina, since they are as great at the ganglion cell level as in the behavioural threshold data.

References

Baker, C.L., Hess, R.F., Olsen, R.T. and Zrenner, E. (1988). Current source density analysis of linear and non-linear components of the primate electroretinogram. J. Physiol. (Lond.) 407: 155–176.

Dick, E., Miller, R.F. and Bloomfield, S. J. (1985). Extracellular K+ activity changes related to electroretinogram components. J. Gen. Physiol. 85: 911-931.

Fiorentini, A., Maffei, L., Pirchio, M., Spinelli, D. and Porcellati, V. (1981). The ERG in response to alternating gratings in patients with diseases of the peripheral retinal pathways. Invest. Ophthal. Vis. Sci. 21: 490–493.

Heynen, H. and van Norren, D. (1985). Origins of the electroretinogram in the intact macaque eye. Vision Res. 25: 709–717.

Maffei, L. and Fiorentini, A. (1981). Electroretinographic responses to alternating gratings before and after section of the optic nerve. Science 211: 953–955.

Marrone, C., Fiorentini, A., Bisti, S., Porciatti, V. and Burr, D. (1994). Pattern-reversal electroretinogram in response to chromatic stimuli. Visual Neurosci. 11: 873–884.

Mills, S.L. and Sperling, H.G. (1990). Red/green opponency in the rhesus macaque ERG spectral sensitivity is reduced by bicuculline. Vis. Neurosci. 5: 217–221.

Seiving, P.A. and Steinberg, R.H. (1985). Contribution from proximal retina to intraretinal pattern ERG. Invest. Ophthal. Vis. Sci. 26: 1642–1647.

Sperling, H.G. and Harwerth, R.S. (1971). Red-green interactions in the spectral sensitivity of rhesus. Science 172:180–184.

Sperling, H.G. and Mills, S.L (1987). ERG and behavioral analysis of spectral sensitivity in normal and blue-blind rhesus monkeys. In: Verriest, E. (ed.) Colour Vision Deficiencies, Vol. VIII, 365–374. Nijhoff, Dordrecht.

Sperling, H.G. and Mishra, J. (1995). Spectral sensitivity by flash flicker and pattern ERG. In: Drum, B. (ed.), Colour Vision Deficiencies XII, Doc. Ophthalmol. Proc. Ser. 57, 381–390. Kluwer, Dordrecht.

Tomita, T. and Yanagida, T. (1981). Origins of the ERG waves. Vision Res. 21: 1703–1707.

The University of Texas, Houston Health Sciences Center
Department of Ophthalmology and Visual Science
Houston, Texas, USA

Present address:
Boston University Medical School
Department of Ophthalmology
30 East Concord St.
Boston, MA 02118-2394, USA

14. Selective stimulation of colour vision in humans

J.J. KULIKOWSKI, A.G. ROBSON and D.J. McKEEFRY
(Manchester, UK)

Abstract

Most cells in the retino-geniculostriate pathway respond to both isoluminant, chromatically modulated and achromatic, luminance-modulated, stimuli. However, the separation of colour-related responses (psychophysical and near-threshold electrophysiological) is possible when the stimuli minimize luminance intrusions caused by chromatic aberration and retinal pigmentation. Under such conditions chromatic (VEP) responses maintain fundamental characteristics of colour mechanisms revealed psychophysically (e.g. low spatial and temporal resolution).

Introduction

Post-receptoral, cone-opponent retinal cells and the parvocellular geniculostriate pathway extract red–green and blue–yellow opponent information, conveniently modelled by the cardinal axes (Krauskopf *et al.*, 1982). However, re-examination of chromatic opponent cells in the striate cortex has revealed that few visual cortical cells conform to this system, and many cells respond to both luminance and chromatically modulated stimuli (Lennie *et al.*, 1990). Although colour information can only be conveyed by the chromatic opponent (parvo) cells, magno-cells also respond to chromatic isoluminant patterns, but in a nonlinear (frequency doubled) manner that is sufficient only to signal a chromatic border (Lee *et al.*, 1990). In general, the magno/parvo streams were originally thought to be completely segregated, at least in V1 and V2, and to subserve different functions (Hubel and Livingstone, 1987). Recent evidence suggests, however, that inputs from these processing streams may be partly mixed (Yoshioka *et al.*, 1994). Both magno and parvo streams are, therefore, likely to be involved in pattern analysis, although at different contrast levels and spatiotemporal ranges of stimuli (Merigan, 1991). Furthermore, some studies (e.g. Gegenfurtner *et al.*, 1994) have challenged the concept of cortical modularity by reporting the existence of cortical cells with multimodal responses, which are not selective for any one particular visual attribute.

Such observations may put in doubt the efforts to isolate magno/parvo functions, particularly in relation to colour mechanisms, but the fact remains

C.R. Cavonius (ed.), Colour Vision Deficiencies XIII, pp. 141–150.

that selective colour deficits of cortical origin do occur (Zeki, 1990), which implies some segregation. Moreover, when V1 cells of high chromatic sensitivity are selected (Kulikowski, 1991; Kulikowski and Walsh, 1993), they form groups with uniform properties, consistent with (threshold) psychophysical responses. In general, suprathreshold responses tend to reveal less specific mechanisms of lower sensitivity. Hence the difficulties in obtaining selective electrophysiological responses may be exacerbated at higher stimulus intensities. Our design for selective stimuli is based on correlation between psychophysical and neuronal responses. The best selectivity in revealing basic mechanisms can be achieved at psychophysical thresholds at which stimulus specificity can be unequivocally verified: if detection and discrimination thresholds are similar the mechanisms involved are labelled (Mullen and Kulikowski, 1990). However, electrophysiological recordings usually require suprathreshold stimulation, and selectivity must be traded off against the response intensity. When global electrophysiological responses such as visual evoked potentials (VEPs) are recorded the stimuli must be optimized not only to maximally activate the mechanism tested, but also to minimally activate the remaining mechanisms. This is because any global recording is a measure of overall activity and provides no information as to how the neural messages are coded. We found that low contrast, low spatial frequency gratings, presented in on–off or reversal modes, dichotomize VEPs. Characteristic colour-specific responses are obtained when gratings are chromatic and isoluminant; qualitatively different, achromatic VEPs are elicited when gratings are luminance modulated (see below).

Global responses tend to increase in magnitude with stimulus size, but isoluminance cannot be accurately specified for spatially extended stimuli. Hence colour-specific recordings are vulnerable to artefacts caused by departures from isoluminance that introduce luminance intrusions to colour stimuli, such as macular pigmentation (Moreland, 1990) and transverse chromatic aberration (Charman, 1991). Thus the rules for optimizing global responses are more stringent than those for single-unit recordings. To avoid recording multimodal and/or artefactual responses, stimulus contrast must be restricted to low or moderate levels (Tootell *et al.,* 1988) and stimulus parameters must be manipulated to allow the greatest discrimination of colour responses from the achromatic mechanisms.

Rationale behind the experiments

Of primary importance in the isolation of colour-specific responses from the cortex is the use of colour stimuli that are near threshold, or of moderate contrast. Under such circumstances VEPs may be related to basic psychophysical colour mechanisms. Five discriminable mechanisms have been isolated at threshold; red, yellow, green, blue and violet (Mullen and Kulikowski, 1990). These appear to correspond with the spectral characteristics of V1 cells (Vautin and Dow, 1985). Secondly, the chromatic system has low spatial and temporal resolution. It is convenient that low-contrast, coarse achromatic gratings stimulate only the 'fast transient' temporal mechanism, because its response is easy to distinguish from the 'slow sustained' colour responses elicited by chromatic gratings. Thirdly, most visual cortical cells have low spontaneous activity, so that the difference between the sustained onset response and the relative silence at offset is accentuated. Conversely, reversal presentations do not effectively reveal sustained components in global responses: each phase of the reversal stimulates roughly the same number of units, and only their transient components are evident.

By exploiting the factors listed above it is possible to distinguish chromatic VEPs from VEPs with achromatic intrusions. Such strategies have been used extensively to verify the specificity of red/green responses. Human VEPs elicited by chromatic gratings (in common with all colour-related responses) must be non-transient. In practice, a substantial difference must exist between onset and reversal presentations and this difference must reach maximum at isoluminance (at which sustained responses are maximal). Chromatic VEPs have different waveforms from achromatic potentials, with a characteristic negative wave (Murray *et al.*, 1987). This negative wave is of maximum amplitude at isoluminance, when low spatial frequencies are used, and disappears at 8 c/deg (c.f. Kulikowski, 1991), consistent with the low resolution of chromatic mechanisms (Hilz and Cavonius, 1970; Mullen, 1985). In addition, fast (12.5 Hz) reversal should generate a minimum VEP at isoluminance as heterochromatic flicker photometry is based upon the sluggishness of the colour system (Kulikowski *et al.*, 1989). However, this holds only at low contrasts, which is not surprising in view of known magno contributions at red/green isoluminance. It is shown that this strategy is also valid for blue/yellow VEPs. The present study aims to compare chromatic and achromatic VEPs elicited by stimuli systematically varying in contrast and spatial extent, in an attempt to empirically define practical limits within which selective stimulation of the chromatic system can be achieved.

Methods

Aberration-controlled apparatus

The apparatus used to generate gratings with minimized chromatic aberration (Mullen, 1985; McKeefry and Kulikowski, 1995) is based on the original design of Regan (1973) (see also Spekreijse in Desmedt, 1977). Sinusoidal gratings generated on two Joyce displays were placed at right angles to each other and viewed at different distances through a semi-silvered mirror. Medium-band filters were placed in the paths of the two displays and blue/yellow (tritan) or red/green gratings were obtained by the optical addition of the two out-of-phase images. The relative distance of each display from the subject depended on the chromatic difference in refraction for the stimulus pair used. This arrangement compensates for both types of chromatic aberration (axial, and more importantly, transverse). The effects of parallax for off-centre bars cannot be entirely eliminated, but were minimized by the use of a bite bar and head restraint.

TV monitor

The apparatus generating standard chromatic gratings was a Grundig TV monitor controlled with a BBC microcomputer and purpose-built interface (Murray *et al.*, 1987; Kulikowski *et al.*, 1989). The main modification was that blue/yellow stimulation was adjustable to follow either the maximum blue/yellow axis (Grigsby *et al.*, 1991) or a tritanopic confusion line. In the case of S-cone specific stimulation, one hue was fixed to the green phosphor ($x = 0.315$; $y = 0.585$) whereas the other, purple, was the mixture of hues of red and blue phosphors. Isoluminance was determined using heterochromatic flicker photometry and tritanopic confusion lines were established for each individual subject using the criterion of minimum distinct border (MacLeod and Boynton, 1979). The number of spatial cycles in a 1 c/deg grating was adjusted between 3 and 15. All gratings were circular (hence their area increased as square of cycles). The mean luminance of the stimulus was 30 cd/m^2.

As a practical guide the contrast of both the chromatic and achromatic stimuli is expressed in terms of the Michelson definition. In the case of the chromatic stimulus, the contrast value expressed is that of its constituent monochromatic elements which are then combined in antiphase. However, a more widely acceptable means of describing the contrast of isoluminant chromatic patterns is in terms of the extent to which they modulate the individual cone types. Employing the Smith and Pokorny (1975) cone fundamentals, cone contrasts for the red/green stimulus are calculated as being 13% of that of achromatic contrast for L cones and 31% for M cones. For the tritan stimulus S cone modulation is in excess of 95% of achromatic contrast.

Recording technique

An Ag/AgCl recording electrode was placed at the mid-occipital position (Oz) and referred to a mid-frontal electrode (Fz) or a linked ears reference. All VEPs were recorded using a Medelec Sensor ER94 averager and an Apple 11E microcomputer. The recordings were averages of a minimum of 128 repetitions. Filter bandwidth was 0.3–30 Hz for low-rate presentations and 1–60 Hz for stimulation rates above 2 Hz. Five subjects were tested. All had normal colour vision and all were corrected myopes.

Results and comments

Figure 1 shows chromatic and achromatic VEPs generated using the aberration-free apparatus. Low contrast (0.05–0.1) gratings elicited blue/yellow and red/green responses which were remarkably similar to each other (reversal VEPs

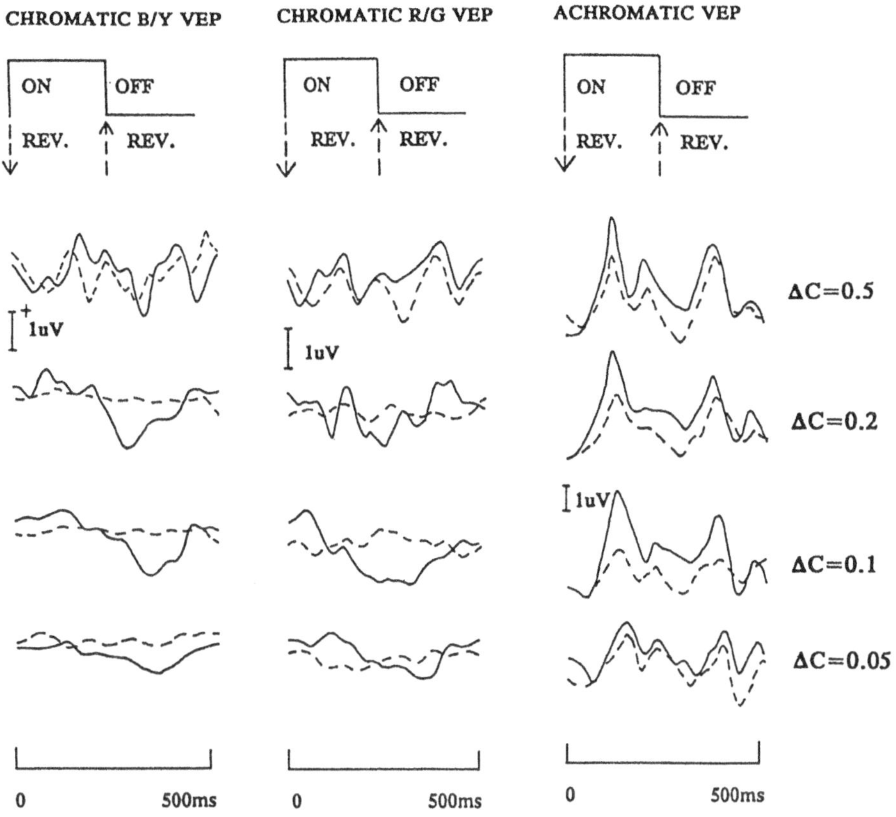

Fig. 1. VEPs elicited by 1 c/deg sinusoidal blue/yellow (close to the tritan axis), red/green and achromatic gratings. The stimuli were presented in either on–off (solid lines) or pattern reversal (dotted lines) modes. Each response is the average of at least 128 repetitions obtained from an electrode at Oz with linked ears as reference, filter bandwidth 0.3–30 Hz, stimulus repetition rate = 2 Hz, Lo = 25 cd/m^2. Note similarity between b/y and r/g VEPs at low contrasts (0.05–0.1), the differences between them at 0.2 and general similarity of all VEPs at 0.5. Subject D.M. (colour normal, corrected myope).

were absent). This is different from achromatic VEPs which are similar for onset, offset and reversal presentations. The achromatic VEPs can be called 'purely transient' since they are solely generated by a change in contrast (from 0 to C, C to 0 or from −0.5C to +0.5C). Conversely, chromatic onset VEPs have a strong negativity and the offset and reversal responses are practically absent. Chromatic onset VEPs can therefore be regarded as 'purely sustained' responses since they are generated by standing contrast.

The prominent negative waves in chromatic onset VEPs peaked around 200–300 ms. In conventional VEP studies (using high contrasts) such VEPs were

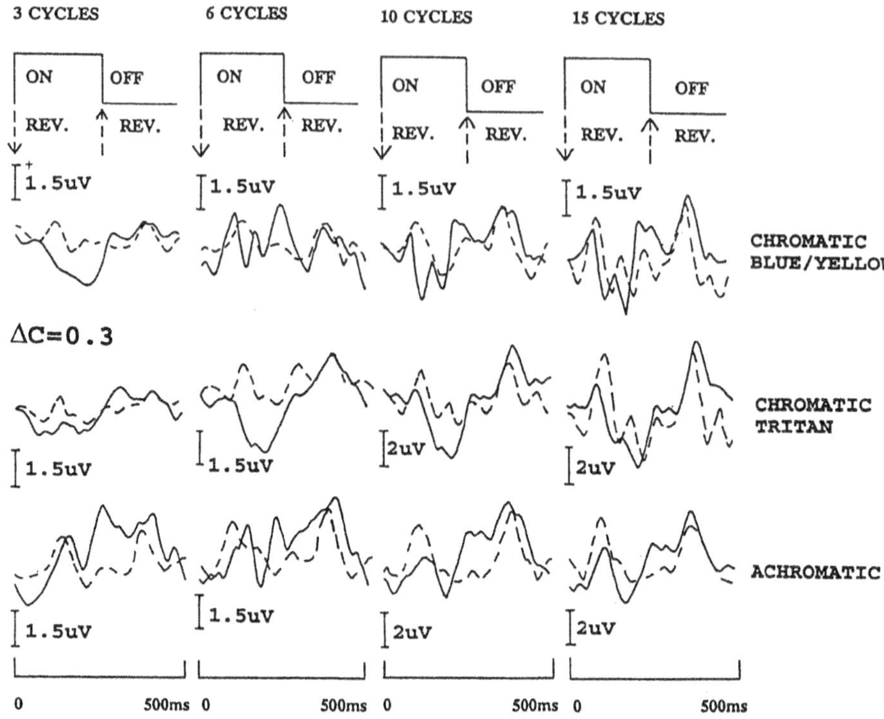

Fig. 2. The effects of field size on VEPs elicited using onset–offset (solid lines) and reversal (broken lines) presentations of blue/yellow and tritan gratings on a TV monitor. Stimulus repetition rate: 2 Hz, spatial frequency: 1 c/deg, Lo = 30 cd/m^2 $\Delta C = 0.3$, subject J.S. (colour normal, corrected myope).

considered to originate from the extrastriate cortical areas. However, microelectrode recordings from monkey V1 revealed that chromatic units gave maximum bursts of spikes to weak suprathreshold stimuli at about 200 ms; the local VEPs (field potentials) recorded with the same electrode had negative waves with peaks which were still more delayed (Kulikowski, 1991). There was no evidence of substantial contribution of a major colour-related area, V4, since monkeys with such lesions had seemingly normal chromatic VEPs (Kulikowski *et al.*, 1994).

At an intermediate contrast (0.2), VEPs to red/green gratings become different from blue/yellow VEPs, mainly by the existence of offset and reversal components. These are consistent with the magno-responses to isoluminant red/green stimuli (Lee *et al.*, 1990). At high contrasts (0.5) the distinct nature of all the chromatic responses was lost, possibly due to some contributions from cells of low selectivity responding to chromatic stimuli in a transient manner.

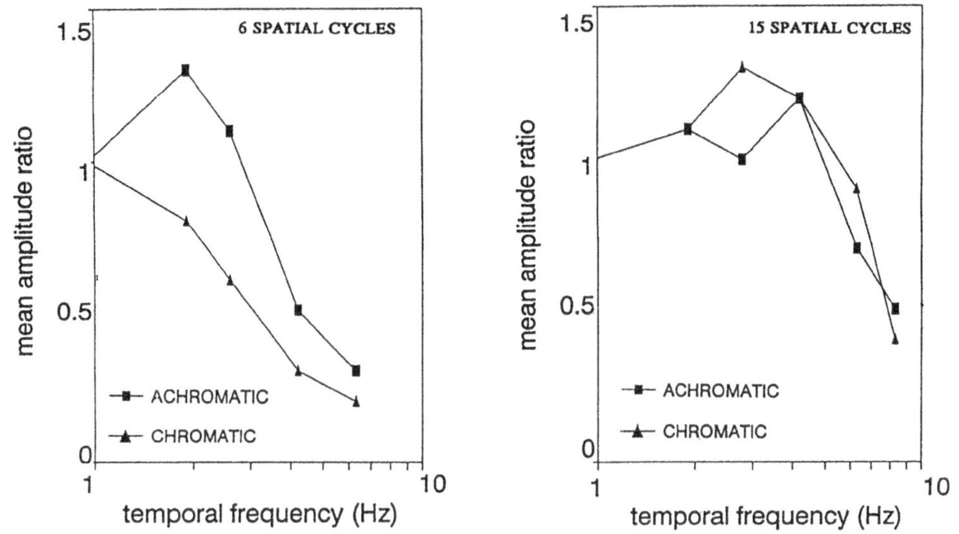

Fig. 3. The effects of field size on VEPs elicited varying temporal frequency of reversal presentations of tritan gratings; spatial frequency: 1 c/deg, Lo = 30 cd/m^2 $\Delta C = 0.3$. Average of 2 subjects: F.E. and J.S. (colour normal, corrected myopes).

Figure 2 compares the VEPs generated by maximum blue/yellow, purple/green (tritanopic pairs) and achromatic stimulation using gratings generated on the TV monitor and containing various numbers of spatial cycles. The most distinct (selective), but small, responses were obtained when gratings contained only 3 spatial cycles. The overall size of the chromatic VEPs increased with the number of cycles, but comparison with reversal indicates that more transient components (which obscure the chromatic component) were generated.

Figure 3 compares the temporal tuning of VEPs. For gratings containing 6 spatial cycles, colour VEP amplitude decreased as temporal frequency increased (low-pass temporal frequency tuning). Conversely, transient responses to achromatic coarse gratings (with 6 spatial cycles) had band-pass characteristics. When larger gratings were used (containing 15 spatial cycles) chromatic VEPs showed a completely different tuning characteristic, with band-pass properties reflecting transient contributions. In view of absent, or negligible, S-cone inputs to the achromatic magno-system (Lee *et al.*, 1990; Nakano and Kaiser, 1992), achromatic contamination (e.g. due to macular pigmentation and chromatic aberration) is the simplest explanation of this result.

Discussion

The present study demonstrates that the generation of chromatically selective VEPs is possible in humans and that these recordings have a neuronal basis, identified previously in primate vision. The main debate is concentrated on the specificity versus the completeness of chromatic responses and their critical properties.

Chromatic VEPs

Chromatic VEPs to red/green gratings have been extensively studied since Carden *et al.* (1985) showed that the gratings provide better distinction of chromatic from achromatic VEPs than checkerboards but only at low contrasts (see discussions in Desmedt, 1977; Kulikowski *et al.,* 1989). This can be explained by reference to unit and local VEP recordings from monkey V1.

In primate visual cortex, units that respond best to low spatial frequency gratings near threshold are only of two types: achromatic cells (concentric, simple and complex) responding transiently up to the temporal resolution limit, and chromatic opponent with prominent sustained responses and with low-pass characteristics. Any stimuli with higher spatial frequencies inevitably activate other cells; even very coarse achromatic checkerboards activate hypercomplex cells, many of which give sustained responses (Kulikowski, 1991). It is interesting to note that the responses to sinusoidal and square-wave gratings are not much different, as though the critical parameter was a check-corner, rather than a sharp edge. The above tests (sustained/transient and low-pass/ band-pass) can further be developed incorporating, for example, Fourier analysis, but the principal distinction in temporal properties remains.

Thus low contrast, coarse gratings seem optimal stimuli to distinguish chromatic from achromatic responses. However, this optimization holds only for low/moderate contrasts, probably because high contrasts activate many units of low specificity (multimodal) and intermediate properties; results obtained by Gegenfurtner *et al.* (1994) should be a cautionary note for over-enthusiastic attempts to isolate parvo/magno, or chromatic/achromatic global responses using high contrast stimuli.

Balance between selective and effective stimulation

These findings also suggest that the price of selectivity is to obtain relatively weak VEPs. It is difficult to argue how far one can depart from selectivity in order to obtain meaningful VEPs since subjects vary in their cortical configuration; so that sometimes a human peripheral representation does not contribute to VEPs, while on other occasions it produces reversed polarity VEPs. However, it must be stressed that the use of spatially extensive stimuli always results in the introduction of achromatic stimulus intrusions, even though it may not give strong artefacts in some subjects. Erroneous conclusions may be drawn, therefore, when extended stimuli are employed to test the limits of selectivity of colour mechanisms (e.g. that the S-cone driven responses are phasic-like: Rabin *et al.,* 1994).

Selective blue/yellow responses

It is often considered important to isolate the S-cone driven mechanism (e.g. according to the tritan axis). However, it is now established that the cortical responses are not organized according to these axes (Lennie *et al.*, 1990). Moreover, Smith and Pokorny (1995) have shown that S-cone sensitivity is predominant at short wavelengths, so small departures from the tritan line may not be critical. The tritanopic confusion line assessed according to a minimally distinct border criterion shows a broad range of inter-subject and intra-subject tolerance. Thus, the purity of the tritan axial stimulation is less important than the control of luminance intrusions which are endemic to polychromatic stimulation (with large differences in wavelengths).

The most effective way of reducing achromatic intrusions is to restrict the number of spatial cycles in a grating to between 3 and 6. Under such conditions human VEP waveforms, elicited by violet-green (tritan) gratings pass all the tests of selectivity, but the price of selectivity may be relatively small VEPs in some subjects.

Acknowledgements

We thank John Simpson for his devoted help in these studies.

References

Carden, D., Kulikowski, J.J., Murray, I.J. and Parry, N.R.P. (1985). Human occipital potentials evoked by the onset of equiluminant chromatic gratings. J. Physiol. 369: 44P.

Charman, W.N. (1991). Limits on visual performance set by the eye's optics and the retinal cone mosaic. In: Kulikowski, J.J., Walsh, V. and Murray, I.J. (eds.), Limits of Vision: 81–96. Macmillan, London.

Desmedt, J.E. (ed.) (1977). Visual Evoked Potentials in Man: New Developments: 3–125, 184–196. Clarendon, Oxford.

Gegenfurtner, K.R., Kiper, D.C. and Fenstemaker, S.B. (1994). Processing of color information in area V2 of the macaque. Perception 23 (suppl): 30–31.

Grigsby S.S., Vingrys, A.J., Benes, S.C. and King-Smith, P.E. (1991). Correlation of chromatic, spatial, and temporal sensitivity in optic nerve disease. Invest. Opthalmol. Vis. Sci. 32: 3252–3262.

Hilz, R. and Cavonius, C.R. (1970). Wavelength discrimination measured with square wave gratings. J. Opt. Soc. Am. 60: 273–277.

Hubel, D.H. and Livingstone, M.S. (1987). Segregation of form, colour and stereopsis in primate area 18. J. Neurosci. 7: 3378–3415.

Krauskopf, J., Williams, D.R. and Heeley, D.M. (1982). The cardinal directions of colour space. Vision Res. 22:1123–1131.

Kulikowski, J.J. (1991). On the nature of evoked potentials, unit responses and psychophysics. In: Valberg, A. and Lee, B.B. (eds.), From Pigments to Perception: 197–209. Plenum, New York.

Kulikowski, J.J. and Walsh, V. (1993). Colour vision: isolating mechanisms in overlapping streams. Prog. Brain Res. 95: 417–26.

Kulikowski, J.J., Murray, I.J. and Parry, N.R.A. (1989). Electrophysiological correlates of chromatic opponent and achromatic stimulation in man. In: Drum, B. and Verriest, G. (eds.), Colour Vision Deficiencies IX. Doc. Ophthal. Proc. Ser. IX: 145–153. Kluwer, Dordrecht.

Kulikowski, J.J., Walsh, V., McKeefry, D.J., Butler, S.R. and Carden, D. (1993). The electrophysiological basis of colour processing in macaques with V4 lesions. Behav. Brain Res. 60: 73–78.

Lee, B.B., Pokorny, J., Smith, V.C., Martin, P.R. and Valberg, A. (1990). Luminance and chromatic modulation sensitivity of macaque ganglion cells and human observers. J. Opt. Soc. Am. A 7: 2223–2235.

Lennie, P., Krauskopf, J. and Sclar, G. (1990). Chromatic mechanisms in the striate cortex of the macaque. J. Neurosci. 10: 649–669.

MacLeod, D.I.A. and Boynton, R.M. (1979). Chromaticity diagram showing cone excitation by stimuli of equal luminance. J. Opt. Soc. Am. 69: 1183–1186.

Merigan, W. (1991). P and M pathway specialization in the macaque: In: Valberg, A. and Lee, B.B. (eds.), From Pigments to Perception: 117–126. Plenum, New York.

McKeefry, D. and Kulikowski, J.J. (1995). Psychophysical and occipital responses to aberration-free blue/yellow and red/green gratings. In: Drum, B. (ed.), Colour Vision Deficiencies XII. Doc. Ophthal. Proc. Ser. XII: 391–398. Kluwer, Dordrecht.

Moreland, J.D. (1990). The clinical utility of anomaloscopy. In: Ohta, Y. (ed.), Color Vision Deficiencies (Proceedings of the Symposium of the IRGCVD): 127–143. Kugler and Ghedini, Amstelveen, NL.

Mullen, K.T. (1985). The contrast sensitivity of human colour vision to red–green and blue–yellow chromatic gratings. J. Physiol. 359: 381–400.

Mullen, K.T. and Kulikowski, J.J. (1990). Wavelength discrimination at detection threshold. J. Opt. Soc. Am. A 7: 733–742.

Murray, I.J., Parry, N.R.A., Carden, D. and Kulikowski, J.J. (1987). Human visual evoked potentials to chromatic and achromatic gratings. Clin. Vision Sci. 1: 231–244.

Nakano, Y. and Kaiser, P.K. (1992). Color fusion and flicker fusion frequencies using tritanopic pairs. Vision Res. 32: 1417–1423.

Rabin, J., Switkes, E., Crognale, M., Schneck, M.E. and Adams, A.J. (1994). Visual evoked potentials in three-dimensional colour space: Correlates of spatiochromatic processing. Vision Res. 34: 2657–2671.

Regan, D. (1973). Evoked potentials specific to spatial patterns of luminance and colour. Vision Res. 13: 2381–2402.

Smith, V.C. and Pokorny, J. (1975). Spectral sensitivity of the foveal cone photopigments between 400 and 500 nm. Vision Res. 15: 161–171.

Smith, V.C. and Pokorny, J. (1995). Chromatic-discrimination axes, CRT phosphor spectra, and individual variations in color vision. J. Opt. Soc. Am. A 12: 27–35.

Tootell, R.B.H., Hamilton, S.L. and Switkes, E. (1988). Functional anatomy of macaque striate cortex. IV. Contrast and magno-parvo streams. J. Neurosci. 8: 1594–1609.

Vautin, R.G. and Dow, B.M. (1985) Color cell groups in foveal striate cortex of the behaving macaque, J. Neurophysiol. 54: 73–292.

Yoshioka, T., Levitt, J.B. and Lund, J.S. (1994). Independence and merger of thalamocortical channels within macaque monkey primary visual cortex. Visual Neurosci. 11: 467–489.

Zeki, S. (1990). A century of cerebral achromatopsia. Brain 113: 1721–1777.

J.J. Kulikowski, A.G. Robson, D.J. McKeefry
Visual Sciences Laboratory, UMIST
PO Box 88
Manchester M60 1QD, UK

15. Topographic differences of spectral sensitivity in rabbit retina. Evidence from small-field ERG and immunohistochemistry

H. REITSAMER, R. PFLUG, P. RÖHLICH and P. AHNELT

(Vienna, Austria; Budapest, Hungary)

Abstract

Recent evidence (Juliusson *et al.*, 1994) has demonstrated a non-uniform topographic distribution of spectral cone types in the rabbit retina. M-cones dominate the major part of the retina with the highest density in the visual streak, while S-cone density increases from the superior to the inferior retina forming a 'blue streak' in the outermost inferior part of the retina. This area lacks M-cones almost completely. Small circular patches of retina (4 mm diameter) obtained along vertical meridians were kept in a superperfusion chamber at 34°C. Intensity–response series of ERG b-waves in response to stimuli between 423 nm and 613 nm were recorded 15 min after adaptation with green light (503 nm). Action spectra calculated from b-wave amplitudes revealed a peak sensitivity in the blue range when patches were cut from the outermost inferior rim of the retina, whereas in patches of the visual streak and the superior retina peak sensitivity was reached at middle wavelengths.

After electrophysiological examination retinal patches were processed for immunohistochemistry using two monoclonal antibodies specific for either middle- or short-wavelength visual pigments. Histological data correlated well with electrophysiological results, thus confirming the different spectral sensitivities in local areas of the rabbit retina.

Introduction

Rabbits are dichromatic mammals. About 5% of all photoreceptors in the retina are cones, with absorption peaks in the middle- and short-wavelength ranges (Dodt and Elenius, 1956: M-cones: 520 nm, S-cones: 460 nm; Caldwell and Daw, 1978: M-cones: 500 nm, S-cones: 460 nm; Lemmer, 1991: M-cones: 515 nm, S-cones: 425 nm). Using recently developed antibodies (COS-1, OS-2) against long- and short-wavelength pigments it has been possible to discriminate between M- and S-cones (Juliusson *et al.*, 1994; Szél *et al.*, 1988; Röhlich and Szél, 1987). A narrow horizontal streak in the outermost inferior retina shows S-

C.R. Cavonius (ed.), Colour Vision Deficiencies XIII, pp. 151–155.

cone densities (10 000/mm^2) that are almost as high as those of the M-cones in the central retina. Because this streak lacks M-cones completely the authors called it the 'blue streak'. This area covers about 5–6% of the whole retina. Just below the penetration area of the optic nerve fibres there is a horizontal band of high cone density called the visual streak which is dominated by M-cones (12 000/mm^2): S-cone density is about 1500/mm^2. In the superior part of the retina densities of both cone types decline towards the superior margin (M-cones 5000/mm^2; S-cones 800/mm^2), but the ratios of M and S-cones are almost the same as in the visual streak.

The purpose of this study was to establish a correlation between local variations of functionally determined spectral sensitivity and immunohistochemical examination of cone density.

Methods

Electrophysiological methods

Twenty rabbits were anaesthetized with urethane (6.8 ml/kg bodyweight) and subsequently sacrificed by air embolism. After enucleation and section of the eye bulb small circular patches (4 mm diameter) of either the superior retina, the visual streak or the inferior retina were cut out. The isolated retinal patches were put into a recording chamber and superperfused with 5 parts Eagle's medium (Sigma) and 3 parts newborn calf serum at 34°C (Hanitzsch and Bornschein, 1965; Lützow and Bornschein, 1966), equilibrated with 5% CO_2/95% O_2 (pH 7.4). ERGs were elicited by Ganzfeld stimuli generated on a two-beam optical bench with a 450 W xenon lamp as light source. Shutters, neutral filters and interference filters (FWHM 7 nm) determined the duration, intensity and colour of the stimuli. Background adaptation with 503 nm (2.75–3.75 log quanta/μm/s) started 15 min before recording, in order to bleach rods. In the test beam spectral stimuli of 9 wavelengths between 423 nm and 613 nm with a duration of 350 ms were presented at different light intensities to the retinal patches.

Action spectra were calculated from intensity–response curves of b-wave amplitudes at constant criteria. The results of various experiments were averaged and normalized at 423 nm.

Immunohistochemical methods

COS-1 and OS-2 are antibodies against long- and short-wavelength absorbing visual pigments, COS-1 being specific for M-cones and OS-2 for S-cones.

Retinal patches and the rest of the bulbus were fixated for 5 h in 4% paraformaldehyde plus phosphate buffer. After brief preincubation with Triton, the pieces were treated with either COS-1 or OS-2 antibodies for 18 h. Bound antibodies were detected with an FITC-conjugated anti-mouse secondary antibody. Retinal pieces were placed on microscope slides and coverslips were mounted with lycerol-PBS. The whole mounts were examined, plot-graphed and counted with a Zeiss epifluorescence microscope.

Results and discussion

Figures 1A–1C show mean spectral sensitivities of ERG b-wave amplitudes recorded in patches of the superior retina, the visual streak and the outermost inferior part of the retina. Each curve is the average of 5–9 recordings, normalized at 423 nm and fitted with Dartnall nomograms. After completion of electrophysiological recordings, S-cones and M-cones in a particular retinal patch from the visual streak and the inferior retina were stained with OS-2 and COS-1 (Fig. 2).

Although the two maxima of each curve show the presence of both cone types in all retinal areas, the regional spectral sensitivities vary. It should also be noted that in some patches the receptor distribution varies considerably within the patch.

Visual streak

The electrophysiological sensitivity of patches from the visual streak peaked at 515 nm (Fig. 1A), with a secondary maximum at 454 nm. Antibody staining revealed a nearly homogeneous receptor distribution, with M-cones outnumbering S-cones by a factor of 7.8 (Fig. 2A).

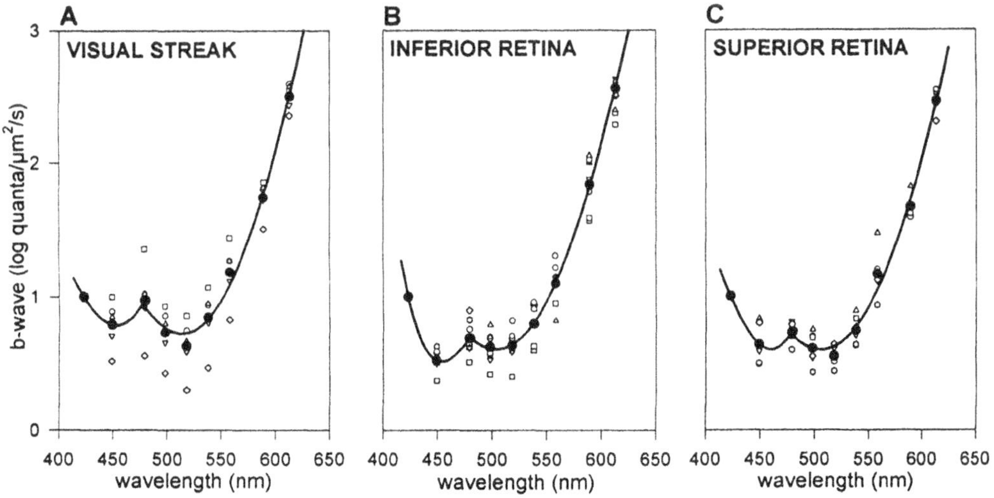

Fig. 1. Individual action spectra (open symbols) and mean action spectra (filled symbols) calculated from ERG b-wave intensity response curves. (A) visual streak ($n = 9$); (B) inferior retina ($n = 6$); (C) superior retina ($n = 5$). Solid lines represent Dartnall nomograms (Wyszecki and Stiles, 1967).

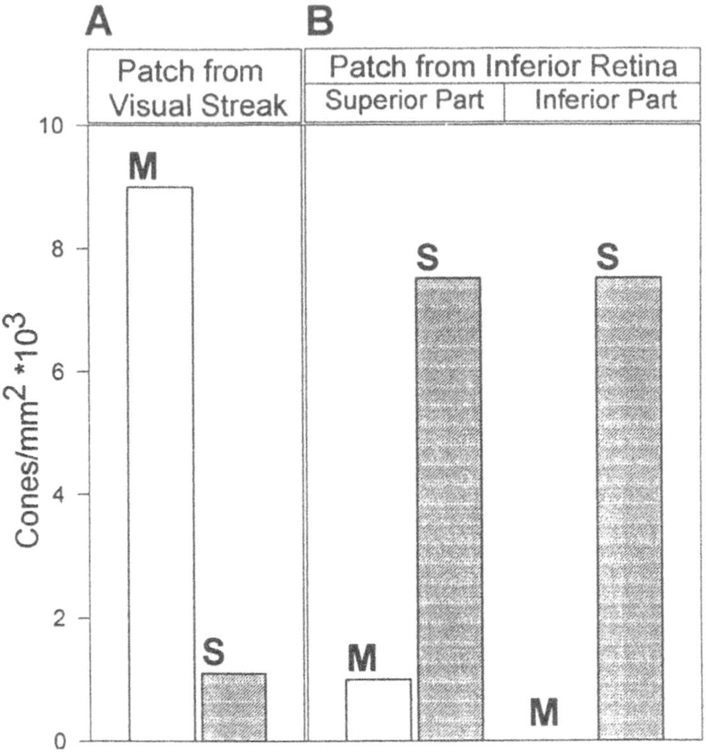

Fig. 2. M-cone (M) and S-cone (S) densities estimated from immunohistochemical staining. Patches were obtained from: (A) visual streak (M: 9000/mm^2, S: 1150/mm^2); (B) inferior retina: superior part of patch (M: 1100/mm^2, S: 7500/mm^2), inferior part of the blue streak (M: 0/mm^2, S: 7500/mm^2)

Inferior retina

Electrophysiologically, the patch cut from the inferior retina was most sensitive to 454 nm light (Fig. 1B), corresponding to the lack of M-cones at its outermost rim, the blue streak (Fig. 2B). Since the patches used (4 mm) were slightly larger than this blue streak (2–3 mm), M-cones at the upper rim of the patches contributed a secondary shallow green peak. The positition of this peak at 505 nm, compared with the sensitivity maximum of M-cones at 515 nm, shows that an influence of rods cannot be completely ruled out despite light adaptation.

Superior retina

Although we did not stain a patch of the superior retina it is known (Juliusson *et al.*, 1994) that there is only a gradient in the absolute density of cones, but not in the ratio of M-cones and S-cones from the visual streak to the superior retina. The blue peak of the action spectrum (Fig. 1c) at 454 nm reveals contributions of S-cones, while the green peak at 505 nm suggests a contribution of rods in addition to M-cones, indicating the increased rod/cone ratio.

The use of patches slightly larger than the blue streak and the rise of rod to cone ratio in the periphery cause electrophysiological data to be less clear than that offered by immunohistochemistry. Nevertheless, the results demonstrate obvious functional differences in colour sensitivity across the rabbit retina. The biological significance of these findings remains to be investigated.

Acknowledgements

Supported by FWF P09 670 - Med.

References

Caldwell, J.H. and Daw, N.W. (1978). New properties of retinal ganglion cells. J. Physiol. (Lond.) 276: 257–276.

Dodt, E. and Elenius, V. (1956). Spektrale Sensitivität einzelner Elemente der Kaninchennetzhaut. Pflüg. Arch. 262: 301–306.

Hanitzsch, R. and Bornschein, H. (1965). Spezielle Uberlebensbedingungen für isolierte Netzhäute verschiedener Warmblüter. Experentia, 21: 484.

Juliusson, B., Bergstöm, A., Röhlich, P., Ehringer, B., van Veen, T. and Szél, A. (1994). Complementary cone fields of the rabbit retina. Invest. Ophtal. Vis. Sci. 35: 811–818.

Lemmer, T. (1991). Stäbchen- und Zapfenprozesse im Elektroretinogramm des Kaninchens. Kyril u. Method-Verlag, München

Lützow, A. and Bornschein, H. (1966). Der Einfluß der Temperatur auf die Charakteristik der b-Welle im ERG der isolierten Kaninchennetzhaut. Graefes Arch. klin. exp. Ophthal. 170: 332–341.

Röhlich, P. and Szél, A. (1987). Color-specific photoreceptors recognized by monoclonal antibodies. Eur. J. Cell Biol. 43 (Suppl. 18): 29.

Szél, A., Diamantstein, T. and Röhlich, P. (1988). Identification of the blue-sensitive cones in the mammalian retina by antivisual pigment antibody. J. Comp. Neurol. 273: 593–602

Wyszecki, G. and Stiles, W.G. (1967). Color Science. Wiley, New York.

H. Reitsamer, R. Pflug, P. Ahnelt
Institut für allgemeine und vergleichende Physiologie
Schwarzspanierstr. 17
1090 Vienna, Austria

PART THREE

Acquired colour vision deficiencies

16. Colour vision in a family with butterfly-shaped pigment dystrophy of the fovea

M. MÄNTYJÄRVI and K. TUPPURAINEN
(Kuopio, Finland)

Abstract

Colour vision was examined in 11 members of a family in which three members had butterfly-shaped pigment dystrophy of the fovea in both eyes. Colour vision was tested with the Standard Pseudoisochromatic Plates part 2, the Farnsworth-Munsell 100-hue (FM 100) test, Nagel anomaloscope, and Color Vision Meter 712 anomaloscope. Eight members of the family had healthy eyes, normal visual fields and normal colour vision. Of the three members with butterfly-shaped dystrophy, one had slightly decreased visual acuity in both eyes (0.8), the two others had normal acuity in one eye but acuity of 0.4 and 0.5, respectively, in the other eye. All three had normal visual fields. During dark adaptation the rod threshold was slightly elevated in two members and normal in one. ERGs were measured in only one of the affected family members and they were normal. In all colour vision tests, the results of the three affected family members were within the normal ranges. However, in the FM 100 test, the error scores increased in all three patients during the 3-year follow-up period: two had error scores at the upper normal limit. The FM 100 test seems to be the best colour vision test for follow-up studies in slowly progressive foveal dystrophies such as butterfly-shaped dystrophy.

Introduction

Butterfly-shaped pigment epithelial dystrophy belongs to a group of genetically transmitted macular disorders called pattern dystrophies of the retinal pigment epithelium. The pattern dystrophies are characterized by a lack of specific symptoms, normal or slightly decreased visual acuity, and normal visual fields, dark adaptation, colour vision and electroretinogram. Electro-oculograms may show abnormal values. Bilateral symmetrical patterned lesions of the retinal pigment epithelium (e.g. reticular, annular, or butterfly-shaped) are found in ocular fundi; they can also be clearly demonstrated by fluorescein angiography, but the fundi are usually otherwise normal. At an advanced stage visual acuity may be reduced, caused by the progressive lesion of the retinal pigment epithelium also affecting the photoreceptors. Inheritance is usually autosomal

C.R. Cavonius (ed.), Colour Vision Deficiencies XIII, pp. 159–168.

dominant (Deutman, 1971; Newsome, 1988; Hsieh *et al.*, 1977; Marmor and Byers, 1977).

Colour vision in butterfly-shaped dystrophy has been previously examined in several studies and generally found to be normal (Pinckers, 1988). Since the first study of butterfly-shaped dystrophy by Deutman *et al.* (1970), we have found reports in the literature of only five patients with different kinds of acquired colour vision abnormalities (Raitta and Erkkilä, 1979; Pinckers, 1988; Cardillo, Piccolino and Zingirian, 1988).

The purpose of the present study was to analyse colour vision in a family in which three members had butterfly-shaped pigment dystrophy of the fovea.

Subjects

Of the 13 family members, we were able to examine 11, aged from 6 to 61 years (Fig. 1), three of whom had butterfly-shaped dystrophy of the fovea. The parents of the proband (Fig. 1) had both died at the age of 40; according to the other members of the family they had had no eye trouble. The only family member who was known to have had poor vision was a deceased paternal uncle of the proband; his eye trouble was thought to have been the result of an explosion during the war. However, the only medical record available for him was from 1976; it showed degenerative pigmented changes in both foveas that might have been a patterned dystrophy. Unfortunately, he had no living descendants available for examination. It is possible that this dominant trait came from the father's side of the proband.

Methods

The eye examination consisted of testing visual acuity, refraction, phorias and tropias, as well as biomicroscopy, ophthalmoscopy and retinoscopy with the aid of tropicamide 0.8% + phenylephrine 5%. Tension was measured with the Goldmann applanation tonometer in those subjects who were 30 years or older.

The visual fields were tested in adults with the full field 120 point three-zone strategy using the Humphrey Visual Field Analyzer (Humphrey Instruments, San Leandro, California, USA), and in children aged 10 years or younger with the Friedman Central Field Analyzer (Clement Clarke Ltd., London, UK).

Colour vision was tested with the Standard Pseudoisochromatic Plates part 2 (SPP2, Ichikawa *et al.*, 1983), Farnsworth-Munsell 100-hue test (FM 100, Farnsworth, 1957), Nagel anomaloscope (Schmidt and Haensch, Berlin, Germany), and Color Vision Meter 712 (CVM, Interzeag, Schlieren, Switzerland). Children ages 10 years or younger were tested with the Farnsworth Panel D-15 (Farnsworth, 1947) and Lanthony Desaturated Panel (Lanthony, 1974) instead of the FM 100 test.

All these tests were administered and interpreted according to the instruc-

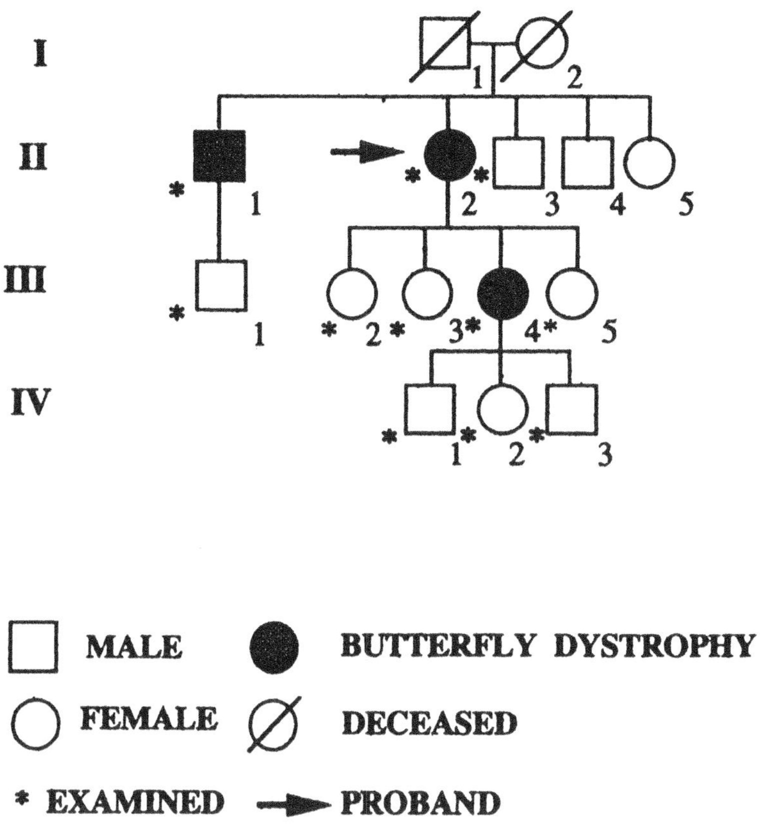

Fig. 1. The family tree of the family with butterfly-shaped pigment dystrophy of the fovea.

tions of the tests. However, in the SPP2 test, subjects were not required to read correctly the left number 2 in the third plate because according to the designers of the test, this number cannot be read by 87.2% of normal subjects and is therefore of no diagnostic value (Ichikawa *et al.*, 1987). Results of the FM 100 test were plotted as suggested by Kinnear (1970). The normal values of the error scores were defined according to Verriest *et al.* (1982), and axis formation according to Smith *et al.* (1985). Nagel anomaloscope testing (the Rayleigh equation) was performed according to the method of Linksz, as described by Birch *et al.* (1979). The CVM examination (for red–green colour vision, the Rayleigh equation, and for blue colour vision, the Moreland equation) was performed, and the results were interpreted in terms of the ranges established by Pelizzone *et al.* (1991, 1993) and Roth *et al.* (1991); the normal values are: for the Rayleigh equation, anomalous quotient (AQ) 0.7–1.3, matching range (MR)

0–7; for the Moreland equation, AQ 0.7–1.4 and MR 0–10.

Dark adaptation (DA; Goldman–Weekers Adaptometer, Haag-Streit, Bern, Switzerland) was performed only on those patients who had a butterfly-shaped dystrophy. Fluorescein angiography (FAG) was performed on two of these dystrophy patients (II/1 and II/2) with the aid of 5 ml 10% fluorescite injected intravenously (Alcon Laboratories Inc., Fort Worth, Texas, USA); fundus pictures were taken with a Canon CFZA. Electroretinogram (ERG) examination was performed only on the proband, in the Neurophysiology Laboratory at the University Hospital in Kuopio; the procedure followed the guidelines for clinical evoked potential studies of the American Electroencephalographic Society (1984). Electro-oculographic testing was not available.

Results

Clinical eye examination with visual fields and colour vision was normal in all the unaffected family members.

Case reports of the affected family members:

Case 1, II/2, the proband

This 58-year-old woman came for an eye examination at the age of 55 years in order to renew her driver's licence. With the best correction (+0.75 c + 1.0 × 0 in the right and +1.0 c + 1.0 × 0 in the left eye), her visual acuity was 0.4 in the right and 1.0 in the left eye. The media were clear and tension was normal in both eyes. A symmetrical yellowish defect with a branch-pattern in the macular area and yellow flecks in the nasal periphery of the fundi was found. Visual fields were normal in both eyes. In DA, the cone threshold and cone-rod break time were normal, but the rod threshold was slightly elevated, 0.6 log units above the upper normal limit. The ERG was normal. In the FAG, a butterfly-shaped defect of the retinal pigment epithelium with no leakages could be seen in both eyes (Fig. 2). These results did not change during the 3-year follow-up period. In the colour vision examination, the patient correctly interpreted all the required SPP2 numbers in 1992 and 1995; other colour vision test results are shown in Tables 1 and 2. During the 3-year follow-up period, the error scores in the FM 100 test increased (within the normal range) fairly uniformly in all directions, giving no indication of a definite axis of error (Fig. 3). In 1995, the error score of the right eye was at the upper normal limit. The anomaloscope test results were normal. The red–green values did not change, but during the follow-up the blue AQ values showed slight changes towards green, but remained within the normal range.

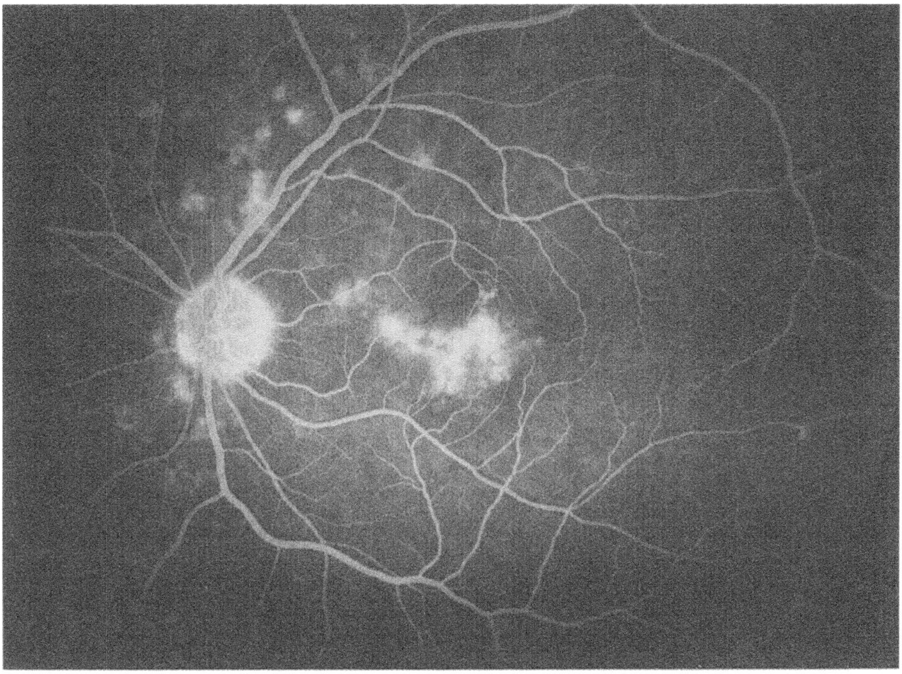

Fig. 2. The fluorescein angiography of the left ocular fundus of the proband

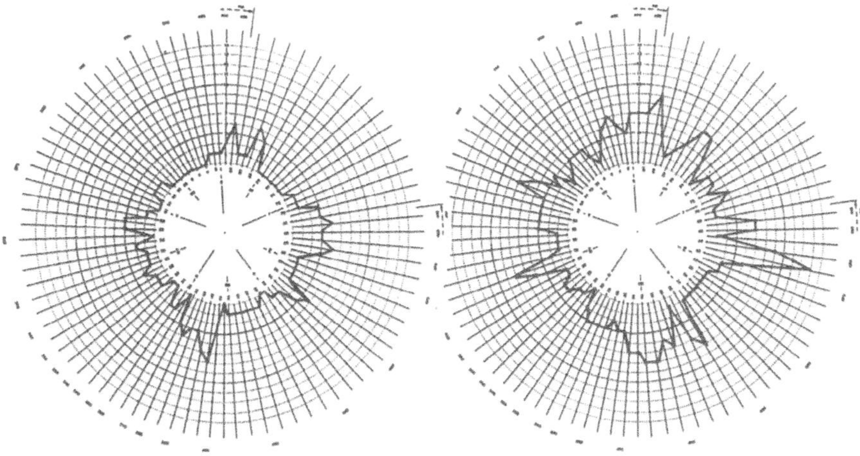

Fig. 3. The change of the FM 100 test results in three years in the right eye of the proband: the result in 1992 (left) and 1995 (right).

Table 1. Square roots of the total error scores on the FM 100 test of the three family members with butterfly-shaped dystrophy during the three years of follow-up from 1992 to 1995.

Family member, age (years)	1992	1993	1995	Normal error scores (Verriest *et al.*, 1982) M ± SD	upper limit: M+2 × SD (95%)
II/1, 61					
Right eye	13.27		15.75	10.93 ± 2.59	16.11
Left eye	13.67		16.43	11.11 ± 2.76	16.63
II/2, 58					
Right eye	10.77	13.34	15.59	10.36 ± 2.43	15.22
Left eye	9.59	10.49	11.31	10.22 ± 1.99	14.20
III/4, 34					
Right eye	4.69	4.47	5.66	7.50 ± 2.68	12.86
Left eye	5.00	4.12	7.81	7.45 ± 2.60	12.65

M = mean; SD = standard deviation; FM 100 = Farnsworth–Munsell 100 hue test.

Case 2, II/1

This 61-year-old brother of the proband had had slightly decreased visual acuity in his left eye since the age of 40. In the eye examination, the best corrected visual acuity (with –1.0 in the right and –0.5 c – 1.5 × 0 in the left eye) was 1.0 in the right eye and 0.5 in the left eye. The media were clear, and tension was normal in both eyes. In the right fundus, an oval-shaped yellowish vertical defect of the retinal pigment epithelium was found, and in the left eye this defect resembled a butterfly. In the FAG, vertical and slightly branching retinal pigment epithelium defects were seen in both eyes. Visual fields were normal. In DA, the cone threshold and cone-rod break time were normal, while the rod threshold was elevated 0.6 log units above the upper normal limit. These results did not change during the 3-year follow-up. In the colour vision examination, the patient correctly interpreted all the required SPP2 numbers in 1992 and 1995; other colour vision results are shown in Tables 1 and 2. During the 3-year follow-up period, the error scores in the FM 100 test increased (within the normal range) evenly in all directions, not giving any indication of a definite axis of error. In 1995, the error score of the left eye was at the upper normal limit and that of the right eye very near to it. The anomaloscope findings were normal and showed no essential changes during the follow-up.

Table 2. Red–green and blue colour vision test results on the anomaloscope of the three family members with butterfly-shaped dystrophy during the 3 years of follow-up from 1992–1995.

	Anomaloscope				
	1992 (Nagel)	1993 (CVM)		1995 (CVM)	
Family member, age (years)	RG	RG	B	RG	B
II/1, 61					
Right eye					
AQ	1.1			1.1	1.0
MR	2			1	11
Left eye					
AQ	1.0			0.9	1.0
MR	1			3	11
II/2, 58					
Right eye					
AQ	0.9	0.9	1.1	1.0	0.9
MR	0	0	4	2	2
Left eye					
AQ	0.9	0.9	1.3	0.9	0.9
MR	0	0	4	1	2
III/4, 34					
Right eye					
AQ	1.0	0.9	1.0	0.9	0.9
MR	5	1	1	2	2
Left eye					
AQ	0.8	0.8	1.1	0.8	1.1
MR	1	1	1	2	1

Case 3, III/4

This 34-year-old daughter of the proband had no visual complaints. With the best correction (+0.5 in both eyes), her visual acuity was 0.8 in both eyes. The media were clear, and tension was normal in both eyes. In the fovea, the beginning of a butterfly-shaped dystrophy pattern was seen. Visual fields were normal in both eyes. In the DA, the cone threshold and cone-rod break time were normal, while the rod threshold was at the upper normal limit. These results did not change during the 3-year follow-up period. The patient read all the SPP2 plates correctly in 1992 and 1995; other colour vision results are shown in Tables 1 and 2. During the 3-year follow-up, the error scores in the FM 100 test increased (within the normal range) evenly in all directions, not giving any indication of a definite axis of error. The error score of the right eye was better than the age-relevant mean, and that of the left eye was at the age-relevant mean. Anomaloscope findings were normal and did not show essential changes during the follow-up.

Discussion

Considering the foveal changes of the patients with a butterfly-shaped dystrophy, it is astonishing that hardly any disturbances of visual function, including colour vision, could be found. All of the colour vision test results were normal: SPP2 plates, FM 100 test scores and the anomaloscope findings. The deterioration of the FM 100 score during the 3-year follow-up period, although within the normal range, was the only sign that this butterfly-shaped dystrophy is progressive.

In earlier studies, colour vision has been examined with various tests: Ishihara plates (Mejia and Gieser, 1981; Starzycka and Bryk, 1976), American Optical and Hardy-Rand-Rittler (HRR) plates (Hsieh *et al.*, 1977; Gutman *et al.*, 1982; Watzke *et al.*, 1982; Nichols *et al.*, 1993), Panel D 15 (Prensky and Bresnick, 1983), FM 100 (Cardillo Piccolino and Zingirian, 1988), Panel D 15 and FM 100 (Raitta and Erkkilä, 1979), Ishihara and HRR plates, Panel D 15 and Nagel anomaloscope (Deutman *et al.*, 1970, Kempeneers *et al.*, 1991). We found in the earlier literature altogether 20 patients with butterfly-shaped dystrophy who had been studied with colour vision tests since Deutman *et al.* (1970) first reported this foveal affection. Only five of them showed acquired colour vision disturbances (in the study of Kempeneers *et al.*, 1991, the patient had a congenital deutan defect). In the follow-up study of Deutman's patients after 11 years, one patient had pseudoprotanomaly and two patients had a blue defect; the age of these patients was at the time 51–56 years (Pinckers, 1988). Another case of protan was found by Raitta and Erkkilä (1979) in a patient of 41 years. A patient aged 42 years with a mild deutan defect was reported by Cardillo Piccolino and Zingirian (1988). Since the youngest patient with butterfly-shaped dystrophy was 8 years old (Prensky and Bresnick, 1983), it can be speculated that this foveal disease progresses very slowly. The possible affection of the photoreceptors probably comes much later, about at an age of 50 years, accompanied by decreasing visual acuity and changes in colour vision. This agrees with the findings of the present study: the error scores in the FM 100 test were at the upper normal limit in the older patients, 58 and 61 years of age, while the younger patient still had error scores better than the mean of her age group in the right eye and at the mean in the left eye. A slight progression of the disease in terms of colour vision could be observed over three years.

In the anomaloscope examination, the AQ of the proband's blue equation showed a slight change towards green; the meaning of this is not clear. It will be interesting to see in future follow-up examinations whether the colour vision of these patients deteriorates further, developing real defects that also show up in the results of other tests.

References

American Electroencephalographic Society (1984). Guidelines for clinical evoked potential studies. J. Clin. Neurophysiol. 1: 3–53.

Birch, J., Chisholm, I.A., Kinnear, P., Pinckers, A.J.L.G., Pokorny, J., Smith, V.C. and Verriest, G. (1979). Clinical testing methods. In: Pokorny, J., Smith, V.C., Verriest, G. and Pinckers, A.J.L.G. (eds.), Congenital and Acquired Color Vision Defects: 83–135. Grune and Stratton, New York.

Cardillo Piccolino, F. and Zingirian, M. (1988). Pattern dystrophy of the retinal pigment epithelium with vitelliform macular lesion: evolution in ten years. Int. Ophthalmol. 11: 207–217.

Deutman, A.F. (1971). The Hereditary Dystrophies of the Posterior Pole of the Eye. Van Gorcum, Assen, The Netherlands.

Deutman, A.F., van Blommestein, J.D.A., Henkes, H.E., Waardenburg, P.J. and Solleveld-van Driest, E. (1970). Butterfly-shaped pigment dystrophy of the fovea. Arch. Ophthalmol. 83: 558–569.

Farnsworth, D. (1947). The Farnsworth Dichotomous Test for Color Blindness, Panel D 15. The Psychological Corporation, New York.

Farnsworth, D. (1957). The Farnsworth-Munsell 100-Hue Test. Munsell Color Company, Baltimore.

Gutman, I., Walsh, J.B. and Henkind, P. (1982). Vitelliform macular dystrophy and butterfly-shaped epithelial dystrophy: a continuum? Br. J. Ophthalmol. 66: 170–173.

Hsieh, R.C., Fine, B.S. and Lyons, J.S. (1977). Patterned dystrophies of the retinal pigment epithelium. Arch. Ophthalmol. 95: 429–435.

Ichikawa, H., Hukami, K. and Tanabe, S. (1983). Standard Pseudoisochromatic Plates part 2. Igaku-Shoin Ltd., Tokyo.

Ichikawa, K., Ichikawa, H. and Tanabe, S. (1987). Detection of acquired colour vision defects by Standard Pseudoisochromatic Plates part 2. In: Verriest, G. (ed.), Colour Vision Deficiencies. VIII. Doc. Ophthalmol. Proc. Ser. 46: 133–140. Martinus Nijhoff/Dr W. Junk Publishers, Dordrecht.

Kempeneers, H.P., Dewachter, A. and Kempeneers, G.M. (1991). Pattern dystrophies of the retinal pigment epithelium. The study of three generations in a family. Doc. Ophthalmol. 76: 261–272.

Kinnear, P.R. (1970). Proposals for scoring and assessing the 100-hue test. Vision Res. 10: 423–433.

Lanthony, P. (1974). Lanthony's Desaturated Panel 15 Hue Test according to Farnsworth-Munsell. Luneau Ophthalmologie, Paris.

Marmor, M.F. and Byers, B. (1977). Pattern dystrophy of the pigment epithelium. Am. J. Ophthalmol. 84: 32–44.

Mejia, J.R. and Gieser, R.G. (1981). Sporadic butterfly macular dystrophy. Ann. Ophthalmol. 13: 1253–1254.

Newsome, D.A. (1988). Pigment epithelial dystrophies. In: Newsome, D.A. (ed.), Retinal Dystrophies and Degenerations: 195–221. Raven Press, New York.

Nichols, B.E., Sheffield, V.C., Vandenburgh, K., Drack, A.V., Kimura, A.E. and Stone, E.M. (1993). Butterfly-shaped pigment dystrophy of the fovea caused by a point mutation in codon 167 of the RDS gene. Nature Genet. 3: 202–207.

Pelizzone, M., Sommerhalder, J., Roth, A. and Hermes, D. (1991). Automated Rayleigh and Moreland matches on a computer-controlled anomaloscope. In: Drum, B., Moreland, J.D. and Serra, A. (eds.), Colour Vision Deficiencies. X. Doc. Ophthalmol. Proc. Ser. 54: 151–159. Kluwer, Dordrecht.

Pelizzone, M., Sommerhalder, J., Roth, A. and Hermes, D. (1993). Automated Rayleigh and Moreland matches. In: Drum, B. (ed.), Colour Vision Deficiencies. XI. Doc. Ophthalmol. Proc. Ser. 56: 345–355. Kluwer, Dordrecht.

Pinckers, A. (1988). Patterned dystrophies of the retinal pigment epithelium. Ophthalm. Paed. Genet. 9: 77–114.

Prensky, J.G. and Bresnick, G.H. (1983). Butterfly-shaped macular dystrophy in four generations. Arch. Ophthalmol. 101: 1198–1203.

Raitta, C. and Erkkilä, H. (1979). Schmetterlingsförmige Dystrophie der Makula bei anomalem Serumlipoprotein. Klin. Mbl. Augenheilkd. 174: 561–566.

Roth, A., Pelizzone, M., Sommerhalder, J., Hermes, D. and Simona, F. (1991). The two equation method III: results in normal subjects above 50 years of age. Correlation with lens opacity. In: Drum, B., Moreland, J.D. and Serra, A. (eds.), Colour Vision Deficiencies. X. Doc. Ophthalmol. Proc. Ser. 54: 353–359. Kluwer, Dordrecht.

Smith, V.C., Pokorny, J. and Pass, A.S. (1985). Color-axis determination on the Farnsworth-Munsell 100-hue test. Am. J. Ophthalmol. 100: 176–182.

Starzycka, M. and Bryk, E. (1976). Schmetterlingsförmige Dystrophie der Makula. Klin. Mbl. Augenheilkd. 169: 454–458.

Verriest, G., Van Laethem, J. and Uvijls, A. (1982). A new assessment of the normal ranges of the Farnsworth-Munsell 100-hue test scores. Am. J. Ophthalmol. 93: 635–642.

Watzke, R.C., Folk, J.C. and Lang, R. M. (1982). Pattern dystrophy of the retinal pigment epithelium. Ophthalmology 89: 1400–1406.

Maija Mäntyjärvi, M.D.
Department of Ophthalmology
University Hospital of Kuopio
P.O.B. 1777
70211 Kuopio, Finland

17. Colour vision deficiencies in two unrelated Sardinian families with butterfly-shaped macular dystrophy

M. FOSSARELLO, I. ZUCCA, C. BERTINI, M. PIRASTU, M.S. GALANTUOMO, G.M. CONTU and A. SERRA
(Cagliari and Tramariglio, Italy)

Abstract

Autosomal dominant butterfly-shaped macular dystrophy (BSMD) has been reported as being associated with different mutations of the peripherin/RDS gene. We describe here the phenotype of two unrelated Sardinian families with a novel mutation in the peripherin/RDS gene, represented by a large deletion which removes exons 2 and 3 and causes macular degeneration, studied in three generations. Four affected members were studied with clinical examinations and measurement of dark adaptation, colour vision (Farnsworth-Munsell 100-hue, Lanthony D-15, Ishihara), automatic perimetry (Octopus G1 and M1 programs), electroretinography and electro-oculography. The results indicate that early in the course of the disease, before the ophthalmoscopic appearance of macular deposits, macular function tests (visual acuity, colour vision and central retinal sensitivity) appear to be normal. After the deposition of material in the macula, test performance is mildly affected. Late in the course of the disease, after the age of 50 years, colour vision shows widespread abnormalities, with an anarchic pattern and high error score on the 100-hue. Retinal sensitivity is also markedly reduced. The age-dependent decay of macular function, together with an abnormal EOG, suggest an important role of retinal pigment epithelium cells in the development of macular degeneration in BSMD.

Introduction

Butterfly-shaped macular dystrophy (BSMD) is an autosomal dominant disease that may accompany mutations in the peripherin/RDS. Several reports of intrafamilial variability of fundus pictures suggest that BSMD is part of a spectrum of various clinical entities (retinal pattern dystrophy (RPD); McKusick, 1990) which should be considered as different cellular responses to a common basic disorder, probably related to the subretinal accumulation of lipofuscin (Watzke *et al.*, 1982; Guiffré and Lodato, 1986; De Jong and

C.R. Cavonius (ed.), Colour Vision Deficiencies XIII, pp. 169–178.

Dellman, 1982; Prensky and Bresnick, 1983; Gutman *et al.*, 1982). This deposition of abnormal material is thought to cause gradual atrophy of the retinal pigment epithelium (RPE) cells and of macular photoreceptors, in which case an significant loss of central vision may be expected.

The human peripherin/RDS gene has been mapped to chromosome 6p21.2-cen (Travis *et al.*, 1991a). The normal gene encodes a photoreceptor-specific glycoprotein containing 346 amino acid residues, located within the outer segment disk membranes of both rod and cone cells (Travis *et al.*, 1991b; Connell and Molday, 1990; Connell *et al.*, 1992; Molday *et al.*, 1987). It is believed to play a role in the morphogenesis and maintenance of disc membrane lamellar structure (Arikawa *et al.*, 1992).

Recent studies have demonstrated that missense or null mutations in the peripherin/RDS gene are associated with different phenotypes, including RPD, autosomal dominant retinitis pigmentosa and flecked retina, with preferential involvement of either the central area or the peripheral retina, or both (for recent references see Nichols *et al.*, 1993; Keen *et al.*, 1994; Kemp *et al.*, 1994; Wroblewski *et al.*, 1994).

In the current study we describe the clinical features exhibited by two unrelated Sardinian families affected by autosomal dominant BSMD, in which the disease was present over three generations. In both families the disease seems to be associated with an identical deletion in the peripherin/RDS gene.

Patients and methods

Subjects and DNA analysis

Patient II-1, a 52-year-old woman, patient III-2, her 18-year-old daughter (pedigree A), patient II-1, a 61-year-old woman, and patient III-1, her 32-year-old daughter (pedigree B), were affected family members from two Sardinian pedigrees with butterfly-shaped macular degeneration. We reported previously that in both pedigrees the disorder is caused by a mutation represented by a large deletion in exons 2 and 3 of the peripherin/RDS gene (Fossarello *et al.*, 1996). In both families the macular dystrophy was observed in three generations: the pattern of inheritance clearly showed autosomal dominant transmission of the disease (Fig. 1).

Visual function tests

Visual acuity was measured with Snellen charts. Colour vision was evaluated with the Farnsworth-Munsell 100-hue and the Lanthony D-15 colour tests using Macbeth illumination. In the same patients automated threshold perimetry was recorded using an Octopus 500EZ, program G1 and M1 (Interzeag, Schlieren, Switzerland). Dark adaptation was measured using a Goldmann-Weekers dark adaptometer (Haag-Streit AG, Berne, Switzerland): baseline dark-adapted

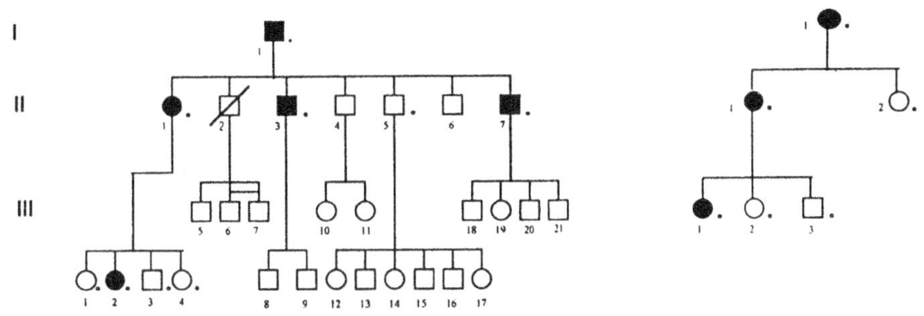

Fig. 1. Family pedigrees with autosomal dominant butterfly-shaped macular dystrophy. Left, Family A. Right, Family B. Individuals are identified by generation and pedigree number. Squares indicate males; circles, females; slashed symbols, deceased; solid symbols, affected; open symbols, unaffected. The black dots to the side of the symbols identify those individuals who were examined ophthalmologically.

thresholds were determined after at least 1 h of dark adaptation, before obtaining a post-bleach full recovery curve (Alexander and Fishman, 1984). Full-field electroretinograms (ERGs) and electro-oculography (EOG) (BM 6000-3maxi, Biomedica Mangoni, Pisa, Italy) were performed using previously described methods that adhere to the international standards for these tests (Marmor *et al.*, 1989; Marmor and Zrenner, 1993). Informed consent was obtained from the patients and from normal subjects involved in the study after the procedures had been explained fully. The research procedures conformed with institutional guidelines and the Declaration of Helsinki.

Results

Table 1 shows the age, sex, fundus picture, corrected visual acuity, age of onset of visual symptoms, Farnsworth-Munsell (FM) 100-hue test total error score (TES), EOG Arden ratio, and automatic perimetry mean sensitivity for the four patients examined.

Patient A II-1 had typical butterfly-shaped macular dystrophy with yellow lesions in the central retina of each eye and yellow dots along temporal arcades and around the optic disc (Fig. 2). The peripheral retina appeared ophthalmoscopically normal. Best corrected visual acuity was 10/10 in both eyes. This patient had a right unilateral paramacular disciform lesion with new subretinal vessels in June 1992; the lesion healed spontaneously within 6 months. Colour vision was tested in June 1993 and again in June 1995: the first FM 100-hue measurement showed a tritan axis with an error score of 221 in RE and 230 in

Table 1. Clinical characteristics of patients with peripherin/RDS deletion.

Family member	Sex/age (years)	Type of macular dystrophy	OD	OS	Age of onset (years)	FMTS[b] OD	FMTS[b] OS	EOG[c] OD	EOG[c] OS	Visual field[d] OD	Visual field[d] OS
A,II-1	F/52	Butterfly+dots	10/10	10/10	39	221	230	174	171	22.8	21.2
A,III-2	F/18	Normal	10/10	10/10		50	68	201	207	26.4	26.6
B,II-1	F/61	Butterfly	10/10	10/10	45	245	266	170	173	21.6	20.8
B,III-1	F/32	Butterfly	10/10	10/10	25	80	115	174	185	27.1	26.5
B,III-2	F/27	Normal	10/10	10/10		48	61			27.4	27.1

[a]Onset of visual symptoms (distortion, blurring, etc.)
[b]Farnsworth–Munsell (FM) 100-hue test: Total error score
[c]Light-dark ratio or Arden ratio: normal, 240; normal lower limit, 180
[d]Mean sensitivity, dB

LE; the second showed an anarchic pattern with an error score of 390 in RE and 264 in LE (Fig.3). Over this 2-year period, the macular lesions enlarged slightly. With automatic perimetry, mean sensitivity was reduced (22.8 RE and 21.2 LE), and mean defect was abnormal in both eyes. Dark adaptometry appeared normal, and the EOG Arden ratio was reduced (174 RE, 171 LE). Patient A, III-1, an 18-year-old girl who was affected on the basis of DNA analysis, showed a completely normal fundus (Fig. 2). All visual function tests were normal, including colour vision (Fig. 3).

In the two patients from family B, II-1 and III-1, a typical butterfly-shaped macular dystrophy was symmetrically present in the macular area (Fig. 2). In patient II-1 corrected visual acuity was 10/10 in each eye; colour vision was abnormal, with values of FM 100-hue TES of 330 in RE and 258 in LE associated with an anarchic axis (Fig. 4). Mean sensitivity was reduced (21.6 db in RE and 20.8 db in LE), as was the EOG Arden ratio (170 in RE and 173 in LE).

In patient III-1 a surface wrinkling maculopathy had been present over the yellow macular deposits in her left eye since her first visit in May 1991. A nearly two-fold enlargement of the butterfly lesion was noted over the next 2 years. In this patient, ICG-A with near-infrared SLO was performed, which in her left eye showed a macular focus of early hypofluorescence followed by intense late hyperfluorescence, not corresponding to the foci of fluorescein hyperfluorescence. The intense ICG hyperfluorescence was probably due to dye staining of fibrotic material located beneath the retinal pigment epithelium, since a similar picture was also detected with infrared light imaging without dye injection (confocal mode). Smaller foci of intense hyperfuorescence were also present in her right eye within the yellow lesion. The exact nature and pathogenesis of such

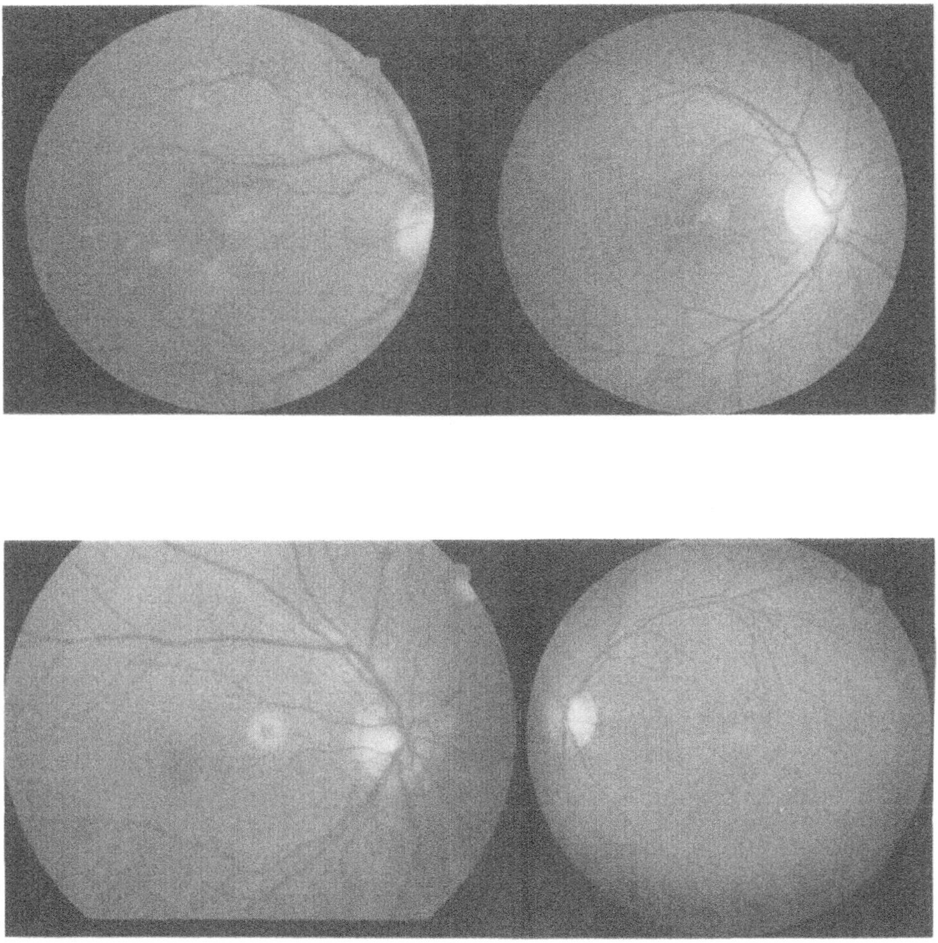

Fig. 2. Top left: right posterior pole of patient A, II-1, at 52 years of age. Besides macular butterfly-shaped yellow deposits and perimacular yellow dots, a small disciform scar is present near the fovea. Top right: right macula of patient B, II-1 at 61 years of age, shows accumulation of yellowish pigment in a butterfly-shaped pattern within the central macular area. Some black pigment is also visible within the lesion. Bottom left: right normal fundus of patient A, III-2, at 18 years of age. Bottom right: left posterior pole of patient B, III-1, at 32 years of age. The accumulation of yellow pigment in the fovea is disposed in three roundish dots and is observed in association with a surface wrinkling maculopathy.

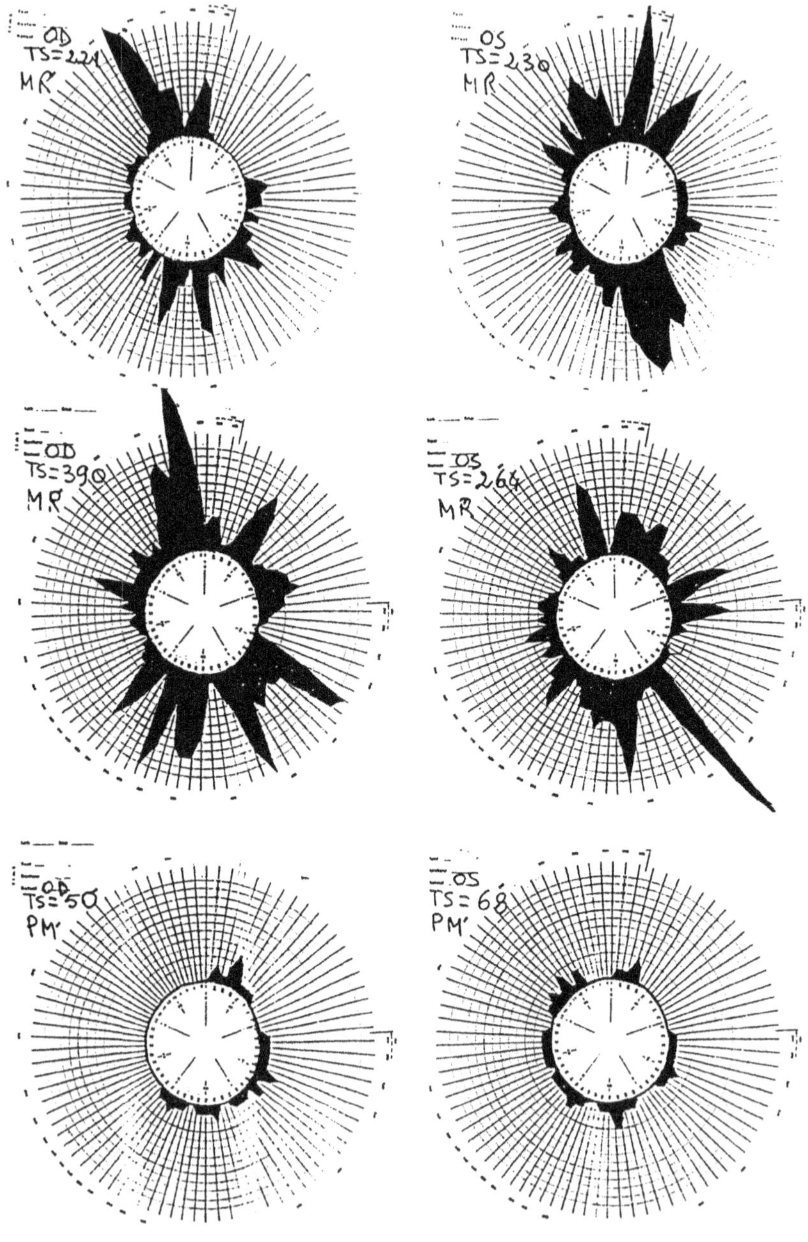

Fig. 3. FM 100-hue plots of patient A, II-1 (top) and 2 years later (centre) and patient A, III-2 (bottom).

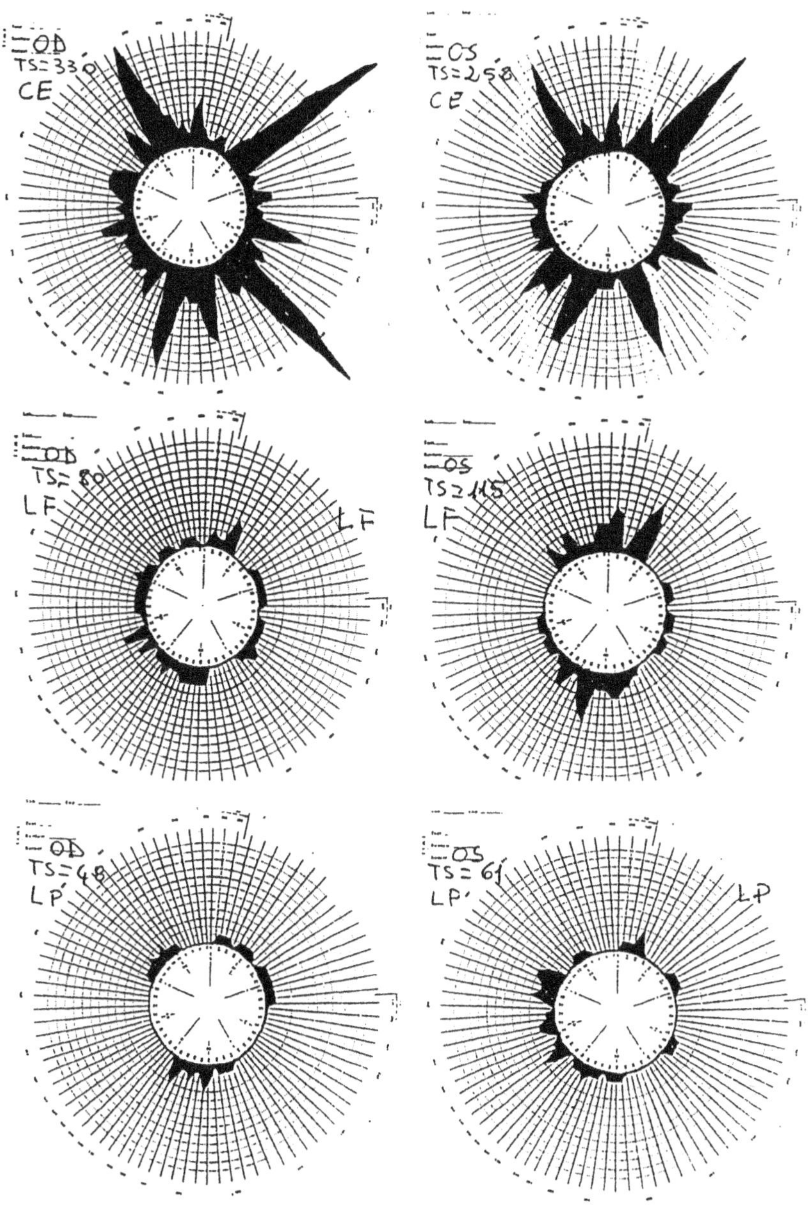

Fig. 4. FM 100-hue plots of patient B, II-1 (top), patient B, III-1 (centre) and subject B, III-2 (bottom).

fibrotic plaque remains to be elucidated. Her visual acuity was 10/10 in each eye, colour vision was slightly affected, with a TES of 80 in RE and 115 in LE. A tritan axis appeared in LE (Fig. 4). For comparison, colour vision of her sister, subject III-2, is also shown (Fig. 4). Retinal sensitivity of III-2 appeared to be normal (27.1 db in RE and 26.5 db in LE), and the EOG Arden ratio (174 in RE and 185 in LE) was slightly reduced. In all patients, dark adaptometry showed a normal cone-rod break and a normal recovery to pre-bleach rod threshold values. Both photopic and scotopic ERG responses were normal.

Comment

In this paper we describe the the clinical phenotype of BSMD in four patients from two unrelated Sardinian families, in which affected individuals show a remarkably homogeneous intrafamilial and interfamilial phenotype. The results of DNA analysis suggest that affected members of both families were heterozygous for an identical mutation, probably shared on a common ancestor chromosome.

Psychophysical and electrophysiological examination revealed normal ERG and slightly subnormal EOG, normal dark adaptometry, normal colour vision and normal retinal sensitivity early in the course of the disease. Later, an abnormal total score on the 100-hue test and a tritan axis defect appear, with reduced central retinal sensitivity. Such defects seem to progress with age.

The clinical phenotype observed in our patients is characterized by a preferential involvement of central photoreceptors and normal activity of the peripheral retina, with a definite indication of disturbance of the RPE activity shown in the EOG. We hypothesize that the availability of peripherin is reduced by 50%, secondary to gene mutation, which triggers an acceleration of the rate of production and shedding of the discs, with the consequence of RPE cells being overcrowded by phagosomes. Early in the course of the disease this might occur without disrupting outer segment metabolism, as suggested by our finding of normal colour vision, normal retinal sensitivity, normal ERG and slightly abnormal EOG in our youngest patient MD2. However, this extraordinary burden of phagocytosis of photoreceptor outer segments should with time lead to the build-up of lipofuscin in RPE cells in the macular area (pattern stage). The accumulation of lipofuscin could result both in RPE cell disruption and atrophy, and in mechanical obstruction to the transport of nutrient materials to the retinal photoreceptors, leading later to atrophy of RPE and the photoreceptors in the macular area. The clinical picture is very similar to senile macular degeneration (atrophic stage).

Compared with more severe phenotypes of BSMD (Kemp *et al.*, 1993; Wroblewski *et al.*, 1994), our patients show a milder type of the disease and have good central visual acuity and central retina structure, even after 60 years of age. Moreover, the yellow plaques appear at a later stage of life, as is demonstrated by case A, III-2, who is genotypically affected but phenotypically normal. Yellow plaques are therefore not a primary feature of the disease.

Acknowledgements

This work was partially supported by Telethon grant No. E.183 and by Assessorato alla Sanità, Regione Sardegna.

References

Alexander, K.R. and Fishman, G.A. (1984). Prolonged rod dark adaptation in retinitis pigmentosa. Br. J. Ophthalmol. 68: 561–569.

Arikawa, K., Molday, L.L., Molday, R.S. and Williams, D.S. (1992). Localization of the peripherin/rds in the disk membranes of cone and rod photoreceptors: Relationship to disc membrane morphogenesis and retinal degeneration. J. Cell. Biol. 116: 659–667.

Connell, G. and Molday, R.S. (1990). Molecular cloning, primary structure and orientation of the vertebrate photoreceptor cell protein peripherin in the rod disc membrane. Biochemistry 29: 4691–4698.

Connell, G.J., Bascom, R., Molday, L., Rei, D., McInnes, R.R. and Molday, R.S. (1992). Photoreceptor peripherin is the normal product of the gene responsible for retinal degeneration in the rds mouse. Proc. Natl. Acad. Sci. USA. 88: 723–726.

De Jong, P.T.V.M and Delleman, J.W. (1982). Pigment epithelial pattern dystrophy. Four different manifestations in a family. Arch. Ophthalmol. 100: 1416–1421.

Fossarello, M., Bertini, C., Galantuomo, M.S., Cao, A., Serra, A. and Pirastu, M. (1996). Deletion in the peripherin/RDS gene in two unrelated Sardinian families with autosomal dominant butterfly-shaped macular dystrophy. Arch. Ophthalmol. (in press).

Giuffrè, G. and Lodato, G. (1986). Vitelliform dystrophy and pattern dystrophy of the retinal pigment epithelium: concomitant presence in a family. Br. J. Ophthalmol. 70: 1526–1530.

Gutman, I., Walsh, J.B. and Henkind, P. (1982). Vitelliform macular dystrophy and butterfly-shaped epithelial dystrophy: a continuum? Br. J. Ophthalmol. 66: 170–173.

Keen, T.J., Inglehearn, C.F., Kim, R., Bird, A.C. and Bhattacharya, S. (1994). Retinal pattern dystrophy associated with a 4 bp insertion at codon 140 in the RDS-peripherin gene. Hum. Mol. Genet. 3: 367–368.

Kemp, C.M., Jacobson, S.G., Cideciyan, A.V., Kimura, A.E., Sheffield, V.C. and Stone E.M. (1994). RDS gene mutations causing retinitis pigmentosa or macular degeneration lead to the same abnormality in photoreceptor function. Invest. Ophtalmol. Vis. Sci. 35: 3154–3162.

Marmor, M.F. and Zrenner, E. (1993). Standard for clinical electro-oculography. Arch. Ophthalmol. 111: 601–604.

Marmor, M.F., Arden, G.B., Nilsson, S.E.G and Zrenner, E. (1989). Standard for clinical electroretinography. Arch. Ophthalmol. 107: 816–819.

McKusick, V. (1990). Mendelian Inheritance in Man, 9th edn. John Hopkins University Press, Baltimore.

Molday, R.S., Hicks, D. and Molday, L. (1987). Peripherin: A rim-specific membrane protein of rod outer segment discs. Invest. Ophthalmol. Vis. Sci. 28: 50-61.

Nakazawa, M., Kikawa, E., Chida, Y. and Tamai, M. (1994). Asn244His mutation of the peripherin/RDS gene causing autosomal dominant cone-rod degeneration. Hum. Mol. Genet. 3: 1195–1196.

Nichols, B.E., Sheffield, V.C., Vandenburgh, K., Drack, A.V., Kimura, A.E. and Stone, E.M. (1993). Butterfly-shaped pigment dystrophy of the fovea caused by a point mutation in codon 167 of the RDS gene. Nature 3: 202–207.

Prensky, J.G. and Bresnick, G.H. (1983). Butterfly-shaped macular dystrophy in four generations. Arch. Ophthalmol. 101: 1198–1203.

Travis, G.H., Christerson, L., Danielson, P. E., Klisak, I., Sparkes, R.S., Hahn, L.B., Dryja, T.P and Sutcliffe, J.G. (1991a). The human retinal degeneration slow (RDS) gene: chromosome assignment and structure of the mRNA. Genomics 10: 733–739.

Travis, G.H., Sutcliffe, J.G. and Bok, D. (1991b). The retinal degeneration slow (rds) gene product is a photoreceptor disc membrane-associated glycoprotein. Neuron 6: 61–70.

Watzke, R.C., Folk, J.C. and Lang, R.M. (1982). Pattern dystrophy of the retinal pigment epithelium. Ophthalmology 1982; 89: 1400–1406.

Wroblewski, J.J., Wells, J.A., Eckstein, A., Fitzke, F., Jubb, C., Keen, T.J., Inglehearn, C., Bhattacharya, S., Arden, G.B., Jay, M. and Bird, A.C. (1994). Macular dystrophy associated with mutations at codon 172 in the human retinal degeneration slow gene. Ophthalmology 101: 12–22.

M. Fossarello, I. Zucca, M.S. Galantuomo, G.M. Contu, A. Serra
Instituto di Clinical Oculistica,
Università degli Studi di Cagliari,
Ospedale S. Giovanni di Dio,
Via Ospedale
I-09100 Cagliari, Italy

C. Bertini
Istituto di Ricerca sulle Talassemie ed Anemie Mediterranee,
Comitato Nazionale delle Ricerche (CNR),
Cagliari, Italy

M. Pirastu
Istituto di Genetica Molecolare,
Comitato Nazionale delle Ricerche (CNR),
Tramariglio (SS), Italy.

18. Spatio-chromatic VEPS in recovered optic neuritis and multiple sclerosis

M.E. SCHNECK, E. LEE, B. FORTUNE, E.W. SWITKES, M. CROGNALE and A.J. ADAMS

(Berkeley, CA, and Santa Cruz, CA, USA)

Abstract

Although colour vision defects are quite common in optic neuritis and multiple sclerosis, there is a lack of general agreement as to the nature of the colour vision defect associated with demyelinating diseases. We investigated whether there is selective involvement of particular channels in demyelinating disease by recording visually evoked cortical potentials (VEPs) elicited by grating patterns that selectively stimulate the chromatic pathways. This paper contains preliminary results on a small number of patients who have essentially recovered from or have suffered no spatial vision loss. None of these patients showed S cone pathway deficits; all deficits involved pathways with L or M cone input.

Introduction

Optic neuritis is an acute, inflammatory, demyelinating disease of the optic nerve that is usually accompanied by ocular pain. It affects women more frequently than men and may be associated with multiple sclerosis. The severity of the visual loss varies from a slight visual field deficit to a loss of light perception; the constellation of visual functions affected also varies greatly among individuals. Although acuity typically returns spontaneously to normal within months, many aspects of visual function remain abnormal. Colour vision defects are quite common in optic neuritis, even after resolution of the defects in spatial vision (e.g. Fleishman *et al.*, 1987; Harrison *et al.*, 1987; Travis and Thompson, 1989; Optic Neuritis Study Group, 1991). However, the nature of the colour defect associated with demyelinating conditions such as optic neuritis and multiple sclerosis is still not agreed upon. Many authors (e.g. François and Verriest, 1957, 1968; Griffin and Wray, 1978) have reported Type II red–green defects in optic neuritis. Others have found a preponderance of blue–yellow defects (e.g. Ohta, 1970; Fleishman *et al.*, 1987; Travis and Thompson, 1989). Non-selective colour defects have also been reported (Marré, 1970; Mullen and Plant, 1986; Dain *et al.*, 1990; Russell *et al.*, 1991).

Chromatic information carried along the optic nerve is segregated into

C.R. Cavonius (ed.), Colour Vision Deficiencies XIII, pp. 179–185.

parallel channels, loosely classified as the blue–yellow and red–green opponent channels and the achromatic channel. In as much as an individual optic nerve fibre subserves only one of these channels, one may ask whether the function of one fibre type is selectively or differentially affected in demyelinating disease. To address this question, we recorded visually evoked cortical potentials (VEPs) elicited by grating patterns that selectively stimulate the chromatic pathways. Using achromatic patterns, many investigators have shown VEPs to be sensitive indicators of visual dysfunction in optic neuritis and multiple sclerosis. We now present preliminary results on a small number of patients who have essentially either recovered or suffered no loss of spatial vision. None of the patients showed S cone pathway deficits; all deficits involved pathways with L or M cone input.

Patients and methods

The subjects were five patients with various diagnoses. Their histories and vision characteristics at the time of testing are summarized in Table 1. Three had resolved optic neuritis (patients 1, 2 and 4), one had no history of optic neuritis but had been diagnosed as having multiple sclerosis (patient 3), and one had multiple sclerosis and had two episodes of optic neuritis in each eye (patient 5). All had near-normal visual acuity and all but patient 5 had normal colour vision as assessed by the FM 100-hue test at the time of the study. This subject had a marginally abnormal square-root-of-total-error score on the FM 100-hue (red/green errors predominating). She was not tested on the desaturated D-15. Patient 3 had marginally abnormal scores on the desaturated D-15 (tritan axis), but normal FM 100-hue results.

Monocular VEPs were recorded from both eyes. Stimuli were 18° circular patches of vertical sine-wave gratings, the contrast of which then tapered to a uniform surround with mean chrominance and luminance (18 cd/m^2) equal to that of the gratings. Five axes in a cone activation space (MacLeod and Boynton, 1978; Krauskopf *et al.*, 1982) were studied: isoluminant stimuli that modulate only short wavelength-sensitive cones (S axis); isoluminant stimuli that countermodulate long wavelength and middle wavelength-sensitive cones (LM axis); stimuli (non-isoluminant) that modulate only M cones (M axis) or L cones (L axis) and achromatic stimuli (A axis). Chromatic gratings were presented for 106 msec, separated by 394 msec periods of mean luminance and chrominance. The achromatic stimulus was 3 cpd with 2 Hz pattern reversal. These stimulus parameters were found to be optimal for the chromatic and achromatic VEP mechanisms. The stimuli and recording methods are given in more detail elsewhere (Crognale *et al.*, 1993; Rabin *et al.*, 1994)

The contrast level used for each subject was the lowest level that produced reliable responses and is indicated in the figures (specified as percentage of maximum contrast available on our display). For achromatic stimuli this maximum corresponded to 100% Michelson contrast (L, M and S cone

Table 1. Characteristics of patients tested.

ID	Age	Sex	Eye	Diagnosis[a]	Duration (MS) or time since episode (ON)	Visual acuity	Adams desat D-15 CCS[b]	Square root Total errors FM 100-hue	Pelli-Robson Log CS[c]	Humphrey visual field (md[d])
1	28	F	OD	–	–	20/15	31.92*	5.3	1.6*[e]	–1.36
			OS	ON	7 months	20/25+	38.35*	6.6	1.6*	–2.96
2	28	M	OD	ON	9 years	20/20+	0	4.0	1.8	–1.44
			OS	ON (×2)	6 years, 2 months	20/20+	0	8.0	1.8	–2.08
3	57	F	OD	MS	8 years	20/15	6.49	7.2	1.9	0.88
			OS	MS	8 years	20/15	6.49	7.5	1.95	1.17
4	40	F	OD	–	–	20/15	6.49	5.3	1.95	–3.24*
			OS	ON	5 years	20/15	6.49	5.3	2.00	–4.52*
5	50	F	OD	ON, MS	2 years, 1 year	20/20–	N/A	14.6*	1.50*	N/A
			OS	ON, MS	2 years, 1 year	20/20+	N/A	12.0	1.45*	N/A

[a]MS: multiple sclerosis; ON: optic neuritis

[b]CCS: colour confusion score

[c]CS: contrast sensitivity

[d]md: mean deviation

[e]denotes abnormal result

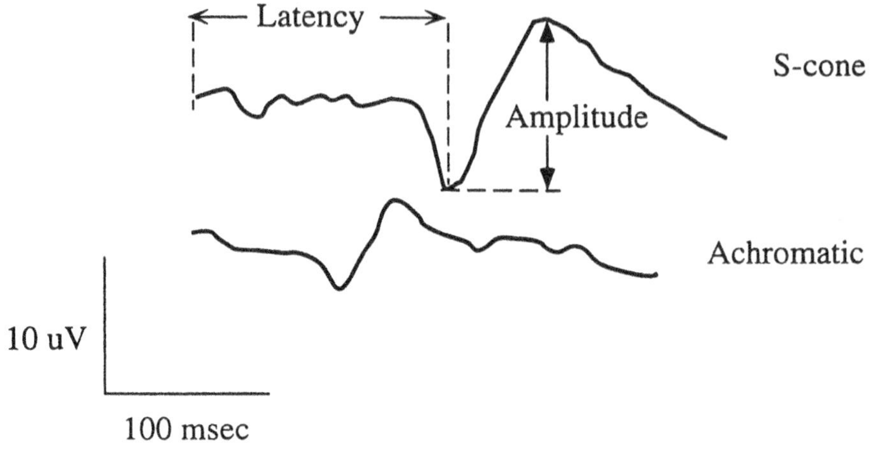

Fig. 1. VEP waveforms for S cone chromatic (top) and achromatic (bottom) modulation.

trasts = 100%). On our display, maximal individual cone contrasts were: for LM axis, L cone contrast 9.3%, M cone contrast 18.0%; for L axis, L cone contrast 25.1%; for M axis, M cone contrast 31.0%; and for S axis, S cone contrast 79.5%.

Results

The latency and amplitude of each VEP response were determined. Examples of chromatic (S axis) and achromatic VEP responses are given in Figure 1 and, as an example, the method of scoring the latency and amplitude of an S axis response is shown. In this study we use latency, rather than amplitude, to compare patients with normal individuals since we have found this to be a more reliable measure, showing better repeatability within individuals, and less intersubject variability among normal individuals.

The five panels in Figure 2 compare the individual subjects' VEP latencies for each eye with those obtained at the same contrast for each axis for a group of normal individuals of similar age. Each subject's affected and unaffected eye and test-contrast levels are indicated in each panel. The error bars around the normal means (squares) represent ± 2 s.d. In every case there was a normal S cone response. Only one subject (patient 5) showed delayed latencies along the achromatic axis; she had moderately reduced contrast sensitivity. All subjects showed abnormalities of the L cone, M cone and/or LM axes. The two subjects affected by MS (patients 3 and 5) showed abnormalities on all three of these

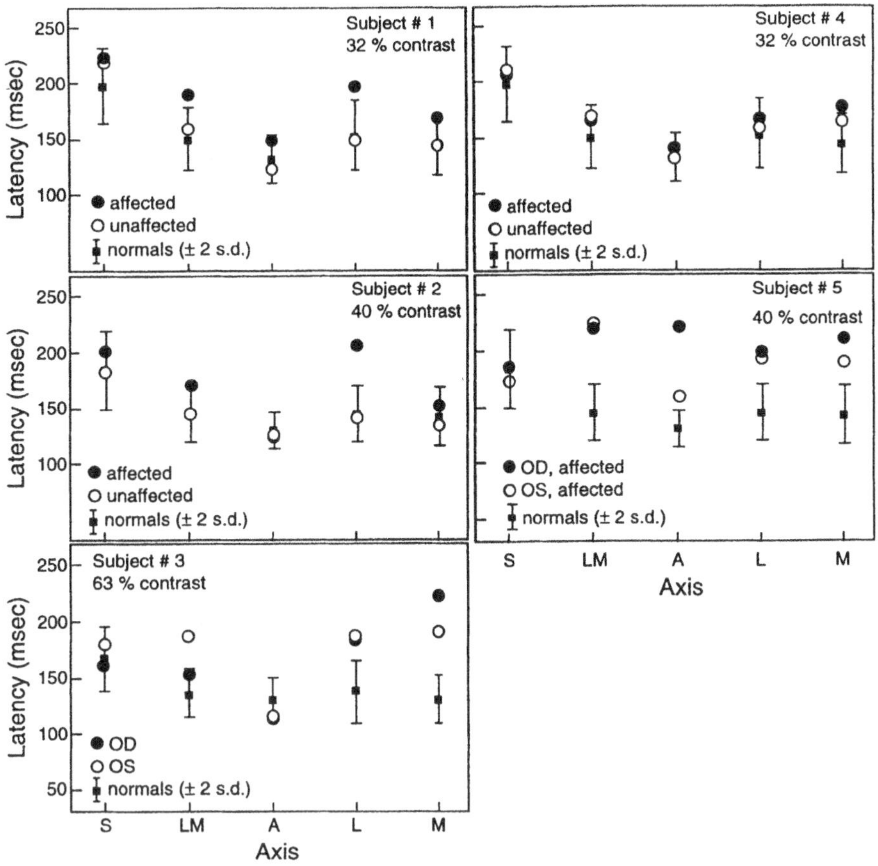

Fig. 2. Patients' VEP latencies (circles) compared with the mean and 95% confidence limits measured in normal individuals of similar age (squares). Axis of stimulus modulation is given along the abscissa. Latency (msec) is given along the ordinate.

axes in one or both eyes. The affected eyes of the other three subjects each show a different abnormality: M cone axis only (patient 4); L cone axis only (patient 2); or LM axis and L cone axis (patient 1).

Discussion

In this small group of patients who have demyelinating disease with normal or near-normal visual acuity and colour vision by standard tests, VEP latencies measured along the L, M and/or LM axes were delayed compared with those measured in normal individuals. S axis responses were normal. Achromatic VEP latencies were affected in only one individual, who had reduced contrast sensitivity. These findings suggest that in recovered optic neuritis and in multiple sclerosis with no visual symptoms, the so-called red–green opponent channel is impaired and the blue–yellow and achromatic channels are spared. These results are in substantial agreement with those of Fallowfield and Krauskopf (1984) who found that discrimination thresholds for stimuli along the LM axis are more abnormal than those along the S cone or achromatic axes in MS and recovered optic neuritis. These findings are also consistent with the report that red–green colour defects are more common than blue–yellow defects in patients with recovered optic neuritis (Schneck *et al.*, 1994), and with clinical wisdom.

Overall this study shows a loss in the M, L, or LM but not in the S pathway. The fact that this agrees with earlier investigations of colour vision in optic neuritis and multiple sclerosis may simply reflect the small sample and the fact that vision was essentially normal. Alternatively, the VEP technique used here, which isolates and individually assesses colour mechanisms, may offer a better method for assessing the selectivity of the defect of demyelinating disease.

References

Crognale, M.A., Switkes, E., Rabin, J., Schneck, M.E., Haegerstrom-Portnoy, G. and Adams A.J. (1993). Application of the spatio-chromatic visual evoked potential to detection of congenital and acquired colour-vision deficiencies. J. Opt. Soc. Am. Am. 10: 1818–1825.

Dain, S.J., Rammohan, K.W., Benes, S.C. and King-Smith P.E. (1990). Chromatic, spatial, and temporal losses of sensitivity in multiple sclerosis. Invest. Ophthalmol. Vis. Sci. 31: 548–558.

Fallowfield, L. and Krauskopf, J. (1984). Selective loss of chromatic sensitivity in demyelinating disease. Invest. Ophthalmol. Vis. Sci. 25: 771–773.

Fleishman, J.A., Beck, R.W., Linares, O.A. and Klein, J.W. (1987). Deficits in visual function after resolution of optic neuritis. Ophthalmology 94: 1029–1035.

François, J. and Verriest, G. (1957). Les dyschromatopsies acquises. Ann. Oculist (Paris) 190: 713–746.

Frañois, J. and Verriest, G. (1968). Nouvelles observations de deficincies acquises de la discrimination chromatique. Ann. Oculist (Paris) 201: 1097–1114

Griffin, J.F. and Wray, S.H. (1978). Acquired colour vision defects in retrobulbar neuritis. Am. J. Ophthalmol. 86: 193–201.

Harrison, A.C., Becker, W.J. and Stell, W.K. (1987). Colour vision abnormalities in multiple sclerosis. Can. J. Neurol. Sci. 14: 279–285.

Krauskopf, J., Williams, D.R. and Heeley, D.W. (1982). Cardinal directions of colour space. Vision Res. 22: 1123–1131.

MacLeod, D.I.A. and Boynton, R.M. (1978). Chromaticity diagram showing cone excitation by stimuli of equal luminance. J. Opt. Soc. Am. 69: 1183–1186.

Marré, M. (1970). Die Darstellung von drei Farbseh-Mechanismen bei erworbenen Farbenseh-störungen. Colour 69, 97–104.

Mullen, K.T. and Plant, G.T. (1986). Colour and luminance vision in human optic neuritis. Brain. 109: 1–13.

Ohta, Y. (1970). Studies on acquired anomalous colour vision. Colour vision anomalies in patients with lesions of the retina, optic chiasma and post-occipital centre. Colour 69: 88–96.

Optic Neuritis Study Group (1991). The clinical profile of optic neuritis. Arch. Ophthalmol. 109: 1673–1678.
Rabin, J., Switkes, E., Crognale, M., Schneck, M.E. and Adams, A.J. (1994). Visual evoked potentials in three-dimensional colour space: correlates of spatio-chromatic processing. Vision Res. 34: 2657–2671.
Russell, M.H.A., Murray, I.J., Metcalfe, R.A. and Kulikowski, J.J. (1991). The visual defect in multiple sclerosis and optic neuritis. A combined psychophysical and electrophysiological investigation. Brain 114: 2419–2435.
Schneck, M.E., Haegerstrom-Portnoy, G., Katz, B., Brabyn, J.A., Fusaro, R. and Beck, R. (1994). Color vision in optic neuritis: Findings in ONTT participants. Invest. Ophthalmol. Vis. Sci. (Suppl.) 35: 1754.
Travis, D. and Thompson, P. (1989). Spatiotemporal contrast sensitivity and colour vision in multiple sclerosis. Brain 112: 283–303.

Marilyn E. Schneck, Ellen Lee, Brad Fortune, Michael Crognale, Anthony J. Adams
School of Optometry
University of California, Berkeley
Berkeley, CA 94720, USA

Eugene Switkes
(as above) and
Department of Chemistry
University of California, Santa Cruz
Santa Cruz, CA 95064, USA

19. The effect of carbamazepine, phenytoin and valproic acid anti-epileptic convulsive drugs on visual perception

W. PAULUS, G. SCHWARZ and B.J. STEINHOFF
(Göttingen, Germany)

Abstract

We investigated the influence of the anti-epileptic drugs carbamazepine (CBZ), phenytoin (PHT) and valproic acid (VPA) on hue perception using the FM 100-hue colour arrangement test and a monitor system generating 24 different gaussian dots and by measuring transient tritanopia. All three drugs significantly altered the error score in the FM 100-hue test. Phenytoin (PHT) produced most errors in the FM 100-hue test, the largest threshold increase for perception of the gaussian dots and an elevated transient tritanopia, at least for the higher adaptation levels. Valproic acid (VPA) also elevated gaussian thresholds, especially for coloured and large stimuli, as well as affecting transient tritanopia.

Introduction

We studied the differential effects of three classic anti-epileptic drugs on colour perception using three different psychophysical methods. FM 100-hue was used as a reference for the description of a colour perception deficit (e.g. Pokorny *et al.*, 1979), gaussian-filtered dots on a monitor system assessed impairment of achromatic and chromatic visual acuity and blue test-light increment and post-adaptation threshold measurements on a yellow background were investigated. Subtracting the increment threshold from the post-adaptation threshold, a transient threshold increase of about 1 log unit occurs for the blue test spot with an exponential decline in about 1–2 sec. This paradoxical phenomenon was named transient tritanopia by Mollon and Polden (1977).

Our aim was to evaluate *in vivo* methods which allow a sensitive and specific test of transmitter or membrane channel function and which, in addition, supply qualitative and quantitative information on the effects of classic anti-epileptic drugs on vision.

C.R. Cavonius (ed.), Colour Vision Deficiencies XIII, pp. 187–191.

Patients and methods

Forty-one epileptic patients receiving treatment with one of three different drugs participated in the study after having given their informed consent according to the Helsinki Declaration. For the gaussian dots we used the computerized Moorfield Vision System to drive a 980 × 600 pixel Barco colour monitor with a frame rate of 90 Hz. Details of the method have been published elsewhere (Haug *et al.*, 1994, 1995). Thirty-five gaussian dots (the four primary colours red, green, yellow, blue and a luminance increment and decrement at four different sizes) were presented to 47 normal individuals (age 28 ± 12 years), 15 patients receiving carbamazepine (CBZ; age 32 ± 9 years), 14 patients receiving phenytoin (PHT; age 31 ± 10 years) and 12 patients receiving valproic acid (VPA; age 23 ± 4 years). We investigated four different dot sizes: 0.13°, 0.43°, 1.32° and 4.1°.

Transient tritanopia, measured by a Maxwellian-view system (Nidek, Japan), was calculated as the difference between the post-adaptation and increment thresholds of a blue test light (1 diameter, 500 msec duration) during adaptation to five different yellow background intensities between 10 and 11000 td (details in Paulus *et al.*, 1993, 1996). In order to account for an age effect, the data are presented with the age plotted on the abscissa. The statistics (F-test; two-dimensional linear regression model with age and disease, details in Paulus *et al.*, 1993) took the age effect into account as well. Fifteen patients treated with CBZ (age 35 ± 13 years), 14 patients receiving PHT (age 27 ± 8 years) and 10 treated with VPA (28.6 ± 13 years) were investigated. Seventy-five subjects of different ages served as a control (mean age 33 ± 14 years).

Results

FM 100-hue test

The total error score in all three patient groups was significantly greater than that of the normal individuals. PHT produced the largest elevation, and VPA a slightly higher score than CBZ (Table 1). No clear colour axis was found in the grand average for any of the three drugs.

Table 1. Total error score of the FM 100-hue.

	Normal individuals	Carbamezepine	Phenytoin	Valproic acid
Total error ± SD	35.6 ± 24.5	66.8 ± 48.29	162.2 ± 118	74.9 ± 51.6
F(T)0		0.0113	0.0001	0.0002
Drug effect		0.0036	0.0001	0.0001

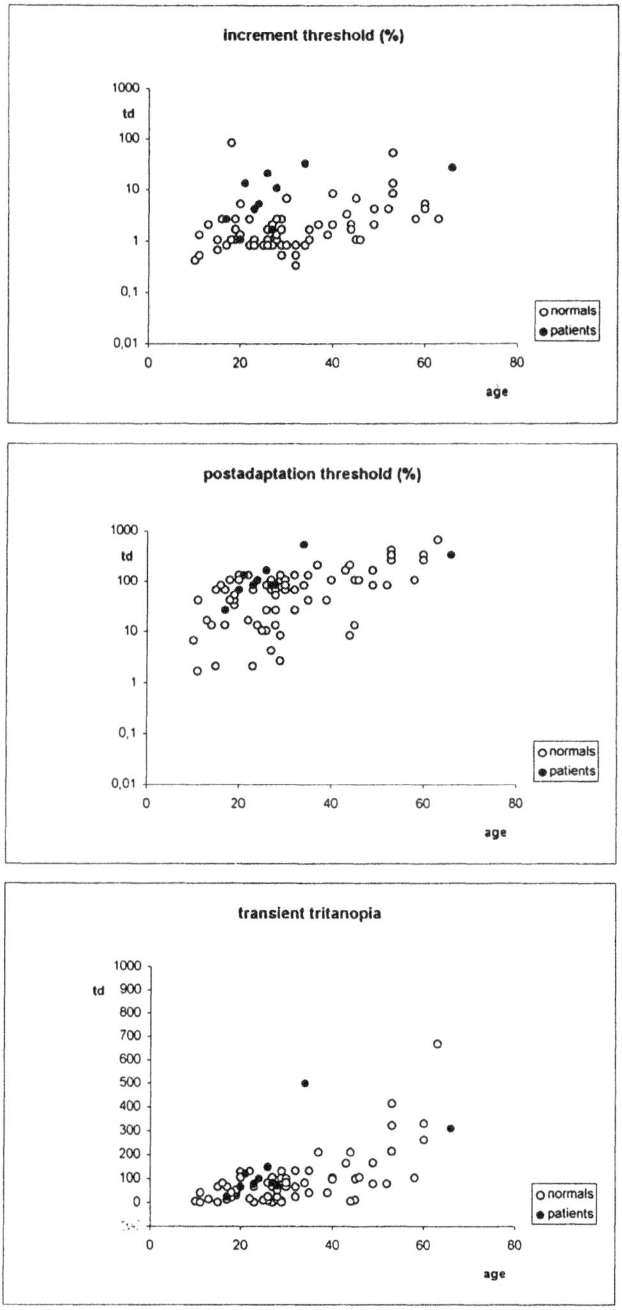

Fig. 1. Increment thresholds (top), post-adaptation thresholds (middle) and transient tritanopia (bottom) under valproic acid.

Gaussian dots

The results of patients treated with CBZ were not different from those of normal individuals in any test. PHT consistently elevated gaussian thresholds for all colours of each size tested. In the VPA group blue gaussian dots provided the greatest deviation from the controls. For blue dots the thresholds of all dot sizes were significantly increased, and for the luminance decrements thresholds were normal. For the remaining colours, and for white increments, the larger dots were detected significantly less well by patients receiving VPA therapy.

Increment threshold, post-adaptation threshold and transient tritanopia

In common with the gaussian dot test, CBZ did not elevate any of the thresholds with the Maxwellian view system. PHT did not raise the increment threshold nor, except for the highest luminance level, the post-adaptation threshold. However, except for the lowest luminance level it consistently elevated transient tritanopia thresholds. Patients receiving VPA showed a consistent increase in increment thresholds at each of the five adaptation levels without a concomitant increase in post-adaptation threshold, except for the highest background luminance level of 11 000 td (Fig. 1). Here increment thresholds (top) and post-adaptation thresholds (middle) were elevated in the patients, whereas the transient tritanopia (the difference between increment and post-adaptation threshold) was not elevated (bottom panel).

Discussion

The FM 100-hue test was the most sensitive test for colour discrimination deficits. In the grand averages of the total error scores no clear colour axis could be detected for the different drugs, which has been proposed previously in a small number group of patients in the FM 100-hue test (Bayer *et al.*, 1991). Non-specific alteration of colour perception was further confirmed by similar gaussian dot threshold elevations for the four main colours red, green, blue and yellow, within each drug group. Gaussian dot thresholds were either not affected (CBZ), were raised for all larger dots (VPA), or were always elevated (PHT). We therefore conclude that PHT impairs red–green and blue–yellow colour vision more than VPA whereas CBZ has a fairly small effect. At four out of five background luminance levels there was a significant increase of increment threshold in patients being treated with VPA. Post-adaptation thresholds were only elevated at the highest luminance level in patients receiving either VPA or PHT. When looking at the difference of their thresholds – the transient tritanopia effect – only the PHT group showed a consistent threshold elevation at the upper four levels, in contrast to the VPA group where elevations occurred at levels three and four.

At present it is difficult to relate the obvious differences in the behaviour of the three drugs to transmitter or channel effects. According to MacDonald and Kelly (1994), PHT and CBZ both act in a rather similar way on Na channels without influencing GABA Receptors or T-Ca channels. Two possibilities arise for the discrepancies found for the two drugs in this study: either further different modes of action exist or the changes found for PHT are due to chronic neurotoxicity. The discrepancies between PHT and VPA are more easy to explain since in addition to Na channel activation sodium valproate may activate GABA receptors and T-Ca channels (MacDonald and Kelly, 1994).

Acknowledgements

This study was supported by grant No. 01 KL 9001 of the German Ministry for Research and Technology (BMFT). We thank J. Baudewig for help with the statistics, R. Kolle for help with the preparation of the figure and Dr. B. Lee for calibrating the Maxwellian view systems.

References

Bayer, A., Zrenner, E. and Paulus, W. (1991). Colour vision deficiencies induced by the anticonvulsants phenytoin and carbamazepine. In: Drum, B., Moreland, J.D. and Serra, A., (eds.), Colour Vision Deficiencies X, Doc. Ophthalmol. Proc. Ser. 54: 599–604. Kluwer, Dordrecht.

Haug, B.A., Trenkwalder, C., Arden, G.B., Oertel, W.H. and Paulus, W. (1994). Visual thresholds to low-contrast pattern displacement, colour contrast and luminance contrast stimuli in Parkinson's disease. Movement Disorders 5: 563–570.

Haug, B.A., Kolle, R., Trenkwalder, C., Oertel, W.H. and Paulus, W. (1995). Predominant affection of the blue cone pathway in Parkinson's disease. Brain 118: 71–778.

MacDonald, R.L. and Kelly, K.M. (1994). Mechanisms of action of currently prescribed abd newly developed antiepileptic drugs. Epilepsia 35 Suppl.4: S41–S50.

Mollon, J.D. and Polden, D.G. (1977). An anomaly in the response of the eye to light of short wavelengths. Phil. Trans. Roy. Soc. Lond. (B) 278: 207–240.

Paulus, W., Schwarz, G., Werner, A., Lange, H., Bayer, A., Hofschuster, M., Müller, N. and Zrenner, E. (1993). Impairment of retinal increment thresholds in Huntington's disease. Ann. Neurol. 34: 574–578.

Paulus, W., Steinhoff, B. and Schwarz, G. (1996) The influence of anti-epileptic drugs on visual perception in patients with epilepsy. Brain 119: 539–549.

Pokorny, J., Smith, V.C., Verriest, G. and Pinckers, A.J.L.G. (1979) Congenital and Acquired Color Vision Defects. Grune & Stratton, New York.

Georg-August-Universität Göttingen
Abteilung Klinische Neurophysiologie
Robert-Koch-Str. 40
D-37075 Göttingen, Germany

20. Colour vision in macular gliosis, cysts, holes and oedema: an approach to depth localization

A. PINCKERS and M.A.D. TILANUS
(Nijmegen, The Netherlands)

Abstract

We investigated retrospectively the results of routine clinical colour vision examination in patients presenting with macular gliosis ($n = 21$), macular holes ($n = 21$), aphakic cystoid macular oedema ($n = 24$), dominant cystoid macular dystrophy ($n = 25$), macular cysts ($n = 21$) or central serous chorioretinopathy ($n = 23$). Special attention was paid to the onset and the presence of anomaloscopic diminished red sensitivity, a symptom of receptor malfunction. It is possible to classify, with the help of routine clinical colour vision examination, the respective diseases according to the primary site of the lesion. This agrees with the histological data, and supports the 'depth localization/eccentrization' theory.

Introduction

The development of vitreoretinal surgery in recent years has resulted in successful treatment of diseases such as macular gliosis (preretinal fibrosis, macular puckering or Jaffe syndrome), and macular holes. With the help of colour vision tests we have been able to contribute to the depth localization of retinal diseases (Pinckers and Marré, 1983). Accurate monitoring of visual function contributes to the diagnosis and enables one to predict the outcome of surgical intervention on vision. This paper is a retrospective study on local foveal alterations: macular gliosis, holes, cysts and oedema. The aim of this study was to determine whether it is possible to achieve depth localization of macular processes with the help of routine clinical colour vision examination.

Material and methods

From the files of the colour vision department we selected comparable groups of patients with uncomplicated macular gliosis without retinal oedema ($n = 21$; mean age 59 years, range 16–82), macular holes ($n = 21$; mean age 59 years, range 29–78), aphakic cystoid macular oedema (CME; $n = 24$; mean age 59

C.R. Cavonius (ed.), Colour Vision Deficiencies XIII, pp. 193–204.

years, range 19–83), dominant cystoid macular dystrophy (DCMD; $n=25$; mean age 23 years, range 6–52), macular cysts ($n=21$; mean age 46 years, range 11–85) or central serous chorioretinopathy (CSC; $n=23$; mean age 42 years, range 32–62). Patients with congenital colour vision defects, nuclear cataract, high myopia or vascular retinal disease were excluded. The patients underwent ophthalmological examination, including measurement of refractive error, corrected visual acuity measurement (distant and near vision), slit-lamp biomicroscopy before and after pupillary dilation, indirect binocular ophthalmoscopy, applanation tonometry and fluoroscein angiography.

Colour vision examination was performed under standardized conditions: 6 Philips fluorescent tubes colour 57, colour rendering index 96, provided 1.750 lux at the level of the test table. A test battery composed of plate tests was used, the FM 100-hue test and 100-hue derivative tests (Pinckers, 1980, 1984). Examination with the Nagel-II or the Neitz OT anomaloscope was performed according to the Linksz procedure (Linksz, 1964). The mid-matching point (MMP) in our laboratory is 42 arbitrary scale units. Pseudoprotanomaly was diagnosed if, for all the anomaloscope equations, the quotient of anomaly was <1.0, and if the fall of luminance from the green end to the red end of the equation zone was at most two arbitrary scale units (Marré and Marré, 1986).

Results

The details of colour vision examination in patients with different diseases of the posterior pole of the eye are listed in Tables 1–6. The results of anomaloscope examination are shown in Table 7. For purposes of comparison, Table 7 includes anomaloscope data on CME in patients with retinitis pigmentosa, obtained in a previous study (Pinckers *et al.*, 1993).

The FM 100-hue error score in macular gliosis was stable, as was the result of anomaloscope testing, when VA falls within the range 0.8–0.3. Acquired red–green defects appeared when VA was 0.25 or less. The MMP was normal. There was no clear loss of red sensitivity, and true pseudoprotanomaly is absent (Tables 1 and 7).

Macular holes were subdivided into lamellar holes (VA 0.8–0.3) and full thickness holes (VA<0.3). In patients with lamellar holes colour vision was normal or blue–yellow defective. The FM 100-hue error score was increased moderately and anomaloscope examination was reliable. The MMP was normal, as was the equation zone. Patients with full-thickness holes frequently showed blue–yellow and red–green defects. The FM 100-hue error score was high. In about half the cases anomaloscope examination was not possible. The MMP may be shifted towards red, and the equation zone enlarged towards red (Tables 2,7).

Patients with aphakic CME showed a combined red–green and blue–yellow defect in the VA range 0.6–0.02. Anomaloscope examination was not usually possible, regardless of the VA. The MMP was within the normal range. No

Table 1. Macular gliosis ($n = 21$).

VA	TMC	HRR	Ish	SPP2	F2	D-15 NCT4	NCT8	D8/2	FM 100-hue type	CI	Anomaloscope (scale units*)
0.8	S/BY	–	2	no	–	no	–	1PD	X	108	33(42)50
0.8	BYRG	S	4	–	–	1ME	–	1D2T	X	132	40(42)49
0.8	S	S	–	no	–	2ME	–	1DT1T	RG	192	–(40)47
0.8	RG	S	–	no	–	no	–	1P4BY	X	132	39(42)45
0.6	RG	S	4	–	–	no	–	3T	DT	100	38(42)46
0.6	S	S	–	–	–	no	–	2ME	–	–	35(42)47
0.5	no	–	no	–	–	no	–	1P3T	X	140	40(46)58
0.5	no	–	no	no	–	no	–	2ME	X	112	39(42)46
0.5	no	–	2	no	–	no	–	1P	no	52	39(43)47
0.4	no	–	no	–	–	2D	1ME	1T	D	152	Imp
0.4	no	–	3	–	–	no	–	2DT	X	132	38(42)47
0.4	Imp	S	–	BY	–	2DT	–	–	–	–	37(41)52
0.3	S	S	1	–	–	no	–	2T	X	80	38(42)45
0.3	BYRG	S/BY	–	no	–	di.BY	1T	–	–	–	30(34)35
0.3	S	no	no	–	–	no	–	2T	X	104	38(42)47
0.25	RG	S	8	–	–	no	–	1P2D	PD	248	39(42)60
0.25	RG	S	2	–	–	1D	–	2P1T	X	140	41(42)44
0.25	RG	–	9	BYRG	–	1DT	1P1DT	–	–	–	37(42)43
0.16	no	–	no	BY	–	3ME	–	2DT7BY	X	488	Imp
0.12	BYRG	S	–	no	–	no	–	–	–	–	–
0.02	BYRG	BYRG	–	BYRG	–	1P	1T	–	–	–	Imp

VA: visual acuity; TMC: Tokyo Medical College test; HRR: Hardy-Rand-Ritler test; Ish: Ishihara test; SPP2: Pseudoisochromatic Plates part 2; F2: Farnsworth Tritan Plate; D-15: Panel D-15; NCT4: New Color Test box 4; NCT8: New Color Test box 8; D8/2: Desaturated Panel D-15; Imp: impossible; CI: error score; No: normal; ME: minor error; S: screening errors; di: dichromatic; RG: red–green; BY: blue–yellow; X: no axis; XX: chaotic; PP: pseudoprotanomaly; –: not examined.
P, PB, D, DT, D, T: confusion direction protan, protan/deutan. deutan, deutan/tritan, tritan.
*scale units: green end of equation zone (midmatching point) red end of equation zone.

Table 2. Macular holes ($n = 21$).

VA	TMC	HRR	Ish	SPP2	F2	D-15 NCT4	NCT8	D8/2	FM 100-hue type	CI	Anomaloscope (scale units*)
0.8	–	no	1	–	–	1ME	–	2P	–	–	36(39)46
0.8	no	–	1	–	–	1ME	–	2BY	T	88	37(39)41
0.8	–	–	–	–	–	1ME	–	1BY	T	76	40(42)46
0.5	–	–	–	–	–	–	–	–	no	68	–
0.4	S	S	–	no	–	1ME	–	3BY	T	140	37(42)45
0.3	no	–	no	no	–	no	–	MET	X	100	41(42)44
0.3	–	–	3	no	–	–	–	–	X	136	42(42)51
0.3	–	no	2	–	no	2BY	–	–	X	324	–(42)–
0.3	BYRG	RG	15	–	–	di.BY	no	–	T	292	–
0.3	BYRG	RG	13	RG	–	di.RG/BY	1ME	–	XX	644	31(42)49
0.25	S	no	1	–	–	no	–	4BY	X	164	43(45)56 PP
0.2	S	no	no	–	–	no	–	2RG	P	188	Imp
0.2	S	S	5	–	–	2DT	–	1RG	PD/T	196	Imp
0.2	RG	S	7	–	–	no	–	5BY	PD	176	Imp
0.16	RG	S	–	no	–	1DT	–	XX	DT	348	43(53)61 PP
0.1	S/BY	S	no	–	–	di.RG/BY	no	–	T	200	Imp
0.06	–	no	no	–	no	4BY	–	4BY	–	–	37(49)56
0.05	RG	S	11	–	–	di.RG	–	–	X	256	0(41)73
0.03	Imp	RGBY	Imp	–	–	–	–	–	–	–	–
0.02	BYRG	RGBY	–	–	RGBY	–	1BY	–	–	–	Imp
0.02	Imp	Imp	Imp	–	–	–	–	–	–	–	Imp

Abbreviations as for Table 1.

Table 3. Irvine-Gass aphakic cystoid macular edema (CME; $n = 24$).

VA	TMC	HRR	Ish	SPP2	F2	D-15 NCT4	NCT8	D8/2	FM 100-hue type	CI	Anomaloscope (scale units*)
0.6	S	no	no	–	–	1ME	2P2BY	–	T	228	39(42)47
0.5	no	no	–	no	–	no	–	1P1T	X	96	37(43)44
0.5	Imp	Imp	–	–	–	–	–	–	–	–	Imp
0.4	RGBY	S/BY	4	–	–	1D	–	–	T	296	Imp
0.4	–	–	–	–	–	–	–	–	T	108	–
0.4	–	RG	32	–	no	1ME	–	–	P	188	–
0.3	S/BY	S/BY	5	–	–	7BY	–	–	X	420	Imp
0.3	–	no	2	RG	–	4ME	–	–	T	116	39(42)50
0.25	Imp	Imp	Imp	–	–	Imp	–	–	–	–	Imp
0.25	S/BY	RG	–	RGBY	–	1D2T	1T	–	T	152	Imp
0.25	–	Rg	7	–	RG	2ME	–	–	X	204	Imp
0.2	Imp	RGBY	–	–	–	XX	2P2T	–	–	–	38(42)43
0.2	–	no	3	–	RG	1ME	–	–	T	148	Imp
0.2	–	no	2	–	RG	2ME	–	–	X	148	Imp
0.2	RG	S	5	–	–	no	–	1T	DT	136	41(42)65
0.16	BYRG	BYRG	–	–	–	2ME	–	–	Imp	–	Imp
0.16	Imp	Rg	6	–	–	1D	–	7BY	P/T	200	Imp
0.12	–	RG	3	–	RG	2P2DT	–	–	DT	400	Imp
0.1	Imp	Imp	Imp	–	–	Imp	1D	–	Imp	–	Imp
0.08	BYRG	BYRG	6	–	–	di.BY	–	RGBY	P/T	268	40(42)52
0.08	Imp	S	7	–	–	no	–	1BY	T	208	41(43)46
0.06	Imp	Imp	Imp	–	–	2ME	–	RGBY	X	264	Imp
0.04	BYRG	BYRG	–	Imp	–	Imp	RGBY	–	Imp	–	Imp
0.02	–	BYRG	17	–	–	2ME	–	2ME	T	136	Imp

Abbreviations as for Table 1.

Table 4. Dominant cystoid macular dystrophy (DCMD, $n = 25$).

VA	TMC	HRR	Ish	SPP2	F2	D-15 NCT4	NCT8	D8/2	FM 100-hue type	CI	Anomaloscope (scale units*)
0.8	no	–	3	–	–	no	–	1T	T	120	40(42)(45
0.8	S	–	1	no	–	4BY	–	1DT1T	X	108	42(46)46
0.8	–	no	no	–	–	no	–	1DT3T	X	228	37(42)47
0.8	–	no	2	–	–	1ME	–	MET	X	168	37(42)46
0.5	BYRG	BYRG	–	BY	–	no	–	1D4BY	T	216	40(45)52
0.5	–	S	–	–	–	2ME	–	3DT1T	–	–	39(42)44
0.4	BYRG	BYRG	16	–	–	1DT1T	1DT	–	DT	284	46(??)54 PP
0.4	BYRG	BYRG	–	BYRG	–	di.BY	–	–	–	–	51(55)55 PP
0.4	BYRG	BYRG	–	BYRG	–	2D	1DT	–	–	–	48(51)56 PP
0.3	–	S	S	–	BY	1DT1T	–	–	T/P	344	Imp
0.3	S/BY	BYRG	–	no	–	1ME	–	1DT4BY	T	272	42(48)48
0.3	BYRG	BYRG	–	no	–	2BY	2BY	–	–	–	47(53)56 PP
0.3	–	no	–	–	–	no	–	1ME	–	–	40(45)–
0.25	S	no	2	–	–	no	–	3RG1T	X	148	40(41)45
0.2	S	no	–	no	–	di.BY	–	6BY	–	–	43(52)56 PP
0.2	–	no	3	–	no	3ME	–	–	RG	236	35(42)49
0.2	–	no	1	–	no	no	–	–	RG	208	32(42)56
0.2	Imp	BYRG	12	–	–	di.BY	–	5RG4BY	D	308	34(54)63
0.16	Imp	BYRG	–	BYRG	–	1DT	no	–	RG	288	34(42)61
0.15	–	S	–	–	–	4ME	–	8RG2BY	–	–	41(42)50
0.1	–	S/BY	7	–	RG	3ME	–	–	T	340	Imp
0.1	Imp	BYRG	12	–	–	di.BY	1ME	–	T/PD	428	Imp
0.08	Imp	BYRG	15	–	–	di.RG/BY	1DT1T	–	T	304	Imp
0.06	–	S	–	–	–	di.BY	1DT	–	T	244	33(42)46
0.04	–	BYRG	–	–	–	1T	–	1DT3T	–	–	40(42)45

Abbreviations as for Table 1.

Table 5. Macular cysts ($n = 21$).

VA	TMC	HRR	Ish	SPP2	F2	D-15 NCT4	NCT8	D8/2	FM 100-hue type	CI	Anomaloscope (scale units*)
0.9	–	no	2	–	no	1BY	–	–	PD	188	47(50)55 PP
0.8	no	–	no	–	–	no	–	2ME	no	60	36(39)41
0.8	–	RG	–	–	RG	2BY	–	–	PD	288	Imp
0.7	no	–	no	–	–	no	–	2ME	no	68	36(41)45
0.6	–	no	2	–	no	no	–	–	T	96	38(42)50
0.5	–	–	–	RGBY	–	1P1DT	–	–	–	–	48(50)51 PP
0.5	–	no	2	–	no	no	–	–	X	88	35(41)47
0.5	–	S	–	BY	–	no	–	1P4BY	X	300	44(46)53 PP
0.4	–	–	–	RGBY	–	1PD	–	–	–	–	43(44)48 PP
0.4	–	no	no	–	RG	2ME	–	–	T	88	40(45)50
0.4	S/BY	S	–	BY	–	di.BY	5BY	–	–	–	39(42)50
0.4	–	no	–	–	no	no	–	–	X	132	35(42)48
0.3	–	no	2	–	RG	no	–	–	P	68	42(49)54
0.3	–	no	1	–	no	1D1T	–	–	DT	316	35(48)58
0.3	–	no	3	–	no	no	–	–	X	164	35(45)58
0.25	–	no	2	–	no	no	–	–	no	44	38(42)48
0.2	–	no	1	–	no	no	–	–	X	92	39(42)??
0.15	–	no	no	–	no	1ME	–	–	T	208	32(40)47
0.15	–	no	no	–	no	2ME	–	–	X	124	–(50)–
0.15	–	no	no	–	no	2ME	–	–	X	212	–(50)–
0.1	–	no	1	–	no	1ME	–	–	X	188	35(42)49

Abbreviations as for Table 1.

Table 6. Central serous chorioretinopathy (CSC, $n = 23$).

VA	TMC	HRR	Ish	SPP2	F2	D-15 NCT4	NCT8	D8/2	FM 100-hue type	CI	Anomaloscope (scale units*)
0.8	–	no	2	–	no	no	–	–	DT	136	32(42)47
0.8	no	–	2	no	–	no	–	1P	no	36	46(48)53 PP
0.8	–	no	no	–	–	no	–	no	no	76	46(47)60 PP
0.8	RG	S	no	–	–	no	–	1P	P	116	50(52)56 PP
0.8	no	–	no	no	–	2ME	–	1ME	X	108	42(44)47
0.8	no	–	no	no	–	no	–	2ME	T	84	43(45)47 PP
0.6	S	–	no	BY	–	1D	–	6BY	DT	116	48(49)53 PP
0.6	no	–	no	no	–	no	–	1P	no	32	40(45)50
0.6	no	–	–	no	–	no	–	1P3BY	X	80	46(46)50 PP
0.6	–	no	S	–	no	1ME	–	–	T	200	37(40)–
0.6	S	no	–	–	–	no	–	1DT7BY	T	148	47(49)61 PP
0.5	S/BY	S/BY	–	BY	–	di.BY	di.BY	–	–	–	42(50)53
0.5	S	S	–	–	–	no	–	1P	PD/T	244	45(49)61 PP
0.5	S	no	no	–	–	no	–	1DT	X	124	37(49)64
0.5	–	S	no	–	–	di.BY	–	1P	T	140	39(42)53
0.4	BYRG	S/BY	–	BY	–	1ME	–	1D4BY	T	124	42(48)50
0.4	–	no	–	–	no	1D3T	–	–	D	260	35(42)50
0.4	BYRG	S	3	–	–	no	–	2D4T	T	240	39(49)50
0.4	BYRG	BYRG	17	–	–	di.BY	2ME	–	XX	604	39(42)54
0.3	–	no	5	–	no	1ME	–	–	T	200	–(43)48
0.3	RG	S	4	–	–	1T	–	4T	T	224	36(42)46
0.3	BYRG	BYRG	–	BYRG	–	di.BY	–	–	Imp	–	Imp
0.2	–	S	10	–	–	2P1T	–	2RG4BY	DT	384	0(45)73

Abbreviations as for Table 1.

Table 7. Anomaloscope in macular diseases.

Mid-matching point

Disease	Eyes (*n*)	MMP <1.0	(%)	MMP >1.0	(%)
Macular gliosis	17	2	(12)	15	(88)
Macular hole	12	3	(25)	9	(75)
Irvine-Gass (CME)	7	2	(28)	5	(72)
RP with CME*	22	9	(41)	13	(59)
DCMD with CME	21	10	(48)	11	(52)
Macular cyst	20	10	(50)	10	(50)
CSC	22	16	(73)	6	(27)

Anomaloscopic diminished red sensitivity

Disease	Eyes (*n*)	Asymmetrical enlarged equation zone	(%)	Pseudo-protanomaly	(%)
Macular gliosis	17	4	(24)	0	(0)
Macular hole	12	4	(33)	2	(17)
Irvine-Gass (CME)	7	3	(43)	0	(0)
DCMD with CME	21	10	(48)	5	(24)
Macular cyst	20	12	(60)	4	(20)
CSC	22	18	(82)	8	(36)

Anomaloscopic diminished red sensitivity related to visual acuity

Disease	Eyes (*n*)	Reliable examination	VA >0.3 (%)	VA 0.2–0.3 (%)	VA <0.2 (%)
Macular gliosis	21	17	12	6	0
Macular hole	21	12	0	17	17
Irvine-Gass (CME)	24	7	0	28	14
DCMD with CME	25	21	19	28	10
Macular cyst	21	20	35	15	15
CSC	32	22	73	5	0

CME: cystoid macular oedema; DCMD: dominant cystoid macular dystrophy; CSC: central serous chorioretinopathy; RP: retinitis pigmentosa; *: Pinckers *et al.*, 1993.

patient with aphakic CME presented with true pseudoprotanomaly. The FM 100-hue error score increased if VA <0.2, but there was no clear relationship between VA and the 100-hue error score (Tables 3,7).

In DCMD a combined blue–yellow and red–green defect appeared when VA was 0.5 or less. The FM 100-hue error score increased with decreasing VA. In about half of the cases the MMP was shifted towards red. Pseudoprotanomaly was not uncommon. Diminished red sensitivity detected by anomaloscopic examination starts at a VA of 0.4. In the group with low VA (0.1 or less) anomaloscope examination became difficult and the diminished red sensitivity disappeared (Tables 4 and 7).

In patients with macular cysts colour vision was remarkably stable over the whole VA range from 0.9 to 0.1, as was the FM 100-hue error score. The MMP was shifted towards red, and the equation zone enlarged towards red. True pseudoprotanomaly was present at an early stage (Tables 5 and 7). In the majority of patients with central serous chorioretinopathy, notably in the early stages, the MMP was shifted towards red. The equation zone was asymmetrically enlarged towards red, often in the form of true pseudoprotanomaly. The FM 100-hue error score increases with decreasing VA. A combined blue–yellow and red–green colour vision defect appears when VA dropped to 0.4 or less (Tables 6 and 7).

Discussion

Using a test battery consisting of electroretinography, electro-oculography, visual evoked potential recording and colour vision testing, it is possible to achieve depth localization and to contribute to the diagnosis of retinal diseases (Pinckers, 1982). Colour vision testing is important for the depth localization of processes in the posterior pole of the eye. We propose two principle rules: (1) central fixation is associated with acquired blue–yellow defects, while eccentric fixation causes red–green defects (Pinckers and Marré, 1983); (2) a diminished red sensitivity of any kind (anomaloscope: MMP shift towards red, asymmetrical enlarged equation zone towards red, pseudoprotanomaly and scotopization) suggests cone damage (Smith *et al.*, 1978; Pokorny *et al.*, 1979; Pinckers and Marré, 1983, 1992).

Anomaloscope examination is a valuable tool in localizing the primary site of a disease. When there is diminished red sensitivity in the early stages of a disease the primary site is likely to be located near the receptor cells. If the diminished red sensitivity appears in later stages the primary site must be another level of the retina, because the cone misalignment here is a secondary symptom. Applying this hypothesis to the patient pool of the present study we should be able to classify the diseases according to the primary site, from the outer retina towards the inner retina (Table 7). The extremes are formed by CSC and macular gliosis, respectively: in CSC diminished red sensitivity is an early event, while it is neglible in macular gliosis. These findings agree with the pathogenesis of the two processes. CSC is a known cause of receptor damage. In CSC there is a serous detachment of the (outer) sensory retina due to altered pumping or barrier functions of the retinal pigment epithelium (Gass, 1967). In macular gliosis contraction of epiretinal membranes produces wrinkling of the inner limiting surface of the retina (Sidd *et al.*, 1982). Both CSC and macular gliosis result in metamorphopsia, a symptom of disturbed mutual relationship of the cones. Obviously, traction on the inner retina may interfere with this mutual relationship without causing permanent cone damage.

A comparable difference exists to a lesser degree among patients clinically diagnosed as having macular cysts or macular holes. In those with lamellar macular holes, colour vision is moderately impaired without distinct signs of receptor damage. Full-thickness holes cause a severe colour vision defect with signs of receptor damage. This natural course reflects the histopathological findings: some factors cause the prefoveal vitreous cortex to shrink and this results in anterior traction at the foveola and, later, foveolar hole formation (Gass, 1988). Once there is a full-thickness hole, subretinal fluid collects and the foveolar region becomes detached. Cone damage can be expected only in advanced stages.

The clinical diagnosis of macular cyst is based on relatively well-preserved VA, a normal Amsler grid, preservation of inner and outer layers of retinal tissue, and no transmission effect in fluorescein angiography (McDonnell *et al.*, 1982). Colour vision in macular cysts is quite different from that in lamellar holes, because there are signs of receptor damage in cysts, but not in lamellar holes. According to Gass the only space available for a centrally located cyst would be the submicroscopic internal limiting membrane and the thin layer of radiating nerve fibres (Gass, 1988). Such a process cannot cause cone damage, unless there is – already in early stages – some leakage of fluid into the outer retinal layers.

Retinal oedema is particularly prominent around the fovea and is called cystoid macular oedema. It originates from leakage through the pigment epithelium or from retinal capillaries. There is potential space between the inner segments of the receptor cells just internal to the outer plexiform layer, and at the inner portion of the bipolar cells just external to the inner plexiform layer. The relative contribution of leakage from pigment epithelium and retinal capillaries cannot be defined by fluorescein angiography since leakage from the retinal capillaries prevents identification of the pigment epithelial contribution to the oedema. Consequently, leakage through the pigment epithelium must be presumed on the basis of knowledge of the primary disorder (Bird, 1984).

In aphakic CME, anomaloscope examination is often not possible. There is some evidence that aphakic CME causes cone damage in eyes in which VA is <0.4, but not in early stages (Table 3). This corresponds with the findings in DCMD patients, in which a diminished red sensitivity, often in the form of pseudoprotanomaly, becomes manifest when VA has dropped to 0.4 or less (Tables 4 and 7). For comparison, Table 7a includes anomaloscope data on CME in retinitis pigmentosa patients from a previous study (Pinckers *et al.*,

1993), and shows that in the three groups of patients CME has a comparable effect on colour vision. The data suggest that aphakic CME, CME in retinitis pigmentosa, and dominantly inherited CME originate primarily from leakage of the retinal capillaries. This concurs with the literature: because angiographic improvement of macular oedema in retinitis pigmentosa patients during acetazolamide treatment is mainly a result of a reduced dye leakage from the retinal capillaries (Fishman *et al.*, 1989). At low VA levels anomaloscope examination in DCMD patients becomes difficult and the diminished red sensitivity disappears. This can be explained by the fact that in advanced stages of DCMD the macular area becomes atrophic.

In conclusion, colour vision examination in the early stages of a disease contributes to establishing the primary site of a process in the posterior pole of the eye. The fact that depth localization with the help of anomaloscopic determination of diminished red sensitivity agrees with the histological data supports the 'depth localization/eccentrization' theory of Pinckers and Marré (1983).

References

Bird, A. (1984). Retina edema. Introduction to the First International Cystoid Macular Edema Symposium. Surv. Ophthal. 28: 433–435.

Fishman, G.A., Gilbert, L.D., Fiscella, R.G., Kimura, A.E. and Jampol, L.M. (1989). Acetazolamide for treatment of chronic macular edema in retinitis pigmentosa. Arch. Ophthal. 107: 1445–1452.

Gass, J.D.M. (1967). Pathogenesis of disciform detachment of the neuroepithelium. II. Idiopathic central serous choroidopathy. Am. J. Ophthalmol. 63: 587–615.

Gass, J.D.M. (1988). Idiopathic senile macular hole. Its early stages and pathogenesis. Arch. Ophthalmol. 106: 629–639.

Linksz, A. (1964). An Essay on Color Vision. Grune & Stratton, New York.

Marré, M. and Marré, E. (1986). Erworbene Störungen des Farbensehens. Diagnostik. VEB Georg Thieme, Leipzig.

McDonnell, P.J., Fine, S.L. and Hillie, A.I. (1982). Clinical features of idiopathic macular cysts and holes. Am. J. Ophthalmol. 93: 777–786.

Pinckers, A. (1980). Color vision and age. Ophthalmologica 181: 23–30.

Pinckers, A. (1982). Retinal dystrophies, a functional classification. Folia Ophthal. 7: 105–111.

Pinckers, A. (1984). Clinical color vision examination. In: Verriest, G. (ed.) Colour Vision Deficiencies VII, Doc. Ophthal. Proc. Ser. 39: 171–179. Junk, The Hague.

Pinckers, A. and Marré, M. (1983). Basic phenomena in acquired colour vision deficiency. Doc. Ophthalmol. 55: 251–271.

Pinckers, A. and Marré, M. (1992). Scotopization and pseudoprotanomaly in blue–yellow colour vision defects. Acta Ophthalmol. 70: 593–596.

Pinckers, A., van Aarem, A. and Keunen, J.E.E. (1993). Colour vision in retinitis pigmentosa. Influence of cystoid macular edema. Int. Ophthalmol. 17: 143–146.

Pokorny, J., Smith, V.C. and Johnston, P.B. (1979). Photoreceptor misalignment accompanying a fibrous scar. Arch. Ophthalmol. 97: 867–869.

Sidd, R.J., Fine, S.L., Owens, S.L. and Patz, A. (1982). Idiopathic preretinal gliosis. Am. J. Ophthalmol. 94: 44–48.

Smith, V.C., Pokorny, J., Ernest, J.T. and Starr J.T. (1978). Visual function in acute posterior multifocal pigment epitheliopathy. Am. J. Ophthalmol. 85: 192–199.

Department of Ophthalmology
University Hospital Nijmegen
PO Box 9101
6500 HB Nijmegen, The Netherlands

21. Juvenile diabetics and the colour vision meter

A. KURTENBACH, U. SCHIEFER, E. ZRENNER and A. NEU

(Tübingen, Germany)

Abstract

We used the colour vision meter, a computer-controlled anomaloscope, to measure pre-retinopathy changes in 17 juvenile patients suffering from diabetes mellitus (Type 1). We find an increased match range (MR) in the Moreland equation compared with an age-matched control group, as well as a smaller increase in the MR for the Rayleigh match. The match midpoint (MMP) for both equations was within normal limits. Eight of the patients were retested after one year and eight patients after 2 years (six patients were common to all 3 years). The results showed on average further increases in the MR for the Moreland equation while the MMP remained unchanged. FM-100 results, on the other hand, showed error scores within the normal limit, which decreased over the 3-year period. The results suggest that in the diabetic eye a decrease in chromatic opponency occurs before a loss of sensitivity at the receptor level.

Introduction

The colour vision meter (CVM) is a computer-controlled electronic anomaloscope that allows determination of two metameric equations, the red–green Rayleigh match and the blue–green Moreland match, using a standardized procedure (Pelizzone *et al.*, 1989, 1991). The aim of this study was to test the sensitivity of the CVM in detecting possible changes in the colour vision of 17 juvenile patients suffering from diabetes mellitus (Type 1), with no, or very mild, retinopathy. The results were compared with those of an aged-matched control group and with the population mean given by the manufacturer.

Subjects and methods

Subjects

Seventeen insulin-dependent diabetics, aged between 11 and 22 years (mean 16.7 ± 2.9 SD) at the beginning of the study, were tested. They had suffered diabetes for between 1.8 and 14.7 (7.74 ± 4.1) years. Fifteen patients showed no

C.R. Cavonius (ed.), Colour Vision Deficiencies XIII, pp. 205–210.

retinopathy and two patients very mild retinopathy (no more than 6 microaneurysms), as evidenced from ophthalmoscopy and colour fundus photographs. There were no significant fundus changes over the period tested.

Visual acuity and intraocular pressure were normal. In addition, blood and urine had been analysed on average every 10 weeks since detection of the disease. Urine was glucose-free and all patients showed normal cholesterin, β2-microglobulin and serum creatine values. Glycosylated haemoglobin (HbA_{1c}) ranged betweeen 5.5% and 8.2% (mean 6.9±0.9%). The results of the diabetic group were compared with those of an age-matched control group of 17 non-diabetic subjects (11–24 years: mean 18.4±4.9) with normal colour vision as tested by the Farnsworth-Munsell 100-hue (FM-100) or 28-hue tests.

Apparatus

The optical system of the CVM comprises 4 independent channels. Interference filters (10 nm bandwidth) produce the primaries, which are combined in pairs to produce two metameric matches: Rayleigh (545 nm + 670 nm = 589 nm) and Moreland (436 nm + 490 nm = 480 nm). The stimulus, a circular bipartite field of 2° diameter, is presented to one eye for 2 sec. A large white field maintains the adaptation level between presentations. The computer algorithm varies the ratio of the two primaries on the left half of the bipartite field and the luminance of the reference colour on the right.

Procedure

Before testing, an ophthalmological examination (visual acuity, slit-lamp ophthalmoscopy) was performed. Fundus examination (indirect and direct ophthalmoscopy) and intraocular pressure measurements were carried out afterwards. Patients were first tested on the FM-100 and then on the CVM.

Results

Figure 1 shows the results for both Rayleigh and Moreland matches for the control and diabetic groups. On the left is the match range (MR), given as a 95% confidence range of the mean. On the right, the match midpoint is shown as mean ±1 SD. The MR of the juvenile diabetic group is larger than the control group for both matches: 8 of the 17 patients showed a significant increase in the Moreland MR, whereas only 2 were marginally significant for the Rayleigh MR. The MMPs, on the other hand, show no difference between control and diabetic groups.

The CVM is provided with population means for these parameters, based on about 50 subjects. The population MR for the Rayleigh match lies at 4, considerably lower than the control value of 8.5 found in this study. For the Moreland match the population MR is 12: our control group gave a value of

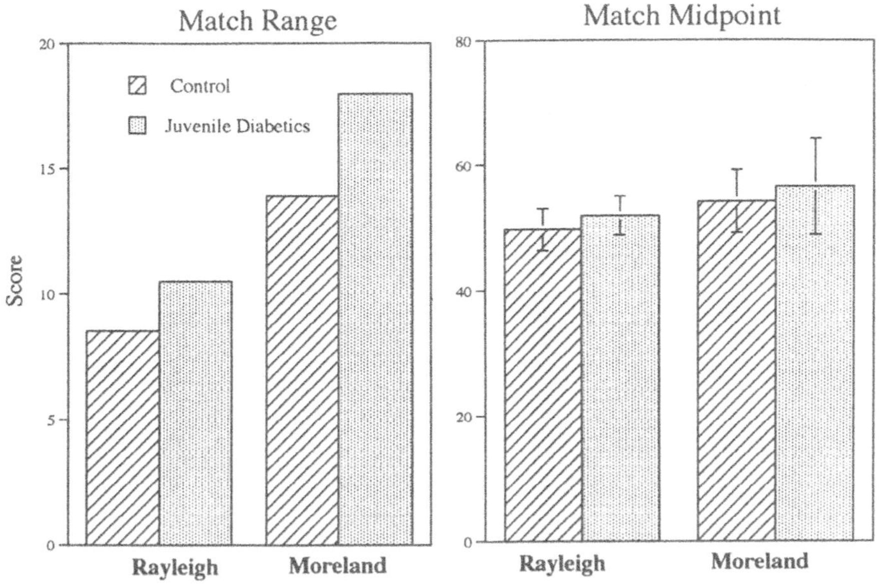

Fig. 1. Comparison of match ranges and match midpoints obtained from 17 juvenile diabetics with those of an age-matched control group of 17 subjects. On the left is the MR, shown as the 95th percentile of the mean. This convention is used by the manufacturers of the colour vision meter, and we have retained it in order to compare our results with theirs, and with those of other clinics; the MMP is shown as the mean ± 1 SD. While the MMPs for the diabetic group lie within the control range, the MRs for both Rayleigh and Moreland matches are raised.

13.1. The population MMP for the Rayleigh match is given as 50 ± 3.5 SD, for the Moreland match the mean is 53 ± 5.0 SD. Both control and diabetic groups lie within these ranges for this parameter.

The results thus show that in the diabetic eye one of the first detectable changes in colour vision is an enlarged MR. This was confirmed by testing 8 of the 17 subjects 1 and 2 years after the first examination. The results are shown in Figure 2. On the left is again shown the MR for both equations, and on the right, the MMP, from 17 subjects for year 1 and eight patients for years 2 and 3. Both the Rayleigh and Moreland MRs were larger than those of the control group (see Fig. 1) for all 3 years. However, the Moreland MR showed a consistent increase over the years. Again, the mean MMPs all lie within the control or population ranges.

Six of these patients were common to all 3 years and if the results for only those patients are examined, the results are similar. Therefore the size of the Moreland range increases with time. In addition, there was a positive correla-

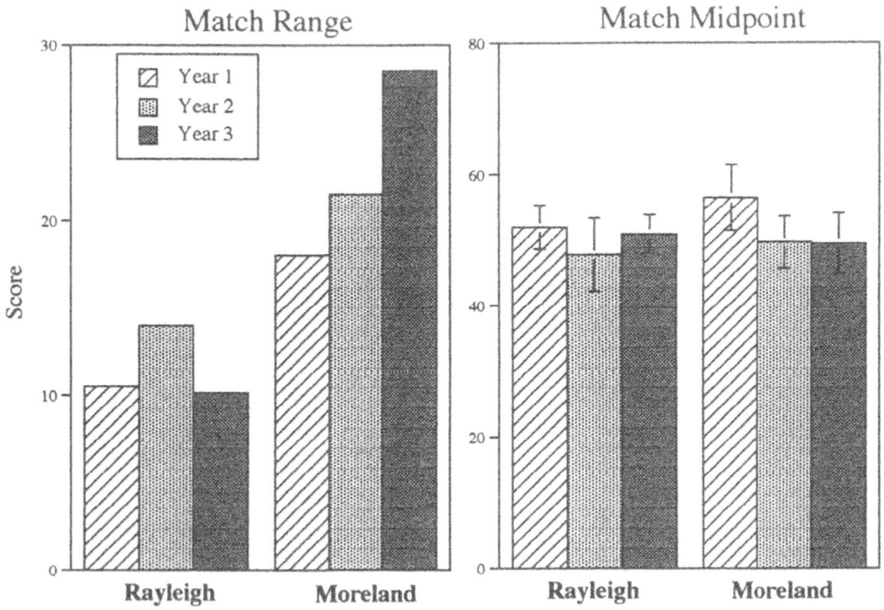

Fig. 2. MR and MMPs obtained after 1 and 2 years for eight of the 17 patients tested in year 1 (6 common to all years). The Rayleigh MR increased in year 2 and decreased in year 3, whereas the Moreland MR showed a continual increase over the 3-year period. MMPs remained within the control range.

tion ($p = 0.002$) between the two MRs for the total pooled data from the 3 years: i.e. an enlarged Moreland MR tends to be accompanied by an enlarged Rayleigh MR.

The results for the FM 100-hue test show error scores below the normal limits. The mean score decreased from 52.29 for the 17 patients in year 1 to 48 for eight patients in year 2 and to 40 for the eight patients in year 3.

Two important factors for the development of retinopathy are the duration of diabetes and blood glucose levels. The relationship between the MR and these two parameters for the pooled data from the 3 years can be seen in Figure 3. The upper panel (Fig. 3a) shows the results for the Rayleigh MR and the lower (Fig. 3b) the Moreland MR. The data have been fitted with linear regression lines. Duration (right) is positively correlated with both Rayleigh and Moreland MRs, although not significantly. Similarly, the rise in glycosylated haemoglobin (HbA_{1c}) values with Moreland MR is not significant.

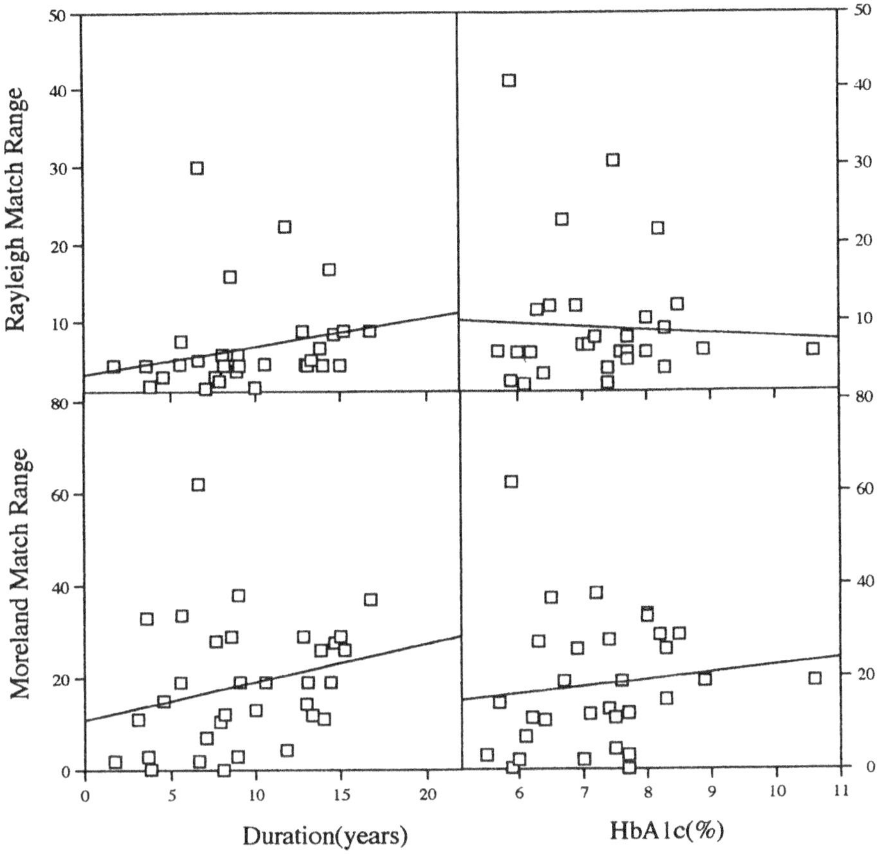

Fig. 3. (a) Rayleigh MRs plotted as a function of HbA_{1c} values (right) and duration of diabetes (left) for the pooled data obtained over the 3 years. Linear regression lines have been fitted to the data. (b) Moreland MRs plotted as in (a).

Discussion

Our results suggest that an increase in the match range, especially of the Moreland equation, occurs in the eyes of juvenile diabetics without retinopathy (Fig. 1). Moreover, the MR becomes larger when retested after 1 and 2 years (Fig. 2), although there are no visible changes in the fundus. In contrast, the match midpoint remains within the normal limits for both metameric matches. The MMP is affected by preretinal structural changes (e.g. lens transmission) as well as the sensitivity of cone receptors, whereas the MR is a measure of wavelength discrimination ability and thus processing at the opponent level. Our data on the CVM therefore suggest that in juvenile diabetics there is an early decrease in chromatic opponency, with no detectable change in lens opacity or sensitivity of the receptors. The results support those of Kinnear *et al.* (1972) and Lakowski *et al.* (1972/73) using the Pickford-Nicolson anomaloscope.

We found a difference between our control group and the population mean provided by the manufacturer, which may reflect the age of our patients. Lakowski *et al.* (1972/73) have shown that the MR on the Pickford-Nicolson anomaloscope decreases on average with age up to 20 years for a green–blue metameric match and up to 25 for the Rayleigh.

The FM-100 error scores of the diabetic group are within the normal limit for this age group. The error scores decrease with retesting. This may be due to a learning effect, which does not occur with the CVM.

Two important factors influencing the development of a retinopathy are duration of diabetes and blood glucose level. The scatter of data points in our results is too large to produce a significant correlation between the MR scores and these parameters, although for the Moreland match both tend to increase with increasing duration of disease (Fig. 3).

References

Kinnear, P. R., Aspinall, P. A. and Lakowski, R. (1972). The diabetic eye and colour vision. Trans. Ophthalmol. Soc. UK 92: 69–78.

Lakowski, R., Aspinall, P. A. and Kinnear. P.R. (1972/73). Association between colour vision losses and diabetes mellitus. Ophth. Res. 4: 145–159.

Pelizzone, M., Roth, A. and Hermes, D. (1989). First trials on an automated examination of *colour* vision. In: Drum, B. and Verriest, G. (eds), Colour Vision Deficiencies IX, Doc. Ophthalmol. Proc. Ser. 52: 151–159. Kluwer, Dordrecht.

Pelizzone, M., Sommerhalder, J., Roth, A. and Hermes, D. (1991). Automated Rayleigh and Moreland matches on a computer-controlled anomaloscope. In: Drum, B., Moreland, J. D. and#38; Serra, A. (eds), Colour Vision Deficiencies X, Doc. Ophthalmol. Proc. Ser. 54: 151–159. Kluwer, Dordrecht.

A. Kurtenbach, U. Schiefer & E. Zrenner
Universitäts-Augenklinik
Abteilung für Pathophysiologie des Sehens und Neuroophthalmologie
Schleichstrasse 12–16
D-72076 Tübingen, Germany

A. Neu
Universitäts-Kinderklinik
Rümelinstrasse 23
D-72076 Tübingen, Germany

22. Chromatic discrimination in subjects with both congenital and acquired colour vision deficiencies

J.L. BARBUR, V.A. COLE and G.T. PLANT
(London, UK)

Abstract

We have isolated the processing of chromatic signals using spatiotemporal background perturbation techniques. Two stimulus configurations were designed and chromatic discrimination thresholds were measured in a subject with both congenital and acquired colour vision deficiency and in subjects with optic neuritis. In the first test, the subject's task is the detection of vertical bars that are presented suddenly in the visual field and can be defined only by means of chromatic signals. The second test involves the detection of colour changes in a large object that is already defined by luminance contrast. The results show that although no great differences exist between these two thresholds in normal trichromats or in dichromats, pattern thresholds, in particular, can be affected significantly in subjects with diseases of the retina and/or the optic nerve.

Introduction

Some 80% of ganglion cells in primates carry and confound information arising from either intensity and/or wavelength changes in the retinal image (Perry *et al.*, 1984). These cells project to the primary visual cortex (area V1) via the four uppermost layers of the dorsal lateral geniculate nucleus (LGN) (Merigan, 1989; Schiller *et al.*, 1990) and account for both colour vision and the discrimination of fine spatial detail (Perry *et al.*, 1984). The separation of chromatic and luminance contrast signals from the firing rate of parvocellular neurones may be achieved by combining appropriately the outputs of pairs of cone-opponent cells (Lennie and D'Zmura, 1988; De Valois and De Valois, 1993). This stage of signal processing can, at least in principle, yield chromatic signals that are independent of associated achromatic luminance contrast components in the image. Parvocellular cells that exhibit only spectral opponency with no trace of centre-surround organization have also been described and labelled type II cells (Wiesel and Hubel, 1966; Dreher *et al.*, 1976). The properties of such cells are ideal for coding chromatic signals (Rodieck, 1991), but they represent only a small percentage of parvocellular neurones (Wiesel and Hubel, 1966). If the extraction and processing of chromatic signals depends

C.R. Cavonius (ed.), Colour Vision Deficiencies XIII, pp. 211–223.

exclusively on the activity of the small number of type II cells, then visual acuity based entirely on chromatic signals would necessarily be low, possibly lower than that observed experimentally when pure chromatic signals are involved (Hilz and Cavonius, 1970; Mullen, 1985). The parallel pathway model also assumes the existence of colour-opponent, spatially co-existent, red–green and blue–yellow cells. The bistratified ganglion cell types that exhibit blue–yellow opponency and have spatially co-extensive on and off regions have been identified recently using combined anatomical and physiological experiments (Dacey, 1996). It is not clear, however, to what extent similar retinal ganglion cells can be found that show red–green opponency. One possibility is that the spatial and spectral properties of some type II, LGN cells are not determined in the retina, but arise from the processing of cone-opponent signals in the LGN. Some parvocellular neurones that are normally described as type I cells could therefore contribute to the generation of pure chromatic signals. This function would be in addition to detecting fine spatial detail defined by either luminance or chromatic contrast. Although this is an attractive hypothesis, it is difficult to test in human vision. Studies of chromatic discrimination in subjects with diseases of the retina and/or the optic nerve that result in acquired deficiencies of colour vision can often yield useful insights into the generation and different uses of chromatic signals (Plant, 1991). In this paper we report the results of chromatic discrimination studies in a subject (GW) with acquired colour vision deficiency that has been linked to the onset of multiple sclerosis (MS). The tests carried out distinguish between the detection of spatially structured patterns and the discrimination of colour changes based on chromatic signals. The results of these tests demonstrate congenital protanopia in addition to pattern-specific acquired colour vision deficiency. Similar tests have been carried out in subjects with optic neuritis (ON) and preliminary findings are also reported here. The results suggest that the use of chromatic signals to detect spatially structured patterns and to discriminate changes in the colour of objects that are defined by luminance contrast can be impaired selectively. The present findings suggest that the two different uses of chromatic signals described here may be mediated by different neural substrates. An alternative, but linked, hypothesis is that diseases of the retina can cause a reduction in the density of parvocellular neurones without significantly affecting magnocellular cells. If this were the case, achromatic sensitivity and visual acuity would not be affected significantly, but discrimination of fine, spatially structured patterns based on chromatic signals would no longer be possible. The retention of almost normal sensitivity for discrimination of colour changes when large stimuli are involved suggests the existence of a separate, low resolution, system (possibly based on type II neurones) that signals only colour changes. Alternatively, the remaining type I parvocellular neurones, although less numerous, may be sufficiently abundant to enable the formation of type II LGN cells that signal only chromatic changes, but have much reduced spatial resolution.

Subjects and methods

Methods

Most dichromats cannot discriminate certain colours because they lack a single class of cone receptor (Wright, 1946; Dartnall *et al.*, 1983), but their ability to see spatially structured objects is seldom impaired. This happens either when the object carries no luminance contrast information for that class of dichromat or when such signals have been made ineffective by a suitable masking technique (Hardy *et al.*, 1954; Reffin *et al.*, 1991). A new colour vision test that effectively masks achromatic signals using spatiotemporal background perturbation techniques (Barbur *et al.*, 1992; 1993) has been modified to investigate the different possible uses of chromatic signals (Barbur *et al.*, 1994). Two versions of the test have been designed and chromatic discrimination ellipses have been measured under conditions that isolate the use of pure chromatic signals. In the first test the subject's task is the detection of vertical bars which appear suddenly in the visual field and can only be defined by means of chromatic signals. The second test involves the detection of colour changes in a large object that is already defined by luminance contrast (Barbur *et al.*, 1994).

The subject views the monitor from a distance of 700 mm, wearing an appropriate spectacle correction. The display subtends a visual angle of $26° \times 21°$, and the test stimulus is always presented in the centre of the screen (CIE (x,y) chromaticity co-ordinates: 0.305, 0.323 and luminance 34 cd/m^2). Luminance contrast masking is provided by presenting the test stimulus embedded in an array of 15×15 achromatic checks (each check subtending a visual angle of about $0.4° \times 0.4°$, giving an array of about $6° \times 6°$). The array of checks appears before, during and after the presentation of the coloured stimulus. The luminance of each check varies randomly within a range defined by a maximum and a minimum value, but the space-averaged luminance over the array remains constant throughout. This range is specified as a percentage of the background luminance (e.g. the random luminance modulation (RLM) range = $\pm 20\%$). The coloured stimulus is formed by a subset of checks, which in addition to the ongoing RLM modulation also undergoes a chromatic displacement in a specified direction in the chromaticity diagram, away from background chromaticity. The checks that form the test pattern are chosen to give the appearance of a set of vertical coloured bars that are separated by 0.4°, against the background of continuously changing RLM. All subjects tested could detect the same vertical bars when defined by luminance contrast.

In order to determine the efficiency of the RLM technique in masking the detection of achromatic contrast changes at stimulus onset, we investigated how thresholds for detection of achromatic vertical bars formed by the same subset of checks that are used in the colour vision test are affected by the RLM amplitude. Contrast thresholds for both increments and decrements measured as a function of RLM are shown in Figure 1. Results are shown for GW and also for a young, normal trichromat. The results show that for both subjects the

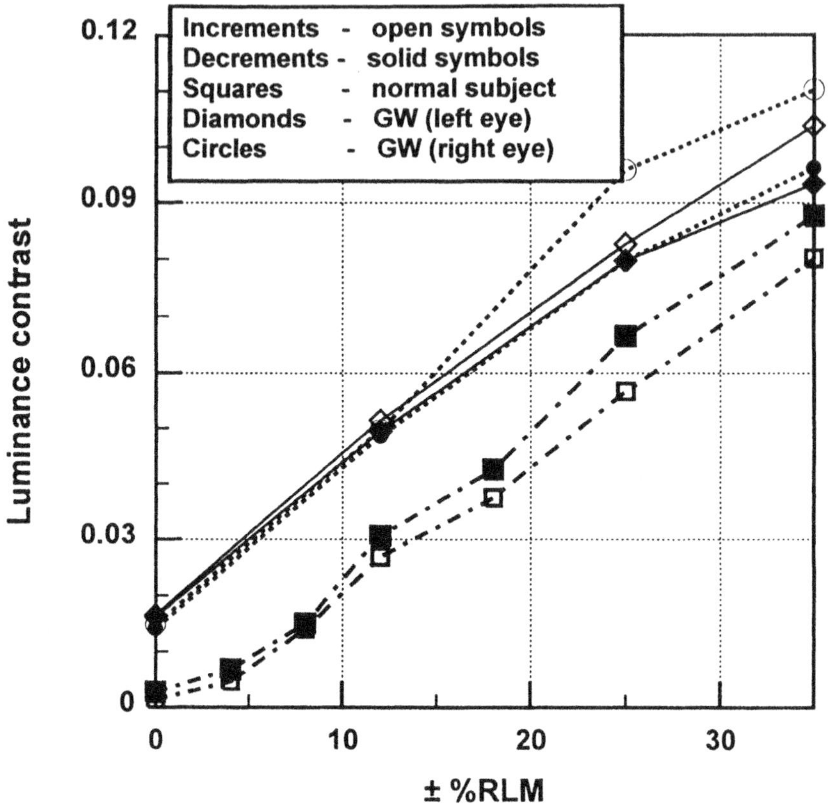

Fig. 1. Threshold contrasts for detection of achromatic vertical bars, plotted as a function of random luminance modulation (RLM) amplitude. The stimulus area consists of a 2-dimensional array of checks. Each check varied randomly in luminance every 0.05 sec within a range specified as a percentage of background luminance. The average luminance of the test pattern and that of the surrounding background field remained constant and equal to 34 cd/m^2 (x,y chromaticity co-ordinates: 0.305, 0.323). Each presentation of the checks lasted for 0.8 sec and consisted of random, spatiotemporal luminance modulation of each check. The bars were generated 0.15 sec after the onset of the RLM stimulus and last for 0.5 sec. When testing for pattern discrimination, the subject's task is to detect vertical bars formed by the introduction of a chromaticity change in some of the checks in the array. This is in addition to the ongoing random luminance modulation. In order to test the effectiveness of the RLM method in masking the detection of achromatic contrast changes, the usual generation of a chromaticity change was replaced by an increase or decrease in mean luminance that affects the same checks and results in the generation of achromatic vertical bars. The results show that both increments in luminance (white symbols) and decrements (dark symbols) yield thresholds that increase almost linearly with RLM modulation for both subjects. The larger thresholds for subject GW are consistent with increased scattered light and the overall reduction in contrast sensitivity observed in both eyes.

thresholds for detection of vertical bars increase almost linearly with RLM amplitude for both increments and decrements. The results for GW indicate somewhat higher thresholds that are caused, at least in part, by increased forward scatter in his eye. During the colour discrimination test, the subject reported after each stimulus presentation whether or not he had been aware of any bars, by pressing one of two response buttons. The chromatic displacement amplitude was then adjusted accordingly, so that thresholds for detection could be measured for each direction of chromatic displacement tested using a staircase procedure with variable step size. The ten directions of chromatic displacement tested were interleaved randomly in each experimental run. Each eye was tested monocularly, and the subject fixated a small cross that was presented continuously in the centre of the stimulus array. A second stimulus configuration was also developed to test the detection of colour changes only. The same array of 15×15 checks is used, but on this occasion the test stimulus was defined as a central block of 5×5 checks (about $2° \times 2°$). The square block is always seen clearly by the subject, since it has a higher mean luminance than the surrounding checks. The subject's task is to detect any change in the colour of the square target when the checks forming the central block all undergo the same shift in chromaticity in addition to the ongoing random luminance modulation of each check. A background of mean luminance $15\,cd/m^2$ was employed in this test, whilst the mean luminance of the central block was $34\,cd/m^2$. The same directions of chromatic displacement were investigated as in the vertical bars test. In this case, the staircase procedure yields the smallest chromaticity shifts that are needed to detect threshold changes in the colour of the test target.

Subjects

The new tests were used to investigate the colour vision of a 69-year-old male subject, GW, and several other subjects with diagnosed optic neuritis (ON). Subject GW is somewhat unusual in that, although he was unaware of it, he is also a congenital protanope. He experienced sudden problems with walking caused by poor balance and he also reported difficulties with his vision in the right eye. Results of clinical ophthalmological investigations failed to distinguish any significant differences between the two eyes. His visual acuity was only slightly worse in this eye (R: 6/9, L: 6/6) and other tests such as fundus examination, perimetry and pupillometry yielded no significant differences that could explain why the vision of his right eye was subjectively worse. He had always considered his colour vision to be 'functionally normal' and had never before been diagnosed as a protanope. Clinical colour vision tests (City University test, Ishihara, HRR and SPP) suggested protanopia, but failed to discriminate between the two eyes. VEPs to achromatic checkerboard patterns were also normal in both eyes. Several months after we completed the chromatic discrimination tests at City University, GW underwent further neuro-ophthalmological tests. MRI scans showed multiple lesions in the white matter that are normally associated with multiple sclerosis of late onset. Analysis of spinal fluid showed evidence of inflammation also compatible with MS.

Results

Increment thresholds for detection of achromatic bars that are spatially identical to those used in the colour vision test are shown in Figure 1 as a function of RLM amplitude. The results show that threshold contrasts for detection of achromatic vertical bars increased almost linearly with RLM amplitude for both GW and for a young subject with normal colour vision. Subject GW had higher thresholds in both eyes, but such an increase is not uncommon in subjects of his age and is caused, at least in part, by increased forward scatter in the eye (Barbur *et al.*, 1995).

GW had no difficulty detecting the achromatic vertical bars or the individual checks that made up the stimulus array: this is consistent with his good achromatic visual acuity in both eyes. Chromatic discrimination contours for detection of vertical bars (i.e. pattern thresholds) and for discrimination of colour changes (i.e. colour thresholds) in a normal trichromat and in a congenital protanope are shown in Figure 2. Data measured for the normal trichromat can be described as elliptical contours similar in orientation to those measured in the past (MacAdam, 1942), with little difference between pattern and colour thresholds. The isochromatic zones for the protanope extend along the corresponding colour confusion line and again show little difference between pattern and colour thresholds. Similar results have also been reported for deuteranopes (Barbur *et al.*, 1994).

In conclusion, the results suggest that corresponding pattern and colour chromatic displacement thresholds are very similar in normal trichromats and in dichromats. Figure 3 shows results obtained using the same chromatic discrimination tests for GW's left eye. The chromatic discrimination thresholds are somewhat unusual, but they reveal the typical colour discrimination loss that is normally associated with congenital protanopia. Although the pattern thresholds are significantly larger than the corresponding colour thresholds, particularly towards the blue and purple regions of the spectrum locus, the orientation of the corresponding colour confusion zone is still consistent with protanopia. Results of identical tests for GW's right eye are presented in Figure 4 and show a large difference between colour and pattern thresholds. Although GW was aware of, and reported correctly, colour changes when measuring pattern thresholds, these changes affect the whole of the stimulus array and he was quite unable to detect the presence of vertical bars. The colour thresholds, on the other hand, were very similar to those observed in congenital protanopes.

The inability of GW to resolve the structure of the stimulus when only chromatic signals are present is the first measurable difference between the two eyes, and correlates with his subjective reports of poor vision in his right eye. Some patients with optic neuritis have been tested for discrimination of vertical

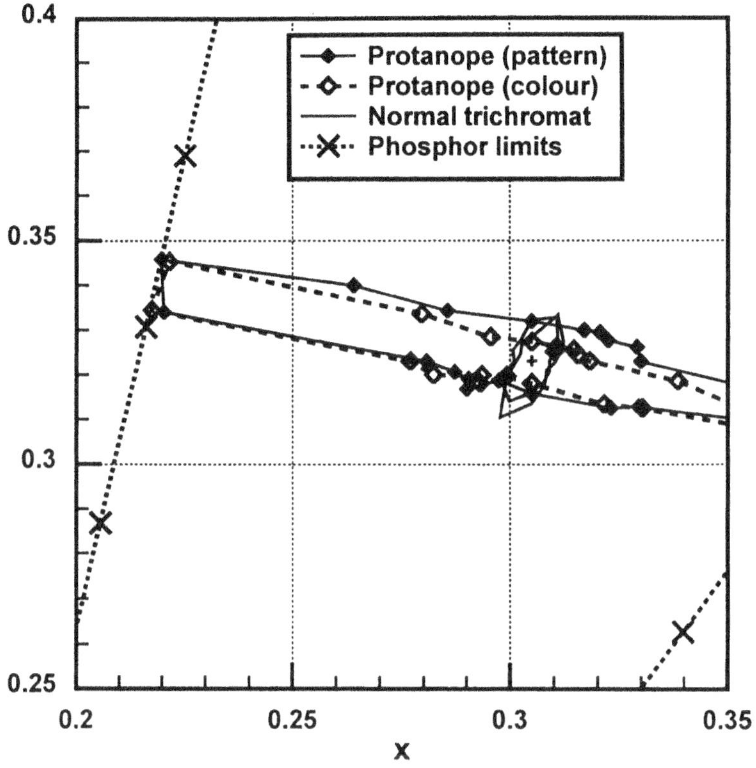

Fig. 2. Direct comparison of chromatic displacement thresholds for detection of vertical bars (i.e. pattern thresholds) or colour changes only (i.e. colour thresholds). The visual stimulus designed for measuring colour difference thresholds consisted of a square target defined by luminance contrast and presented in the centre of the randomly modulated checks that form the 2-dimensional array (Barbur *et al.*, 1994). Each check varied randomly in luminance every 0.05 sec, but the centre 5 × 5 region of the array was of higher mean luminance. This luminance difference defines a square target of 30% contrast. Since a shift in the chromaticity of the elements that form the square target can only be detected as a change in the perceived colour of the square, this has been labelled a colour threshold. The black, continuous lines that form elliptical contours at the centre of the diagram (surrounding the single cross that defines the chromaticity of the achromatic background) show data measured in a normal trichromat. Pattern thresholds (solid symbols) and colour thresholds (open symbols) are also shown for a congenital protanope. The results are consistent with those observed in deuteranopes and show no significant differences between structure and colour thresholds (Barbur *et al.*, 1994).

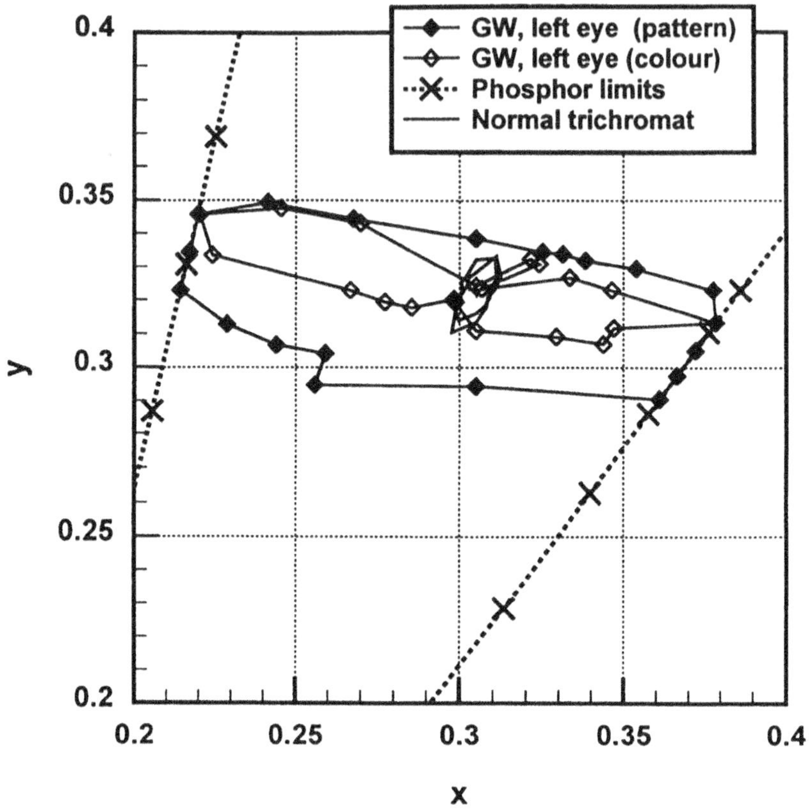

Fig. 3. Chromatic discrimination contours measured using both pattern and colour thresholds are shown for subject GW's left eye. The results show data similar to those measured in a protanope, although there are some systematic differences between colour and structure thresholds that are particularly evident for some directions of chromatic displacement.

bars using the same RLM test that we employed with GW. Data obtained for the normal and the affected eyes of one of the ON patients tested are shown in Figure 5: there was a massive loss of chromatic sensitivity for discrimination of vertical bars in the affected eye (Fig. 5B). Although the corresponding thresholds in the unaffected eye were somewhat larger, they followed the typical pattern of a normal trichromat (Fig. 5A). Several ON subjects have been tested in this way and, although the results obtained were similar in every subject, the size of the chromatic discrimination contours for pattern detection in the affected eye varied within the ON group. Although we have not measured colour thresholds in this group, the inability of these subjects to detect vertical

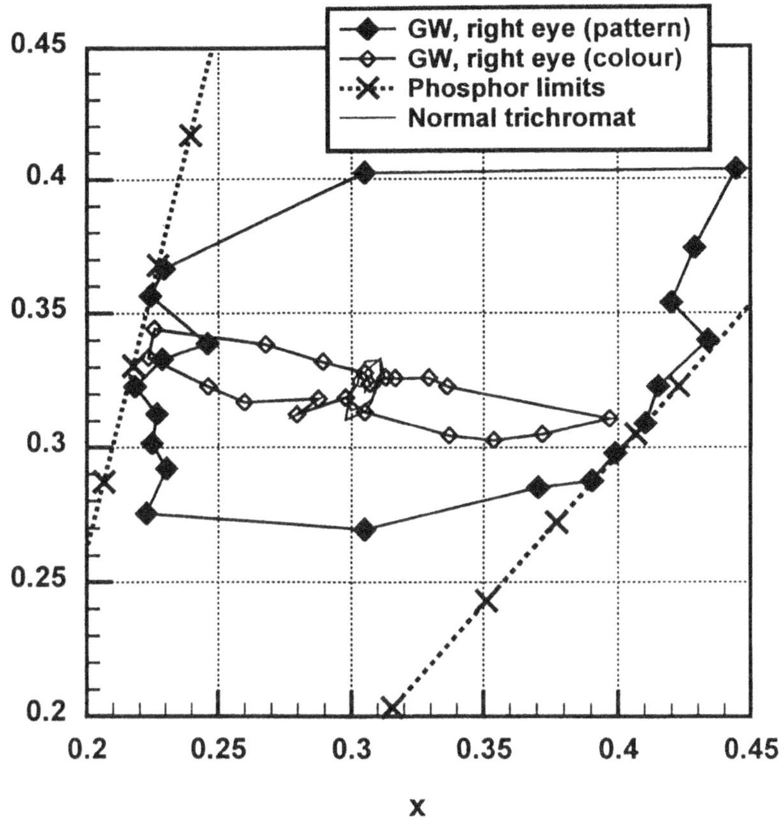

Fig. 4. Chromatic discrimination contours for GW's right eye. When the stimulus involved the detection of colour changes, the results were similar to those observed in a protanope. The chromatic discrimination contour for the pattern detection task was, however, greatly enlarged compared with that expected for a protanope. The results demonstrate a clear loss of pattern detection when only chromatic signals are involved. GW had no difficulty in detecting the same bars when defined by luminance contrast (see Fig. 1).

bars in spite of being able to discriminate correctly colour changes that affected the whole of the stimulus array, match very well both the subjective reports and the visual performance observed in GW. The generality of these findings remains to be established, but the results shown in Figure 5C suggest that loss of pattern discrimination when one can only make use of chromatic signals can be demonstrated even when thresholds for discriminating colour changes are within the normal range.

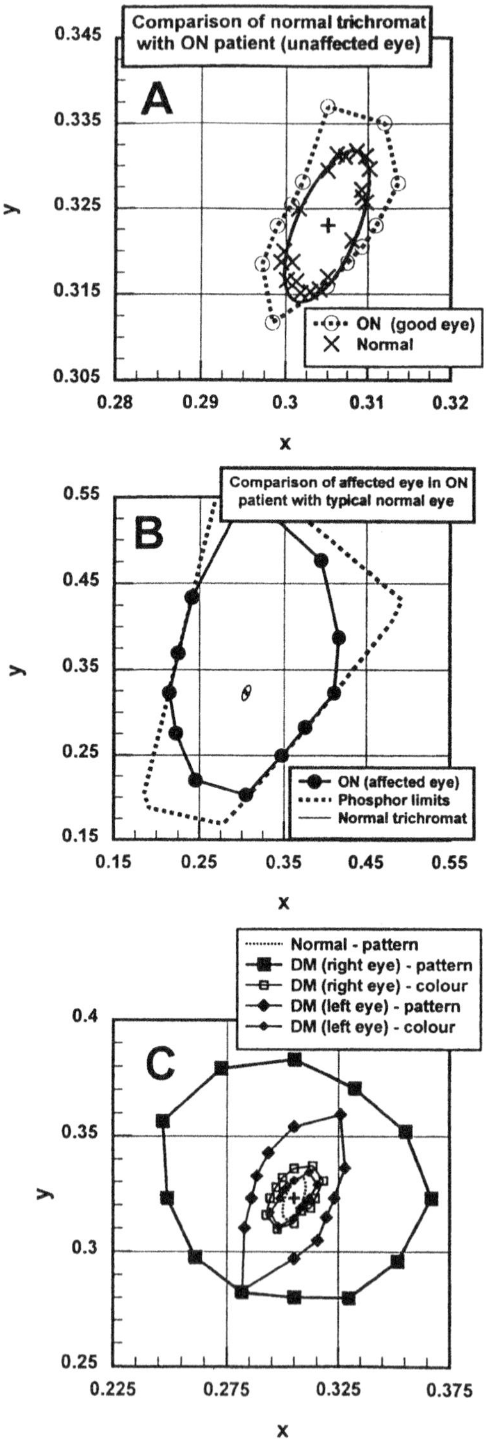
Comparison of normal trichromat with ON patient (unaffected eye)
A
0.345
0.335
0.325
0.315
0.305
y
0.28
0.29
0.3
0.31
0.32
x
ON (good eye)
Normal
Comparison of affected eye in ON patient with typical normal eye
B
0.55
0.45
0.35
0.25
0.15
0.15
0.25
0.35
0.45
0.55
ON (affected eye)
Phosphor limits
Normal trichromat
Normal - pattern
DM (right eye) - pattern
DM (right eye) - colour
DM (left eye) - pattern
DM (left eye) - colour
C
0.4
0.35
0.3
0.25
0.225
0.275
0.325
0.375

Conclusions

The chromatic discrimination tests employed in this study make it possible to isolate two different uses of chromatic signals. One is the extraction of boundaries and edges that contribute to the formation of spatially structured patterns in the absence of luminance contrast signals, and the other is the detection of colour changes when the test target is defined by luminance contrast. Although no great differences exist between these two thresholds in normal trichromats or in dichromats, pattern thresholds in particular can be affected significantly in subjects with diseases of the retina and/or the optic nerve. The results obtained in GW and the optic neuritis patients examined in our studies are consistent with a significant loss of spatial resolution when the stimulus conditions isolate the use of pure chromatic signals. No equivalent loss in the discrimination of colour changes was observed when chromatic signals were not needed to define the spatial structure of the target This was particularly evident for subject DM (Fig. 5C) who has 6/5 visual acuity in both eyes, but a massive loss of chromatic discrimination for detection of vertical bars. Thresholds for the detection of colour changes in a pattern defined by luminance contrast are only marginally affected (Fig. 5C). How can we account for these findings? The difference between GW's left and right eyes suggests an acquired abnormality of colour vision that affects the discrimination of spatial detail at an early stage in the visual pathway. The cone-opponent parvocellular pathways must somehow be affected in GW's right eye without significant changes in the magnocellular pathways. This hypothesis would account for the subject's own subjective report of difficulties with the vision in his right eye, in spite of

Fig. 5. Example of measurements of chromatic discrimination in the affected and unaffected eyes of two subjects with optic neuritis (ON). The first subject (AB) was tested 1 year after the first attack of ON. The data show chromatic discrimination thresholds for detection of vertical bars generated in an array of checks that vary randomly in luminance (Barbur *et al.*, 1994). The RLM amplitude was $\pm 20\%$ and the luminance of the background field was $34 cd/m^2$; x,y chromaticity was 0.305, 0323. (A) Data for a normal trichromat and the unaffected eye of the ON subject. The extent to which the sensitivity for discrimination of vertical bars based on chromatic signals has been reduced in the affected eye is shown in (B). Although chromatic displacements along some directions hit the phosphor limits, the subject was unable to discriminate vertical bars. The subject did, however, report correctly a change in the colour of the whole square array for each of the large chromatic displacements shown in section. Colour thresholds were not measured when this subject's chromatic discrimination was tested. (C) Comparison of thresholds for detection of vertical bars (filled symbols) and for discriminating colour changes (open symbols) in a 56-year-old subject (DM) 6 months after the first subjective observation of colour loss in the right eye. DM's Snellen visual acuity was 6/5 in both eyes. The chromatic discrimination results show decrease sensitivity for detection of vertical bars with the right eye (squares) being more affected. The subject failed most Ishihara plates with the right eye, but no specific pattern emerged. GM passed all the Ishihara plates when viewing with the left eye, in spite of reduced left eye sensitivity for detection of chromatic bars (solid diamonds).

relatively normal visual acuity. The immediate hypothesis that accounts for the sparing of colour discrimination ability when chromatic signals are not needed to convey spatial pattern information assumes the existence of different neural substrates at an early stage in the visual system for handling the coding and use of chromatic signals. Type II LGN cells are less numerous, but their large, spatially co-existent and chromatically opponent receptive fields are ideal for signalling chromatic information (Rodieck, 1991). The type I cone-opponent LGN cells are more numerous and exhibit centre-surround receptive field organization. Such cells respond well to fine spatial detail defined by either luminance contrast, chromatic boundaries or both. A decrease in the density of type I cells or a change in their centre-surround, chromatically opponent organization could account for the much reduced sensitivity for detection of chromatic borders in GW's right eye. Type I cells have identical counterparts in the retina (DeMonasterio and Gouras, 1975), but this may not be the case with all type II cells. If the colour-coding properties of some type II LGN cells reflect the processing of signals derived from cone-opponent type I cells that are presumably affected in GW's right eye, how can we account for GW's relatively normal discrimination of colour changes when large stimuli are involved? One possibility is that most type II LGN cells have counterparts in the retina and that these cells have not been affected to the same extent by the disease.

Alternatively, it is highly likely that not all cone-opponent parvocellular cells have been impaired and that the remaining type I neurones, although less numerous, can still contribute to the formation of type II LGN cells that are fewer and have reduced spatial resolution. If this hypothesis is correct then the poor spatial sampling of the remaining type I neurones can account for the selective loss of sensitivity for detection of vertical bars in GW and in subjects with optic neuritis. The findings of this investigation suggest that much remains to be learned about the coding and use of chromatic signals in human vision from studies of chromatic discrimination in subjects with retinal/optic nerve disease.

Acknowledgements

We thank our subjects for their interest and experimental participation. We acknowledge the Wellcome Trust for support with some of the equipment used in this study. This investigation was also supported by a BCO research scholarship awarded to V.A. Cole.

References

Barbur, J.L., Birch, J. and Harlow, A.J. (1992). Threshold and suprathreshold responses to chromatic stimuli using psychophysical methods. Opt. Soc. Am., Tech. Digest 1: 51–54.

Barbur, J.L., Birch, J. and Harlow, J.A. (1993). Colour vision testing using spatiotemporal luminance masking. In Drum, B. (ed.): Colour Vision Deficiencies XI: 417–426. Kluwer, Dordrecht.

Barbur, J.L., Harlow, A.J. and Plant, G.T. (1994). Insights into the different exploits of colour in the visual cortex. Proc. Roy. Soc. (Lond.) B. 258: 327–334.

Barbur, J.L., Edgar, D.F. and Woodward, E.G. (1995). Measurement of the scattering characteristics of the eye in relation to pupil size. Opt. Soc. Am., Tech. Digest Ser. 1: 250–253.

Dacey, D.M. (1996). Circuitry for colour coding in the primate retina. Proc. Natl. Acad. Sci. USA. 93: 582–588.

Dartnall, H.J.A., Bowmaker, J.K. and Mollon, J.D. (1983). Microspectrophotometry of human photoreceptors. In Mollon, J.D. and Sharpe, L.T. (eds.): Colour Vision: 69–80. Academic Press, London.

De Monasterio, F.M. and Gouras, P. (1975). Functional properties of ganglion cells of the rhesus monkey retina. J. Physiol. (Lond.) 251: 167–195.

De Valois, R.L. and De Valois, K.K. (1993). A multi-stage colour model. Vision Res. 33: 1053–1065.

Dreher, B., Fukada, Y. and Rodieck, R.W. (1976). Identification, classification and anatomical segregation of cells with X-like and Y-like properties in the lateral geniculate nucleus of old world primates. J. Physiol. (Lond.) 258: 433–452.

Hardy, L.H., Rand, C. and Rittler, M.C. (1954). H-R-R Polychromatic Plates. J Opt. Soc. Am. 44: 509–523.

Hilz, R.L. and Cavonius, C.R. (1970). Wavelength discrimination measured with square-wave gratings. J. Opt. Soc. Am. 60: 273–277.

Lennie, P. and D'Zmura, M. (1988). Mechanisms of colour vision. Crit. Rev. Neurobiol. 3: 333–400.

MacAdam, D.L. (1942). Visual sensitivities to colour differences in daylight. J. Opt. Soc. Am. 32: 247–274.

Merigan, W.H. (1989). Chromatic and achromatic vision of macaques: role of P pathway. J. Neurosci. 9: 776–783.

Mullen, K.T. (1985). The contrast sensitivity of human colour vision to red–green and blue–yellow chromatic gratings. J. Physiol. 359: 381–400.

Perry, V.H., Oehler, R. and Cowey, A. (1984). Retinal ganglion cells that project to the dorsal lateral geniculate nucleus in the macaque monkey. Neuroscience 12: 1101–1123.

Plant, G.T. (1991). Disorders of colour vision in diseases of the nervous system. Vision Visual Disorders 17: 173–198.

Reffin, J.P., Astell, S. and Mollon, J.D. (1991). Trials of a computer-controlled colour vision test that preserves the advantages of pseudoisochromatic plates. In: Drum, B., Moreland, J.D. and Serra, A. (eds.): Colour vision deficiencies X: 69–76. Kluwer, Dordrecht.

Rodieck, R.W. (1991). Which cells code for colour? In: Valberg, A. and Lee, B.B. (eds.) From Pigments to Perception: 83–93. Plenum Press, New York.

Schiller, P.H., Logothetis, N. and Charles, E. (1990). Functions of the colour-opponent and broad-band channels of the visual system. Nature 343: 68–70.

Wiesel, T.N and Hubel, D.H (1966). Spatial and chromatic interactions in the lateral geniculate body of the rhesus monkey. J. Neurophysiol. 29: 1115–1156.

Wright, W.D. (1946). Researches on Normal and Defective Colour Vision. Henry Kimpton, London.

J.L. Barbur, V.A. Cole
Applied Vision Research Centre
City University
311–321 Goswell Road
London EC1V 7DD, UK

23. Acquired tritanopia in diabetic maculopathy

J. BIRCH
(London, UK)

Abstract

Ten patients with diabetic maculopathy and VA of 6/12 or better (20 eyes) were examined with a battery of colour vision tests consisting of pseudoischromatic plates, the Farnsworth D15, the Farnsworth–Munsell 100-hue test (FM 100-hue) and a modified TNO test. Severe tritan defects were diagnosed by the Birch tritan plates in all 20 eyes and by the D15 test in 18 eyes. Acquired type 3 dichromatism (tritanopia) was confirmed by the TNO test in 17 eyes. Six patients (12 eyes) failed to interpret red–green screening plates in the Ishihara test. FM 100-hue error scores were generally very high and a tritan axis of confusion could be discerned in only 5 plots by inspection. Further analysis in terms of colour difference vectors, angle and S index, failed to identify significant polarity. It is concluded that red–green deficits occur in type 3 dichromatism, which results in poor overall hue discrimination. Other clinical tests are therefore recommended to diagnose the type of acquired colour deficiency when FM 100-hue error scores are very large.

Introduction

Acquired type 3 (tritan) colour deficiency occurs in a number of eye diseases and can be used to monitor the disease process and to assess the effects of treatment (Birch *et al.*, 1980). Classification of the defect is usually made from the results of clinical colour vision tests, such a pseudoisochromatic plates, the Farnsworth D15 test and the Farnsworth-Munsell 100-hue test. An estimate of severity is made from the cumulative test results and from error scores (Pokorny *et al.*, 1979). The TNO test measures a short-wavelength increment threshold and was designed to identify and diagnose congenital tritan defects (van Norren and Went, 1981). A modified version of the TNO test has been used previously to diagnose acquired type 3 colour deficiency in patients with proliferative diabetic retinopathy (Birch *et al.*, 1987). The aim of the present study was to establish a diagnosis of acquired colour deficiency in patients with diabetic maculopathy and to link this diagnosis to performance on clinical tests.

C.R. Cavonius (ed.), Colour Vision Deficiencies XIII, pp. 225–229.

Patients and methods

Ten diabetic patients with maculopathy and clear optic media were examined with the TNO test and a battery of clinical colour vision tests. Monocular acuities were 6/12 (0.5) or better and each eye was examined separately (20 eyes; Table 1). The modified TNO test used in this study has a 14° yellow adapting field (wavelengths greater than 560 nm) on which a rectangular blue 462 nm test field subtending 0.6° × 1.6° is superimposed (Birch *et al.*, 1987). The blue test field is modulated at 1 Hz. The luminance of the yellow adapting field is 3700 cd/m^2 and the maximum luminance of the blue test field is 1035 cd/m^2. The luminance of the test field is altered by a continuously variable neutral density wedge. Increment thresholds at 462 nm are influenced by variations in macular pigment density. The mean normal threshold for 39 normal observers (78 eyes) was found to be 2.85 ND units (range 2.2–3.6) and that of a congenital tritanope (2 eyes) to be 1.7 (range 1.6–1.8).

The clinical test battery consisted of the Ishihara plates (38-plate, 1987 edition), the City University tritan plates (Birch tritan plates), the Farnsworth D15 test and the Farnsworth-Munsell 100-hue test. These tests were illuminated by natural daylight from a high level north-facing window giving approximately 500 lux. The Birch tritan plates consist of 5 plates with vanishing designs in which the contained figure is a geometric symbol. There are three screening plates for identifying tritan colour deficiency and two plates for diagnosing severe tritan defects. FM 100-hue error scores were analysed using colour difference vectors (Vingrys *et al.*, 1992). An angle, C and S index were calculated for each monocular FM 100-hue plot.

Table 1. Diabetic patients with maculopathy.

Patient	M/F	Age	Duration of diabetes (years)	V/A RE	V/A LE
1	M	59	28	6/9	6/9
2	M	53	12	6/12	6/9
3	F	21	7	6/5	6/9
4	M	54	1	6/6	6/6
5	M	53	11	6/6	6/6
6	M	46	18	6/6	6/9
7	M	47	19	6/9	6/9
8	M	25	12	6/6	6/9
9	M	50	20	6/12	6/9
10	F	22	17	6/12	6/12

Results

The screening and diagnostic plates of the Birch tritan test were failed by all the 20 eyes examined indicating severe type 3 (tritan) colour deficiency. Only one patient (patient 3) was able to pass the Farnsworth D15 test; the remaining 18 eyes yielded tritan patterns. The results for the TNO, Ishihara and FM 100-hue tests are summarized in Table 2. The results obtained for a congenital tritanope are included for reference. Of the 20 eyes examined, three had short wavelength

Table 2. Test results.

Patient	TNO threshold	Ishihara plates	FM 100-hue error score	Angle	C index	S index
1 RE	0.8	Pass	400	69.77[b]	4.12	1.79
LE	1.4	Pass	304	71.49[b]	3.20	1.74
2 RE	1.6	Fail	344	86.34[b]	3.33	1.50
LE	1.2	Fail	312	61.89	3.23	1.61
3 RE[a]	1.6	Pass	80	45.96	1.66	1.50
LE[a]	1.2	Pass	144	46.31	1.83	1.23
4 RE	1.5	Fail	300	68.32[b]	3.28	1.88
LE	1.5	Fail	292	66.50	3.10	1.61
5 RE	1.4	Fail	316	–89.58[b]	3.52	1.95
LE	1.2	Fail	428	69.78	4.03	1.77
6 RE	0.9	Pass	676	78.29	5.18	1.28
LE	0.9	Pass	496	37.54	4.18	1.18
7 RE	0.6	Fail	736	61.46	5.84	1.24
LE	1.2	Fail	676	81.24	5.20	1.22
8 RE	2.2	Fail	412	65.91	3.73	1.35
LE	1.8	Fail	456	77.82	3.63	1.79
9 RE	1.2	Pass	460	70.00	4.37	1.79
LE	1.9	Pass	444	82.10	3.81	1.31
10 RE	1.2	Fail	680	71.79	4.87	1.26
LE	1.4	Fail	560	68.75	4.55	1.46
Congential tritanope	1.7	Pass	144	–84.65[b]	2.68	2.07

[a]Pass D15; [b]Tritan axis of confusion by inspection

thresholds between that of the congenital tritanope and the normal range, the thresholds of four eyes were similar to that of a congenital tritanope and thresholds were raised in the remaining 13 eyes. The Ishihara test was failed by 12 eyes. Failure of the Ishihara plates consisted of the inability to distinguish a numeral in vanishing designs or failure to see a numeral in transformation designs on 3 or more plates. Failure to see a 'confusion' numeral in transformation designs is characteristic of acquired colour deficiency and distinguishes acquired colour deficiency from congenital red–green deficiency. One patient (patient 3) obtained an FM 100-hue error scores less than 150; otherwise error scores were extremely high. A possible tritan axis of confusion was found by inspection in five plots with the remaining results diagrams showing loss of overall hue discrimination. The angle by colour difference vector analysis, which defined the polarity of the FM 100-hue plot was similar to that of the congenital tritanope in only one case (patient 5, RE). In the remaining four plots, perceived to be tritan, the regions of poor hue discrimination were not diagrammatically opposite in the hue circle. This lack of symmetry had a significant effect on both the angle and on the S index. The S index was not found to be reliable in identifying polar plots. The large values for the C index are a consequence of the large error scores rather than polarity.

Discussion

Results for the TNO test showed acquired type 3 (tritan) colour deficiency in all 20 eyes with diabetic maculopathy. The short wavelength increment threshold obtained for 17 eyes was similar to, or greater than, that of a congenital tritanope and it was concluded that the defect was acquired type 3 dichromatism (tritanopia). In the remaining 3 eyes the increment threshold was outside the normal range but smaller than that of a congenital tritanope, and it was concluded that the defect was acquired type 3 colour deficiency in the trichromatic stage (acquired tritanomalous trichromatism). Results for the Birch tritan plates and the Farnsworth D15 test confirmed acquired type 3 colour deficiency in all patients. The results obtained for the TNO test may be influenced by variations in macular pigment density. However, the background luminance of the modified TNO test does not entirely eliminate a small contribution from the red–green mechanism, and the raised short wavelength thresholds found in 13 of the eyes with diabetic maculopathy show a deficit in red–green vision, described by Marré (1973) as characteristic of acquired type 3 colour deficiency in the dichromatic stage. Poor red–green vision is confirmed by failure on the Ishihara plates and by the loss of overall hue discrimination, with very large error scores, on the FM 100-hue test. No direct relationship was found between the TNO threshold and the results of hue discrimination tests. For example, patient 3 was found to have raised thresholds on the TNO test but obtained good results on the D15 and FM 100-hue test while patient 8, who had only slightly raised thresholds, performed poorly on both tests.

The results of this study show that patients with diabetic maculopathy can have very poor colour vision even when visual acuity is relatively good. This deterioration in colour vision was not directly related to the age of the patient or to the duration of the disease. In most cases acquired type 3 dichromatism (tritanopia) was present. This type of colour deficiency is not identical to congenital tritanopia since red–green hue discrimination is also affected. Similar results with clinical tests alone have been obtained with diabetic patients with proliferative retinopathy, before and after laser photocoagulation, and with other groups of patients with maculopathy (Birch-Cox, 1978a, 1978b; Birch *et al.*, 1980; Ohta *et al.*, 1982; Robertson and Moreland, 1980). The FM 100-hue test is often the test of choice for assessing acquired colour deficiency but fails to diagnose the type of defect in this case even after colour difference vector analysis. The Birch tritan plates and the D15 were both found to be useful diagnostic tests and are recommended when the FM 100-hue error score is very large.

References

Birch, J., Hamilton, A.M. and Gould, E.S. (1980). Colour vision in relation to the clinical features and extent of field loss in diabetic retinopathy. In: Verriest, G. (ed.), Colour Vision Deficiencies V, 83–88. Adam Hilger, Bristol.

Birch, J., Tertuik, K. and North, F. (1987). Tritan screening with a modified TNO test. In: Verriest, G. (ed.), Colour Vision Deficiencies VIII, Doc. Ophthalmol. Proc. Series 46; 227–232. Junk, Dordrecht.

Birch-Cox, J. (1978a). A comparison between congenital and acquired tritanopia due to diabetic retinopathy. Mod. Probl. Ophthalmol. 19: 296–298.

Birch-Cox, J. (1978b). Defective colour vision in diabetic retinopathy before and after laser photocoagulation. Mod. Probl. Ophthalmol. 19: 326–329.

Marré, M. (1973). The investigation of acquired colour deficiencies. Colour 73: 99–135.

Ohta, Y., Miyamoto, T., Seki, A. and Kajigaya Y. (1982). On spectral sensitivity of the retinal receptors in central serous chorioretinopathy. In: Verriest, G. (ed.), Colour Vision Deficiencies VI, Doc. Ophthalmol. Proc. Series 33: 389–398. W. Junk, The Hague.

Pokorny, J., Smith, V.C., Verriest, G. and Pinckers, A.J.L.G. (1979). Congenital and Acquired Color Vision Defects. Grune & Stratton, New York.

Robertson, K.M. and Moreland, J.D. (1980). Study of familial retinitis pigmentosa sine pigmento combined with cystoid macular oedema. In: Verriest, G. (ed.), Colour Vision Deficiencies V, 285–298. Adam Hilger, Bristol.

van Norren, D. and Went, L.N. (1981). New test for the detection of tritan defects evaluated in two surveys. Vision Res. 21, 1303–1306.

Vingrys, A.J., Atchison D.A. and Bowman K.J. (1992). The use of colour difference vectors in diagnosing congenital colour deficiencies with the Farnsworth-Munsell 100 hue test. Ophthal. Physiol. Opt. 12: 38–35.

Applied Vision Research Centre
Department of Optometry and Visual Science
City University
London EV1V 0HB, UK

24. A protocol for monitoring treatment with synthetic anti-malarial drugs

J. LEID, J. PLANTEY and V. LEID

(Pau, France)

Abstract

We have developed a protocol for monitoring visual function during the course of treatment with synthetic anti-malarial drugs, with the goal of combining efficiency in detection of subclinical anomalies with a simple and rapid test. Such a test is indispensable for accurate monitoring, which should ideally be available in any ophthalmology practice.

We combined a colour vision test using Lanthony's desaturated D-15 with a detailed study of the paramacular visual field by automated static perimetry. For the latter, we have developed a novel perimetric test, rapid in execution, which allows exploration of extra-foveal areas between 2° and 8° eccentricity. Following evaluation of the clinical efficiency of this pair of tests, a recommended timetable for monitoring the ophthalmological consequences of treatment with synthetic anti-malarial drugs is proposed.

Introduction

Prolonged ingestion of synthetic anti-malarial drugs (SAMD), resulting in a sufficiently large cumulative dose, can cause a maculopathy accompanied by a visual field defect and a dyschromatopsia (Nozik *et al.*, 1964; Babel, 1966; Hart *et al.*, 1984).

Aetiology and pathogenesis

The substances involved are derivatives of 4-aminoquinolone (chloroquine) marketed in France as Nivaquine (chloroquine sulphate 100 mg) and Plaquénil (hydroxychloroquine sulphate 200 mg). Two conditions are treated with these drugs: Nivaquine is essentially used for the prevention of malaria, while collagen vascular diseases and inflammatory rheumatism are both treated with Plaquénil.

The pathogenesis of the resulting maculopathy can be divided into two stages: in the first stage, dilation of the choroidal vessels and pigment epithelial cell swelling result in an acquired blue–yellow axis defect; later, photoreceptor

C.R. Cavonius (ed.), Colour Vision Deficiencies XIII, pp. 231–238.

atrophy and ganglion cell injury result in a red–green axis defect. Once established, there is no effective treatment for this type of maculopathy. The cumulative effect of the doses absorbed during a patient's life is thought to underlie this toxicity.

Different authors (Carr, 1966; Nylander, 1967; François *et al.*, 1972) have suggested different estimates of a toxic cumulative dose, and it is rarely indicated whether chloroquine or hydroxychloroquine is involved. Although chloroquine is less efficient than hydroxychloroquine as an anti-inflammatory drug, the toxic side effects of the former on the retina can be much greater. A range of 100 to 500 g has been proposed by different authors as the threshold of toxicity. The oldest publications refer most frequently to choloroquine.

We are concerned only with hydroxychloroquine doses, as this is currently used almost exclusively. As a safety measure, we selected only one type of threshold for our recommended dose: this applies to both drugs since those thresholds are sufficiently low. According to Larricart (1985), 4% of treated patients developed a subclinical thesaurismosis maculopathy and 1.6% a confirmed maculopathy. Cambiaggi (1957) described the first case of retinopathy. The associated acquired dyschromatopsia was later described by Carr (1966, 1968), Grützner (1969), François *et al.* (1972), Lakowski (1972), Pinckers and ven der Eerden (1973) and Jaeger (1977).

Clinical manifestations

The maculopathy

There are three stages of progression: in the preclinical stage, both visual acuity and the optic fundus remain normal. However, early signs of primarily blue–yellow axis defects can be detected in colour vision and the visual field shows a pericentral scotoma with decrease of foveal sensitivity. The electroretinogram may reveal a few defects (Rigaudière, 1990), especially if performed with fluorescent light stimulus (Alfieri and Sole, 1969). If the patient remains on medication confirmed maculopathy then develops, characterized by a substantial loss of visual acuity, the characteristic 'bull's eye' image of the fundus (Fig. 1), severe acquired blue–yellow and red–green dyschromatopsia, a typically 2°–6° absolute pericentral annular scotoma (Hart *et al.*, 1984) and a grossly abnormal ERG with increased a-wave and depressed b-wave amplitude (Schmitt *et al.*, 1989).

In the final stage visual acuity is dramatically reduced, the fundus shows a perimacular diffuse depigmentation, colour vision exhibits dyschromatopsia with no clear axis of confusion, and both visual fields and the ERG are dramatically affected. Once established, the maculopathy is irreversible and may worsen despite cessation of therapy.

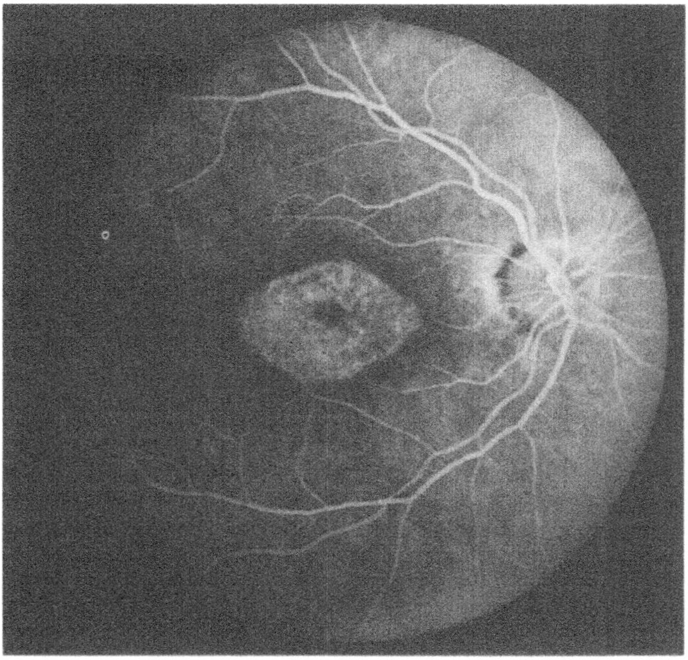

Fig. 1. Typical SAMD retinopathy.

The dyschromatopsia

Acute intoxication results in total achromatopsia. Chronologically, acquired dyschromatopsia appears early in the subclinical stage. It follows in a concomitant fashion the course of the maculopathy, worsening in a variable manner depending on the patient, and seems to be reversible during the early stage. There are various means of studying this dyschromatopsia. Grützner (1969) and François *et al.* (1972) used Ishihara and AO HRR, while more recent authors preferred SPP2. Grützner (1969) and Lakowski (1972) also tested their patients with the anomaloscope, and Zanen (1971) measured relative luminous efficiency curves. However, arrangement tests now appear to be more widely used: 15-hue (François *et al.*, 1972), 100-hue (Lakowski, 1972), desaturated D15 (Lagerlöf, 1980).

This dyschromatopsia, besides being an acquired form, typically resembles a Verriest type I red–green defect in the advanced stages (Fig. 2). It is actually rather variable, often showing a blue–yellow axis in the early stage (D15, NCT), and at times a red–green axis at an early stage, appearing as a pseudoprotanomaly (Jaeger and Krastel, 1987).

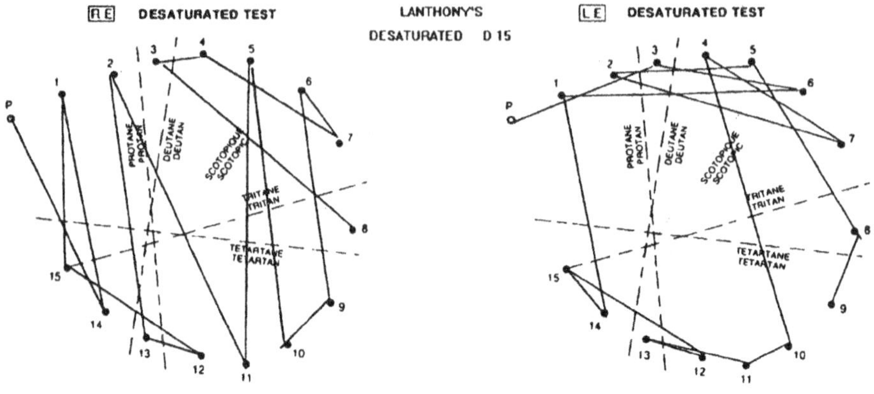

Fig. 2. Typical advanced SAMD color vision defect, Type I R–G axis.

Correlation with visual fields

Only Friedmann (1969) has claimed that threshold measures are more sensitive than colour vision tests. In the initial stage, one can simultaneously observe the onset of dyschromatopsia, a relative annular scotoma and, eventually, a loss of foveal sensitivity.

The manner in which visual fields are studied has changed from kinetic perimetry, which is more or less useless in the perimacular area, through the Friedmann visual field analyser, then profile static perimetry, and finally to automated static perimetry, although the commercial equipment does not have any ready-to-use specific programs. Schmitt (1989) proposed specifically studying the central 12°.

Our specific program

Our goal was to evaluate rapidly but carefully the area likely to sustain damage, specifically 2°–6° around the fovea. Our program was designed for the Humphrey automated perimeter, consisting of 36 points between 2° and 7°, paired with measurement of the foveal threshold (Fig. 3). The suprathreshold test requires less than 3 min/eye and the threshold test requires about 6 min/eye. We extensively evaluated the position and spacing of points that seemed to us the most logical and efficient. This established a set of normative thresholds based on subjects without ocular pathology and undergoing no treatment. These norms obviously depend on the age of the subject.

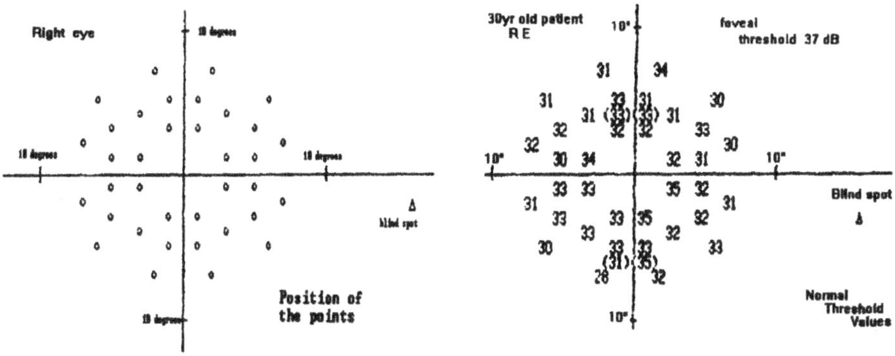

Fig. 3. Leid's test of paracentral visual field. (A) Position of the points tested. (B) Normal thresholds.

Monitoring protocol

In the absence of a consensus on this subject, we propose a protocol for monitoring treatment by SAMD which can be easily followed in any reasonably equipped ophthalmology practice. The protocol provides for rapid, reliable and non-invasive testing and aims to detect subclinical retinopathy. It combines standard ophthalmological examination with visual acuity and fundus examinations, colour vision tests that we have ourselves devised with Lanthony's desaturated D15, and an examination of the paracentral visual field as described above.

The initial examination consists of the following evaluations: visual acuity, fundus examination, Ishihara for inherited colour deficiencies, a desaturated arrangement test (D15 or New Color Test) to test for acquired anomaly antedating SAMD therapy (default SPP2), and finally a visual field with the specific SAMD test using the suprathreshold procedure.

During the 2 years following the beginning of treatment (for moderate doses: 2 tablets/day), visual acuity, fundus examination and colour vision testing should be performed every 6 months, and visual field should be examined with the specific suprathreshold test every year. After 2 years, or when a cumulative dose of 300 g has been reached, visual acuity, fundus examination, colour vision testing and specific threshold visual field should be monitored every 6 months. When the cumulative dose exceeds 500 g, the same monitoring should be performed every 4 months, but unfortunately this is usually not practical.

If an anomalous result is obtained on one of the tests, more frequent monitoring is required. If the anomaly is moderate, the battery of tests should be repeated 2 months later. If it is more severe, retesting is required even earlier, and the slightest doubt is an indication for ERG and fluorescein angiography.

Whatever the cause, if two consecutive tests show anomaly drug treatment should be stopped.

Validation of the clinical protocol

Methods

We tested 17 patients (ages 12–73 years, mean age 55), who had ingested a total of 45–1265 g chloroquine (mean 680 g). The diseases for which they were being treated were: rheumatoid arthritis (10), systemic lupus erythematosus (1), juvenile rheumatoid arthritis (2), Behçet's disease (1), pseudorheumatoid arthritis (1), Sjögren's syndrome (1) and severe allergy to sunlight (1).

There were 16 females and 1 male, which reflects the well-known prevalence of these conditions among females. Patients treated for malaria were intentionally excluded, since the absorbed doses are overall very small and thus present the least risk. For these patients we recommend the same type of monitoring, basing the frequency on the cumulative ingested dose.

Results

Thirteen patients, four of whom had ingested a cumulative dose greater than 900 g, showed normal results. These cases allowed us to establish the normality of perimacular threshold in patients with SAMD. Two patients showed a discrete anomaly of colour vision with errors along a tetartan axis and a normal visual field. We have repeated the examinations conforming to the protocol and resolved the problem with the aid of the ERG in one and angiography in the other. These were false positives on the colour vision testing. One patient seemed to have a small anomaly of the visual field. Repeat examination with angiography identified this as a false positive result. Two patients presented advanced maculopathies. They had absorbed around 600 g of chloroquine and 800 g of hydroxychloroquine, respectively, when their retinopathy began. These cases, being well documented, allow us to identify the expected pathological aspects of the visual field (Fig. 4).

Desired improvements

The specific test proposed has been implemented with respect to the standard possibilities offered on the Humphrey perimeter at our disposal. Interpretation of the threshold test is limited by the fact that the data are displayed inconveniently as the threshold value at each point tested. It would be desirable to obtain from the different perimeter manufacturers the possibility of representing threshold as grey levels and above all in the manner used for the classical tests, in order to make statistical comparisons with published numerical norms

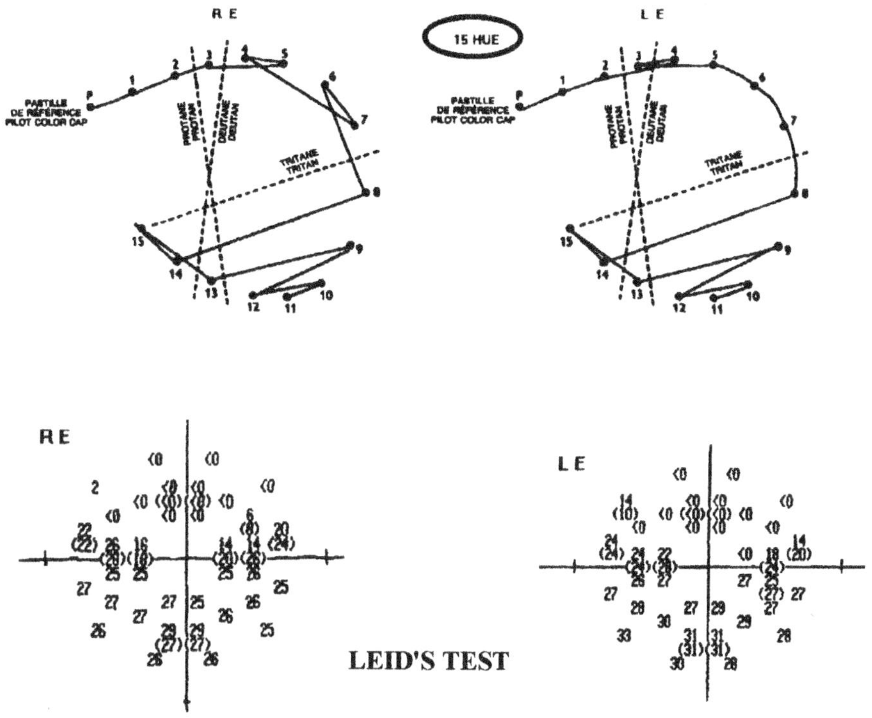

Fig. 4. Desaturated D15 results of patient shown in Fig. 1. Thresholds for the patient shown in Figure 1.

as a function of age and sex. We would be interested in collaborating with manufacturers toward this goal.

References

Alfieri, R. and Sole, P. (1969). On very short wavelength electroretinography. An original technique of diagnosis in infraclinical synthetic antimalarial retinopathies. In: François, J. (ed.), Occupational and Medicative Hazards in Ophthalmology, 661–668. Karger, Basel.

Babel, J. (1966). Rétinopathie dûe à la chloroquine. Ophthalmologica, 152: 74–79.

Cambiaggi, A. (1957). Unusual ocular lesions in a case of systemic lupus erythematous. Arch. Ophthalmol. 57: 453–451.

Carr, R.E., Gouras, P. and Gunkel, R.D. (1966). Chloroquine retinopathy. Arch. Ophthalmol. 75: 171–178.

Carr, R.E., Henkind, P., Rothfield, N. and Siegel, I. (1968). Ocular toxicity of antimalarial drugs. Am. J. Ophthalmol. 66: 738–744.

François, J., De Rouck, A., Cambie, E. and De Laey, J.J. (1972). Rétinopathie chloroquinique. Ophthalmologica 165: 81–99.

Friedmann, A. (1969). The early detection of chloroquine retinopathy with the Friedmann visual field analyser. Ophthalmol. Addit. 158: 583–591.
Grützner, P. (1969). Acquired colour vision defects secondary to retinal drug toxicity. Ophthalmologica 158: 592–604.
Hart, W.M., Burde, R.M., Hohnston, G.P. and Drews, R.C. (1984). Static perimetry in chloroquine retinopathy perifoveal patterns of visual field depression. Arch. Ophthalmol. 102: 377.
Jaeger, W. (1977). Erworbene Farbsinnstörungen als Nebenwirkungen von Medikamenten. Klin. Mbl. Augenheilkd. 170: 460.
Jaeger, W. and Krastel, H. (1987). Colour vision deficiencies caused by pharmacotherapy. In: Verriest, G. (ed.), Colour Vision Deficiencies VIII, Doc. Ophthal. Proc. Ser. 46: 39–52. Nijhoff/Junk, Dordrecht.
Lagerlöf, O. (1980). Drug-induced colour vision deficiency. In: Verriest, G. (ed.), Colour Vision Deficiencies V: 317–319. Adam Hilger, Bristol.
Lakowski, R. (1972). The P-N anomaloscope as a test for acquired dyschromatopsias. Mod. Prob. Ophthalmol. 11: 25–33.
Larricart, P. (1985). Les affections rétiniennes. In: Raspiller, A. Les effets indésirables des médicaments en ophtalmologie. Bull. Soc. Opht. Fr. Rapport Annuel: 193–211.
Nylander, U. (1967). Ocular damage in chloroquine therapy. Acta Ophthal. Suppl., 92: 1–71.
Nozik, R.A., Weinstoc, F.J. and Vignos, P.J. Jr. (1964). Ocular complications of chloroquine. A series of case presentations with a simple method for early detection of retinopathy. Am. J. Ophthalmol. 58: 774–778.
Pinckers, A. and van der Eerden, J. (1973). La polyarthrite rhumatismale et les antipaludéens de synthèse. Ann. Oculist 206: 305–309.
Rigaudière, F., Pizzato, M., Albuisson, E., Le Fargasson, J.F. and Grall, Y. (1990). Bilan statistique de 700 examens électrophysiologiques chez des patients sans signe ophtalmologique traités par antipaludéens de synthèse pour maladie rhumatologique ou dermatologique. Ophtalmologie 4: 254–259.
Schmitt, D., Parent De Curzon, H., Zamfirescu, F. and Campinchi, R. (1989). Protocole de surveillance du traitement par antipaludéens de synthèse. Ophtalmologie 3: 239.
Zanen, J. (1971). La méthode des seuils en pathologie oculaire. Conc. Ophthalmol. Acta 3: 1750–1756.

4, Place Royale
F-64000 Pau, France

25. Visual function in patients affected by multiple sclerosis: visual field and colour vision

A. SERRA, I. ZUCCA, A. TANDA, V. PIRAS, S. MULAS, C.M. DESSY and M. FOSSARELLO

(Cagliari, Italy)

Abstract

Twenty-one patients aged 20–45 years suffering from multiple sclerosis (MS), with or without optic neuritis (ON), were examined ophthalmologically. Visual field (automatic perimetry) was tested as well as accommodation. There was a significant correlation between colour vision defect, visual field and range of accommodation.

Introduction

In a previous paper (Serra *et al.*, 1990) we reported data from 16 MS patients; in particular we compared the FM 100-hue responses with the deviations from norm of the profiles recorded along the horizontal meridian on the Goldmann perimeter. Automated perimetry now allows measurement of sensitivity loss, corrected for age, for every point examined in the visual field.

We frequently observed sensitivity loss in parts of the central area of the visual field, in spite of normal visual acuity in MS patients, whether or not they had suffered an episode of optic neuritis. We investigated whether there is any correlation between colour vision defect and the areas in which there is a loss of sensitivity. We also asked whether there was an association between colour vision defects and reduced accommodation.

Subjects and methods

For this study we selected 21 patients (7 males and 14 females), aged between 20 and 45 years, suffering from MS h or without a history of ON (41 eyes because one amblyopic eye was not considered). The patients were tested for visual acuity (Snellen chart), colour vision (FM 100-hue at 100 lux C-illuminant, Lanthony D15), accommodative amplitude by means of modified version of Badal's optometer, where the subject's task consisted of recording the blur-to-sharp transition for a set of targets each 6.5 × 6.5 mm (the targets are seen through a lens and the eye of the observer is placed at the focal distance of the lens; thus, the target appears a constant size independent of the distance) and

C.R. Cavonius (ed.), Colour Vision Deficiencies XIII, pp. 239–242.

visual field (Octopus 500 EZ automatic perimeter, Interzeag, G1 program). We tested three areas in the visual field (26 central degrees, G1 program): CI, CII, CIII, as shown in Figure 1, so that each area had equal number of examined points; CI (10), subdivided into four sectors of 5 points each: C1 higher nasal, C2 higher temporal, C3 lower temporal and C4 lower nasal (Fig. 2).

We calculated the correlation between total score on the FM 100-hue test and the sum of sensitivity loss for each area of visual field for every patient. We then calculated the correlation between accommodative amplitude and total score and the sensitivity loss of each of the areas, both for the group data for all

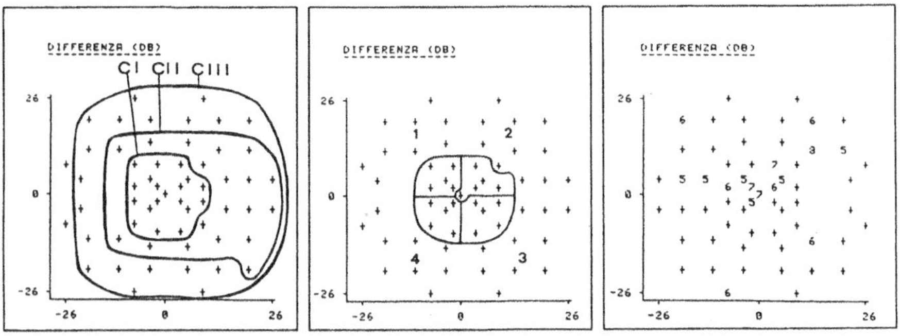

Fig. 1. Left: visual field (represented by difference in retinal sensitivity, expressed in decibels, between observed patient and a standard normal subject) divided into three areas: CI, CII, CIII, of an equal number of points each in order to compare central to peripheral areas. Middle: central area of visual field (represented by difference in retinal sensitivity, expressed in decibels, between observed patient and a standard normal subject) divided into four sectors of five points each in order to evaluate the distribution of defects. Right: a sample of visual field (Octopus 500 EZ, G1 program) of MS patient with reduced sensitivity in the central area.

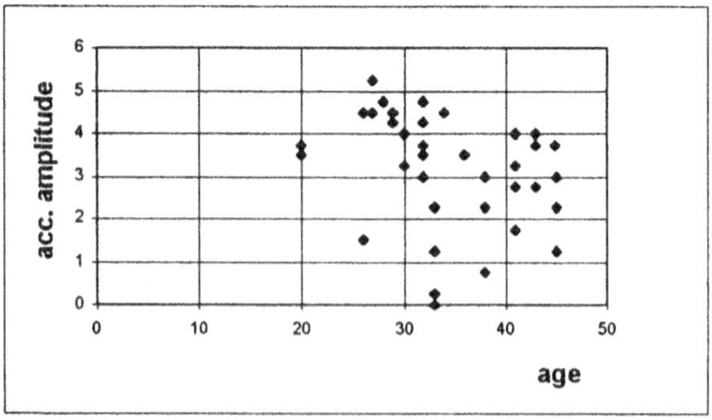

Fig. 2. Accommodation amplitude (ordinate) plotted versus age (abscissa) for all patients.

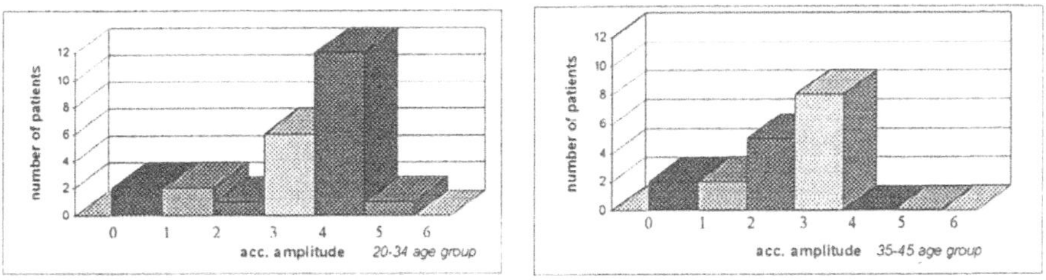

Fig. 3. Left: frequency histogram of accommodative amplitude for the 20–34 year age group. Right: frequency histogram of accommodative amplitude for the 35–45 year age group.

patients and after division of the patients into two age groups: 20–34 years and 35–45 years.

Results and discussion

All patients included in this study, after proper dioptric correction, had normal acuity (10/10) for distant vision. Colour vision was impaired a in majority of patients with or without optic neuritis, as reported previously (Pokorny *et al.*, 1975, Serra and Mascia, 1981; Serra *et al.*, 1990). Automated visual fields showed an increased mean loss in the majority of our patients, in spite of good visual acuity; in particular we found a loss of sensitivity in the central area (Fig. 1, right panel). Accommodative amplitude was reduced when patients were considered as a single group, and for the patients in the 35–45 years age group (Figs 2 and 3).

Table 1 shows the correlation of total FM 100-hue score versus sensitivity loss in each area of visual field for each patient. The correlation coefficient $r = 0.70$ ($p < 0.05$) is significant for the central area at 10°, but there are also significant correlations for the other areas.

Table 2 shows the correlation between accommodative amplitude and FM 100-hue score and sensitivity loss in the visual field. The correlation coefficient was significant for FM score ($r = -0.47$ at $p < 0.05$) and for the central area 10°, CI ($r = -0.41$ at $p < 0.05$) for the entire group. When patients were divided in two age groups the correlation was more significant for the young group ($r = -0.53$; $p < 0.05$) than for the older patients ($r = -0.34$ at $p < 0.05$).

Conclusion

Our study shows that a majority of patients with MS and normal visual acuity have a significant reduction of accommodative amplitude associated with impairment of colour vision and central retinal sensitivity. The basic mechanisms giving rise to poor accommodation have yet to be studied. However, if we exclude a reduced function of the oculomotor fibres of accommodation, which

Table 1. *r* value for TS (FM 100-hue) versus visual field at $p < 0.05000$.

	CI	CII	CIII	C1	C2	C3	C4
TS	0.70	0.64	0.50	0.60	0.56	0.61	0.54

Table 2. *r* value for accomodative amplitude versus TS (FM 100-hue) and visual field areas at $p < 0.05000$.

	TS	CI	CII	CIII	C1	C2	C3	C4
Accommodative amplitude	–0.5	–0.4	–0.3	–0.1	–0.3	–0.5	–0.4	–0.2

remains to be shown, we can presume that the disease affects the optic nerve in a selective way, impairing the transmission of certain signals, such as colour discrimination, which in turn affects the accommodative system.

Acknowledgements

This work was supported in part by Research Grant from the Italian Ministero della Pubblica Istruzione.

References

Bruscaglioni, R. (1976). On a modified version of Badal's Optometer. Atti Fond. G. Ronchi 31: 835–844.

Pokorny, J., Smith, V.C., Verriest, G. and Pinckers, A.J.L.G. (1979). Congenital and Acquired Color Vision Defects. Grune & Stratton, New York.

Serra, A. and Mascia, C. (1981). Quantitative perimetry in optic subatrophy from previous optical neuritis in multiple sclerosis. In: Greve, E.L. and Verriest, G. (eds.), Fourth International Visual Fields Symposium. Doc. Ophthalmol. Proc. Ser. 26: 85–90. W. Junk, The Hague.

Serra A, Zucca, I. and Mulas, S. (1990). Some remarks on colour discrimination in multiple sclerosis. In: Ohta, Y. (ed.), Colour Vision Deficiencies: 251–254. Kugler and Ghedini, Amstelveen.

Ogden, N.A., Raymond, J.E. and Seland, T.P. (1992). Visual accommodation and sustained visual resolution in multiple sclerosis. Invest. Ophthalmol. Vis. Sci. 33: 274–2753.

Clinica Oculistica Universitaria
Ospedale S. Giovanni di Dio
I-09100 Cagliari, Italy

PART FOUR

Congenital colour deficiencies

26. Features of foveal dichromacy illustrated by deuteranopia

H. SCHEIBNER and A. ORAZEM
(Düsseldorf, Germany)

Abstract

Using a Guild-Bechstein visual tristimulus colorimeter, the missing colour, the alychne trace and the neutral zone of a male deuteranope were measured using a 2° diameter visual field. The resulting data were processed by means of linear homogeneous forms. An outline of dichromatic opponent-colour vision is given not only for deuteranopia but also for protanopia and tritanopia.

Introduction

Making the transition from trichromacy to dichromacy – in other words reducing trichromacy to dichromacy – can be made by means of the dichromat's missing colour (in German, *Fehlfarbe*, Schrödinger, 1920; cf. also König and Dieterici, 1892; Nuberg and Yustova, 1958; Scheibner, 1968). The tool is the *dichromatic* colour match within the framework of a *trichromatic* colour space. Trichromatic as well as dichromatic colour matches imply that the two halves of a visual field are made 'indistinguishably equal'. Whereas a trichromatic colour match results in a point locus in the chromaticity chart, dichromatic colour matches against a constant test stimulus result in a straight line with a well defined point on it, the so-called copunctal point, which is common to all other line loci of the dichromat. In terms of projective geometry, points and straight lines are equivalent elements of the chromaticity chart (e.g. Coxeter, 1974; Rees, 1983). We ask whether this approach can simplify or unify our views on dichromacy.

We wish to show how to deal with, in addition to the colour match, perceptual criteria leading to straight-line loci, how to deal with dichromatic opponent-colour vision by means of straight line loci and how to unify the views on dichromacy by such means.

Measuring the missing colour through dichromatic colour matches

Figure 1 shows the (r,g) chromaticity chart derived from Wright's B, G, R

C.R. Cavonius (ed.), Colour Vision Deficiencies XIII, pp. 245–259.

system (Wright, 1946). The **C** = B**B** + G**G** + R**R** reference system was realized on a visual tristimulus colorimeter (Beck and Richter, 1958), in particular by its spectral optical output, consisting of the wavelengths 460 nm (blue, B), 530 nm (green, G) and 650 nm (red, R).

The three deuteranopic confusion lines shown in Figure 1 are derived from

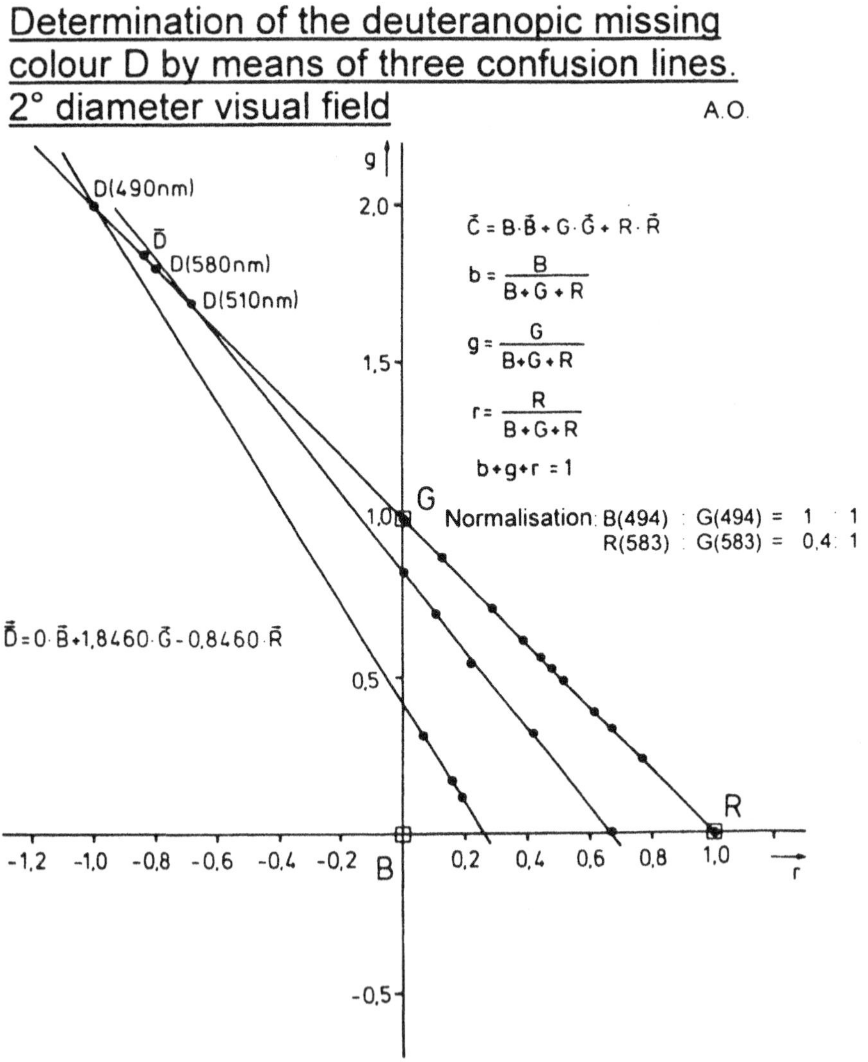

Fig. 1. Instrumental (r, g) chromaticity chart. The points within the triangle BGR are chromaticity loci of dichromatic colour matches against spectral test stimuli of 490 nm, 510 nm and 580 nm, each test stimulus producing a confusion line. The points D(490 nm), D(580 nm and D(510 nm) each pertain to a confusion line and are chromaticity loci of averaged difference vectors or missing points; $\overline{\mathrm{D}}$ is the grand average.

dichromatic colour matches made by a deuteranopic observer. The green component (**GG**) in the mixture was kept constant, but the blue (**BB**) and red (**RR**) components could be varied at will (Scheibner, 1976; Kröger and Scheibner, 1977). The mixtures were matched against three spectral test stimuli of 490 nm, 510 nm and 580 nm, as shown in Figure 1. The vector difference of two matches made against an identical test stimulus, where the matches contain different fixed green components, is a representative of the deuteranopic missing colour vector. Thus, a single difference vector results in a chromaticity locus called *missing point*, and one confusion line may produce many missing points. A copunctal point, on the other hand, is an intersection of (at least) two different confusion lines. In Fig. 1, D(490 nm) is the mean of the missing points that are associated with the confusion line produced by the monochromatic stimulus of 490 nm. The loci D(580 nm) and (510 nm) have an analogous meaning. The grand average of these three mean missing points is indicated by $\overline{\text{D}}$. Its vector representation is

$$\mathbf{D} = 0\mathbf{B} + 1.8460\mathbf{G} - 0.8460\mathbf{R}, \qquad (1)$$

where the tristimulus values B, G, and R are normalized to chromaticity coordinates b, g, r. It should be noted that none of these loci was derived as a copunctal point. According to equation (1) **D** will be taken as the common copunctal point in constructing the deuteranopic ray pencils shown in Figures 4 and 5. Equation (1) is a non-zero vector for a trichromat (non-match), but a zero vector for the deuteranope (match).

Following Wright (1946) the usual red–green normalization is 1:1, similar to the blue–yellow normalization. In Figure 1 and the subsequent figures, however, the red–green normalization is 0.4:1.0, as indicated. By this modification, the chromaticity locus of the deuteranopic missing points is moved closer to the instrumental primary **G**. With a red–green normalization of 1:1, deuteranopic missing points would lie very distant from **G**; in limiting cases even at infinity.

Reduction to a deuteranopic colour space by means of the deuteranopic missing colour

A linear mapping that transforms the trichromatic colour space into a deuteranopic colour space must consist of a matrix that contains two rows and three columns, i.e. six constants. Figure 2 explains the equation of the mapping:

$$\begin{bmatrix} B' \\ \\ R' \end{bmatrix} = A \begin{bmatrix} B \\ G \\ R \end{bmatrix} \qquad (2)$$

where A is the desired 2 × 3 matrix. As the missing colour is a null vector to the deuteranope, it is mapped onto the origin of the deuteranopic colour space. Figure 2 shows this equation by making use of the notion of kernel. A kernel is a subspace that is mapped to zero, i.e. annihilated (Kostrikin and Manin, 1989). This equation gives two constants of the desired matrix. Fixing the two

<u>Deuteranopic missing colour $\vec{D}$:</u>

$$\vec{D} = B_D \cdot \vec{B} + G_D \cdot \vec{G} + R_D \cdot \vec{R}$$

$$\begin{pmatrix} B' \\ R' \end{pmatrix} = A \cdot \begin{pmatrix} B \\ G \\ R \end{pmatrix}$$

B, G, R	tri-stimulus values
B′, R′	di-stimulus values

Missing colour as the kernel of a reduction mapping:

$$\begin{pmatrix} 0 \\ 0 \end{pmatrix} = A \cdot \begin{pmatrix} B_D \\ G_D \\ R_D \end{pmatrix}$$

Additionally

$$\begin{pmatrix} 1 \\ 0 \end{pmatrix} = A \cdot \begin{pmatrix} 1 \\ 0 \\ 0 \end{pmatrix}$$

$$\begin{pmatrix} 0 \\ 1 \end{pmatrix} = A \cdot \begin{pmatrix} 0 \\ 0 \\ 1 \end{pmatrix}$$

Fig. 2. Scheme of how to determine the 2 × 3 matrix A, if the missing colour vector **D** is known and the two deuteranopic instrumental primaries **B′** and **R′** are fixed.

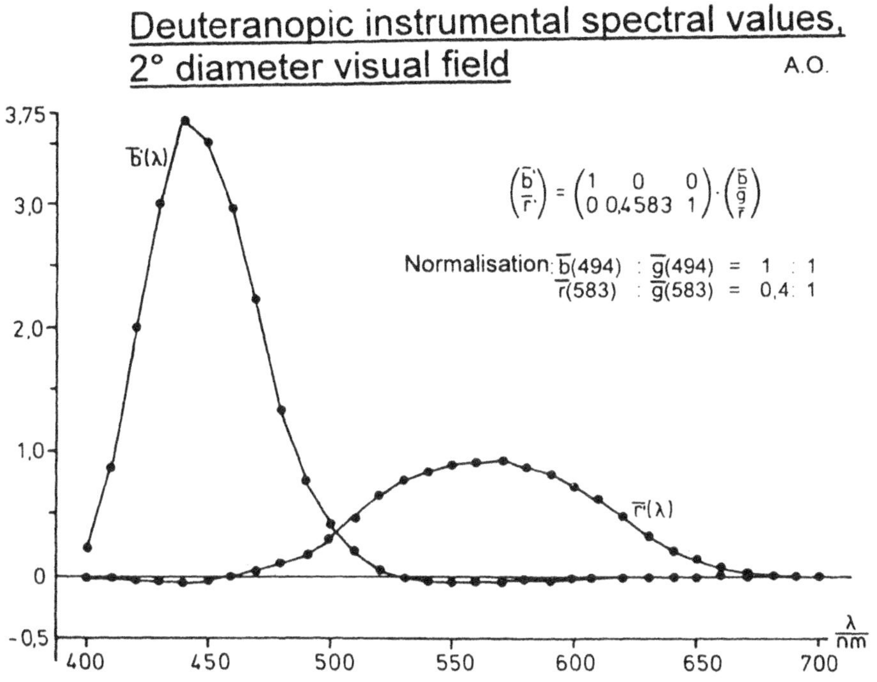

Fig. 3. By means of the numerically given 2 × 3 matrix A (Fig. 2), the trichromatic instrumental spectral values $\bar{b}(\lambda)$, $\bar{g}(\lambda)$, $\bar{r}(\lambda)$ (Estévez, 1982) are transformed to deuteranopic instrumental spectral values $\bar{b}'(\lambda)$, $\bar{r}'(\lambda)$. Due to the green–red normalization shown in Fig. 1, the values of $\bar{r}'(\lambda)$ are reduced to 40% of the usual values.

deuteranopic primaries, i.e. basis vectors **B**′, **R**′, gives us the remaining four constants of the matrix. This is shown in the two equations at the bottom of Figure 2. This choice for fixing the deuteranopic basis vectors is arbitrary, but is the simplest one.

Figure 3 shows the numerical results following from the numerical vector **D**, equation (1), given in Figure 1. The mapping is applied to trichromatic instrumental spectral values $\bar{b}(\lambda)$, $\bar{g}(\lambda)$, $\bar{r}(\lambda)$ (Estévez, 1982), which is why they appear in Figure 3 instead of tristimulus values B, G, R:

$$\begin{aligned} \bar{b}'(\lambda) &= \bar{b}(\lambda) \\ \bar{r}'(\lambda) &= 0.4583\bar{g}(\lambda) + \bar{r}(\lambda) \end{aligned} \qquad (3)$$

Note that the deuteranopic curve $r'(\lambda)$ has been reduced to 40% of the normal value due to the modified green–red normalization in Fig. 1. It may be seen that the collapse of trichromacy to deuteranopia happens in the green–red region and has no effect on the blue mechanism.

Equations (2) and (3) map the trichromatic instrumental colour space (B,G,R) onto the deuteranopic instrumental colour space (B′,R′). Within the chromaticity chart (considered as a projective plane), this is equivalent to a central projection of trichromatic chromaticities (loci lying in a plane) to deuteranopic chromaticities (loci lying on a straight line, the target line). Figure 4 shows the unique copunctal point **D**, in which four straight lines intersect. Point **D** is the projection centre and the intersecting lines are projection rays. The choice of the target line is arbitrary with one restriction: the projection centre must not lie on the target line. For deuteranopia, a convenient choice is the abscissa (axis) r (Fig. 4). Such a projection gives, e.g., the following results: **B**′ = **B**; **R**′ = **R**. **G**′ is a linear combination of **B**′ and **R**′ and, therefore, is not independent within the system (B′,R′), so that **G**′ can be discarded; **D** disappears within the system (B′,R′).

Reduction to deuteranopia in dual terms

A confusion line can be determined from repeated dichromatic colour matches against an identical colour stimulus (Fig. 1), i.e. through the perceptual criterion of indistinguishably equal fields (Scheibner, 1976; Kröger and Scheibner, 1977). This property makes it possible for a straight line in the chromaticity chart to serve as an expression of a dichromatic chromaticity.

As a first step, such a straight line may be expressed by an equation of the type

$$g = m\,r + h,$$

where the variables r and g are chromaticity coordinates in Wright's chromaticity chart (Wright, 1946). By changing from chromaticity coordinates r, g to tristimulus values B, G, R by the relationships indicated in Figure 1, the straight lines assume the following homogeneous linear form:

$$\beta B + \gamma G + \rho R = 0, \qquad (4)$$

where the constant β, γ, ρ are the *coefficients.*

Equation (4) is the general incidence condition of a point and a straight line, and expresses the principle of duality (von Staudt, 1847; Plaumann and Strambach, 1981): (1) keep the triple (β,γ,ρ) constant, vary the triple (B,G,R) and you obtain a fixed straight line; (2) keep the triple (B,G,R) constant, vary the triple (β,γ,ρ) and you obtain a pencil of straight lines through a fixed point. For this reason, (B,G,R) are called point coordinates, (β,γ,ρ) are called line coordinates in the projective plane, which here is identified with the chromaticity chart. The aspect of incidence condition inherent in equation (4) means that if a point described by (B,G,R) lies on the straight line described by (β, γ, ρ),

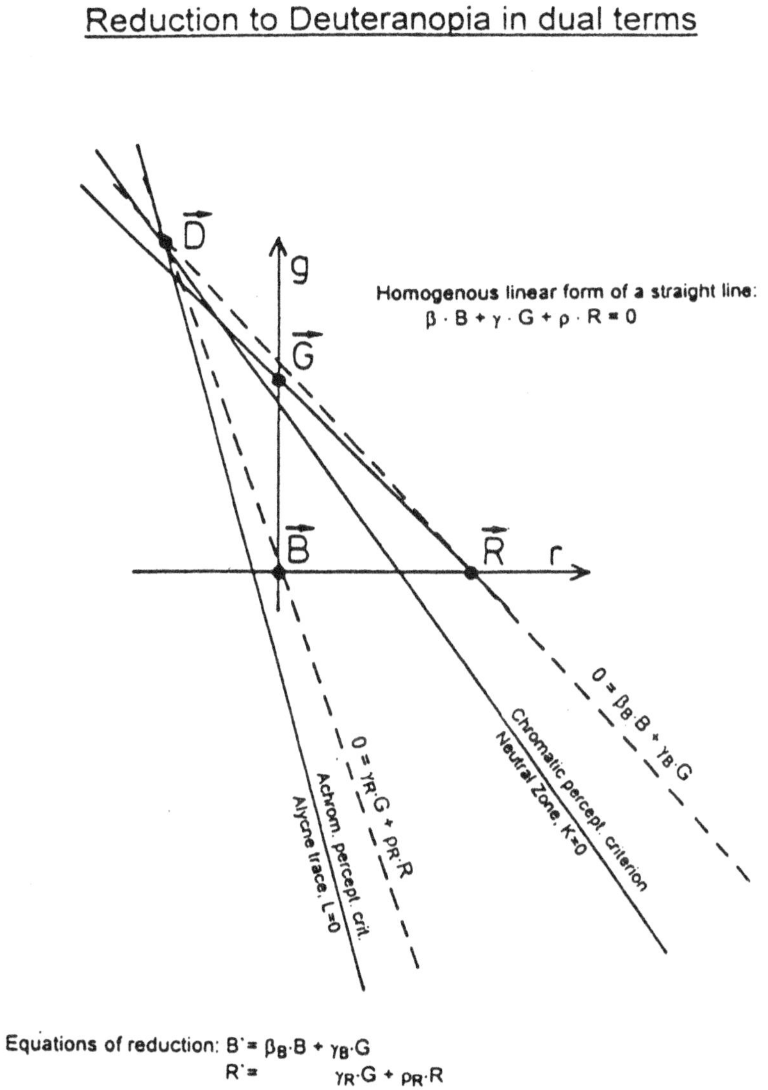

Fig. 4. Instrumental chromaticity triangle in the modified normalization of Fig. 1. The two dashed lines through **B** and **R** are described by linear forms that consist of only two terms. They are essential for the reduction to the deuteranopic instrumental system (B′,R′), where **B**′ = **B** and **R**′ = **R**. The neutral zone and the alychne trace are essential for a reduction to a deuteranopic opponent-colour system (K,L).

equation (4) has the value zero; otherwise, equation (4) has a value other than zero. If the point crosses the line, equation (4) changes sign. This change of sign reflects the blue–yellow opponency present in deuteranopic colour perception.

If a set of straight lines possessing one unique copunctal point is known (i.e. a ray pencil) there is an alternative dual way of deriving the reduction equation (2): Figure 4 shows a plot of the two dashed lines joining the copunctal point **D** with **B**′ = **B** and **R**′ = **R**. Their linear forms contain only two terms:

$$0 = \beta_B B + \gamma_B G \qquad (6a)$$
$$0 = \gamma_R G + \rho_R R. \qquad (6b)$$

We now change our point of view by considering the two forms (equations 6a and 6b) as explicit transformations that map the (B,G,R) space onto a very specialized restricted range of the deuteranopic space, namely the origin. In other words, equations (6a) and (6b) describe mapping kernels in the form of two-dimensional subspaces, i.e. planes containing the origin. These two subspaces are mapped to zero, i.e. are annihilated. The generalization that opens this limited range says: all chromaticity point loci in Figure 4, in particular those that are not incident on the two dashed lines, are mapped onto the full deuteranopic space according to

$$B' = \beta_B B + \gamma_B G \qquad (6c)$$
$$R' = \gamma_R G + \rho_R R. \qquad (6d)$$

This derivation of the reduction to deuteranopia by means of straight lines and linear homogeneous forms seems more straightforward than that shown in Figure 2. Because of the more general site of **D** assumed in Figure 4, equations (6c) and (6d) are slightly more general than equation (3), since a small green component $\gamma_B G$ may also be attributed to the blue deuteranopic component **B**′. Its magnitude can be estimated from the small distance of the dashed straight line (equation 6a; Fig. 4) from the green instrumental primary **G**.

Immediate reduction to a deuteranopic opponent-colour space

It is possible to single out two straight lines by means of perceptual criteria that are quite different from indistinguishably equal, namely by means of the criterion 'neither blue nor yellow' and 'equally bright' (Scheibner, 1987; Orazem and Scheibner, 1992).

For the (chromatic) criterion 'neither blue nor yellow', the deuteranopic observer adjusted various binary colour mixtures so that they appeared neither blue nor yellow. This was done in the left side of the bipartite visual field of the colorimeter. When satisfied with this setting, the observer matched it to the superposition of the three instrumental primaries, presented in the right side of the bipartite visual field. Thus, the colours obeying the criterion were specified

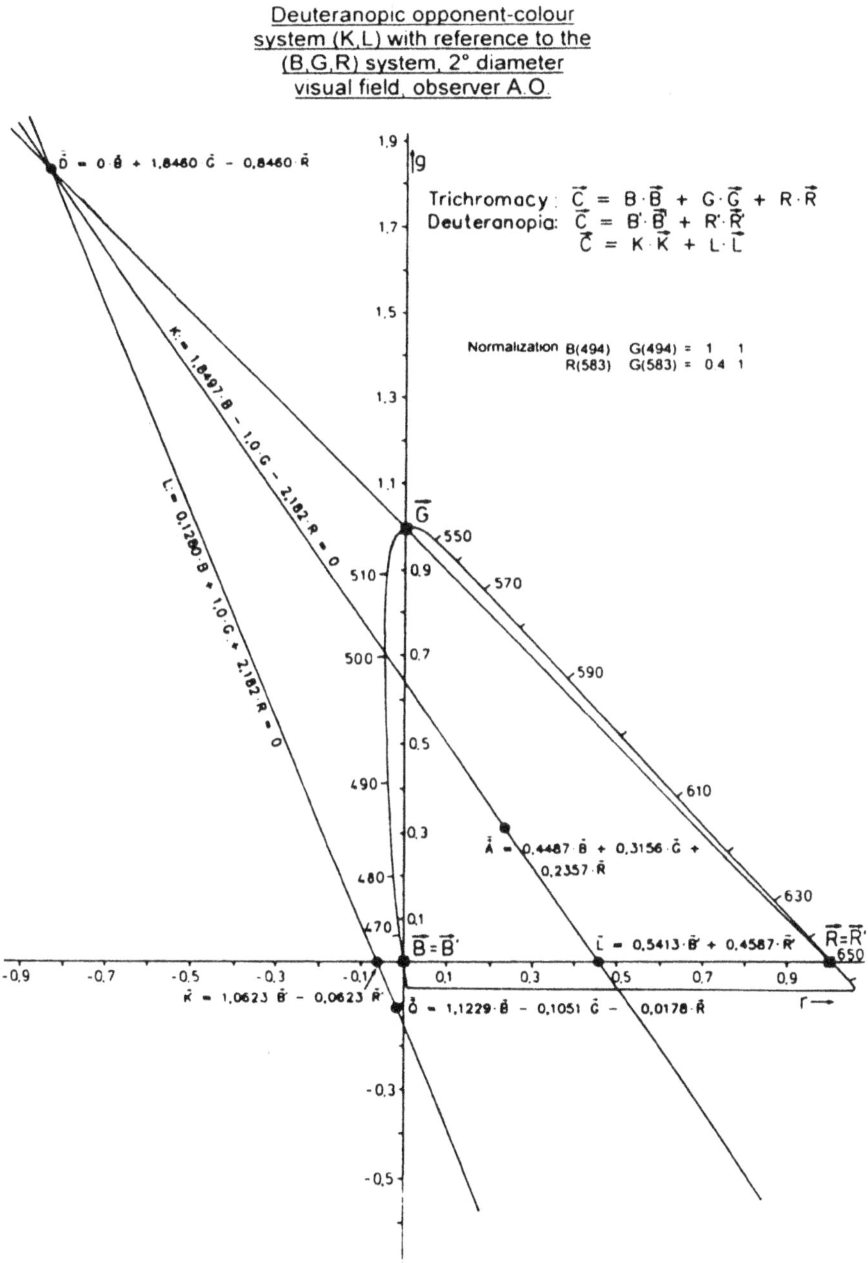

Fig. 5. Graphical and numerical description of (a) the missing colour **D**; (b) the neutral zone K = 0, (c) the alychne trace L = 0; and, additionally, (d) the transition from (**B**′, **R**′) to (**K**, **L**).

within the (B,G,R) system and could be plotted in the (r,g) chromaticity chart. They were well approximated by a straight line – the so-called neutral zone. In Figures 4 and 5, this is shown as the straight line that runs through the central area of the colour triangle (**BGR**).

In applying the (achromatic) criterion 'equally bright', the deuteranopic observer equated pairs of colours, which differed in hue to him, in brightness. For that purpose, he used the criterion of minimally distinct border (Boynton, 1978; Thoma, 1982). By matching the equated colours to a superposition of the three experimental primaries, the colours were specified within the (B,G,R) system. Pairwise vectorial differences were then calculated. If one accepts the minimally distinct border method, such difference vectors are by definition luminance-free. Their chromaticity loci lay in the imaginary region of the chromaticity chart and were well approximated by a straight line. In Figures 4 and 5, this is the straight line that passes the left side of the colour triangle. It is the alychne (i.e., without light) of Schrödinger (1925; cf. also Lang, 1995). The final averaging process for constructing the neutral zone and the alychne trace took the missing point **D** into account so that the two lines intersected in locus **D**.

The two straight lines were expressed as homogeneous linear forms as explained in the preceding section. The result, shown in Figure 5, was

$$\begin{aligned} 1.8497B - 1.0G - 2.182R &= 0 \\ 0.1280B + 1.0G + 2.182R &= 0 \end{aligned} \qquad (7)$$

The interpretation of these equations as incidence condition (von Staudt, 1847) implies that a sign change takes place when a colour crosses these lines, in agreement with the chromatic and achromatic perceptual criteria. The generalization from the two linear forms (equation 7) to a corresponding mapping is as follows: all trichromatic colours, in particular also those whose loci are not incident on the neutral zone and the alychne trace, are mapped to a deuteranopic colour space according to

$$\begin{aligned} K &= 1.8497B - 1.0G - 2.182R \\ L &= 0.1280B + 1.0G + 2.182R. \end{aligned} \qquad (8)$$

We call this (K,L) space the deuteranopic opponent-colour space, in which a colour is presented by

$$\mathbf{C} = K\mathbf{K} + L\mathbf{L}\,,$$

where the opponent di-stimulus value K is the blue–yellow chrominance and the opponent di-stimulus value L is the luminance. The neutral zone and the alychne trace represent the reference chromaticities of deuteranopic opponent-colour vision, in which the neutral zone is chrominance-free and the alychne trace is luminance-free.

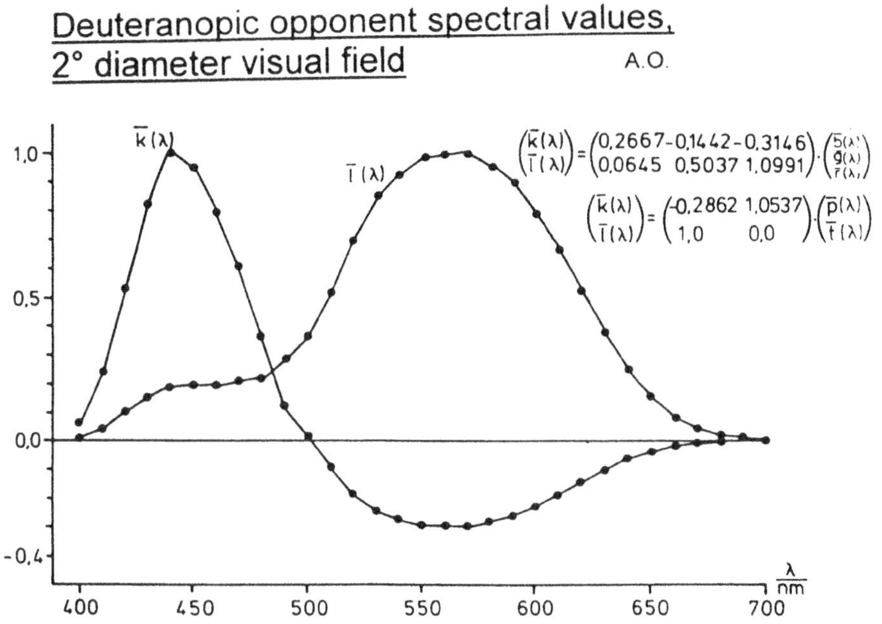

Fig. 6. The linear forms describing the neutral zone and the alychne trace in Fig. 5 provide, after renormalization, the 2 × 3 matrix that transforms the trichromatic spectral functions $\bar{b}(\lambda)$, $\bar{g}(\lambda)$, $\bar{r}(\lambda)$ (Estévez, 1982) to deuteranopic opponent spectral functions $\bar{k}(\lambda)$, $\bar{l}(\lambda)$. The connection of the same functions $\bar{k}(\lambda)$, $\bar{l}(\lambda)$ to the fundamental spectral functions $\bar{p}(\lambda)$, $t(\lambda)$ is also shown.

In Figure 6, the transformation equations (equation 8) were applied to trichromatic instrumental spectral values $\bar{b}(\lambda)$, $\bar{g}(\lambda)$, $\bar{r}(\lambda)$, the so-called colour-matching functions (Estévez, 1982). Figure 6 shows the resulting opponent spectral curves $\bar{k}(\lambda)$, $\bar{l}(\lambda)$, where $\bar{k}(\lambda)$ describes the spectral blue–yellow opponent function and $\bar{l}(\lambda)$ describes the (photopic) spectral luminous efficiency. The transformation equations (8) were renormalized row-wise to make the maxima of the curves $\bar{k}(\lambda)$ and $\bar{l}(\lambda)$ unity.

The second set of equations shown in Figure 6 gives the transformation of the so-called fundamental curves $\bar{p}(\lambda)$ (sensitivity of long-wavelength cones) and $\bar{t}(\lambda)$ (sensitivity of short-wavelength cones) into the opponent spectral curves $\bar{k}(\lambda)$ and $\bar{l}(\lambda)$:

$$\begin{aligned} \bar{k}(\lambda) &= -0.2802\bar{p}(\lambda) + 1.0537\bar{t}(\lambda) \\ \bar{l}(\lambda) &= 1\bar{p}(\lambda) \end{aligned} \qquad (9)$$

Their derivation – omitted here – is effected by a matrix concatenation (e.g. Scheibner, 1988/89; Scheibner and Lochner, 1991). Equation (9) expresses a linear excitation transfer from the deuteranope's retinal cones onto his post-receptoral pathways and centres. Since $\bar{l}(\lambda) = \bar{p}(\lambda)$, deuteranopic brightness is signalled by the long-wavelength cones alone.

A survey of dichromacy

Although the use of an instrumental colour-reference system is indispensable for experimental investigations – we worked with the instrumental (B, G, R) system of W. D. Wright (1946) – such a system involves an arbitrary choice of working primaries. However, the excitation transfer given by equation (9) is not arbitrary.

Figure 7 shows all three types of dichromacy with reference to the trichromatic fundamental system (P, D, T). P stands for the protanopic missing colour, D for the deuteranopic and T for tritanopic missing colour. These fundamental primaries (basis vectors) are arranged in an equilateral triangle (Fig. 7).

The excitation transfer may be written in the following way, where fundamental and opponent di-stimulus values are used (instead of fundamental and opponent spectral values as in equation 9):

Protanopia $$\begin{bmatrix} K_p \\ L_p \end{bmatrix} = A_p \begin{bmatrix} D \\ T \end{bmatrix} \quad (10)$$

Deuteranopia $$\begin{bmatrix} K_d \\ L_d \end{bmatrix} = A_d \begin{bmatrix} P \\ T \end{bmatrix} \quad (11)$$

Tritanopia $$\begin{bmatrix} K_t \\ L_t \end{bmatrix} = A_t \begin{bmatrix} P \\ D \end{bmatrix} \quad (12$$

In these equations, (D, T), (P, T), (P, D) are fundamental di-stimulus values, (K_p, L_p), (K_d, L_d), (K_t, L_t) are opponent di-stimulus values, K_p, K_d meaning blue–yellow chrominance, K_t green–red chrominance, and L_p, L_d, L_t luminance. A is a 2×2 matrix. A numerical example for equation (11) is equation (9).

As is shown in linear algebra (e.g. Boseck, 1984; Kostrikin and Manin, 1989), the various dichromatic primaries can be transformed contragrediently to equations (10), (11) and (12):

$$(\mathbf{D}, \mathbf{T}) = (\mathbf{K}_p, \mathbf{L}_p)\, A_p \quad (10a)$$
$$(\mathbf{P}, \mathbf{T}) = (\mathbf{K}_d, \mathbf{L}_d)\, A_p \quad (11a)$$
$$(\mathbf{P}, \mathbf{D}) = (\mathbf{K}_t, \mathbf{L}_t)\, A_t \quad (12a)$$

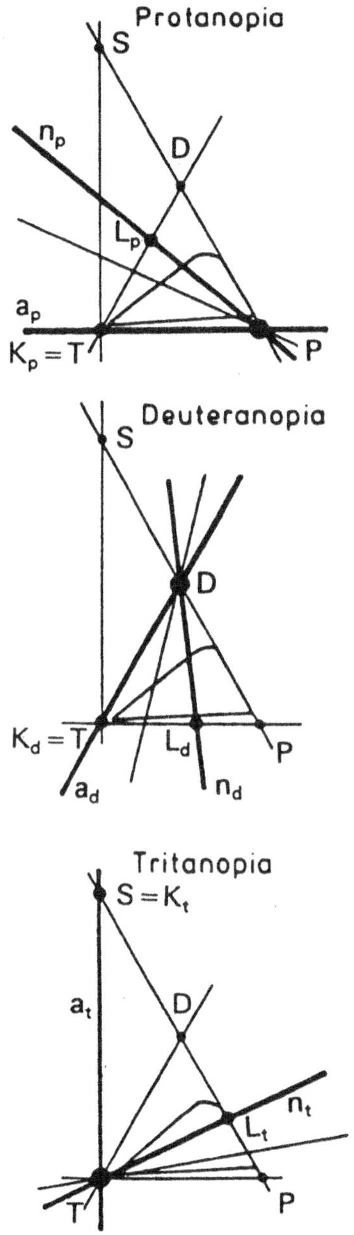

Fig. 7. A triangle assumed to be equilateral is plotted three times. P, D, T are the copunctal points of the protanope, deuteranope and tritanope, n_p, n_d, n_t are their neutral zones, a_p, a_d, a_t are their alychne traces. K_i, L_i where i = p, d, t, are the loci of the dichromatic opponent primaries. The large filled circles are the vertices of the dichromatic chromaticity pencils. The point S may be interpreted as the copunctal point of a hypothetical fusion deuteranope exhibiting the luminosity function of the normal trichromat (and the tritanope) (modified from Scheibner and Bruckwilder, 1989).

The loci of the fundamental primaries **P, D, T** are the vertices of the fundamental triangle (Fig. 7). The luminance primaries $\mathbf{L_p}$, $\mathbf{L_d}$, $\mathbf{L_t}$ lie on their respective neutral zones n_p, n_d, n_t and are, therefore, chrominance-free. A missing colour and a chrominance primary are by definition luminance-free and lie, therefore, on the appropriate alychne trace: in protanopia, **P**, $\mathbf{K_p}$ and a_p coincide ('are incident'), in deuteranopia, **D**, $\mathbf{K_d}$ and d_d coincide, in tritanopia, **T**, $\mathbf{K_t}$ and a_t coincide. The six straight lines n_p, n_d, n_t, a_p, a_d, a_t are shown in Figure 7 as heavy rays. In protanopia and deuteranopia, the contributions of the blue retinal cones (T) to luminance can be neglected, so **T** and a_p as well as **T** and a_d are also coincident. Therefore, the rays a_p and a_d become sides and $\mathbf{K_p}$ and $\mathbf{K_d}$ become vertices of the fundamental triangle. These important features do not hold for tritanopia: a_t is not a side of the triangle nor is $\mathbf{K_t}$ the apex of the triangle. One reason for these tritanopic properties may be the greater similarity of tritanopia to normal trichromacy: the tritanopic alychne trace a_t is approximately identical with the trichromatic alychne trace, and the tritanopic neutral zone n_t resembles the hue *yellow* of the normal trichromat.

The reduction of trichromacy to a dichromacy implies the degeneration of the fundamental triangle to a single side, which is left when the missing colour is removed: line DT for protanopia, line PT for deuteranopia and line PD for tritanopia (Fig. 7). It may appear strange that all these dichromatic chromaticity charts are outside the gamut of real colours, i.e. they are imaginary. The solution of this problem may be that one should consider the rays of pencils whose vertices are the missing colours **P, D, T**. The reduction to dichromacy, then, consists not of removing the missing primary, but of removing the triangle side opposite to the missing primary. In Figure 7, the two dichromatic fundamental reference chromaticities are plotted as thin rays, while the opponent ones are plotted as heavy rays.

Figure 7 demonstrates the importance of the alychne (Schrödinger, 1925): it is an imaginary confusion line. One primary of a dichromatic opponent-colour space is real, whereas the other imaginary. The extension to a trichromatic opponent-colour space adds one imaginary primary (Scheibner and Wolf, 1985). It is beyond the scope of this paper to go into details of (dual) straight line coordinates by which dichromatic colour vision may be handled. Some initial attempts have been reported elsewhere (Scheibner, 1993; Scheibner and Orazem, 1996).

References

Beck, J. and Richter, M. (1958). Neukonstruktion des Dreifarbenmeßgerätes nach Guild-Bechstein. Die Farbe 7: 141–152.

Boseck, H. (1984). Einführung in die Theorie der linearen Vektorräume: 152–153. Deutscher Verlag der Wissenschaften, Berlin.

Boynton, R.M. (1978). Ten years of research with the minimally distinct border. In: Armington, J.C., Krauskopf, J.G. and Wooten, B.R. (eds.), Visual Psychophysics and Physiology: 193–207. Academic, New York.

Bruckwilder, R. and Scheibner, H. (1988/89). Spektralwerte des Protanopen, insbesondere ihre Bestimmung bei minimaler Buntsättigung. Die Farbe 35/36: 215–258.

Coxeter, H.S.M. (1974). Projective Geometry. University of Toronto Press, Toronto.

Estévez, O. (1982). A better colorimetric standard observer for color-vision studies: The Stiles and Burch 2 color matching functions. Color Res. Appl. 7: 131–134.

König, A. and Dieterici, C. (1892). Die Grundempfindungen und ihre Intensitätsverteilung im Spektrum. Z. Psychol. Physiol. Sinnesorg. 4: 221–347.

Kostrikin, A.I. and Manin, Y.I. (1989). Linear Algebra and Geometry. Gordon and Breach, New York.

Kröger, A. and Scheibner, H. (1977). Reduktion der Deuteranopie aus der Trichromasie. Ber. Dtsch. Ophthalm. Ges. 75: 515–517.

Lang, H. (1995). Bemerkungen zu der Arbeit *Über das Verhältnis der Vierfarben- zur Dreifarbentheorie* von Erwin Schrödinger. Die Farbe 41: 169–197.

Nuberg, H.D. and Yustova, E.N. (1958). Researches on dichromatic vision and the spectral sensitivity of the receptors of trichromats. In: Stiles, W.S. (ed.), Visual Problems of Colour: 477–486. HMSO, London.

Orazem, A. and Scheibner, H. (1992). Gegenfarbensehen eines Deuteranopen. Pflügers Archiv, Eur. J. Physiol. 420 (Suppl. 1): R48.

Plaumann, P. and Strambach, K. (eds.) (1981). Geometry – von Staudt's Point of View. Reidel, Dordrecht.

Rees, E.G. (1983). Notes on Geometry. Springer, Berlin.

Scheibner, H. (1968). Dichromasie als Homomorphismus der Trichromasie. Optica Acta 15: 339–349.

Scheibner, H. (1976). Missing colours (Fehlfarben) of deuteranopes and extreme deuteranomalous observers. In: Verriest, G. (ed.), Colour Vision Deficiencies III: 21–26. Karger, Berlin.

Scheibner, H. (1987). Opponent-color vision in relation to perceptual criteria. Die Farbe 34: 243–252.

Scheibner, H. (1993). Transformation of luminance coefficients. J. Opt. Soc. Am. A10: 1392–1395.

Scheibner, H. and Bruckwilder, R. (1989). Protanopic opponent colour vision. In: Drum, B. and Verriest, G. (eds.), Colour Vision Deficiencies IX, Doc. Ophthal. Proc. Ser. 52: 125–130. Kluwer, Dordrecht.

Scheibner, H. and Lochner, D. (1991). Subspaces of an opponent-color space define chrominance and luminance channels. In: Blum, B. (ed.), Channels in the Visual Nervous System: Neurophysiology, Psychophysics and Models: 125–137. Freund, London and Tel Aviv.

Scheibner, H. and Orazem, A. (1997). Linear models of dichromacy. In: Dickinson, C.M, Murray, I.J. and Carden, D. (eds.), John Dalton's Colour Vision Legacy. Taylor and Francis, London (in press).

Scheibner, H. and Wolf, E. (1985). Psychophysik und Physiologie des Farbensehens . In: Bodmann, H.-W. (ed.), Aspekte der Informationsverarbeitung, Funktion des Sehsystems und technische Bilderdarbietung: 1–65. Springer, Berlin.

Schrödinger, E. (1920). Grundlinien einer Theorie der Farbenmetrik im Tagessehen. Ann. Physik (IV) 63: 397–456.

Schrödinger, E. (1925). Über das Verhältnis der Vierfarben- zur Dreifarbentheorie. Sbr. Akad. Wiss. Wien (IIa) 134: 471–490.

Staudt, K.G. Chr. von (1847). Geometrie der Lage. Bauer und Raspe, Nürnberg.

Thoma, W. (1982). Trennlinien-Deutlichkeit und tritanopische Buntsättigung. Die Farbe 30: 167–197.

Wright, W.D. (1946). Researches on Normal and Defective Colour Vision. Henry Kimpton, London.

Physiologisches Institut der Universität Düsseldorf
Abt. für Neuro- und Sinnesphysiologie
Postfach 101007
D-40001 Düsseldorf, Germany

27. The relative salience of the cardinal axes of colour space in normal and anomalous trichromats

B.C. REGAN and J.D. MOLLON
(Cambridge, UK)

Abstract

By imposing rival perceptual organizations on a regular spatial array, we have measured the relative salience of colour differences along the two cardinal axes of colour space. Results are reported for normal and anomalous trichromats. For most, but not all, anomalous trichromats S cone modulations are relatively more salient than they are for normal trichromats. The spatial separation of the elements of the array has a major effect on its perceptual organization.

Introduction

One important role of colour vision is to segregate the visual field. Pseudoisochromatic plates such as those of Ishihara plainly demonstrate the power of colour to impose organization on a stimulus array: the digits in these plates are composed of discrete patches of varying lightness, and can be identified only by the hue that is common to their component parts.

Accumulated evidence suggests that our colour vision depends upon two subsystems, which evolved at different times and for different purposes. The ancient mammalian subsystem of colour vision compares the signal of the short wavelength cones with some combined signal of the long and middle wavelength cones, whereas the phylogenetically recent subsystem compares the signals of the long and the middle wavelength cones (Gouras, 1984; Mollon and Jordan, 1988; Mollon, 1991). The question of the relative roles of the two subsystems in perceptual segregation is largely unexplored.

In the normal observer, the two subsystems correspond to the two axes of the chromaticity diagram of MacLeod and Boynton (1979): at equiluminance, variations along the ordinate correspond only to variations in the short wavelength cone signal, whereas variations along the abscissa correspond to changes in the ratio of long- and middle-wavelength signals. It is convenient to refer to these 'cardinal directions' (Krauskopf *et al.*, 1982) as the S (or tritan) and the L-M (or constant-blue) axes, respectively.

Our experiments were designed to measure the relative organizing power of the signals of the two subsystems. The principle was to pit the organizing effect

C.R. Cavonius (ed.), Colour Vision Deficiencies XIII, pp. 261–270.

of a fixed colour difference along one cardinal direction against that of a variable colour difference along the other axis. By a staircase procedure, the magnitude of the variable colour difference was adjusted so as to estimate the point at which each subsystem determined the reported organization on an equal number of trials. The method is conceptually not unlike that of Hochberg and Silverstein (1956) or Ben-Av and Sagi (1995), who pit the proximity of stimulus elements against other perceptual dimensions.

We have also applied this technique to anomalous trichromats. Although the Rayleigh matches of anomalous observers have been studied extensively, only occasional studies have attempted to penetrate the suprathreshold world of the anomal (Müller *et al.*, 1991). A natural expectation might be that a given red–green variation would be much less salient, i.e. it would enjoy less organizing power for the anomal than for the normal: the anomal is commonly thought to depend for his residual red–green discrimination on two photopigments that have very similar spectral sensitivities and these can offer only a limited range of difference signals to any channel that compares their outputs. There is, however, a plausible alternative to this prediction. During development, or indeed by relatively short-term adaptation, the dynamic range of each colour channel may match itself to the range of chromaticities in the subject's visual world. In the case of the anomal, the signals in his residual long/middle wavelength channel might expand to fill the neural space available. A process of this kind known as 'contrast adaptation' is, in fact, known to occur in the colour-normal observer (Webster and Mollon, 1995): a given colour modulation will look more salient if the observer has adapted to an array in which there are only small modulations on that axis. Laughlin (1994) has discussed the biological utility of matching the dynamic range of visual neurons to the range of contrasts available in natural scenes.

In order to compare normal trichromats and anomalous trichromats in their response to external stimuli, we have used the same stimulus dimensions for both. Since the colour matching functions of anomals differ from those of normals, the MacLeod-Boynton chromaticity diagram is not valid for anomalous observers. Moreover, there is probably a range of anomalous phenotypes (Regan *et al.*, 1994). Our unpublished calculations suggest that modulations along the S axis for normals also very nearly isolate the short wavelength cone signal for anomalous trichromats, whereas modulations along the L-M axis of normal trichromats may introduce some additional modulation of the short wavelength signal (when our stimuli are equated in luminosity for the anomal). Therefore, in discussing the results of anomalous trichromats below, we use the term 'red–green modulation' to make it clear that we refer to the external stimulus.

Design and stimuli

The stimulus consisted of small circles, 0.21° in diameter, presented on a graphics monitor (Sony GDM-1936), and arranged so that their nearest neighbours lay along diagonal lines (Fig. 1, right). The mean chromaticity of the circles was fixed, at Illuminant E (x,y = 0.333,0.333), but the colour along one diagonal of the array was modulated along the S cone axis, and along the other was modulated along the L-M axis of colour space. Thus, the two chromatic modulations imposed on the array suggest alternative perceptual organizations. The subject saw a Gestalt of diagonal lines orthogonal to the direction of the more salient modulation. On each trial he or she was asked to indicate, by pressing one of two buttons, whether the predominant organization was from upper left to lower right or upper right to lower left. The computer program randomly paired the directions of chromatic modulation with the two spatial organizations.

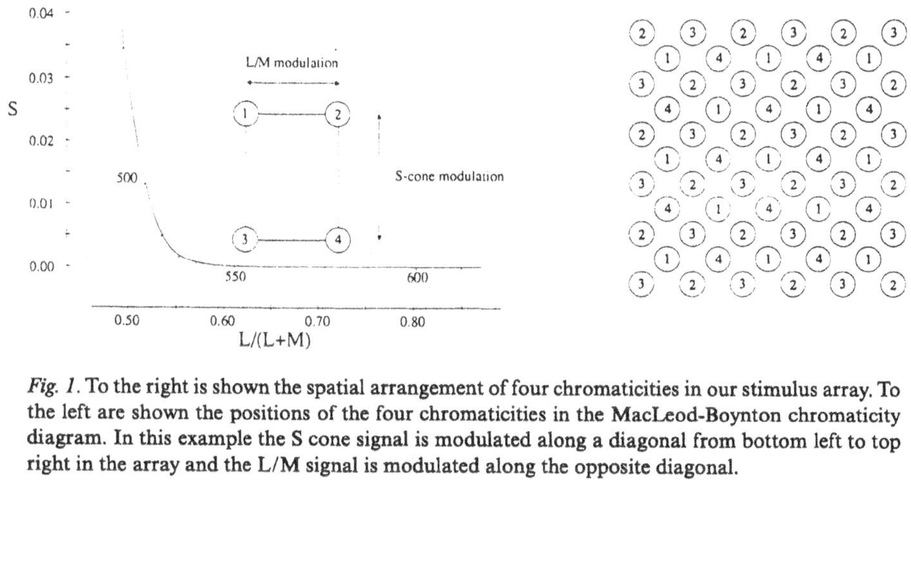

Fig. 1. To the right is shown the spatial arrangement of four chromaticities in our stimulus array. To the left are shown the positions of the four chromaticities in the MacLeod-Boynton chromaticity diagram. In this example the S cone signal is modulated along a diagonal from bottom left to top right in the array and the L/M signal is modulated along the opposite diagonal.

To ensure that the perceptual organization was driven by chromatic modulation and not luminance modulation, we varied randomly the luminances of the elements in the array, so that the maximum luminance was 1.6 times the minimum. In the case of all protanomalous and all but two deuteranomalous subjects, we additionally adjusted the mean luminances of the stimuli to allow for the altered spectral sensitivity functions of anomalous trichromats. These equations were obtained for each subject before the experiment began, using heterochromatic flicker photometry on the same monitor. For normal subjects, this procedure was not performed, and the mean luminances of differently coloured stimuli were held constant.

Chromatic modulations were computed, and are reported, in the units of the MacLeod-Boynton chromaticity diagram. Along each cardinal axis we chose five fixed modulations. For each we adjusted the variable colour modulation along the other axis, using a staircase procedure, so that the competing modulations were each equally likely to determine the organization of the array. The ten staircases used were randomly intermingled in each block of trials. Subjects repeated the staircases several times over a number of days, and the decisions made by the subjects on every trial on each repetition of a particular staircase were combined to produce a psychometric function: the probability of the variable modulation imposing perceptual organization was plotted against the depth of that modulation. Cumulative normal distribution curves were fitted to these functions and were used to estimate the depth of the variable modulation that was equal in salience to a given fixed modulation on the other cardinal axis.

We repeated the measurements for different centre-to-centre separations of the array elements, from 0.21° to 1.0° of visual angle. Initial results showed little effect of spatial separation beyond about 0.40°; therefore some of the subjects were not tested with separations greater than this.

Subjects

Of the 20 subjects in this experiment, three were normal trichromats (ages 20, 23 and 50 years), five were simple protanomalous (ages 14–25, mean 21.0 years) and 12 were simple deuteranomalous (ages 15–29, mean 21.25 years). Colour vision was diagnosed with the Nagel anomaloscope (model I, Schmidt and Haensch). Subjects were also tested on the Ishihara plates (10th edition) under the Illuminant C of a Macbeth easel lamp.

Results and discussion

Figure 2a shows results for a normal trichromat. Each subset of data points corresponds to a particular separation of the elements in the array, and each individual point shows the depth of modulation on one axis of colour space that has the same perceptual salience as a particular modulation on the other axis. We have fitted regression lines through each subset of points, indicating the locus of equal organizing power for modulations along the two cardinal axes. The data seem to be fitted quite well by straight lines. This is a curious result, for there is much evidence that the signals of chromatically opponent channels are increasingly compressed as they depart from an equilibrium value (e.g. Le

Figs. 2–4. The upper graph in each pair shows the loci of equal organizing power for modulations along the two cardinal axes, at different spatial separations of the elements in the array. The lower graph shows the slopes of these loci plotted as a function of the separation of the elements in the array. The units are those of the MacLeod-Boynton chromaticity diagram. Note that the ordinate scale in Figures 3a and 4a differs from that of Figure 2a.

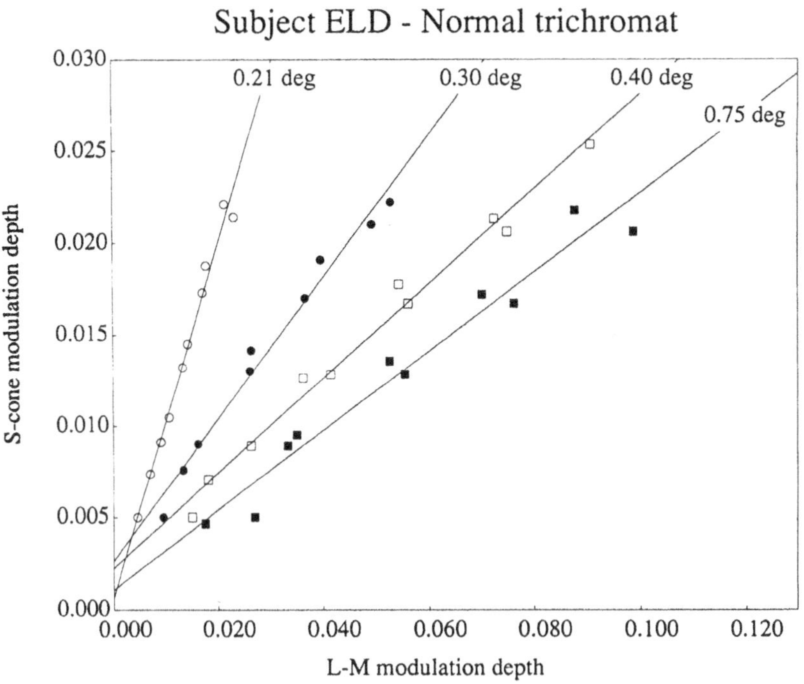
Subject ELD - Normal trichromat
0.21 deg
0.30 deg
0.40 deg
0.75 deg
S-cone modulation depth
0.000
0.005
0.010
0.015
0.020
0.025
0.030
0.000
0.020
0.040
0.060
0.080
0.100
0.120
L-M modulation depth

(a)

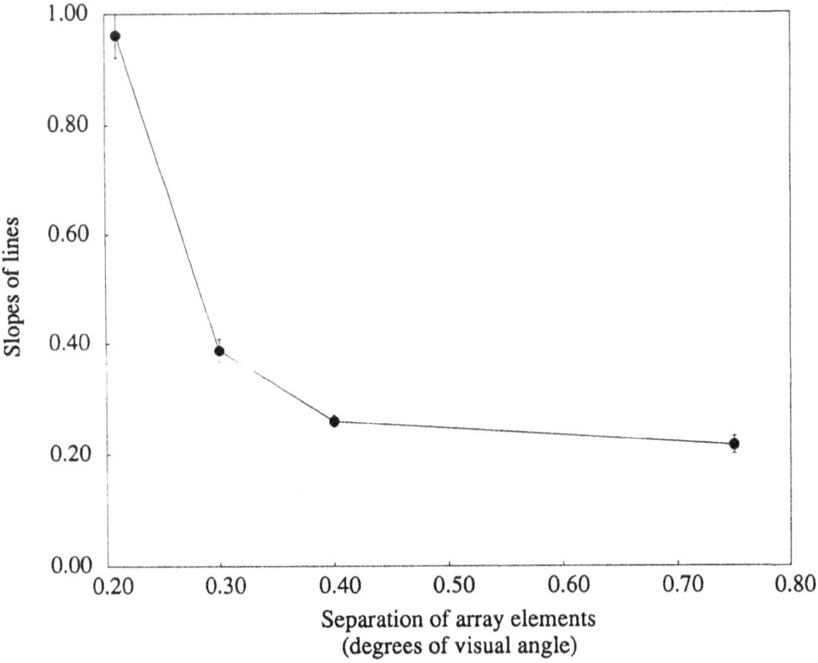
Slopes of lines
0.00
0.20
0.40
0.60
0.80
1.00
0.20
0.30
0.40
0.50
0.60
0.70
0.80
Separation of array elements
(degrees of visual angle)

(b)

Fig. 2

Grand, 1949; Stiles, 1972; Polden and Mollon, 1980; Mollon and Estévez, 1988). Our data suggest either that there is a concomitant and equal compression of the signals of the two subsystems, or that the chromaticities we used were not saturated enough to drive the chromatic channels into non-linear regions of their response ranges.

A second feature of Figure 2a is that the slopes of the lines decrease as the separation of the array elements increases. In Figure 2b these slopes are plotted as a function of the separation of the array elements. The slope at 0.21° of visual angle is over four times as great as at 0.75°, indicating that the relative salience of a given S cone modulation is more than four times greater at 0.75° than at 0.21°. This may be due to (i) the low spatial resolution of the S cone system (Stiles, 1949), which must be constrained by the sparsity of the corresponding receptors (Mollon and Bowmaker, 1992), and (ii) to chromatic aberration, which would tend to blur S cone modulation more than L-M modulation. There are implications here for the construction of perceptually uniform colour spaces: such a space can be valid only for a specified spatial configuration.

Figures 3a and 3b show corresponding results for a typical protanomalous subject, and Figures 4a and 4b for a deuteranomalous subject. The S cone axis has been scaled by a factor of three in Figures 3a and 4a in order to present the results more clearly. The data still show the same effect of increasing the separation of the array elements. However, the slopes of all the lines are much less than for the normal subjects, showing that for these subjects, red–green modulation is much less salient relative to S cone modulation than for normal trichromats. For all but two anomalous trichromats in our sample, the relative salience of red–green modulation was reduced. One exception was a simple deuteranomalous subject whose results were similar to those of the normal trichromats. His setting on the Nagel anomaloscope was 18.5 units with a matching range of 2 units, but he misread only three of the Ishihara plates and obtained a 'superior normal' score on the Farnsworth-Munsell 100-hue test. A second deuteranomalous subject resembled normals at small separations, but gave greater weighting to the S cone signal at larger separations.

Figure 5 summarizes the results from all the subjects. Amongst normal trichromats, there was a two-fold variation in the locus of equal salience of modulations on the two cardinal axes, a result that has a further implication for perceptually uniform colour spaces: such spaces will not necessarily be uniform for all normal observers. Among anomalous trichromats, there is a contrast between the one or two subjects whose results are similar to normal, and the majority, for whom the salience of red–green modulation relative to the S cone modulation is much reduced.

We mentioned above the Ishihara plates as an example of the importance of colour in perceptual segregation. It is a feature of this test that eight of the plates are so-called transformation plates: they do not test directly the threshold for chromatic discrimination, but use a different stratagem, offering two alternative readings of the numbers (Lakowski, 1969). In the normal reading of these plates, the digit is distinguished from the background by a red–green difference,

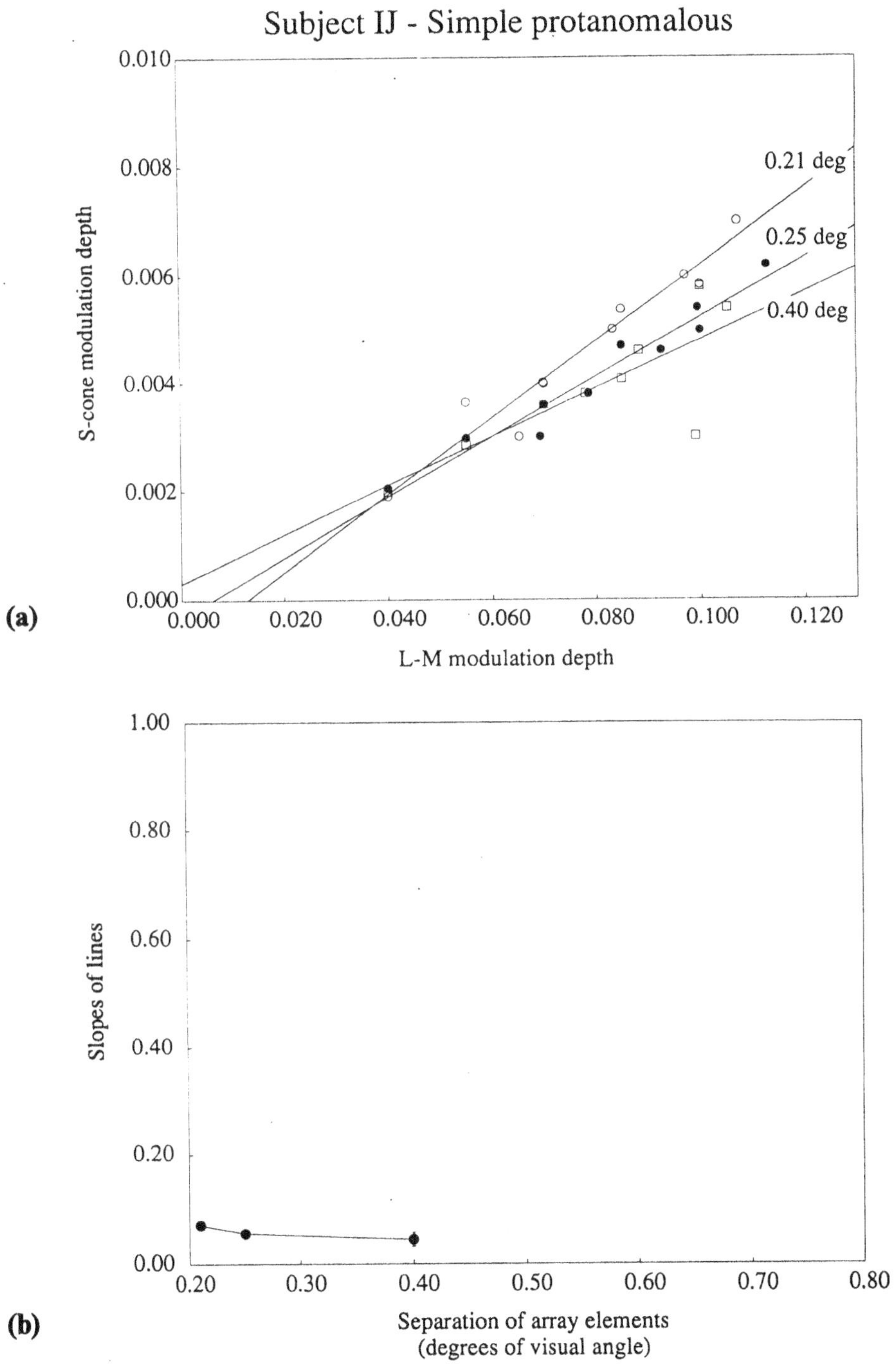

Fig.3. (see legend for Figure 2).

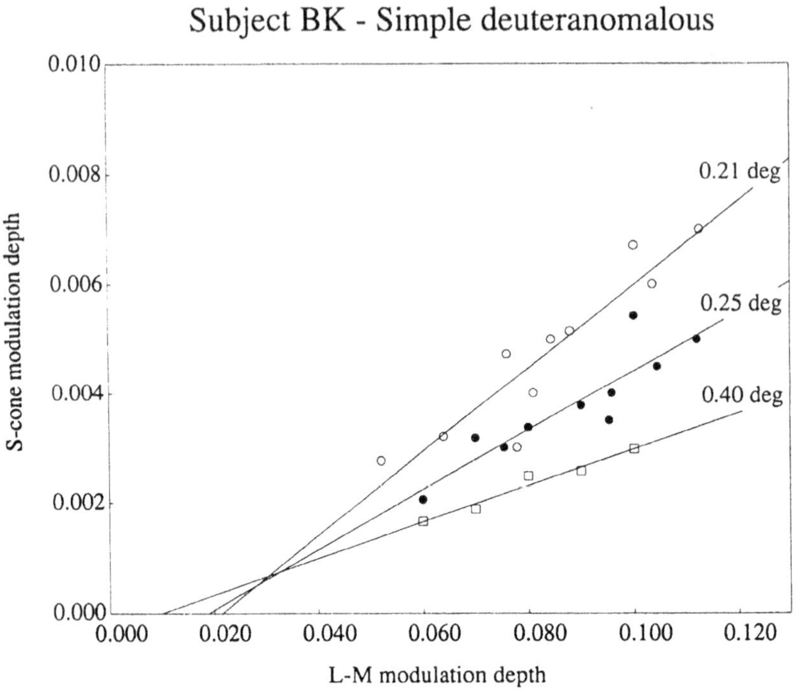

(a)

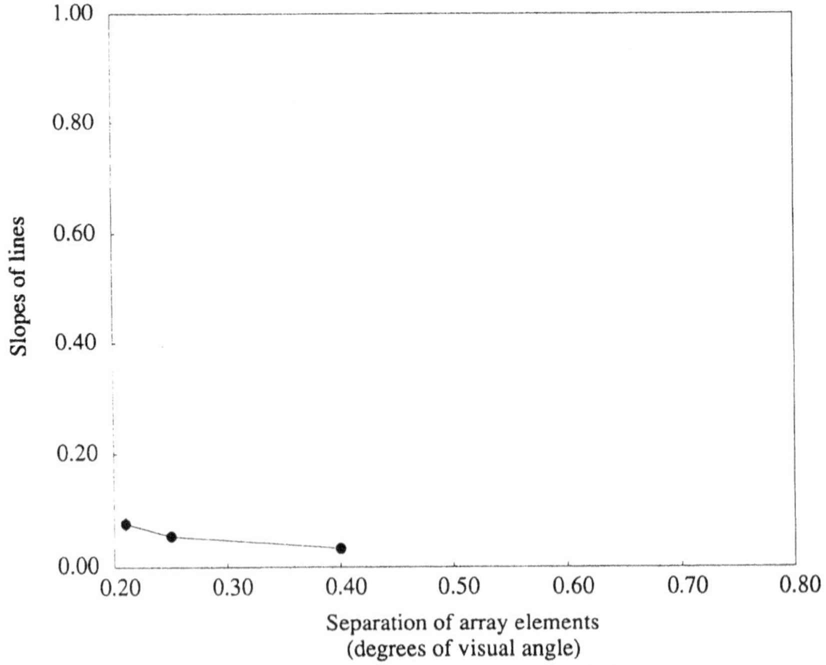

(b)

Fig. 4. (see legend for Figure 2).

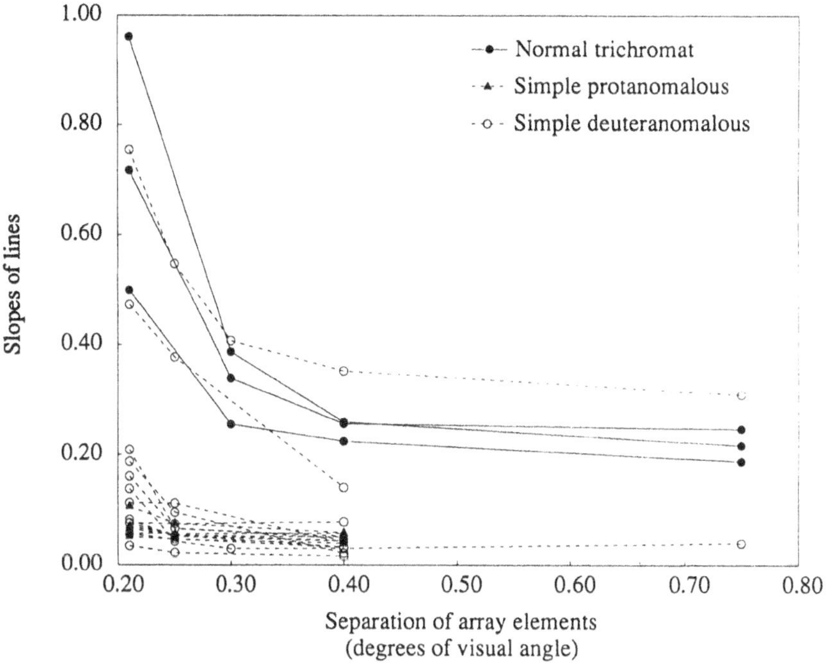

Fig. 5. Combined results from all the subjects in our sample. The slopes of the loci of equal power of the two cardinal axes are plotted as a function of the separation of the elements in the array.

whereas in the reading favoured by most anomalous trichromats the digit differs from the background in S cone excitation. It would appear that these plates depend upon the salience of the S cone difference being greater than the red–green difference for anomalous trichromats, but not for normal individuals (Reffin *et al.*, 1991). If this is the case, then for each anomalous trichromat, the relative salience of the two cardinal axes of colour space should be related to the number of misreadings made in the Ishihara transformation plates. In our sample of anomalous trichromats we found a significant correlation between the relative salience of red–green modulation and the number of Ishihara transformation plates passed. This was true for all spatial separations of the array. (With a sample size of 17, the Spearman rank order correlation coefficient was 0.456, 0.625, and 0.622 for 0.21°, 0.25° and 0.40° centre-to-centre separation of the elements in the array, significant at the 5%, 0.5% and 0.5% levels, respectively.) However, two of these eight plates were misread even by the deuteranomalous subject who resembled the normals at all separations. The broad-band stimuli of the pigments in the Ishihara plates may provide a weaker red–green signal for anomalous trichromats than the relatively narrow-band stimuli of television phosphors.

Acknowledgements

Supported by MRC grant G911363ON. We are grateful to Dr J.P. Reffin for contributions to the early stages of these experiments and to Prof. C.R. Cavonius for discussion.

References

Ben-Av, M.B. and Sagi, D. (1995). Perceptual grouping by similarity and proximity: experimental results can be predicted by intensity autocorrelations. Vision Res. 35: 853–866.

Gouras, P. (1984). Colour vision. In: Osbourne, N.N. and Chader, G.J. (eds.) Progress in Retinal Research: 227–261. Pergamon, Oxford.

Hochberg, J. and Silverstein, A. (1956). A quantitative index of stimulus-similarity: Proximity vs. differences in brightness. Am. J. Psychol. 69: 456–458.

Krauskopf, J., Williams, D.R. and Heeley, D.W. (1982). Cardinal directions of color space. Vision Res. 22: 841–856.

Lakowski, R. (1969). Theory and practice of colour vision testing: a review. Part 2. Br. J. Industr. Med. 26: 265–288.

Laughlin, S.B. (1994). Matching coding, circuits, cells, and molecules to signals: general principles of retinal design in the fly's eye. Prog. Retinal Eye Res. 13: 165–196.

Le Grand, Y. (1949). Les seuils différentiels de couleurs dans la théorie de Young. Rev. Optique Théor. Exp. 28: 261–278.

MacLeod, D.I.A. and Boynton, R.M. (1979). Chromaticity diagram showing cone excitation by stimuli of equal luminance. J. Opt. Soc. Am. 69: 1183–1186.

Mollon, J.D. (1991). Uses and evolutionary origins of primate colour vision. In: Cronly-Dillon, J.R. and Gregory, R.L. (eds.), Vision and Visual Dysfunction, vol. 2: 306–319. Macmillan, London.

Mollon, J.D. and Bowmaker, J.K. (1992). The spatial arrangement of cones in the primate fovea. Nature, 360: 677–679.

Mollon, J.D. and Estévez, O. (1988). Tyndall's paradox of hue discrimination. J. Opt. Soc. Am. A5: 151–159.

Mollon, J.D. and Jordan, G.J. (1988). Eine evolutionäre Interpretation des menschlichen Farbensehens. Die Farbe, 35/36: 139–170.

Müller, M., Cavonius, C.R. and Mollon, J.D. (1991). Constructing the colour space of the deuteranomalous observer. In: Drum, B., Moreland, J.D. and Serra, A. (eds.), Colour Vision Deficiencies X, Doc. Ophthalmol. Proc. Ser. 54: 377–387. Kluwer, Dordrecht.

Polden, P.G. and Mollon, J.D. (1980). Reversed effect of adapting stimuli on visual sensitivity. Proc. Roy. Soc. Lond. B 210: 235–272.

Reffin, J.P., Astell, S. and Mollon, J.D. (1991). Trials of a computer-controlled colour vision test that preserves the advantages of pseudoisochromatic plates. In: Drum, B., Moreland J.D. and Serra, A. (eds.) Colour Vision Deficiencies X, Doc. Ophthalmol. Proc. Ser. 54: 69–76. Kluwer, Dordrecht.

Regan, B.C., Reffin, J.P. and Mollon, J.D. (1994) Luminance noise and the rapid determination of discrimination ellipses in colour deficiency. Vision Res. 34: 1279–1299.

Stiles, W.S. (1949). Increment thresholds and the mechanisms of colour vision. Doc. Ophthal. 3: 138–163.

Stiles, W.S. (1972). The line element in colour theory: a historical review. In: Vos, J.J., Friele, L.F.C and Walraven, P.L. (eds.), Color Metrics: 1–25. Institute for Perception TNO, Soesterberg.

Webster, M.A. and Mollon, J.D. (1995). Colour constancy influenced by contrast adaptation. Nature, 373: 694–698.

University of Cambridge
Department of Experimental Psychology
Downing Street
Cambridge CB2 3EB, UK

28. Age-matched comparisons of chromatic thresholds between normal and strabismic observers

K. KNOBLAUCH, F. VITAL-DURAND and J.L. BARBUR
(St Etienne, France; Bron, France; London, UK)

Abstract

Chromatic thresholds in three directions in an equiluminant plane were obtained from strabismic observers (both amblyopic and not amblyopic) and were compared with those obtained from normal observers. Two methods were used to factor out the variations due to age. Despite a tendency for the thresholds from the amblyopic group to be higher along protan and deutan axes, the magnitude of the thresholds fell within the statistical norms under every condition.

Introduction

In a classic paper, Wald and Burian (1944) demonstrated that the absolute thresholds and spectral sensitivities of amblyopic observers do not differ from those of normal observers. Since acuity could be severely degraded with no loss of sensitivity to coloured lights, they concluded that there was a dissociation of the visual pathways subserving form and light perception.

More recent hypotheses on the functional pathways in the primate visual system (Shapley, 1990) also lead us to conclude that these stimuli were detected within pathways different from those used in an acuity task. For example, one might argue that the large, short-duration, flashes presented on a dark background that they employed would be best detected by means of signals propagating along the magnocellular pathway. Conversely, the acuity targets would be resolved by means of signals propagating along the parvocellular pathway. There is growing evidence that cells in the magnocellular pathway form the substrate for a luminance-sensitive pathway (Lee *et al.*, 1988). The parvocellular pathway, on the other hand, carries high spatial frequency luminance information and low frequency chromatic information (Derrington *et al.*, 1984). To the extent that amblyopia is a defect of the parvocellular system, we might expect to see losses of chromatic discrimination associated with acuity losses. On the other hand, the acuity loss in amblyopia might arise at a stage after the low frequency chromatic signal has been separated out.

We measured chromatic thresholds in a sample of strabismic observers with

C.R. Cavonius (ed.), Colour Vision Deficiencies XIII, pp. 271–276.

and without amblyopia. If amblyopia were associated with a defect in the parvocellular pathway, we would expect to see differences between these two groups. In addition, we evaluated the thresholds with respect to recently collected age norms (Knoblauch *et al.*, 1995) using the same apparatus, so that age variation of the thresholds would be factored out of the analyses.

Subjects and methods

All subjects were tested binocularly with the same stimulus configuration. The stimuli were presented on a colour TV display (EIZO) under the control of a TIGA graphics card running in an 80486 CPU. We used a simplified version of the technique published by Barbur *et al.* (1993). Two fields of spatio-temporal random luminance noise of 16% contrast were presented simultaneously for a 5 sec duration. Each field subtended $7 \times 7°$ at the viewing distance of 57 cm and fields were separated by 7°, edge-to-edge. A set of three stripes, equiluminant with the background, appeared on one of the noise fields, chosen randomly by the computer. The bars subtended 0.43 c/deg and alternated between vertical and horizontal orientations at a rate of 1 Hz. In a given condition, the chromaticity coordinates of the bars were chosen to be along one of the three dichromatic confusion axes passing through the background colour ($x = 0.305$, $y = 0.323$, $Y = 34\,cd/m^2$) in CIE x, y space.

All subjects (normal and strabismic) were recruited in a non-selective fashion from the patients, staff and visitors at the Ophthalmology Unit of Jules Courmont Hospital of Lyon. Consent to participate in the experiments was given by each subject (or if a minor, by the accompanying parent) after being informed that the procedure was experimental and of the risks involved. Observers younger than 2 years of age were tested using a preferential looking technique (Teller, 1979; Vital-Durand, 1992). Observers older than 3 years responded verbally or by pointing. Between trials, the colour difference was adjusted according to a staircase procedure along the axis being tested, as a function of the observer's response. Adults with defective colour vision were excluded if they failed a standard colour vision test such as the Roth 28. Male infants were excluded if there was a maternal history of colour deficiency and the ratio of protan to deutan axis thresholds fell outside of the 95% confidence limits for this measure. These criteria were similarly applied to strabismic observers, although no strabismic observers were excluded on this basis. Strabismus was diagnosed using the cover/uncover test. Observers were classified as amblyopic if the difference in visual acuity between the two eyes was greater than or equal to one half of an octave. This difference corresponds to about 40% or equivalently 0.15 (a line and a half) on a LogMAR scale.

Results

In a previous study (Knoblauch *et al.*, 1995), we measured chromatic thresholds of observers whose ages ranged from 3 months to 86 years along three colour directions in an equiluminant plane. We found that when plotted on log–log axes, the thresholds decreased linearly as a function of age up to adolescence (developmental phase) and thereafter increased linearly (ageing phase). When fitted with two linear segments (a bilinear function), thresholds during the developmental phase declined by almost a factor of two with each doubling of age. The ageing phase was more gradual, thresholds increasing by slightly less than a factor of the square root of two with each doubling of age. These trends were nearly identical along the three axes tested. The principal difference was that thresholds along the tritan axis were higher than along the protan and deutan axes.

The three panels of Figure 1 show plots for each of the colour directions of the residual difference between the chromatic thresholds measured from 28 strabismic observers and the bilinear function fit to the normal data as a function of age. The filled symbols represent amblyopic observers and the unfilled symbols represent observers without amblyopia. The solid horizontal line indicates a residual of zero. A point falling on this line lies exactly on the bilinear function that was fit to the original normal data. The dashed lines indicate the 95% confidence intervals for an individual point; these were computed using a bootstrap technique (Efron and Tibshirani, 1993). The confidence intervals computed in this way were found to be nearly symmetrical about the bilinear function and approximately independent of age.

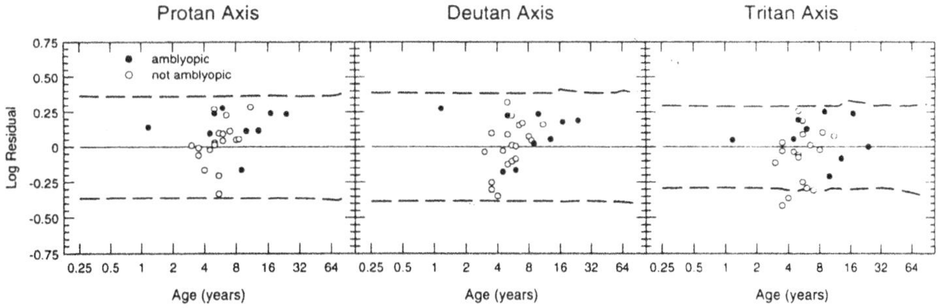

Fig. 1. The difference of the logarithms of the threshold of individual strabismic observers and a bilinear function that predicts the logarithm of normal thresholds as a function of age. Data were collected along three axes in the CIE diagram, from left to right, protan, deutan and tritan. The filled circles represent the data of strabismic amblyopes. The unfilled circles are from strabismic observers without amblyopia. The dashed curves indicate the 95% confidence intervals calculated for individual points.

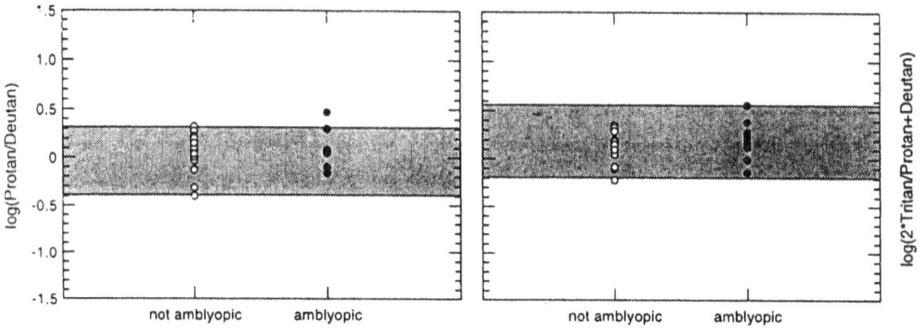

Fig. 2. Left. The ratio between thresholds measured along the protan and deutan axes is indicated for strabismic observers without and with amblyopia. The grey region indicates the 95% confidence bands for an individual point calculated from a sample of 172 normal observers. Right. The ratio between the threshold along the tritan axis and the mean of the thresholds along the protan and deutan axes is indicated for strabismic observers without and with amblyopia. The grey region indicates the 95% confidence bands for an individual point.

There was a tendency for the amblyopic residuals to lie above zero in the protan and deutan directions and less so in the tritan direction. This tendency is confirmed by a sign test (protan: $p<0.02$; deutan: $p<0.09$; tritan: $p<0.25$). Observers without amblyopia showed no such trend. However, all of the thresholds lie within the 95% confidence bands. A χ^2 test confirmed that the magnitudes of these residuals were not elevated significantly for either group along any of the axes.

The similarity of threshold variation throughout the lifespan of normal observers suggests that an alternate way to factor out age variation would be to examine the ratio of thresholds across colour directions. Such a procedure is insensitive to an overall sensitivity loss (i.e. one that is equal in all colour directions). It is effective, however, in detecting losses specific to an axis in colour space. We have elsewhere shown that such a procedure is capable of identifying colour-defective individuals independently of the age variation in their thresholds (Knoblauch *et al.*, 1995).

The ratio between protan and deutan thresholds is plotted in the left panel of Figure 2. The grey area indicates the 95% confidence interval computed from the normal data. With the exception of a single point, none of the strabismic observers demonstrates a significant imbalance with respect to the thresholds along these two axes. In the right panel of Figure 2 we have plotted the ratio between thresholds obtained along the tritan axis to the mean of the thresholds along the protan and deutan axes. Here, again, we see no evidence for an imbalance in relative sensitivity across colour directions.

Discussion

We found no influence of strabismus on chromatic sensitivity despite taking into account variations due to age. One criticism of this study is that because we tested all observers binocularly, the good eye may have always dominated. Binocular thresholds are typically lower than monocular thresholds, however. Probability summation predicts that binocular thresholds should be on the order of 1.4 times lower than monocular ones (Arditi, 1986). Thus, if one eye suffers reduced sensitivity, it is not necessarily unreasonable to expect reduced binocular sensitivity. The normal eyes of amblyopes have previously been shown to function more poorly than those of normal observers on some measures of visual function (Barbur *et al.*, 1994). Indeed, there is a suggestion of a slight reduction in sensitivity of strabismic amblyopes along protan and deutan axes, but not along a tritan direction. This difference is detected by a non-parametric, but not by a parametric, statistical test. Such a selective reduction in sensitivity is consistent with a recent report by Earley *et al.* (1993) who found reductions in chromatic sensitivity along red–green but not yellow–blue directions in strabismic amblyopes. The most numerous cell class in the parvocellular pathway responds only to differences between L and M cone signals (Derrington *et al.*, 1984) and would theoretically be the most responsive to the stimuli we presented along the protan and deutan axes. If we were to suppose further that this class of cells in the parvocellular pathway is primarily affected in strabismic amblyopia, then the differences that we observe could be taken as weak support for the hypothesis of parvocellular involvement in amblyopia. Further testing is obviously needed to clarify this issue. It would be interesting, for example, to compare the non-amblyopic and the amblyopic eye of each observer, as Wald and Burian (1944) did.

Previous studies (Pokorny *et al.*, 1979) have demonstrated that monocular losses in chromatic sensitivity in amblyopes can largely be accounted for by eccentric fixation, i.e. the normal reduced chromatic sensitivity of the peripheral retina. We found no differences, however, between the normal sample and the strabismics without amblyopia. Thus, the tendency that we observed, if confirmed in further studies, should be attributed to the amblyopia and not simply to the misalignment of the eyes. Such a result would be in accord with the results of Mtanda *et al.* (1986) who concluded that in strabismic and anisometropic amblyopes 'visual function is worse than predicted on the basis of the area used for fixation'.

An additional criticism that can be raised is that it is likely that the criterion for threshold varies across age. The thresholds based on preferential looking are probably based on a higher criterion than those based on verbal or pointing responses. Adult observers can be motivated to search for very small colour differences that might not be sufficiently salient to engage an infant's attention. Since all analyses were performed relative to the norms for a given age, however, we only need to assume that the criterion is constant within a given age group.

Note added in proof

While this paper was under review, Mullen *et al.* [Mullen, K.T., Sankeralli, M.J. and Hess, R.F. (1996). Colour and luminance vision in human amblyopia: Shifts in isoluminance, contrast sensitivity losses, and positional deficits. Vis. Res. 25: 645–653.] reported 5 of 8 amblyopic observers that they examined required more red in a red/green isoluminance equation with their amblyopic eyes, consistent with the elevation along a protan direction reported here.

References

Arditi, A. (1986). Binocular vision. In: Boff, K., Kaufman, L. and Thomas; J. (eds.), Handbook of Perception and Human Performance: 1–41. Wiley, New York.

Barbur, J.L., Birch, J. and Harlow, A.J. (1993). Colour vision testing using spatiotemporal luminance masking. In: Drum, B. (ed.), Colour Vision Deficiencies XI: Doc. Ophthal. Proc. Ser. 56: 417–426. Kluwer, Dordrecht.

Barbur, J.L., Hess, R.F. and Pinney, H.D. (1994). Pupillary function in human amblyopia. Ophthal. Physiol. Opt. 14: 139–149.

Derrington, A.M., Krauskopf, J. and Lennie, P. (1984). Chromatic mechanisms in lateral geniculate nucleus of macaque. J. Physiol. 357: 241–265.

Earley, M.J., Pierce, G.E. and King-Smith, P.E. (1993). Analysis of spatial, temporal and chromatic vision in human amblyopes. Invest. Ophthalmol. Vis. Sci. (Suppl.) 34: 711.

Efron, B. and Tibshirani, R.J. (1993). An Introduction to the Bootstrap. Chapman & Hall, New York.

Knoblauch, K., Barbur, J.L. and Vital-Durand, F. (1995). Development and aging of chromatic sensitivity. Invest. Ophthalmol. and Vis. Sci. (Suppl.) 36: 910.

Lee, B.B., Martin, P.R. and Valberg, A. (1988). The physiological basis of heterochromatic photometry demonstrated in the ganglion cells of the macaque retina. J. Physiol. (Lond) 404: 323–347.

Mtanda, A.T., Cruysberg, J.R.M., Pinckers, A. and van der Werf, S. (1986). Evaluation of colour vision, mesopic vision, visual evoked potentials and lightness discrimination in adult amblyopes. Doc. Opthalmol. 62: 247–264.

Pokorny, J., Smith, V.C., Verriest, G. and Pinckers, A.J.L.G. (1979). Congenital and Acquired Color Vision Defects: 339–348. Grune & Stratton, New York.

Shapley, R. (1990). Visual sensitivity and parallel retinocortical channels. Annu. Rev. Psychol. 41: 635–658.

Teller, D.Y. (1979). The forced-choice preferential looking procedure: A psychophysical technique for use with human infants. Infant Behav. Dev. 2: 135–153.

Vital-Durand, F. (1992). Acuity card procedures and the linearity of grating resolution development during the first year of human infants. Behav. Brain Res. 49: 99–106.

Wald, G. and Burian, H.M. (1944). The dissociation of form vision and light perception in strabismic amblyopia. Am. J. Ophthalmol. 27: 950–963.

K. Knoblauch
Institut de l'Ingénierie de la Vision
Site GIAT-Industries
3, rue Javelin Pagnon – BP 505
42007 Saint-Etienne Cedex 1, France

29. Detecting colour vision deficiency in 4- and 8-week-old human infants

M.L. BIEBER, K. KNOBLAUCH and J.S. WERNER

(Boulder, CO, USA; St. Etienne, France)

Abstract

Using silent substitution we have measured visually evoked potentials (VEPs) from 4- and 8-week-old infants to L and M cone-isolating stimuli, while the rod response was nulled. Data are shown from six male infants, four of whom were classified as candidates for defective colour vision. These results demonstrate that infants as young as 4 weeks show clear responses to L and M cone-isolating stimuli. This technique may be useful in identifying infants with sex-linked colour deficiencies.

Introduction

It is generally accepted that infants as young as 2 months of age possess some ability to discriminate stimuli on the basis of wavelength. For example, previous studies have demonstrated the ability of infants as young as 2 months to make both tritan (Varner *et al.*, 1985) and Rayleigh discriminations (Hamer *et al.*, 1982; Packer *et al.*, 1984). Based on the results of these studies, one might conclude that all three cone types (S, M and L cones) and the post-receptoral circuitry necessary for wavelength discrimination are functional by this age. However, under the conditions in which infants are often tested, it is difficult to be certain which mechanisms are involved. The use of large fields, poorly controlled fixation, and low luminance levels all favour contributions from rod signals. Rod signals combined with one functional cone type may explain some of the Rayleigh discriminations reported from infants.

In an attempt to verify the functioning of both L and M cones in early infancy, we used 'silent substitution' (Estévez and Spekreijse, 1982) which allows one to isolate the response of a single receptor type while maintaining a constant output from the other receptors. If this method is valid it should allow one to examine the ontogenesis of both normal and abnormal visual mechanisms. We have collected data from three adults and 22 infants to M and L cone-isolating stimuli. The data presented here are from six male infants, four of whom yielded weak responses to only one of the cone-isolating stimuli; these were subsequently classified as candidates for defective colour vision, based on

C.R. Cavonius (ed.), Colour Vision Deficiencies XIII, pp. 277–282.

verbal reports from their parents of a familial history of colour deficiency on the maternal side of the family.

Methods

Three adults (1 with normal colour vision, 1 protanope, and 1 deuteranope) and six full-term infants (three 4-week-olds and three 8-week-olds) served as subjects. Each participated in 1–4 testing sessions that each lasted approximately 45–60 min.

To simplify the stimulus, we worked in the Rayleigh region of the visible spectrum (wavelengths $\geqslant 540$ nm), thereby minimizing any S cone contribution. Assuming that there are three functional photoreceptor types (rods, L and M cones) operating in the Rayleigh region, a variable mixture of three lights is needed to silence any two of them at a time. Our three-primary stimulus, presented in a 6° disc, consisted of a 570 nm light alternating in square-wave counterphase at 7.5 Hz with a mixture of 610 and 540 nm. The intensity of the 570 nm light and the ratio of the 610/540 nm mixture can be set so as to modulate only the L or M receptors. For example, L cone isolation results when the number of quantal absorptions produced by the 570 nm light is the same as that produced by the 610/540 nm mixture for the rods and M cones, but not for the L cones. In two separate conditions (L and M cone isolation) the amplitude of the fundamental component of the VEP was measured while continuously increasing the mean luminance of the receptor-isolating stimulus for 30 sec, over a 2 log unit range.

Electrodes were placed over the occipital cortex with a ground on the forehead. The EEG signal was amplified and filtered by passing it through an isolation amplifier and then into a battery-powered pre-amplifier. The amplified and filtered EEG signal was then fed into a vector voltmeter, which selectively amplifies a particular frequency component of the EEG signal. Our vector voltmeter was synchronized to the 7.5 Hz modulation of our stimulus, in order to selectively amplify the phase-locked amplitude of the fundamental component (7.5 Hz) of the VEP signal. Because the vector voltmeter extracts the VEP in real time, continuous changes in amplitude of the 7.5 Hz VEP response could be plotted as a function of the intensity of the receptor-isolating stimulus.

To estimate the level of background noise and artefacts during testing, separate VEP measurements were made at a frequency 10% above the fundamental (8.25 Hz) and at the second harmonic in 15 full-term infants between 1 and 2 months of age. The amplitude of the VEP response was maximal at the fundamental frequency.

Results

Figure 1 shows mean VEP data obtained using L and M cone-isolating stimuli from two 4-week-old infants and two 8-week-old infants. VEP functions represent the mean of 2–8 VEP records (the total number of VEP records collected from each infant under each condition). Mean VEP data have been collected under both L and M cone isolation from 18 candidate colour-normal infants (see Knoblauch *et al.*, 1994; Werner *et al.*, 1995). Candidates for normal colour vision were classified based on strong responses to both L and the M cone isolation and no reported family history of colour deficiency. The data for two of these infants are presented in the upper panels of Figure 1. Data from two infants who were classified as candidate protans are shown in the lower panels. Both these infants showed a stronger response to the M cone-isolating than to the L cone-isolating stimulus. For both of these infants the VEP responses obtained using the L cone-isolating stimulus were no greater than

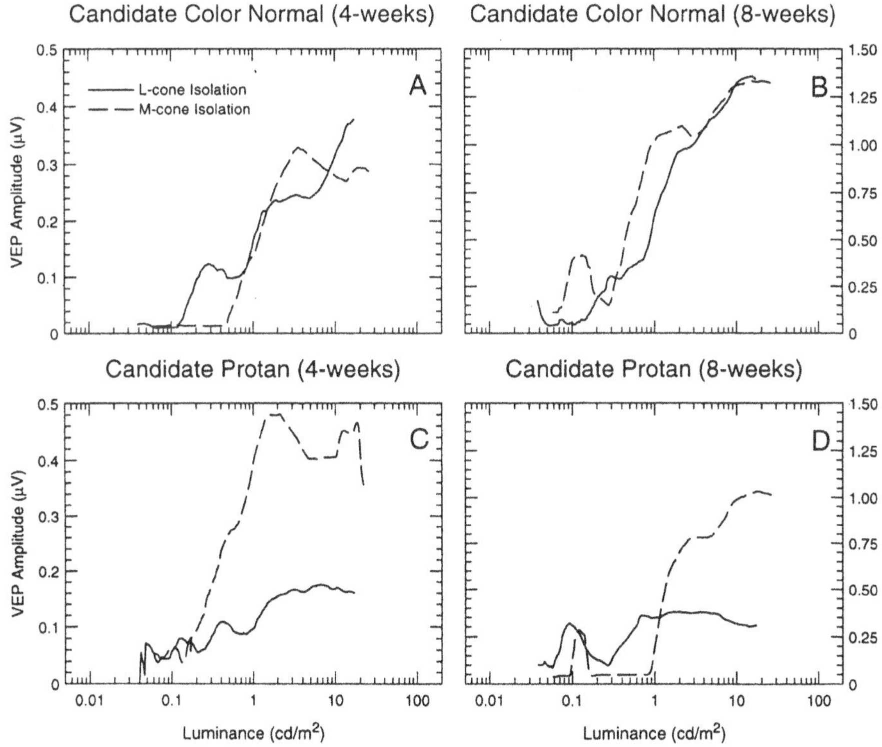

Fig. 1. Mean VEP records obtained from two 4-week-old (A, C) and two 8-week-old (B, D) infants. The upper and lower panels present data from candidate colour-normal infants and candidate protans, respectively. Solid and dashed functions represent mean VEP functions obtained using L and M cone-isolating stimuli, respectively. Amplitude (µV) of the VEP response is plotted as a function of the luminance (cd/m^2) of the receptor-isolating stimulus. Note that the amplitudes measured from 4- and 8-week-old infants are scaled differently.

the noise level. Two other infants (one 4-week and one 8-week, data not shown) were classified as candidate deutans on the basis of a strong response to the L cone-isolating stimulus with little or no response to the M cone-isolating stimulus. The parents of all four of these infants subsequently reported a familial history of colour deficiency on the maternal side. In addition, red/green flicker photometric data were obtained from the mother of the 4-week-old candidate protan. Females, heterozygous for a protan deficiency, often demonstrate Schmidt's sign, a reduced sensitivity at longer wavelengths (Schmidt, 1955). A 530 nm standard was alternated at 15 Hz in square-wave counterphase with a 650 nm test field, in a 2° disk. The intensity of the test field was varied, using the method of adjustment, to minimize flicker. Consistent with Schmidt's sign, the sensitivity of the mother of the 4-week-old protan suspect fell between the normal and protan luminosity function. Her flicker minimum required more than twice as much light at 650 nm as two presumably non-heterozygous females tested at the same time.

Discussion

One necessary assumption of this approach is that the spectral sensitivities of the photoreceptors are known. It is reasonable to assume that the spectral sensitivities of infant photoreceptors are the same as those of the adult in the Rayleigh region, with the exception of differences that arise from variations in photopigment optical density and pre-retinal absorption. The shorter outer segments of infant cones and consequent reduced photopigment optical density manifests itself as a narrowing of the photopigment absorption spectrum. However, we have shown earlier (Knoblauch *et al.*, 1996), that over a range of cone optical densities that might be reasonably assumed for human vision, an incorrect assumption is not likely to upset the receptor isolation conditions enough to produce a false-positive VEP response.

In order to validate further our receptor isolation conditions we performed several control experiments using both colour-normal and dichromatic adult observers. Individuals with normal colour vision should demonstrate a strong VEP response to both the L and M cone-isolating stimuli, whereas dichromats should demonstrate a strong response to only one of the isolating stimuli, giving a negligible response to the isolated mechanism that they lack (i.e. protanopes should not respond to L cone isolation and deuteranopes should show no response to M cone isolation). Figure 2 shows mean VEP data collected from a colour-normal observer to L cone and M cone-isolating stimuli. VEP data from a protanope and a deuteranope are also shown. VEP functions represent the mean of 2–7 records. The colour-normal adult showed a strong response to both the L and M cone-isolating stimuli. The protanope, on the other hand, produced a strong response only to the M cone isolation condition, and the deuteranope produced a strong response only to the L cone isolation condition.

While the measurements made with dichromatic individuals provide strong

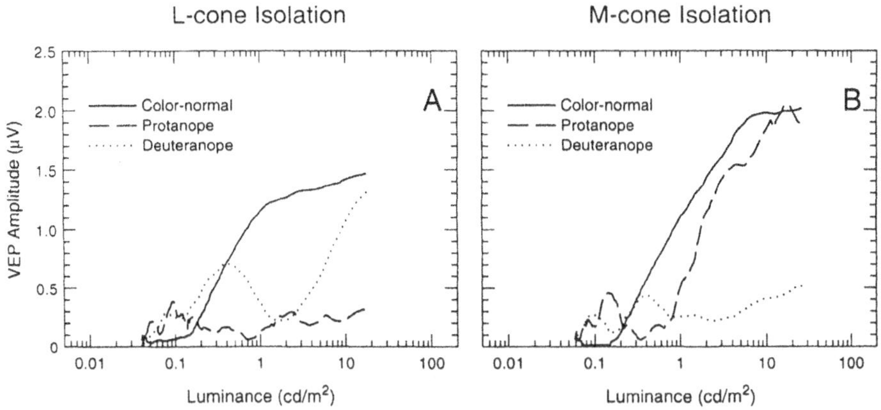

Fig. 2. Mean VEP records obtained from an adult with normal colour vision (bold function), deuteranope (dotted function) and protanope (dashed function) for L (A) and M (B) cone-isolating stimuli, respectively. Amplitude (μV) of the VEP response is plotted as a function of the luminance (cd/m^2) of the receptor-isolating stimulus.

support for our isolation conditions, a more direct test would be to measure the action spectra of isolated mechanisms. For example, the action spectrum measured under L cone isolation should correspond to the L cone fundamental and that measured under M cone isolation should correspond to the M cone fundamental. Psychophysically derived action spectra, not shown here, collected from two colour-normal adults using L cone- and M cone-isolating stimuli are in good agreement with the Smith and Pokorny L and M cone fundamentals, consistent with a previously reported study using similar methods (Estévez *et al.*, 1975).

Compared with previous methods used to study the ontogeny of visual mechanisms, the silent substitution technique shows considerable promise. The VEP data obtained using this technique demonstrate that infants (with the exception of those classified as candidate colour defectives) as young as 4 to 8 weeks of age show clear responses to L and M cone-isolating stimuli. These results are consistent with previous studies using different methods (Hamer *et al.*, 1982; Packer *et al.*, 1984). However, because these previous studies did not adequately control for rod response, it is difficult to be sure which mechanisms are functional in early infancy. While holding rod output constant we have demonstrated that L and M cones are functional and that their signals can be measured in the cortex as early as 4 weeks of age. In addition, this technique appears to be able to identify sex-linked colour deficiencies in early infancy.

References

Estévez, O. and Spekreijse, H. (1982). The 'silent substitution' method in visual research. Vision Res. 22: 681–691.

Estévez, O., Spekreijse, H., Van den Berg, T.J.T.P. and Cavonius, C.R. (1975). The spectral sensitivities of isolated human color mechanisms determined from contrast evoked potential measurements. Vision Res. 15: 1205–1212.

Hamer, R.D., Alexander, K.R. and Teller, D.Y. (1982). Rayleigh discriminations in young human infants. Vision Res. 22: 575–587.

Knoblauch, K., Bieber, M.L. and Werner, J.S. (1994). VEP responses to receptor-isolating stimuli from 8-week-old infants. Invest. Ophthalmol. Vis. Sci. (Suppl.) 35: 1571.

Knoblauch, K., Bieber, M.L. and Werner, J.S. (1996) Assessing dimensionality in infant color vision. In Vital-Durand, F., Braddick. O. and Braddick, J. (eds.), Infant Vision: 51–62. Oxford, New York.

Packer, O., Hartmann, E.E. and Teller, D.Y. (1984). Infant color vision: The effect of test field size on Rayleigh discriminations. Vision Res. 24: 1247–1260.

Schmidt, I. (1955). A sign of manifest heterozygosity in carriers of color deficiency. Am. J. Opt. 32: 404–408.

Smith, V.C. and Pokorny J. (1975). Spectral sensitivity of the foveal cone photopigments between 400 and 500 nm. Vision Res. 15: 161–71.

Varner, D., Cook, J.E., Schneck, M.E., MacDonald, M.A. and Teller, D.Y. (1985). Tritan discriminations by 1- and 2-month old human infants. Vision Res. 25: 821–831.

Werner, J.S., Bieber, M.L. and Knoblauch, K. (1995). Isolated M- and L-cone responses in the VEPs of 4-week-old human infants. Invest. Ophthal. Vis. Sci. (Suppl.) 36: S910.

Michelle L. Bieber, J.S. Werner
Department of Psychology
University of Colorado
Boulder, CO 80309-0345, USA

K. Knoblauch
INSERM Unite 371
18 ave. du Doyen Lepine
Bron 69675, France

30. Cone monochromacy: a case report

L. VAJOCZKI and P.L. PEASE
(Houston, TX, USA)

Abstract

Spectral sensitivities and results of clinical colour tests were obtained for a 28-year-old male who exhibited a complete absence of colour perception, normal visual acuity, no nystagmus and no photophobia. Increment-threshold spectral sensitivities showed multiple peaks, suggesting the presence of more than one photopigment in the middle to long wavelength region. The absolute values of the increment-threshold spectral sensitivities were lower than those in individuals with normal colour vision, with a marked sensitivity loss at short wavelengths, suggesting reduced or complete absence of the short wavelength sensitive cones. Spectral sensitivity to rapid flicker (25 Hz) was spectrally broad, indicating the presence of more than one cone type in the middle and long wavelengths, but again with reduced sensitivity.

Introduction

Cone monochromacy is an extremely rare type of complete achromatopsia, with an estimated incidence of one in one hundred million (0.000001%; Pitt, 1944). There are two types of complete achromatopsia, one with reduced visual acuity (typical or rod monochromacy) and one with normal visual acuity (atypical or cone monochromacy). In addition to complete achromatopsia, two types of incomplete achromatopsia are distinguished: X-chromosome-linked incomplete achromatopsia (blue mono-cone monochromacy) and autosomal recessive incomplete achromatopsia (Pokorny *et al.*, 1979). Complete achromats have no ability to discriminate hues of equal brightness, whereas incomplete achromats exhibit dichromatic colour vision under some viewing conditions.

Cone monochromats have normal visual acuity and demonstrate a Purkinje shift. They do not have the associated symptoms found with rod monochromats. Rod monochromats have reduced visual acuity, photophobia, may show nystagmus, they exhibit no Purkinje shift, and their spectral sensitivity agrees with the absorbance spectrum of rhodopsin. Spectral sensitivity of the cone monochromat may have one of two forms: it may resemble that of the protanope as shown by Pitt (1944) for one cone monochromat and by Weale (1953) for three, or may resemble that of the deuteranope as shown by Alpern (1974) for one observer. In

C.R. Cavonius (ed.), Colour Vision Deficiencies XIII, pp. 283–290.

terms of the number of receptor photopigment mechanisms, Alpern provided evidence for two cone mechanisms, the long-wavelength sensitive (L) and the short-wavelength (S) sensitive cones. Pitt suggested a single cone photopigment, the mid-wavelength sensitive (M) cones and described his observer as a tritanopic protanope. The spectral sensitivity curves for each of Weale's three observers are too broad to match exactly that of a protanope or a tritanope and the results therefore suggest that all three cone photopigments are present. There is general agreement that the colour defect of the cone monochromat is of post-receptoral origin (Pokorny *et al.*, 1979). The term cone monochromacy is a misnomer if it is understood to imply that the defect is in the photopigment or that a cone monochromat has only one cone type.

The observer who participated in this study is a complete achromat with normal visual acuity who made himself available to us for testing on three different occasions. A number of colour vision tests were administered and spectral sensitivity was determined by measurement of the increment-threshold for both 1° and 4° targets superimposed upon dim white background and by measurement of thresholds for detection of 25 Hz flicker. Dark adaptation was assessed with broad-band test stimuli using a modified Goldmann–Weekers dark adaptometer and the results showed elevated cone thresholds but a normal level for the rod branch. This report is about the results of colour testing and the measures of spectral sensitivity.

Methods

Subject

A 28-year-old Caucasian male (MR) presented to the clinic inquiring about his vision, which had previously been diagnosed as rod monochromatism. He reported a complete absence of colour vision since birth. His medical history included a diagnosis of Crohn's disease in 1991 for which he is currently taking prednisone 20 mg/day. Family ocular history revealed red–green colour deficiencies in his maternal grandfather and cousin. Entering acuities through −8.00D Acuvue lenses were 20/20 OD, OS, OU. He exhibited no nystagmus or photophobia. Anterior ocular assessment and fundus examination were unremarkable OU.

Colour vision tests

The tests of colour vision included the Ishihara (24 Plate Ed., 1976) and the AO-HRR pseudoisochromatic plates, the Farnsworth Dichotomous Test (Panel D15), a large version of the Panel D15, and the H16, all of which were administered monocularly under standard conditions with the Macbeth easel lamp. The Dvorine Nomenclature test, the Sloan Achromatopsia test and the Nagel anomaloscope were also administered.

Spectral sensitivity

Spectral sensitivity was determined with a free-viewing rear projection apparatus with tungsten-halogen lamps (Thorn THE-335: 12 V, 100 W from Gilway Technical Lamps stock no. L9390) and heat-absorbing glass in place for both monochromatic test and achromatic adaptation fields. An interference wedge (Verils 200 continuous line filter 200 × 25 mm, Schott Glass Technologies, Duryea, PA) provided monochromatic light with a half-bandwidth of 18 nm throughout the applicable spectrum. Spectral thresholds were obtained with a random sequence of wavelengths at intervals of 25 nm between 425 nm and 675 nm. Circular monochromatic targets of either 1° or 4° subtense were superimposed and centred upon an 8° diameter achromatic adaptation field having a colour temperature of 3148K and, depending upon protocol, a luminance of either 40 cd/m^2 or 2.2 cd/m^2. Cross-hairs were used for fixation and data were obtained with monocular viewing. Test stimuli were either on continuously or flickered at 25 Hz. The mean of six thresholds for the detection of the 25 Hz flicker was determined (3 ascending and 3 descending series). Increment thresholds were determined by increasing the radiance of the stimulus from below threshold to threshold and then lowering the radiance approximately 0.75 log units below threshold before the next determination. For the increment-threshold spectral sensitivities, the average result (log geometric mean of radiance) of three trials is reported at each wavelength. Data are reported for the cone monochromat (MR) and an age-matched control (RG) with normal colour vision. Each of the subjects was a first-time observer for the type of experiments described here. Threshold values are expressed in terms of the relative stimulus radiance (watts/ster*m^2/nm).

Results

Colour vision testing revealed that MR was unable to identify any of the Ishihara or the AO-HRR pseudoisochromatic plates with either eye except for plate three of the AO-HRR screening series where he could faintly see the 'x' (this symbol has the least contrast of any of the symbols on the test). When asked to identify the colours on the Dvorine Nomenclature test, he reported only shades of grey, though when pressed to use colour names, he consistently labelled red as 'black' and blue as 'red'. The results of the large (4° disc) and standard D15 and the H16 tests were anarchic. With the Sloan Achromatopsia test, he was able to find an exact match of each of the six colours with a grey paper and the values were equivalent to those of a colour normal. With the Nagel anomaloscope (Model I), he was capable of making exact colour matches over the entire range of red–green mixtures by adjusting the luminosity of the yellow. Pure red (664 nm) and pure green (549 nm) were matched with nearly the same low yellow setting of about 1–2 Nagel scale units, with a peak value of 13 scale units in the middle (40) of the red–green range. The Nagel results are shown in Figure 1. A diagnosis of cone monochromacy was made based on the above colour vision results and normal visual acuity.

Figure 2 shows the spectral sensitivity to rapid flicker (25 Hz) obtained with 4° monochromatic stimuli superimposed on a 40 cd/m^2 background. The results

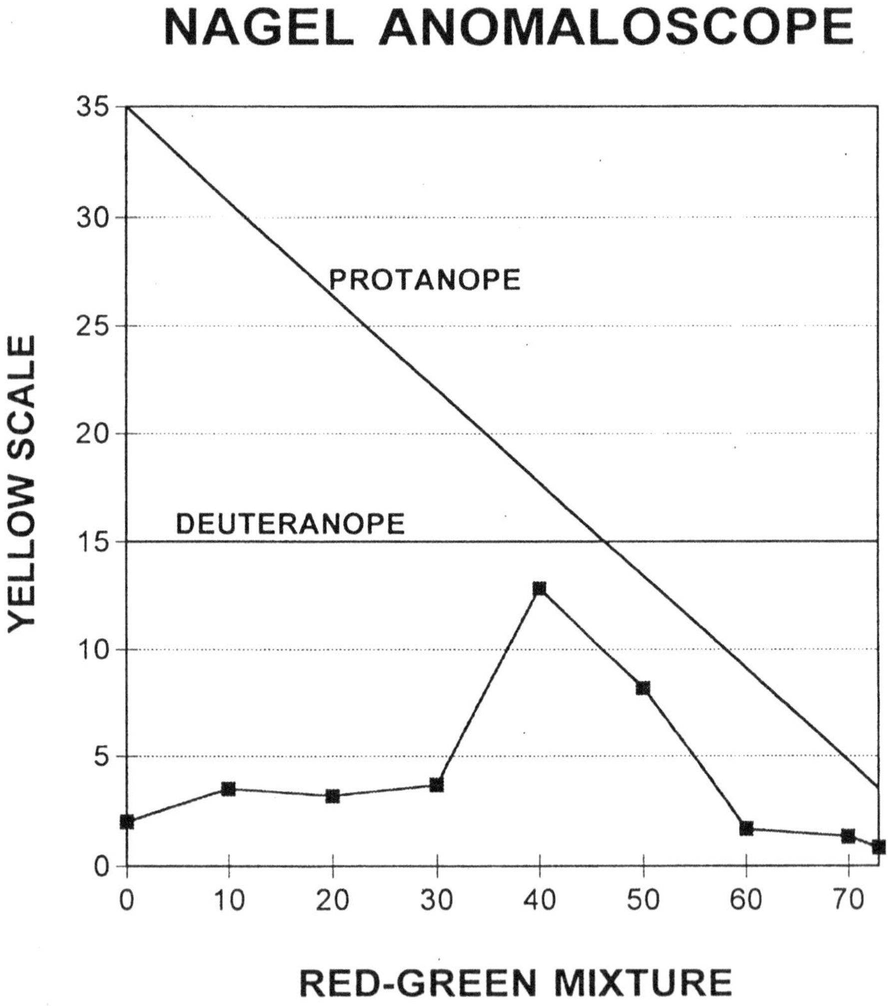

Fig. 1. Results of matching the brightness of yellow to various mixtures of red and green from pure green (0) to pure red (73) on the Nagel anomaloscope. Lines for the protanope and deuteranope are from Pokorny *et al.*, 1979.

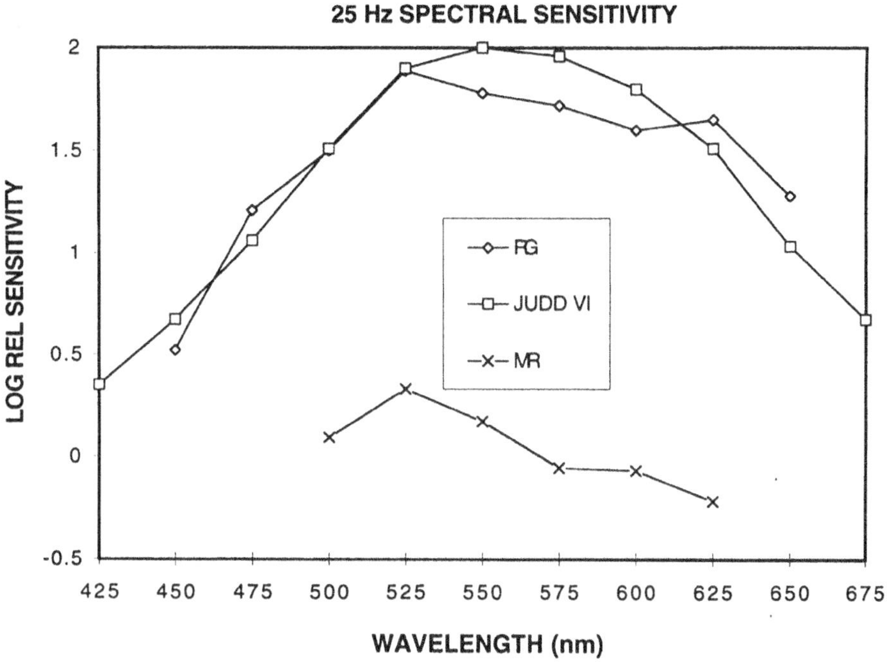

Fig. 2. Log relative sensitivity (log 1/watts/ster*m^2/nm) for detection of flicker for the cone monochromat (MR) and a colour normal (RG). Open squares show the Judd modification of the V_λ function (Wyszecki and Stiles, 1967). Data are the means of six thresholds obtained with 4° test targets flickered at 25 Hz and superimposed upon an 8° achromatic background having a luminance of 40 cd/m^2.

are similar to that of the colour normal but depressed by about 1.6 log units. Data for MR could not be obtained at wavelengths shorter than 500 nm or longer than 625 nm because of the limited radiance of the apparatus.

Increment-threshold spectral sensitivities for both observers are shown in Figure 3. These data were obtained with a background luminance of 2.2 cd/m^2 for both 1° and 4° targets. The absence of a more pronounced peak at short wavelengths for the colour normal with the 4° target is likely due to the low luminance of the background (Gunji *et al.*, 1995). Compared with the colour normal, the cone monochromat shows reduced sensitivity to all wavelengths and, even with the dimmer background, thresholds exceeded the stimulus levels that could be obtained with the apparatus at wavelengths shorter than 500 nm. The shapes of the curves for the different size targets are fairly similar in the mid- to long wavelengths for the colour normal but there are obviously very different shapes for each size target for the cone monochromat.

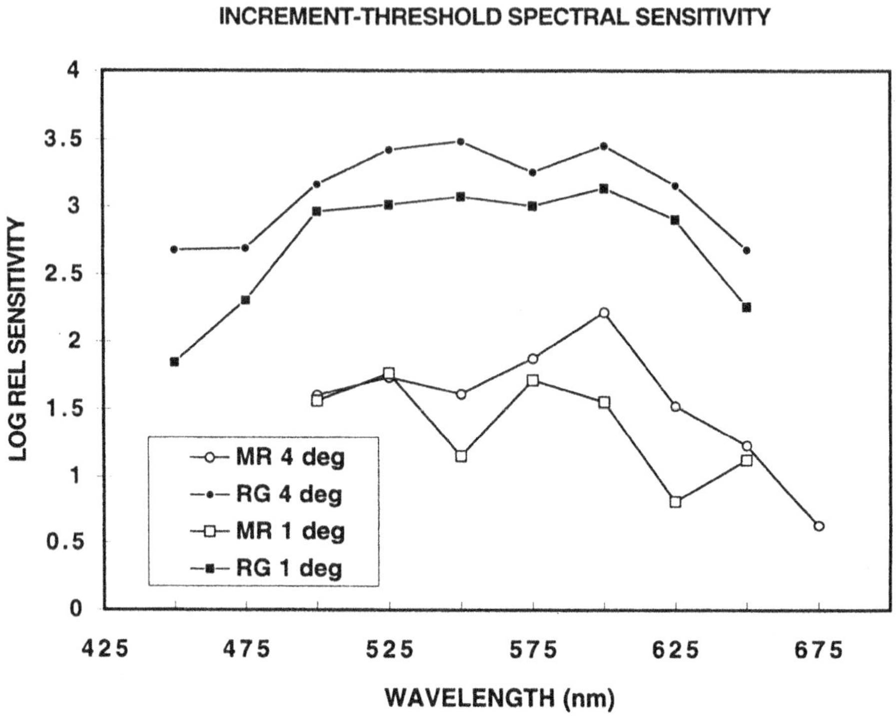

Fig. 3. Increment-threshold spectral sensitivity for MR (open symbols) and RG (filled symbols). Sensitivity was determined for 1° (open and filled circles) and 4° (open and filled squares) test targets superimposed upon an 8° achromatic background having a luminance of 2.2 cd/m^2. Data points are the average of three threshold determinations at each wavelength. Repeat determinations of thresholds for each subject at each wavelength did not differ by more than 0.15 log units.

Discussion

The incidence of cone monochromacy was estimated by Pitt (1944) at one in one hundred million. This estimate was based upon the premise that his observer was a double dichromat, i.e. a protanope who was also tritanopic. The incidence figures that Pitt used for the different forms of dichromacy, protanopia (1%), deuteranopia (1.1%), and tritanopia (0.0001%) were multiplied to estimate the incidence for each of the two types of double dichromat. Hence, the prevalence for a tritanopic deuteranope is the product of the prevalence of both tritanopia and deuteranopia (0.000001×0.011) and the tritanopic protanope (0.000001×0.01). Since the incidence figures that Pitt used for protanopia and deuteranopia are those for males alone, the estimate should be revised for the total population (males and females), that is 1 out of

200 hundred million not 1 out of 100 million as Pitt estimated for each type of double dichromat. Further revision is necessary in this context since more recent figures (Pokorny *et al.*, 1979) provide an estimate of tritanopia at 0.002–0.007%; hence, to the nearest million, the likelihood for a double dichromat would be either 1 in 10 million or 1 in 3 million. The data reported here and that given by both Weale (1953) and Alpern (1974) provide evidence that some cone monochromats have more than one cone type and are not, therefore, double dichromats. To estimate the prevalence of cone monochromats who have more than one cone type requires a mechanism different to that proposed by Pitt.

The comparison of the different measures of spectral sensitivity of MR to the colour normal show some interesting differences and similarities. The results obtained for detection of rapid flicker (25 Hz) showed a nearly normal luminosity function, though depressed by about 1.6 log units. The increment-threshold spectral sensitivity curves for the colour normal showed multiple peaks, as expected from inhibitory interaction between two cone types in the mid- to long wavelengths (Sperling and Harwerth, 1971). The decrease in sensitivity for the colour normal at 450 nm for the 1° target is expected. The cone monochromat is less sensitive to all wavelengths and his thresholds are higher than those obtainable with the apparatus for wavelengths shorter than 500 nm. The positions of the peaks in the 4° curve for MR are similar to those of the colour normal, but MR shows a very different profile for the 1° stimulus. These results may indicate that there are abnormal interactions in different regions of the retina that result from changing the stimulus size, or that the cone types or cone pigment densities are different in different regions of the retina. The data are insufficient to select a viable model, but the breadth of the spectral sensitivity curves clearly indicates the presence of more than one cone type. The lack of sensitivity in the short wavelengths indicates that the S cones for this patient are either absent or present in very few numbers.

This achromat has at least two cone mechanisms and is likely to lack the S cones. The low luminosity that MR shows with the yellow setting on the Nagel anomaloscope to match pure red and pure green, 665 nm and 550 nm on our instrument, are roughly in agreement with the increment-threshold sensitivities with the 1° target (Fig. 3). Previous measures of spectral sensitivity of cone monochromats were obtained by colour matching with 2° or smaller fields (Pitt, 1945; Weale, 1954; Alpern,1974). Because of the different methods, our data are not truly comparable to these earlier measures, though we have begun to collect colour matching data using a 4° bipartite field and the results are similar in shape to the 4° increment-threshold spectral sensitivity curve. We conclude that MR is an achromat with normal visual acuity and has at least two cone photopigment mechanisms.

Acknowledgments

Technical services for the design and construction of the optical system were provided by Chris Kuether and Ming Qi and supported in part by Core Center Grant (P30-EY07551).

References

Alpern, M. (1974). What is it that confines in a world without colour? Invest. Ophthalmol. 13: 648–674.

Gunji, H., Kitahara, K., Noji, J. and Kubo, A. (1995). The necessary intensity of the white background when measuring the response of the S cone system. In: Drum, B. (ed), Colour Vision Deficiencies XII, Doc. Ophthalmol. Proc. Ser. 57: 317–320. Kluwer, Dordrecht.

Pitt, F.H.G. (1944). Monochromatism. Nature 154: 466–468.

Pokorny, J., Smith, V.C. and Verriest, G. (1979). Congenital colour defects. In: Pokorny, J., Smith, V.C., Verriest, G. and Pinckers, A.J.L.G. (eds.), Congenital and Acquired Colour Vision Defects: 232–241. Grune & Stratton, New York.

Sperling, H.G. and Harwerth, R.S. (1971). Red–green cone interactions in the increment-threshold spectral sensitivity of primates. Science 172: 180–184.

Weale, R.A. (1953). Cone-monochromatism. J. Physiol. (Lond.) 121: 548–569.

Wyszecki, G. and Stiles W.S. (1967). Colour Science: Concepts and Methods, Quantitative Data and Formulas: 436. John Wiley, New York.

University of Houston
College of Optometry
Houston, TX 77204-6052, USA

31. A new phenotype? Nagel 'dichromats' who reject 'red matches green' for a 2° field

J.D. MORELAND and C. BRADSHAW
(Keele, UK)

Abstract

Population statistics on the relative frequencies of different red–green deficiencies are based on diagnoses using the Nagel anomaloscope with a 2° field. However, the colour range of the Nagel anomaloscope is insufficient for a secure diagnosis of true dichromacy. Tests with a 2° field have revealed, in a group of 12 classical 'Nagel dichromats', three deutans who cannot match the Rayleigh red and green stimuli with each other. This subgroup is characterized by a matching range intermediate between the widest available on Nagel's anomaloscope and the greatest achievable range for dichromacy. The incidence of complete dichromacy may be overestimated by Nagel anomaloscopy.

Introduction

Several frequently cited large population studies (Le Grand, 1968) used the Nagel anomaloscope (Nagel, 1907) to determine the relative frequencies of the various red–green deficiencies. This instrument presents narrow-band spectral lights for metameric colour matching in the Rayleigh equation. The stimuli are a yellow test (Y, 589 nm) and a comparison mixture of red (R, 671 nm) and green (G, 546 nm) presented in two juxtaposed halves of a circular 2° field.

Subjects with shifted matches (too red or too green) but with a narrow range of accepted R+G mixtures are classified as simple anomalous trichromats. Those with wide matching ranges which include the normal match and possibly one of the primaries are classified as extreme anomalous trichromats, while those accepting all R+G mixtures, including R and G, as colour matches are classified as dichromats.

Following Lakowski's (1965) procedure, chromatic distances between stimuli may be expressed in JND (normal just noticeable difference) units. The chromatic distance between R and Y is about 50 JND and that between Y and G is about 34 JND (Fig. 1). Both of these distances are significantly smaller than the highest potential threshold (about 100 JND) for red–green dichromacies. Moreover, Nagel's anomaloscope does not allow the comparison of R with G, which represents a chromatic range of at least 84 JND. Consequently,

C.R. Cavonius (ed.), Colour Vision Deficiencies XIII, pp. 291–295.

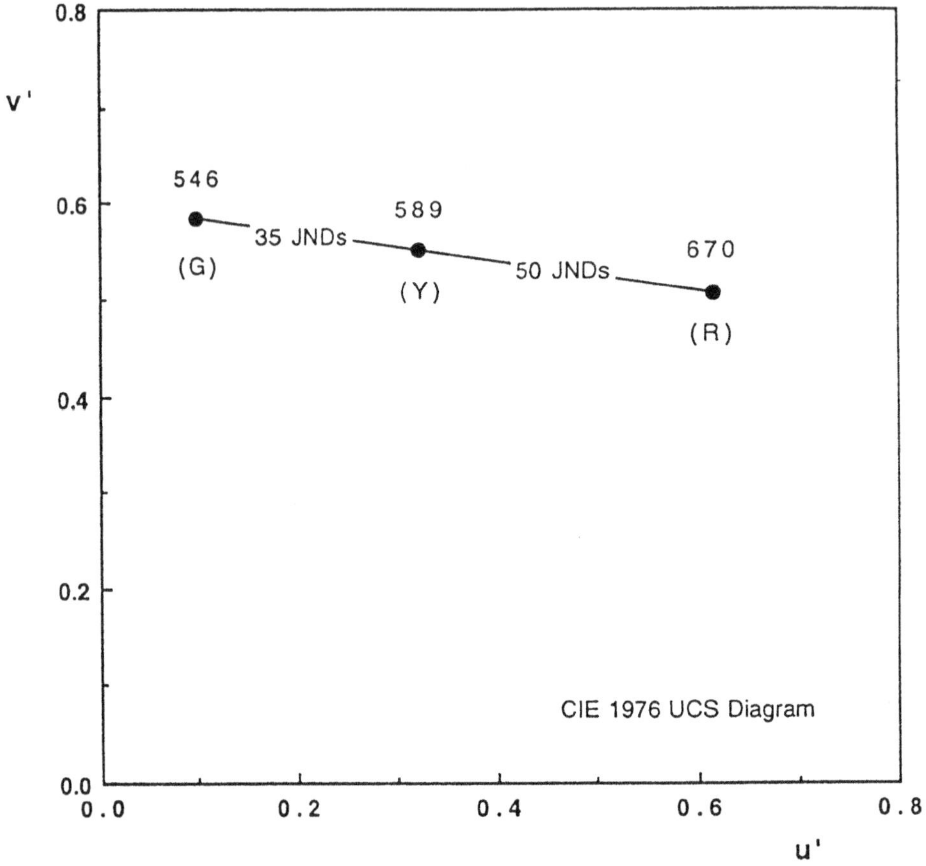

Fig. 1. The red (R), yellow (Y) and green (G) stimuli of Rayleigh's equation shown in the approximately uniform CIE 1976 chromaticity diagram. Since this diagram retains significant non-uniformity, the chromatic separations of the stimuli were calculated in normal JND units from Wright's (1946) classic wavelength discrimination data. The diagnosis of dichromacy on Nagel's anomaloscope requires the separate acceptance of 'R matches Y' and 'G matches Y', representing matching ranges of 50 and 35 JND respectively. This diagnosis is, however, uncertain since Nagel's anomaloscope cannot present larger chromatic differences: for example, the two primaries, separated by 85 normal JND, cannot be compared in an 'R matches G' equation.

acceptance or rejection of the extreme R–G comparison cannot be tested and, as has been noted by Alpern and Wake (1977) on other grounds, the classical diagnosis of dichromacy based upon Nagel 2° anomaloscopy is uncertain. Indeed, if anomalous trichromats with extremely poor hue discrimination (>50 JND) exist they would be classified erroneously as dichromats and the currently accepted published incidence of dichromacy would be an over-estimate.

Materials and methods

A group of self-identified volunteers with defective colour vision were screened using the Ishihara plates (38-plate, 1972 edition under a North Sky daylight fluorescent lamp, 6127K). Some 30 red–green defectives were identified and selected for further tests. The Moreland anomaloscope (Moreland and Kerr, 1979) was set for Rayleigh matching with the standard 2° field. Subjects were presented with the two extremes of the Nagel anomaloscope, 'R matches Y' and 'G matches Y', and then with the more exacting comparison of the two primaries, 'R matches G' (Fig. 2). Subjects were allowed to adjust the radiance of the yellow field to match the comparison field for luminance. The key instruction was 'think of the two light patches as windows and judge whether the amounts of light coming through them are the same'. This instruction is quite effective in helping inexperienced subjects to discount the disturbing effect of large chromatic differences between the fields, which has been documented by Schmidt (1974) and others. The luminance match was repeated five times and, on each occasion, the subject was asked to describe any colour differences between the fields. Finally, the luminance of the yellow stimulus was fixed at the mean of the subject's five settings and the question repeated.

Moreland matches (Moreland, 1993) were made to exclude blue–green deficits. All observations, for both Rayleigh and Moreland equations, were made using a short glance (1–2 sec) technique to minimize 'tuned' (*Umstimmung*) responses (Krastel and Moreland, 1991).

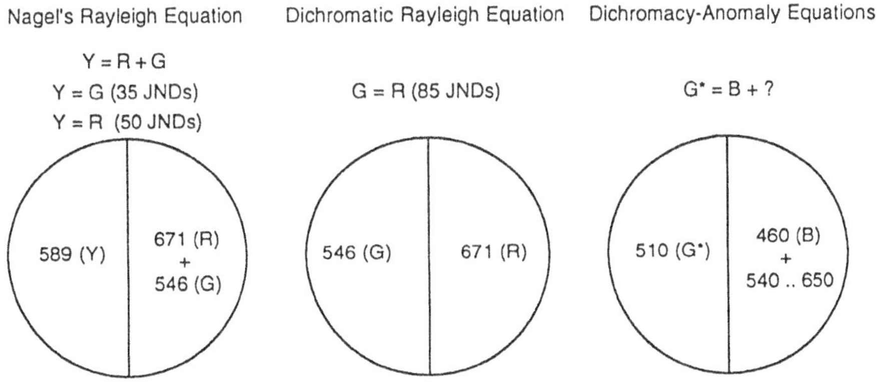

Fig. 2. Matching fields with 2° diameter. Left: simulation of the Nagel anomaloscope's Rayleigh equation. Centre: direct comparison of the Rayleigh primaries. Right: comparison of a green test light (510 nm) and mixtures of blue (460 nm) with various longer wavelength lights (540–650 nm). This provides a choice of matching ranges from about 12–90 normal JND (see Fig. 3).

Results and discussion

Colour naming was found to be generally very consistent and this was taken as evidence of reliable performance. The occasional inconsistency that did occur was usually attributable to a luminance mis-match. Attention to this point is particularly important in establishing large matching ranges.

We found 12 classical 'Nagel dichromats' (1 protan, 11 deutan), three of whom were deutans who rejected 'R matches G'. Two of these three were related (grandfather and grandson). Since we used a 2° field it is unlikely that this rejection could have been due to the large-field anomaly performance exhibited by many classical dichromats (Moreland, 1993).

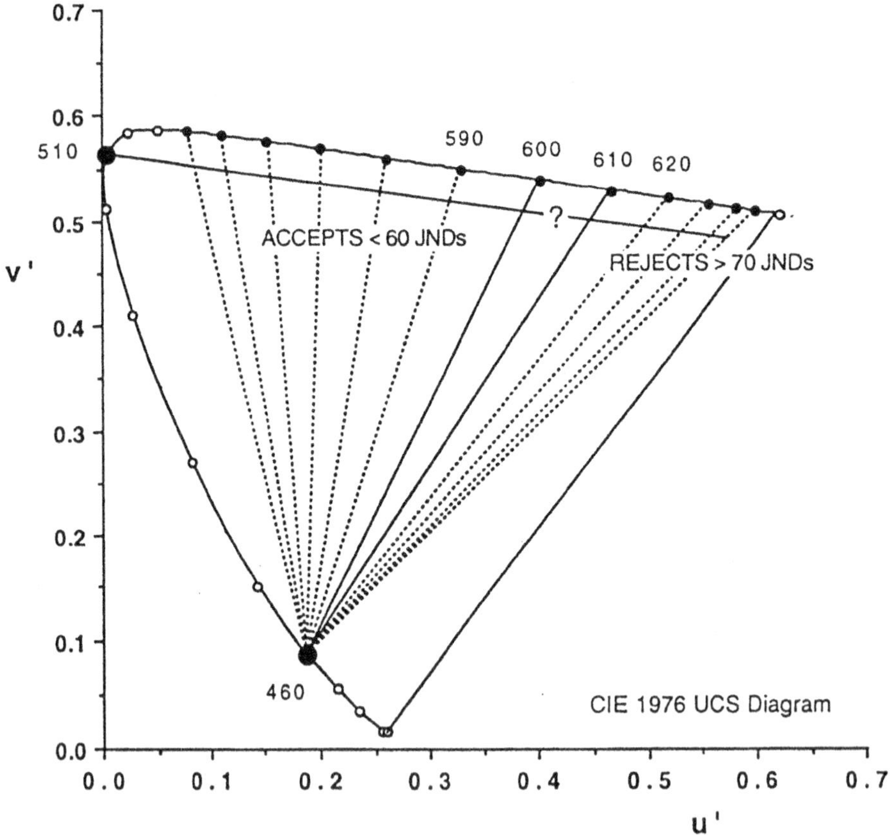

Fig. 3. Comparison of a green test light (510 nm) and mixtures of blue (460 nm) with various longer wavelength lights (540–650 nm) in a 2° field. The line drawn through the 510 nm point characterizes the direction of deutan confusions and the intersections of that line with the various mixture lines are used to determine the matching range limits in JND units. One classical 'Nagel 2° dichromat', a deutan who rejected 'R matches G' (85 JND), accepted matches made with mixtures containing lights of wavelength less than 600 nm but rejected mixtures with lights of wavelength greater than 610 nm. This establishes his matching range as 60–70 JND.

Further tests on one of these three, using colour matches between a green test light (510 nm) and mixtures of blue (460 nm) with various longer wavelength lights (540–650 nm), showed a matching range of some 60–70 normal JND (Fig. 3). Consequently, extending the terminology describing anomalous trichromats as simple or extreme, these three deutans could be described as very extreme deuteranomalous trichromats.

The sample is small and so it is too early to claim the existence of a distinct phenotype. However, the grandfather–grandson relationship for two of the subjects does point in that direction.

Conclusions

The colour range of the Nagel anomaloscope is insufficient for a secure diagnosis of dichromacy. Tests with a 2° field and a wider colour gamut have revealed classical 'Nagel dichromats' who cannot match the Rayleigh red and green stimuli with each other and whose matching range is intermediate between the widest available on Nagel's anomaloscope and the greatest achievable range for dichromacy.

References

Alpern, M. and Wake, T. (1977). Cone pigments in human deutan colour vision defects. J. Physiol. (Lond.) 266: 595–612.

Krastel, H. and Moreland, J.D. (1991). Colour vision deficiencies in ophthalmic diseases. In: Foster, D.H. (ed.), Inherited and Acquired Colour Vision Deficiencies: 115–172. Macmillan, Basingstoke.

Lakowski, R. (1965). Testing of colour vision in prospective printer's apprentices and the problems this represents in selection. Br. J. Physiol. Opt. 22: 10–32.

Le Grand, Y. (1968). Light, Colour and Vision. Chapman and Hall, London.

Moreland, J.D. (1993). Design criteria for a clinical anomaloscope. In: Drum, B. (ed.), Colour Vision Deficiencies XI. Doc. Ophthalmol. Proc. Ser. 56: 335–344. Kluwer, Dordrecht.

Moreland, J.D. and Kerr, J. (1979). Optimization of a Rayleigh-type equation for the detection of tritanomaly. Vision Res. 19: 1369–1375.

Nagel, W.A. (1907). Zwei Apparate für die augenärzliche Functionsprüfung: Adaptometer und kleines Spektralphotometer (Anomaloskop). Zeitschr. Augenheilkd. 17: 201–208.

Schmidt, I. (1974). Brightness matches on the Nagel anomaloscope. Mod. Prob. Ophthalmol. 13: 19–25.

Wright, W.D. (1946). Researches on Normal and Defective Colour Vision. Kimpton, London.

Department of Communication and Neuroscience
Keele University
Staffordshire, ST5 5BG, UK

PART FIVE

Plate tests, arrangement tests, and automated tests

32. Influence of stimulus luminance on automated Moreland matches

J. SOMMERHALDER, M. PELIZZONE, J.D. MORELAND and A. ROTH

(Geneva, Switzerland and Keele, UK)

Abstract

Stimulus luminance is an important parameter in colour matching procedures. In particular, we do not know exactly what optimal luminance conditions are for automated blue–green Moreland matches. Automated Moreland matches were made at different luminance levels in a group of 8 normal subjects (age range: 25–64 years). Stimulus luminance 'at match' was set to 0.2, 0.8, 3, 11 or 5 cd/m^2. Two field sizes were used: 2° and 7°. The subject's pupil diameter was measured for each stimulus condition during the test with an infra-red camera. Pupil size and the estimated influence of age on lens transmission were used to compute retinal illuminance. In the range 40–1000 trolands, matching ranges and match midpoints of the Moreland equation were not significantly influenced by retinal illuminance. Below 10 trolands, matching ranges were significantly enlarged and match midpoints tended to shift towards the blue primary. These effects were very similar for both field sizes. These results have practical implications since they indicate that suitable stimulus luminance for Moreland matches are in the range 10–30 cd/m^2, and that these values are suitable for subjects of all ages.

Introduction

Most congenital and acquired colour vision defects can be diagnosed and classified with the so-called 'two equation method', combining the red–green Rayleigh equation with the blue–green Moreland equation (Roth *et al.*, 1989, 1990; Roth, 1990; Pokorny and Smith, 1984, 1990; Moreland, 1990). Our routine clinical examination of colour vision uses the two equation method presented on an automated, computer-controlled anomaloscope (Pelizzone *et al.*, 1991, 1995). A commercial version is available from Interzeag (Schlieren, Switzerland).

Several stimulation parameters can influence the results of a colour vision examination. Moreland recently summarized design criteria for a clinical anomaloscope (Moreland, 1993b), and we investigated the effects of several stimulus parameters on the result of colour vision examinations (Pelizzone *et al.*,

C.R. Cavonius (ed.), Colour Vision Deficiencies XIII, pp. 299–305.

1993; Sommerhalder *et al.*, 1995). Our objective is to develop a standard colour vision examination for routine clinical use. To achieve this goal, optimal conditions need to be determined.

One of the most important stimulation parameters in the Moreland equation for which the optimal value has not yet been determined is the luminance of the observation field or the retinal illuminance (for the Rayleigh equation 175 trolands is recommended by the German Industry Norm DIN 6160). The experiments reported in this paper were designed to determine the range of stimulus luminance over which match midpoint and matching range of automated Moreland matches are stable.

Methods

Eight normal trichromats (aged 64, 55, 49, 42, 33, 30, 26 and 25; 4 females and 4 males) made Moreland matches at five different luminances of the observation field and on two different field sizes (2° and 7°). Measurements at each luminance were replicated three times on each subject (therefore each subject made $5 \times 2 \times 3 = 30$ Moreland matches). The lengthy data collection (approximately 5 min per match) was distributed over two or more experimental sessions. Only one eye of each subject was tested. During the matching procedure, the exposure time of each coloured stimulus was 2 sec (Pelizzone *et al.*, 1993).

In order to achieve different field luminance levels, the laboratory anomaloscope was set to 45 cd/m^2 at 50% of the blue primary. Lower field luminance levels of 11, 3, 0.8 and 0.2 cd/m^2 (at 50% of the blue primary) were made by introducing neutral filters in the optical path, just behind the observation screen (see Pelizzone *et al.*, 1989). The spectral transmission of each neutral filter was measured and used to calculate luminance levels as well as minor corrections of the match midpoint. Since the observation field was viewed directly, retinal illuminance was influenced by the size of the natural pupil. Pupil diameter was measured during the test by a small infra-red (IR) camera installed on top of the tube through which the subject viewed the field. During testing, the eye of the subject was illuminated by an annular set of IR LEDs, emitting radiation at about 920 nm. This radiation was not seen by the subject and presumably does not influence the measurement. Horizontal pupil diameters (magnified 5.7 times) were measured on a black and white monitor with an accuracy of ± 0.3 mm.

For each subject, tests were made first on the 7° observation field, then on the 2° field. Field luminances were tested in random order. Prior to each examination static heterochromatic luminance matches between the primary and the test field were established for 5 different mixture values (20, 40, 50, 60 and 80% of green primary), which allowed us to determine individual luminance match settings for all test field mixtures. The pupil diameter was recorded at each luminance match.

Results

Matching range (MR) and match midpoint (MMP) versus luminance of the observation field

Figure 1a shows group mean values of the log MRs versus field luminance. For both field sizes MRs increased at the two lowest luminance levels of 0.8 and 0.2 cd/m^2. These increases were significant for both field sizes (paired *t*-tests between 0.2 and 0.8 cd/m^2: $p = 0.0005$ for the 7° field, $p = 0.0002$ for the 2° field; paired *t*-tests between 0.8 and 3 cd/m^2: $p = 0.02$ for the 7° field, $p = 0.05$ for the 2° field). The mean MRs on the 7° field were systematically about 0.2 log units smaller than on the 2° field. As reported in a previous study (Rossillion *et al.*, 1995), this reflects the fact that the test is much easier on the larger field.

Figure 1b shows group mean values of the MMPs versus field luminance. There was no significant influence of luminance within the range 0.8–45 cd/m^2, but the MMPs were shifted towards blue at the lowest luminance level (0.2 cd/m). These shifts were significant for both field sizes (paired *t*-tests between 0.2 and 0.8 cd/m^2: $p = 0.004$ for the 7° field, $p = 0.01$ for the 2° field). Mean MMPs on the 2° field were systematically about 2 units higher than on the 7° field. As reported in a previous study (Rossillion *et al.*, 1995), this effect can be attributed to the retinal distribution of macular pigment: Moreland matches require more green primary for the smaller field size.

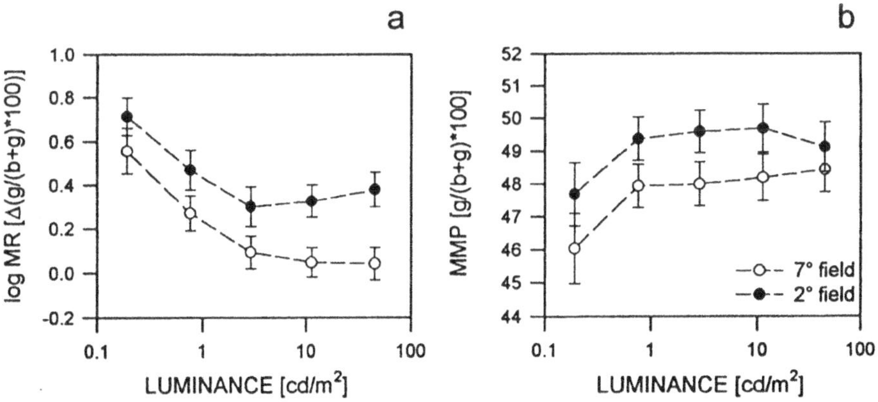

Fig. 1. (a) Log of matching range versus observation field luminance. Mean values and standard errors for eight subjects on the 7° observation field (open circles) and on the 2° observation field (filled circles). (b) Match midpoint versus observation field luminance. Mean values and standard errors for eight subjects on the 7° observation field (open circles) and on the 2° observation field (filled circles).

For most subjects, the examination was (quite) difficult to perform at very low luminances (notice the larger error bars for the 0.2 cd/m^2 results). Some younger subjects complained that the highest value of 45 cd/m^2 was too luminous; they preferred the 11 cd/m^2 value. Younger subjects have lenses with higher transmission and (usually) larger pupils; therefore their retinal illuminance is higher, possibly making them more sensitive to higher luminances.

Pupil diameter versus luminance of the observation field

Table 1 gives measured pupil diameter at match for the eight subjects. We found no difference for the pupil diameter between the two field sizes. As expected, the pupil diameter decreased with increasing field luminance and younger subjects tended to have larger pupils.

Table 1. Pupil diameter at match (mm).

		Luminance (cd/m^2)									
		7° Field					2° Field				
Subject/Age		0.19	0.77	2.93	11.3	45.0	0.19	0.77	2.93	11.3	45.0
S1	64	4.2	4.2	4.2	4.2	3.5	4.2	4.2	4.2	4.1	3.8
S2	55	6.0	6.2	6.0	5.8	4.8	6.0	6.2	5.7	5.9	5.4
S3	49	4.3	3.6	3.5	3.0	2.8	4.3	3.9	3.3	3.9	3.3
S4	42	4.7	4.0	3.6	3.6	3.2	5.2	4.6	4.5	3.9	3.4
S5	33	6.2	6.0	5.7	5.8	5.5	6.1	6.2	6.2	6.2	5.9
S6	30	4.5	4.0	3.8	3.5	3.1	4.0	3.4	3.1	2.9	2.7
S7	26	5.8	5.3	4.9	5.1	4.6	5.4	5.3	4.9	4.5	3.9
S8	25	7.0	6.7	6.8	6.5	6.7	6.9	6.6	7.0	6.4	6.0
Average		5.3	5.0	4.8	4.7	4.3	5.3	5.1	4.9	4.7	4.3

*at 50% of blue primary

Discussion

The mean MR of the Moreland match increases and the mean MMP shifts towards blue at the lowest luminance. Inspection of individual data suggested that these trends were more pronounced for older subjects. The amount of light actually reaching the retina depends on pupil diameter and lens transmission, which are both related to age and different for each subject. Therefore retinal illuminance is the more appropriate parameter to assess the influence of stimulation luminance.

Retinal illuminance was calculated for each subject and for each condition, using measurements of pupil diameter at match and by estimating lens transmission. Age-dependent lens transmission was estimated, following the procedure proposed by Moreland (1993a) which is based on Weale's data on the transmission of the human crystalline lens (Weale, 1988). These computations gave individual retinal illuminances that were markedly different across subjects. Therefore, in order to present group means similar to those of Figure 1, we averaged data by groups. Normalized – compensating for normal interindividual differences – log(MR) were used because they best exhibit the intrinsic influence of retinal illuminance on the data.

Figure 2a shows normalized log(MR) versus retinal illuminance for both field sizes. While MRs were not affected by retinal illuminance above 40 trolands, there was a significant increase of the MR at low retinal illuminances (paired *t*-tests between 4 and 14 trolands: $p = 0.03$ for the 7° field, $p = 0.003$ for the 2° field). The same data analysis was performed on MMP data. MMPs were also corrected for the age dependency of lens transparency by computing MMPs for

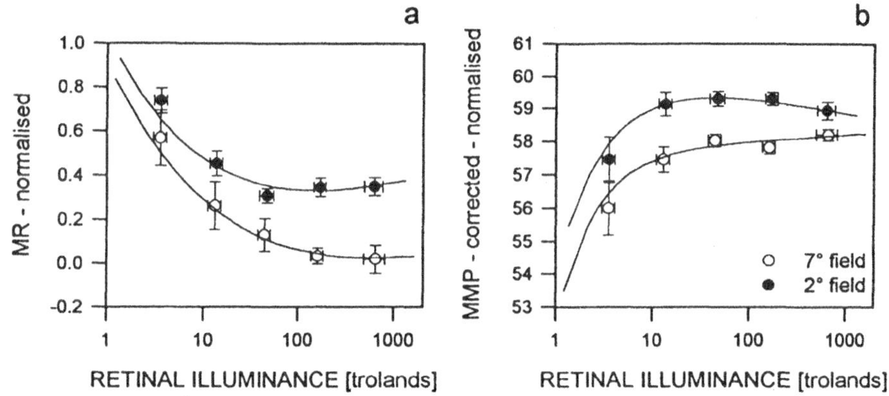

Fig. 2. (a) Normalized log of matching range versus retinal illuminance. Mean values and standard errors on the 7° observation field (open circles) and on the 2° observation field (filled circles). The set of 40 data-points (8 subjects × 5 retinal illuminances) were split into 5 groups (according to increasing retinal illuminance) and the mean retinal illuminance and mean MR were computed for each group. Data were normalized by shifting the mean value of the three highest luminances, 45, 11 and 3 cd/m^2, of each subject to the mean value of the same luminances of all subjects. The sets of 40 data points were fitted to a combination of a straight line and an exponential function. (b) Age corrected and normalized match midpoints versus retinal illuminance. Mean values and standard errors on the 7° observation field (open circles) and on the 2° observation field (filled circles). In addition to data analysis similar to that performed for the MRs, MMPs were corrected for the age dependency of lens transmission. MMPs were corrected to age 0 years on the basis of the estimated lens transmission used to compute retinal illumination (see text).

age. Figure 2b shows mean values of age-corrected, normalized MMP versus retinal illuminance for both field sizes. As for MRs, MMPs were not affected by retinal illuminance above 40 trolands, but there was a significant shift towards the blue primary at low retinal illuminances on the 2° field (paired *t*-tests between 4 and 14 trolands: $p = 0.05$). Although this shift is not significant for the 7° field, it is interesting to note that these shifts appear to be of similar magnitude for both field sizes. We do not know why the MMP shifts towards the blue primary at very low levels.

These results have practical implications since they indicate that suitable stimulus luminance for Moreland matches are in the range 10–30 cd/m^2, and these values are suitable for subjects of all ages (enough light for old subjects and no glare for young ones). The upper limit can probably be extended if a neutral adaptation field is shown to the subject between two colour presentations.

Conclusion

Pupil size and the estimated influence of age on lens transmission were used to compute retinal illuminance. In the range 40–1000 trolands, matching ranges and match midpoints of the Moreland equation were quite stable and were not significantly influenced by retinal illuminance. Below 10 trolands, matching ranges were significantly increased and match midpoints tended to shift towards the blue primary. These effects were very similar for both field sizes.

Acknowledgement

This work was supported by the Swiss Science Foundation (grant no. 32-33931.92).

References

Moreland, J.D. (1978). Optimisation of stimuli for tritanomaloscopy. Mod. Probl. Ophthalmol. 19:162–166.

Moreland, J.D. (1990). The clinical utility of anomaloscopy. In: Y. Ohta (ed.), Colour Vision Deficiencies: 127–143. Kugler and Ghedini, Amsterveen.

Moreland, J.D. (1993a). Matching range and age in a blue-green equation. In: Drum, B. (ed.), Colour Vision Deficiencies XI, Doc. Ophthalmol. Proc. Ser. 56: 129–134, Kluwer, Dordrecht.

Moreland, J.D. (1993b). Design criteria for a clinical anomaloscope. In: Drum, B. (ed.), Colour Vision Deficiencies XI, Doc. Ophthalmol. Proc. Ser. 56: 335–344. Kluwer, Dordrecht.

Pelizzone, M., Sommerhalder, J., Roth, A. and Hermès, D. (1991). Automated Rayleigh and Moreland matches on a computer controlled anomaloscope. In: Drum, B., Moreland, J.D. and Serra, S. (eds), Colour Vision Deficiencies X, Doc. Ophthalmol. Proc. Ser. 54: 151–159. Kluwer, Dordrecht.

Pelizzone, M., Sommerhalder, J., Roth, A. and Hermès, D. (1993). Automated Rayleigh and Moreland matches. Optimization of stimulation parameters for normal observers. In: Drum,

B. (ed.), Colour Vision Deficiencies XI, Doc. Ophthalmol. Proc. Ser. 56: 345–355. Kluwer, Dordrecht.
Pelizzone, M., Sommerhalder, J. and Roth, A. (1995). Détermination des seuils de discrimination dans les épreuves d'égalisation colorée. Ophtalmologie 9: 229–234.
Pokorny, J. and Smith, V.C. (1984). Metameric matches for assessment of colour vision. I. Theoretical considerations. In: Verriest, G. (ed), Colour Vision Deficiencies VII, Doc. Ophthalmol. Proc. Ser. 39: 83–94. Junk, The Hague.
Pokorny, J. and Smith, V.C. (1990). Colour matching as a clinical tool. Theory of modification by disease. In: Ohta, Y. (ed.), Colour Vision Deficiencies: 255–267. Kugler and Ghedini, Amstelveen.
Rossillion, B., Pelizzone, M., Sommerhalder, J. and Roth, A. (1995). Automated Moreland equations on 7° and 2° fields. In: Drum, B. (ed.), Colour Vision Deficiencies XII, Doc. Ophthalmol. Proc. Ser. 57: 473–480. Kluwer, Dordrecht.
Roth, A. (1990). The power of metameric colour equations in testing colour vision. In: Ohta, Y. (ed.), Colour Vision Deficiencies: 181–190. Kugler and Ghedini, Amstelveen.
Roth, A., Pelizzone, M., Hermès, D., Sommerhalder, J. (1989). Neuere Ueberlegungen und Entwicklungen zur klinischen Untersuchung des Farbensehens – die zwei-Gleichungsmethode. Fortschr. Ophthalmol. 86: 374–379.
Roth, A., Pelizzone, M., Hermès, D. and Sommerhalder, J. (1990). L'examen de la vision colorée par la méthode des deux équations métamériques. Ophtalmologie 4: 197–205.
Sommerhalder, J., Pelizzone, M., Rossillion, B. and Roth, A. (1995). Heterochromatic luminance matches in automated Rayleigh and Moreland equations. In: Drum, B. (ed.), Colour Vision Deficiencies XII, Doc. Ophthalmol. Proc. Ser. 57: 473–480. Kluwer, Dordrecht.
Weale, R.A. (1988). Age and the transmittance of the human crystalline lens. J. Physiol. (Lond.) 395: 577–587.

J. Sommerhalder, M. Pelizzone and A. Roth
Clinique d'ophtalmologie
Hôpital Cantonal
Rue Micheli-du-Crest 24
CH-1211 Genève 14, Switzerland

J. D. Moreland
Department of Communication and Neuroscience
Keele University
Staffordshire ST5 5BG, UK

33. The clinical use of saturated and desaturated Munsell panel tests

J. BIRCH
(London, UK)

Abstract

215 people with red–green colour deficiency were examined with the Farnsworth D15 test. The 88 subjects who passed completed a desaturated version of the D15, the Adams test. 48 of the 88 subjects passed this test. The 127 subjects who failed the D15 were given a panel test containing more saturated colours, the Paulson H16 test. 19 of the 127 subjects passed the H16. The 4 grades of severity obtained by the combined use of these three tests were compared with Nagel anomaloscope matching ranges and error profiles obtained with the Farnsworth-Munsell 100-hue test. The mean matching ranges and error scores for the 4 categories have a rank order but there is considerable overlap between the individual results in Grades 1 and 2 and Grades 3 and 4. In particular, the H16 is not failed exclusively by dichromats. These data show that the 2-category system of severity based on pass or fail of the Farnsworth D15 is strongly based and is the optimum that can be applied clinically.

Introduction

The Farnsworth D15 test is a shortened version of the dichotomous test described by Farnsworth in 1943. The prototype test had 20 colours whereas the D15 has one reference colour and 15 moveable colours, making 16 in all. The test is composed of Munsell samples selected from a complete hue circle with Munsell value 5 and chroma 4 (Paulson, 1971). The colours are held in small caps which subtend 1.5° at a test distance of 50 cm. The subject has to arrange the colours in what is perceived to be a natural colour order beginning with the reference colour. Isochromatic colour confusions are demonstrated when colours from opposite sides of the hue circle are placed together in the subject's arrangement.

The aim of the D15 test is to identify people with 'significant' colour deficiency. The test divides people into two groups. The first group consists of those with normal colour vision or slight colour deficiency who pass the test. The second group consists of people with moderate or severe colour deficiency who fail. Farnsworth considered that people who passed the test would be able

C.R. Cavonius (ed.), Colour Vision Deficiencies XIII, pp. 307–312.

to set the electrical colour codes of the time safely and that people who failed the test would be unsafe. The D15 is normally used together with a screening test, such as the Ishihara plates, to establish the severity of colour deficiency for vocational purposes (Birch, 1993; Cole, 1993). Desaturated D15 tests, having the same Munsell hues but different Munsell values and chroma, were developed initially for the assessment of acquired colour deficiency (Lanthony, 1974; Adams and Rodic, 1982; Birch, 1989). These tests have different pass/fail levels compared with the standard D15 and can be used to grade colour deficiency into a larger number of categories (Adams and Haegerstrom-Portnoy, 1987; Birch, 1993). However, even desaturated D15 tests do not provide accurate colour vision screening (Vingrys *et al.*, 1993). Two desaturated tests are in general use. The Adams desaturated D15 test has Munsell value 5 and chroma 2 and is preferred to the Lanthony desaturated D15 (Munsell value 8 and chroma 2) for the level of illuminance provided by the Macbeth easel lamp (Birch *et al.*, 1987). A similar panel test having more saturated Munsell colours, with the aim of identifying protanopia and deuteranopia only, was described by Paulson in 1971. The hues of the H16 test are not the same as in the D15 and there is no provision for identifying tritan colour deficiency. Munsell value is kept at 5 but chroma is either 6 or 8 (Birch, 1993). There is one reference colour and 16 moveable colours, making 17 colours in all.

The aim of the present study was to assess the combined function of these three tests in grading the severity of congenital colour deficiency into 4 categories; very slight (Grade 1), slight (Grade 2), moderate (Grade 3) and severe (Grade 4). Grade classification is compared with the matching range found with the Nagel anomaloscope and the error score obtained with the Farnsworth-Munsell 100-hue test.

Subjects and methods

The D15 test was given to 215 subjects with congenital red–green colour deficiency identified by failure on the Ishihara plates. Diagnosis of the type of colour deficiency was made with a Nagel Type I anomaloscope for each subject (Birch, 1982). There were 41 protanopes, 13 protanomalous trichromats, 50 deuteranopes and 111 deuteranomalous trichromats. For the anomaloscope used in this study pure green is at 0 on the red–green mixture scale and pure red is at 73. The mean normal match was established with 100 normal trichromats as 42 ± 2 scale units on the red–green mixture scale. Dichromats were able to match all red/green ratios. The matching range was recorded for the anomalous trichromats. All the subjects completed the Farnsworth D15 test and the FM 100-hue test. The Adams desaturated D15 test was given to all the subjects who passed the D15 test, and the H16 test was given to all the subjects who failed the D15. The tests were illuminated with the Macbeth easel lamp, standard source C, providing approximately 350 lux and were viewed binocularly. The correct sequence or one transposition error of adjacent hues was considered to

Table 1. Grading red–green colour deficiency with panel tests.

	Grade 1 (pass D15 test; pass Adams test)	Grade 2 (pass D15 test; fail Adams test)	Grade 3 (fail D15 test; (pass H16 test)	Grade 4 (fail D15 test; (fail H16 test)	No. of subjects
P	0	1	0	40	41
PA	1	6	6	0	13
D	0	0	0	50	50
DA	47	33	13	18	111
Total	48	40	19	108	215

P = protanopia, PA = protanomalous trichromatism, D = deuteranopia, DA = deuteranomalous trichromatism

constitute a 'pass' of each of the panel tests. More than one transposition was considered as 'failure'. The S index obtained by vector analysis has been used to indicate polarity of FM 100-hue plots (Vingrys *et al.*, 1992).

Results

Of the 215 people examined, 88 passed and 127 failed the D15 (Table 1). All 50 deuteranopes failed the D15 test and all but one of the 41 protanopes also failed. All the dichromats who failed the D15 also failed the H16 and were categorized as having the most severe type of colour deficiency (Grade 4). The protanope who passed the D15 failed the Adams test and was identified as having slight colour deficiency (Grade 2). This subject was an experienced colour matcher and achieved the very low error score of 64 on the FM 100-hue test.

Of the 124 anomalous trichromats examined 87 (70%) passed the D15 and 37 (30%) failed. The grade of colour deficiency together with the mean anomaloscope matching range and the mean FM 100-hue error score for all the anomalous trichromats is shown in Table 2. The results show that both the mean Nagel matching range and the mean FM 100-hue error score increase from Grade 1 to Grade 4. However, the standard deviations for both measurements are very large, showing that there is a considerable overlap of the results obtained in adjacent groups.

A smaller anomaloscope matching range was found for protanomalous trichromats than for deuteranomalous trichromats. Of the 13 protanomalous subjects examined, seven passed the D15 and 6 failed. Only one protanomalous trichromat passed the Adams test and was categorized as having very slight colour deficiency (Grade 1). Six failed the Adams test and were classified as

Table 2. Grading 124 anomalous trichromats with saturated and desaturated D15 tests.

Grade	Percentage of subjects	Nagel		FM 100-hue	
		Mean range	SD	Mean error score	SD
1	38.7	11.40	6.15	71.25	45.24
2	31.5	16.12	11.55	102.52	53.40
3	15.3	23.73	13.51	127.95	82.47
4	14.5	32.88	13.89	174.22	60.37
1 + 2	70.2	13.63	8.93	86.88	41.02
3 + 4	29.8	28.31	13.19	151.08	42.72

Table 3. Grading 111 deuteranomalous trichromats with saturated and desaturated D15 tests.

Nagel grade	Percentage of subjects	Mean range	SD	FM 100-hue		
				Mean error score	SD	Mean S index
1	42.3	11.19	5.81	73.44	41.25	1.31
2	29.8	17.93	11.64	103.57	48.98	1.31
3	11.7	28.15	14.13	162.46	52.37	1.44
4	16.2	31.22	14.11	182.22	54.18	1.49
1 + 2	72. 1	13.07	8.90	87.71	47.24	1.31
3 + 4	27.9	31.87	11.84	162.96	63.32	1.32

having slight colour deficiency (Grade 2). All six protanomalous trichromats who failed the D15 passed the H16 test and were classed as having moderate colour deficiency (Grade 3). The number of protanomalous subjects was too small for further analysis.

The results obtained by 111 deuteranomalous trichromats, including the S index for the FM 100-hue test, are shown in Table 3. Eighty deuteranomalous trichromats passed the D15 and 31 failed. Of the 80 deuteranomalous trichromats who passed the D15, 47 were categorized as very slight (Grade 1) and 33 as slight (Grade 2) by the Adams test. Of the 31 deuteranomalous trichromats who failed the D15, 13 passed the H16 and were categorized as having moderate colour deficiency (Grade 3) and 18 failed, indicating severe colour deficiency (Grade 4). A similar systematic increase in the mean Nagel matching range and the mean FM 100-hue error score, both with large standard deviations, was found from Grade 1 to Grade 4 as that shown in Table 2. The S index was the same in Grade 1 and Grade 2 and had a similar value in Grades 3 and 4.

Discussion

Combined use of the D15 and the H16 tests was effective in showing that dichromats have severe colour deficiency (Grade 4). The protanope who passed the D15 may have an advantage because of his previous experience as a colour matcher in industry and his result may be due to training. Not only dichromats failed; 18 anomalous trichromats (24.5% of the anomalous trichromats in this study) also failed. All these subjects were deuteranomalous trichromats who tended to have larger anomaloscope matching ranges and higher FM 100-hue error scores than other anomalous trichromats in this study.

The mean values for the Nagel matching range and for the FM 100-hue error scores shown in Tables 2 and 3 suggest that the combined use of three panel tests that differ in saturation (Munsell chroma) may provide a useful 4-category grading system. However, the large standard deviations show that there is considerable overlap between grades and that it is difficult to apply a 4-category grading system accurately. The overlap of results was particularly marked between Grades 1 and 2 and between Grades 3 and 4, suggesting that these categories can be recombined. Two categories of severity, based on pass or fail of the standard D15 only, have a much stronger relationship with the severity of colour deficiency, as shown by the Nagel matching range, and practical colour discrimination ability, as shown by the FM 100-hue error score (Table 2). The S indices, which indicate polarity of the FM 100-hue plots, also showed two categories of colour deficiency. These results emphasize that the current system of offering vocational advice on the basis of pass or fail in the D15 is sound and that additional saturated or desaturated D15 tests offer little benefit in grading congenital colour deficiency into additional categories. A further problem with the Adams test is that some of the desaturated hues do not lie along the same isochromatic lines as those of the standard test (Dain and Adams, 1990). In the present study, failure on the Adams test was often based on multiple single transformation errors, between adjacent hues, rather than isochromatic colour confusions across the hue circle, so that protan/deutan classification could not be obtained from visual inspection of the results diagram. Additional qualitative scoring of D15 panel tests has been shown to be ineffective unless isochromatic confusions are made and, in consequence, classification of the type of colour deficiency cannot be obtained from these results (Atchison *et al.*, 1991).

References

Atchison, D.A., Bowman, K.J. and Vingrys, A.J. (1991). Quantitative scoring methods for D15 panel tests in the diagnosis of congenital color deficiencies. Optom. Vis. Sci. 68, 41–48.

Adams, A.J. and Haegerstrom-Portnoy, G. (1987). Colour Deficiency. In: Amos, J.F. (ed.), Diagnosis and Management in Vision Care: 671–711. J.F. Butterworths, Boston.

Adams, A.J. and Rodic, R. (1982). Use of desaturated versions of the D15 test in glaucoma. In: Verriest, G. (ed.) Colour Vision Deficiencies VI. Doc. Ophthalmol. Proc. Ser. 33: 419–423. Junk, The Hague.

Birch, J. (1982). Diagnosis of defective colour vision with the Nagel anomaloscope. In: Verriest, G. (ed.), Colour Vision Deficiencies VI. Doc. Ophthalmol. Proc. Ser. 33: 231–235. Junk, The Hague.

Birch, J. (1989). Farnsworth tests in the study of acquired colour vision defects in diabetic retinopathy. In: Drum, B. and Verriest, G. (eds.), Colour Vision Deficiencies IX, Doc Ophthalmol. Proc. Ser. 52: 433–440. Kluwer, Dordrecht.

Birch, J. (1993). Diagnosis of Defective Colour Vision. Oxford University Press, Oxford.

Birch, J., Tertuik, K. and North, F. (1987). Tritan screening with a modified TNO test. In: Verriest, G. (ed.), Colour Vision Deficiencies VIII, Doc. Ophthalmol. Proc. Ser. 46: 227–232. Junk, The Hague.

Cole, B.L. (1993). Does defective colour vision really matter? In: Drum, B. (ed.), Colour Vision Deficiencies XI, Doc. Ophthalmol. Proc. Ser. 56: 67–86. Kluwer, Dordrecht.

Dain, S.J. and Adams, A.J. (1990). A comparison of the standard and Adams desaturated D15 tests with congenital colour deficiency. Ophthal. Physiol. Opt. 10: 40–44.

Farnsworth, D. (1943). The Farnsworth-Munsell 100-hue test and dichotomous test for colour vision. J. Opt. Soc. Am. 33: 567–587.

Lanthony, P. (1974). Etude de la saturation au cours des dyschromatopsies acquises au moyen de l'album de Munsell. Ann. Oculist. (Paris) 207: 741–751.

Paulson, H.M. (1971). Colour vision testing in the US Navy. In: Pierce, J.R. and Levene, J.R. (eds.), Visual Science: 164–170. World Press, Bloomington, IL.

Vingrys, A.J., Atchison, D.A. and Bowman, K.J. (1992). The use of colour difference vectors in diagnosing congenital colour deficiencies with the Farnsworth-Munsell 100-hue test. Ophthal. Physiol. Opt. 12: 38–45.

Vingrys, A.J., Atchison, D.A. and Bowman, K.J. (1993). Using panel tests in screening for congenital colour vision defects. In: Drum, B. (ed.), Colour Vision Deficiencies XI, Doc. Ophthalmol. Proc. Ser. 56: 267–284. Kluwer, Dordrecht.

Applied Vision Research Centre
Department of Optometry and Visual Science
City University
London EC1V OHB, UK

34. The pseudoisochromatic plates of E.N. Yustova

M.V. DANILOVA and J.D. MOLLON
(St. Petersburg, Russia and Cambridge, UK)

Abstract

The new pseudoisochromatic plates of Yustova have been tested in various illuminants for a well characterized panel of colour-defective observers. Under the same conditions, colorimetric measurements were made of the target and ground patches of each plate. In a subset of plates (1, 5 and 9) the chromaticities of target and ground did not lie on the intended confusion lines and these were the plates that were diagnostically unsatisfactory. The most successful plates (e.g. 7 and 8) are those whose chromaticities lie close to the intended confusion line. There is clear value in making concurrent spectroradiometric measurements when field testing colour tests.

Introduction

The pseudoisochromatic plates of E.N. Yustova (Alekseeva and Yustova, 1978) were described to the International Research Group at its Dresden meeting (Yustova *et al.*, 1986). A production version of the test has recently been introduced and is offered as a replacement for the Rabkin pseudoisochromatic plates, which are widely used in the former Soviet Union but have been out of print for many years. Since the principle of the Yustova plates most closely resembles that of the HRR plates (Hardy *et al.*, 1954), themselves long out of print, we have made a preliminary examination of the new Russian plates. Each plate consists of a mosaic of patches, 9 mm square, with 2 mm gaps. The target is defined by a subset of patches that form three sides of a square (Fig. 1). The examinee must indicate the orientation of the gap in the square. The examiner is thus free to present a given plate repeatedly in random orientations, and it would be difficult for a candidate to learn specific responses for particular plates, as he might in the case of the Ishihara or Rabkin tests. In this property the Yustova plates resemble those of Schaaff (1925; illustrated in Oblath, 1929) and those of Ishihara and Okuma (1975).

The plates are colorimetrically based, in that the target and background are in each case intended to lie on the confusion line of one type of dichromat. There are four protan plates, four deutan plates and three tritan. Within each subset the chromaticity difference between target and background is intended to

C.R. Cavonius (ed.), Colour Vision Deficiencies XIII, pp. 313–322.

Fig. 1. The achromatic demonstration plate of the Yustova test. The coloured test plates have the same geometry.

increase in successive plates. (According to the test manual, these chromaticity differences have been adjusted to correspond to fixed numbers of threshold units, on the basis of discrimination measurements with a Maxwell disc.) Like the HRR plates, the Yustova plates are intended to detect and classify daltonians entirely on the basis of chromatic difference thresholds. And like Hardy *et al.* (1954), the authors of the present plates introduce a classification scheme corresponding to their graded plates. Thus plates 1–3 correspond to 'protodeficits' of the first, second and third degrees, while failure on plate 4 is intended to indicate protanopia. Plates 5–7 correspond to 'deutodeficits' of the first, second and third degree, while failure on plate 8 indicates deuteranopia.

A particular feature of the Yustova plates is that the target and background are designed to be equated not only in colour for the corresponding dichromat but also in lightness. Having made this equation, the authors of the plates took the daring step of foregoing the luminance noise that has been a feature of most pseudoisochromatic plates since their invention by Stilling in 1877 (see Regan *et al.*, 1994). An effective (but not commercially available) plate without luminance noise has been designed by Birch (1975), but here spatial noise is introduced.

Methods

We tested the Yustova plates on a panel of 31 colour-defective observers who had been classified in detail by means of the Nagel anomaloscope (Model I; Schmidt & Haensch) and the computerized test of Regan *et al.* (1994). Since the plates are promoted as being colorimetrically based and yet the recommended illumination is specified only as 'fluorescent light or natural daylight', we tried more than one illuminant. All subjects were tested using the Macbeth easel daylight lamp that is typically used for clinical colour testing in the West; most subjects were also tested in natural northern daylight (in Cambridge in the period July–October) and under white fluorescent lighting (GE F58W/35).

The test manual calls for a rather high level of illumination of 500–1000 lux. This range is higher than provided either by our Macbeth easel lamp (330 lux) or by our normal office fluorescent lighting or by natural daylight when the subject sat with his back to a north-facing window (350–500 lux). We therefore equipped our testing room with additional lighting to give 750 lux of illumination from fluorescent lamps. Of the conditions we explored, these are the ones that best match those called for by the test manual.

Using a PhotoResearch 650 spectroradiometer, we measured the chromaticities of target and ground patches of each plate in each of the three illuminants used for testing. All measurements were repeated with a Minolta Chroma Meter. The chromaticities obtained with the Chroma Meter showed small systematic shifts from those measured with the PhotoResearch instrument, but the vectors defined by the chromaticities of target and ground were very similar.

Results

Test performance

Table 1 summarizes the performance of our subjects. One deuteranomalous trichromat passed all plates in all illuminants; and three protanomalous subjects passed all plates in natural daylight. Otherwise, the test successfully detected red–green defectives (including all dichromats) in all illuminants used. However, it was plate 9, nominally a tritan plate, that caught most red–green defectives, whereas plates 1 and 5, intended to detect mild protans and mild deutans respectively, were failed only by one elderly deuteranope. Deuteranopes were usually identified as such (by plate 8), but all the protanopes passed the 'protanope' plate (no. 4) in at least one illumination. Anomalous trichromats were not well classified, often failing both the second protan and the second deutan plates. No protanomalous trichromats were misclassified as dichromats, but some deuteranomalous subjects were classified as deuteranopes, in that they failed plate 8. A patient suffering from a complete acquired tritanopia (subject DC in the study of Regan *et al.*, 1994) passed all the plates, including the three graded tritan plates.

Colorimetric measurements

The foregoing test results can more readily be understood by reference to our concurrent colorimetric measurements (Fig. 2). In the case of the protan plates 2, 3, and 4, the colorimetric measurements suggest that the designers' intentions

Table 1. Performance of colour-defective subjects on individual plates of the Yustova test.

	Plate											
	Protan				Deutan				Tritan			
	1	2	3	4	5	6	7	8	9	10	11	Nagel
Subject												
WL									xxx			PA
BJ									/ / x			PA
JI		xxx				xx /			xxx			PA
RJ									xxx			PA
TR		x /										PA
LC		xx	/ x			xx			xx			PA
LL									x /			PA
WS		xxx	x / /	xx /		x / /			xxx			P
WM						xx			xx			P
HT		xxx		xx /		xxx			/ xx			P
McCA		x										DA
MC		xx				xx	x /		xx			DA
PD		xx				xx			xx			DA
BM						x / /			x / x			DA
KB		x							x			DA
KS		xx /				xx /	x / /	xx /	xxx			DA
MD		xx /				x / /	xxx	xx /	/ xx			DA
HC		xxx				xx /	/ / x	xxx	xxx			DA
SR												DA
SJ		x				x			x			DA
OD		xx /				x / /	/ / x	xxx	xxx			DA
HI		xxx				xx /	xxx	xxx	xxx			DA
WC		x				x	x	x	x			D
LW		xx				x /	xx	xx	xx			D
BK	x/ /	xxx			xx /	xx /	xxx	xxx	xxx	/ xx	/ x /	D
KM		/ x						/ x	xx			D
VS		xxx				xx /	xxx	xxx	xxx			D
BB		xx				xx	x/	xx	xx			D
BS								xx /	/ xx			D
CD												T
CL			x				x		x			Cone dystrophy

An empty cell indicates that the subject passed that plate in all illuminants tried for that subject. Cells containing symbols indicate plates that were failed under at least one illuminant. In these cases the first symbol in the cell refers to the Macbeth illuminant (used in all cases), the second to natural daylight, and the third to fluorescent light. x indicates failure and / indicates a pass. In the case of the congenital red–green defectives the rightmost column shows the diagnosis according to the Nagel anomaloscope (PA protanomalous; P protanopic; DA deuteranomalous; D deuteranopic). T indicates a case of acquired total tritanopia.

Protan plates

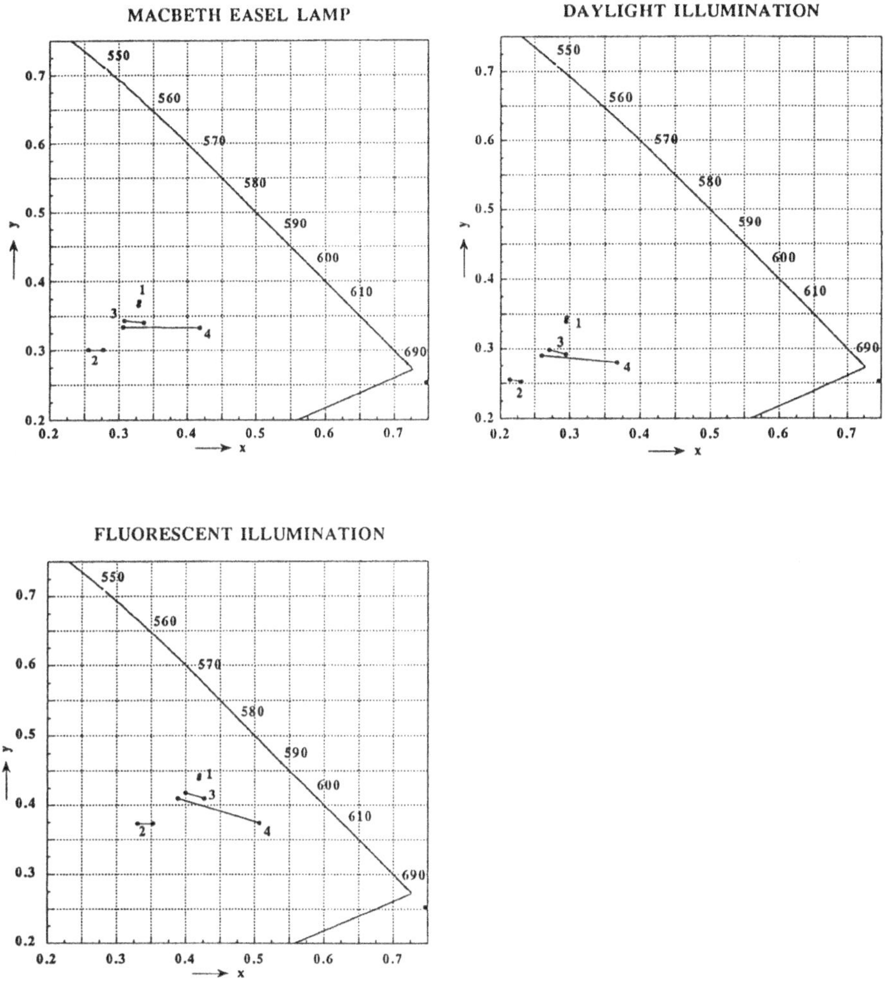

Fig. 2. The chromaticities of target and ground for the four protan plates of the Yustova test, plotted in the 1931 CIE diagram. Measurements are shown for three different illuminants. The protan copunctal point is shown at the bottom right of each diagram. The numbers by the data points indicate plate number.

are reasonably well incorporated into the production version of the test: in all the illuminations tried, the chromaticity differences in these three plates lie approximately along protan lines. In the case of the deutan plates 6, 7, and 8 the pattern of chromaticities depends on the illuminant: under a Macbeth easel

Deutan plates

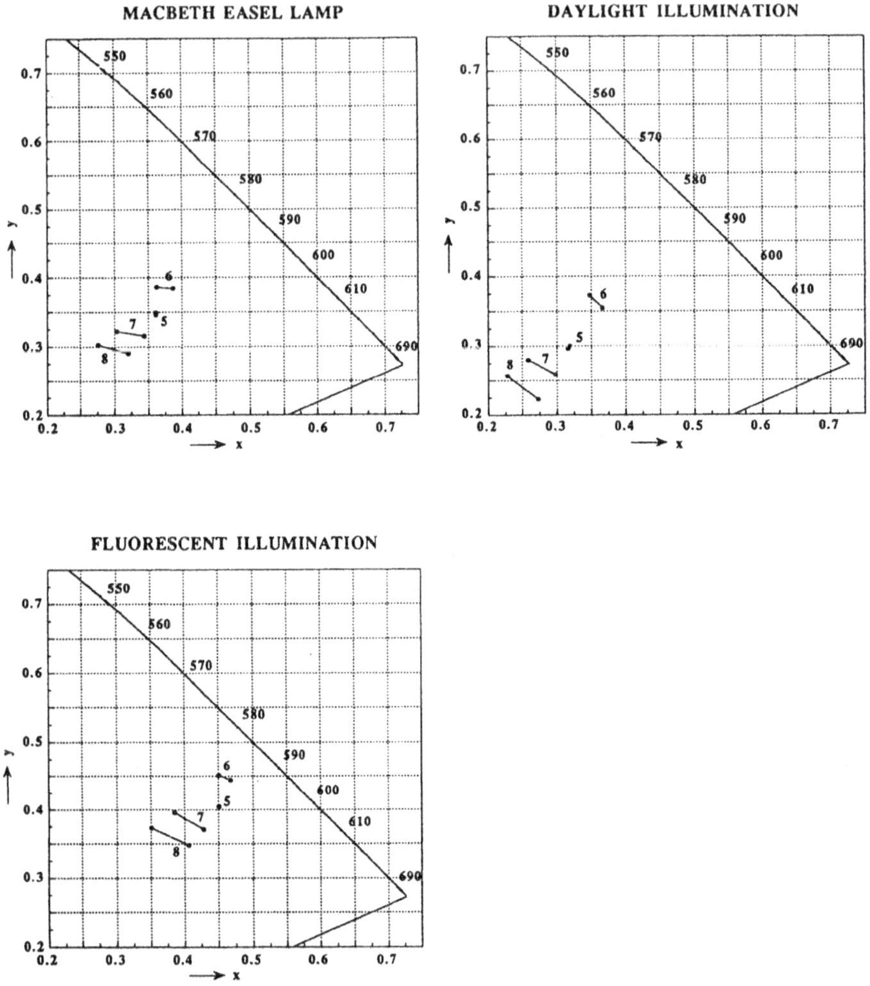

Fig. 3. The chromaticities of target and ground for the four deutan plates, plotted in the 1931 CIE diagram. Measurements are shown for three different illuminants.

lamp the confusion lines are predominantly protan – this is especially true for plate 6, which many protans fail. In true daylight, the chromaticity differences lie much more nearly along deutan lines, whereas the fluorescent lamps give a pattern of chromaticities that is less clearly deutan or protan.

The lack of diagnostic power of the mild protan and mild deutan plates (1 and 5), so manifest in Table 1, is easily understood from Figure 2: in neither plate do

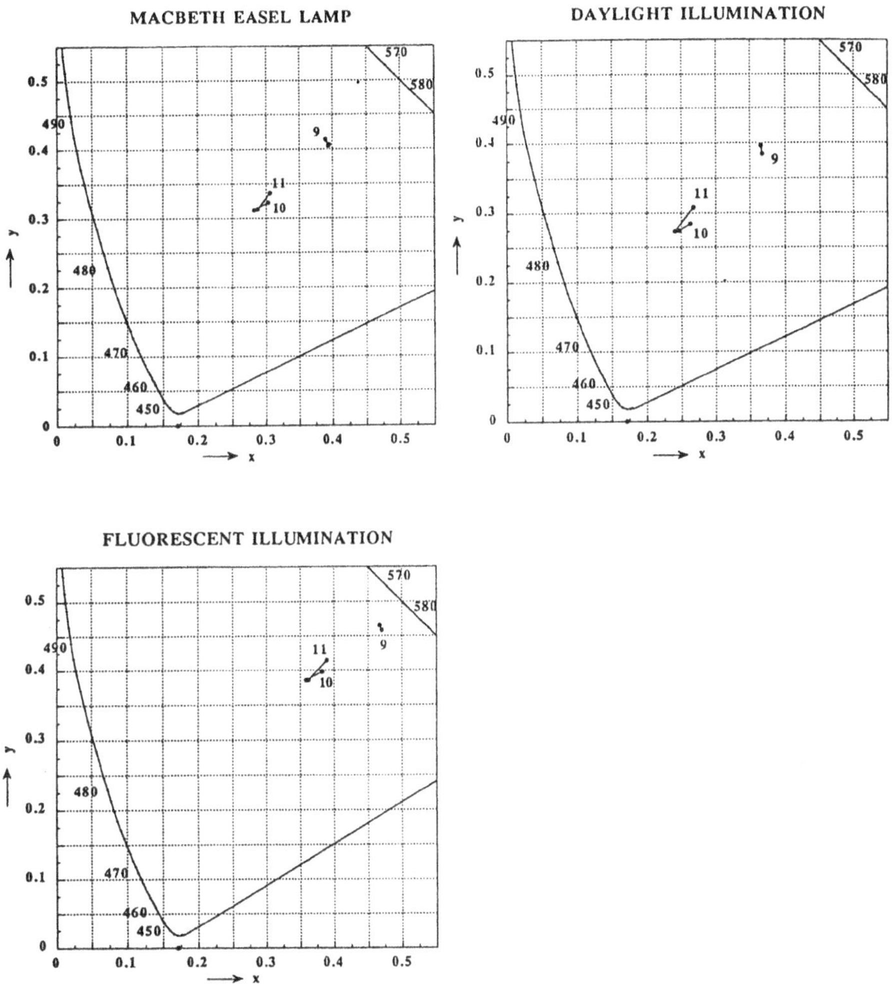

Fig. 4. The chromaticities of target and ground for the three tritan plates, plotted in the 1931 CIE diagram. Measurements are shown for three different illuminants.

the chromaticities lie along the required confusion line. The tritan plates 10 and 11 are the least satisfactory of the set, being read easily by a complete tritanope. Two factors contribute here. First, the chromaticities of target and ground lie along a yellow–blue line rather than a tritan line; and second, in both plates a clear lightness difference between target and ground is apparent to the normal eye (and to the spectroradiometer). Yet the luminous efficiency functions of congenital tritanopes do not differ systematically from those of normal observers (Wright, 1952). Plate 9, intended for detecting tritan defects of the mildest degree, exhibits small chromaticity differences that lie more in a protan or deutan direction (Fig. 2) and thus it is not remarkable that this plate serves well as a screening plate for red–green deficiencies. The user might do well to cross out the '9' on the back of this plate and substitute '1'.

Discussion

It has proved instructive to test subjects on a new set of plates and concurrently to measure the chromaticities of the plates. The subjects behave rather lawfully when their passes and failures are compared to the colorimetric data. It seems unlikely that the chromaticity vectors in plates 1, 5, and 9 are those intended by the designers. It is possible that errors in reproduction have occurred between the hand-made prototypes and the sets now circulated. Even if reproduction of the chromaticities of the prototype were perfect, there would be a second reason to expect differences in the behaviour of the field-tested prototype and that of the production version: spectroradiometrically different stimuli that are perfectly metameric for a normal observer will not be metameric for anomalous trichromats. Thus data gathered on a prototype are valid for a production version of a colour test only if the stimuli are spectroradiometrically, and not merely colorimetrically, identical to the original.

Equation of lightnesses

By eschewing the use of luminance noise in their plates, Yustova and her colleagues deny themselves one of the hard-won devices of colour vision testing (Regan *et al.*, 1994). Even if lightnesses could be reproduced perfectly, we today have reason to doubt that lightness equations can be made *a priori* for all daltonians: both the long and middle wavelength photopigments are genetically polymorphic, varying in their spectral position (Merbs and Nathans, 1992; Winderickx *et al*, 1992), and so there are good grounds for expecting the variations in psychophysical sensitivities that are in practice seen in populations of dichromats and anomalous trichromats (Alpern and Wake, 1977; Cavonius and Kammann, 1984; Jordan and Mollon, this volume). Lens and macular pigment are additional sources of variance. Yustova's idea of adjusting the lightnesses to compensate for the altered spectral sensitivities of daltonians is an excellent one, but the test might gain considerably by the addition of luminance noise centred on the mean match for each type of dichromat.

Anomalous trichromats

Yustova and her colleagues recognize that a standard colorimetric system, based on normal observers, cannot be applied to anomalous trichromats, whose colour matching functions differ from those of normals. The Yustova plates are explicitly intended only to detect impairments of colour discrimination. Clearly, however, in healthy young populations most of the subjects diagnosed as having protan or deutan deficits of the first, second or third kinds are likely to be anomalous trichromats. Insofar as the Yustova or HRR plates detect anomalous trichromats, they depend on the fact that the discrimination ellipse of the anomal is typically elongated in the direction of the confusion line of the corresponding dichromat. There does exist a subset of anomalous trichromats whose discrimination ellipses resemble those of normal individuals and whose spectral saturation function shows little loss at the wavelength of the neutral point of the corresponding dichromat (Chapanis, 1944; Regan *et al.*, 1994). Members of this subgroup of anomalous trichromats with good discrimination are usually caught by the transformation plates of the Ishihara test. The latter plates work not by measuring chromatic thresholds but by a quite distinct and little recognized principle (Reffin *et al.*, 1991): they measure the relative salience of the signals derived from the phylogenetically older and younger subsystems of colour vision. Those anomals who passed the Yustova plates in the present study are all known to fail the Ishihara.

Conclusions

Although the most sensitive plate is not the one intended by the designers, the Yustova test will serve adequately to screen for actual impairment of colour discrimination, while passing a small proportion of anomals who have good discrimination. Plates 7 and 8 are useful in distinguishing deutans from protans, and the Yustova test probably could be used to make this distinction as well as other plate tests in common use (Frey, 1958; Walls, 1959; Lakowski, 1966). Plates 1 and 5 are ineffective and should be omitted. Plate 9, intended as a tritan plate, can be usefully retained as a red–green screening plate. It is clear how the weaknesses of the test could be removed without sacrificing its strengths.

Acknowledgements

Supported by MRC Grant G9113630N, and by a Physiological Society travel grant to M. Danilova. We are grateful to E.N. Yustova and K.A. Alekseeva for information about the test and to our colleagues G. Jordan and B. Regan for information on the phenotypes of many members of our subject panel.

References

Alekseeva K.A. and Yustova E.N. (1978). Novye tablicy dlya ispytaniya cvetovogo zreniya. Trudy metrologiheskih institutov SSSR. 144: 74–85.

Alpern, M. and Wake, T. (1977). Cone pigments in human deutan colour vision defects. J. Physiol. (Lond.) 266: 595–612.

Birch, J. (1975). New diagnostic pseudoisochromatic plates for dichromats based on subtractive color matches. Am. J. Physiol. Optics 52: 398–404.

Cavonius, C.R. and Kammann, J. (1984). A clinical evaluation of the 'OSCAR' color vision set. In Verriest, G. (ed.), Colour Vision Deficiencies VII, Doc. Ophthal. Proc. Ser. 39: 275–279. Junk, The Hague.

Chapanis, A. (1944). Spectral saturation and its relation to color-vision defects. J. Exp. Psychol. 34: 24–44.

Frey, R.G. (1958). Welche pseudoisochromatischen Tafeln sind für die Praxis am besten geeignet? Graefe's Arch. Ophthalmol. 160: 301–320.

Hardy, L.H., Rand, G. and Rittler, M.C. (1954). H-R-R polychromatic plates. J. Opt. Soc. Am. 44: 509–523.

Ishihara and Okuma (1975). Plates for Detecting Colour Abnormality and Determining Its Degree [In Japanese]. Kanehara Shupper, Tokyo.

Lakowski, R.A. (1966). A critical evaluation of colour vision tests. Br. J. Physiol. Opt. 23: 186–209.

Merbs, S.L. and Nathans, J. (1992). Absorption spectra of human cone pigments. Nature 356: 433–435.

Oblath, O. (1929). Colour Vision Tests. International Labour Office (League of Nations), Studies and Reports, Ser. F, No. 12, Geneva.

Reffin, J.P., Astell, S. and Mollon, J.D. (1991). Trials of a computer-controlled colour vision test that preserves the advantages of pseudoisochromatic plates. In: Drum, B., Moreland, J.D. and Serra, A. (eds.), Colour Vision Deficiencies X, Doc. Ophthal. Proc. Ser. 54: 69–76. Kluwer, Dordrecht.

Regan, B.C., Reffin, J.P. and Mollon, J.D. (1994). Luminance noise and the rapid determination of discrimination ellipses in colour deficiences. Vision Res. 34: 1279–1299.

Schaaff (1925). L'examen du sens chromatique dans le service des chemins de fer. Bull. Acad. Méd. 94: 952–955.

Stilling, J. (1877). Die Prüfung des Farbensinnes beim Eisenbahn- und Marine-personal. Theodor Fischer, Cassel.

Walls, G.L. (1959). How good is the H-R-R test for color blindness? Am. J. Optom. Monogr. No. 249.

Winderickx, J., Lindsey, D.T., Sanocki, E., Teller, D.Y., Motulsky, A.G. and Deeb, S.S. (1992). Polymorphism in red photopigment underlies variation in colour matching. Nature 356: 431–433.

Wright, W.D. (1952). The characteristics of tritanopia. J. Opt. Soc. Am. 42: 509–521.

Yustova E.N., Volkov V.V., Alekseyeva, K.A. and Sergenyev, V.P. (1987). New tables for the testing of colour vision. In: Marré, E., Tost, M. and Zenker, H.-J. (eds.), Normal and Pathological Colour Vision: 25–35. Martin-Luther-Universität, Halle-Wittenberg.

M.V. Danilova
Visual Physiology Laboratory, I.P. Pavlov Institute of Physiology
Academy of Sciences of Russia
Nab. Makarova 6
St. Petersburg, 199034, Russia

J.D. Mollon
Department of Experimental Psychology, University of Cambridge
Downing St., Cambridge, CB2 3EB, UK

35. Evaluation of the desaturated Roth 28 HUE colour test – preliminary results

C. ERB, M. ADLER, N. STÜBIGER, M. WOHLRAB and H.-J. THIEL
(Tübingen, Germany)

Abstract

In order to quantify colour vision deficiencies, we have evaluated the desaturated Roth 28 HUE colour test for error scores in 95 normal subjects (71 males, 24 females) with no ocular or general chronic disease who were tested monocularly under standard conditions. The background was a black cardboard; illumination consisted of two fluorescent lamps (Osram L18W/11, Germany). The subjects were divided into four age groups: 0–19, 20–39, 40–59 and 60–79 years. Quantitative analysis was performed using specially developed software. The mean error score in the colour test was 66 ± 35 (minimum 0, maximum 174; median 60). There were no differences between the right or left eye ($p = 0.6$, Mann–Whitney test), or between males and females ($p = 0.64$, Mann–Whitney test). No correlation was found between age and error score (regression coefficient $= -0.027$; $p = 0.94$). Based on these data, error scores on the desaturated Roth 28 HUE test appear to be stable over age.

Introduction

Colour vision and its disturbances are important in ophthalmological examinations, because it has been shown that drugs and systemic diseases, such as diabetes mellitus, can cause acquired colour vision deficiencies. The detection of such disturbances, however, is difficult because most of them appear very slowly and begin with slight disturbances in colour vision. They are therefore often overlooked or underestimated. Second, in middle-aged and elderly people there is a relationship between acquired colour vision deficiencies and disturbances in the optic system, such as cataract, and in the macula (increased deposits of lipofuscin). Variation in the patient's power of concentration also has to be taken into account. An appropriate colour test must, therefore, be easy to understand, rapidly administered and has to show the disturbed colour axis and demonstrate the degree of disturbance of colour vision, in order to obtain quantitative and qualitative information. Many colour tests do not fulfil these criteria, because they are not exact enough (e.g. the Ishihara plates: there is no colour combination for the detection of a blue–yellow anomaly), too complicated (e.g.

C.R. Cavonius (ed.), Colour Vision Deficiencies XIII, pp. 323–329.

Farnsworth-Munsell 100 (FM-100)) or do not give satisfactory qualitative information (e.g. Farnsworth Panel D15). Therefore, we tested the cap-sorting test Roth 28 HUE desaturated, to determine its practical use and to obtain error scores for a normal population.

Methods

Patients

95 subjects (F: 24; M: 71) underwent an ophthalmological examination at the University Eye Clinic Tübingen between 1993 and 1995. Preconditions for participating in this study were: absence of systemic diseases, such as hypotension, hypertension, diabetes mellitus or stroke, no drug treaments, non-smokers and no history of ophthalmological surgery (e.g. cataract extraction) or ophthalmological diseases.

The ophthalmological examination consisted of a slit-lamp examination, test of visual acuity, including refraction and, after colour testing, measurement of the interocular pressure with Goldmann applanation tonometry. Best corrected visual acuity was at least 0.8 and the intraocular pressure was within normal limits (right eye: 15 ± 2 mmHg, left eye: 14 ± 2 mmHg).

Colour test

The cap-sorting test Roth 28 HUE (Lumeau, Paris) was first described by Roth (1966a, 1966b) and also exists in a desaturated version. The background used was a black cardboard, illuminated by two fluorescent lamps (Osram L18W/11, Germany) at 4800 lux. The subjects' visual acuity was corrected, if necessary, and gloves were used to protect the surface of the colour caps. Each subject was tested right eye first, followed by the left eye. After an explanation of the colour test all colour caps were removed and mixed. There was no time limit for the test, and the subject was allowed to correct each of her or his arrangements.

Calculation and graphic demonstration of the Roth 28 HUE desaturated colour test

All data were sorted into four age groups: <20, 20–39, 40–59 and >59 years. The Roth 28 HUE desaturated is a subset of the Farnsworth-Munsell 100-hue test. Therefore, the colour chips of this test are numbered from 1 to 82, with only each third number used. The colour chip number 82 is fixed and defined as the starting and end point. Since calculation of the error score resembles the method used in the FM 100-hue test, the error scores are similiar to those of the FM 100-hue. For each of the 28 colour chips, the difference of the chip number to both adjacent colour chips was calculated (value X). Values X and 84[=(82-1)+3]–X were then compared and the lower was chosen as the distance.

The shortest distance between two caps within the circular arrangement is thus calculated, e.g. the distance 82–1 is calculated as 3, which is the standard value for error-free arrangement. The values of distance on both sides were added, and the value 6 [2 × 3] (corresponding to error-free handling of the test) was subtracted from this sum and the resulting value stored as a local error score. The sum of the local error scores of the 28 colour chips is the resulting global error score. The graphic representation shows the 28 local error scores.

Results

The distribution of error scores (Fig. 1) was not Gaussian. Therefore, for our calculations we used the Wilcoxon signed-rank test, a paired, non-parametric test. The mean error score of all subjects was 66 ± 35 (Table 1). The error scores differed slightly across the four age classes, but no correlation was found between age and error scores (Fig. 2; regression coefficient −0.027, $p = 0.94$). No differences could be detected between the right and the left eyes, although all subjects started with the right eye. The median of the error score of the right eye, the left eye and of both eyes was 60 ($p = 0.5825$, Wilcoxon signed-rank test). No differences were found between males and females (Table 2; $I = 0.6468$; Mann–Whitney test).

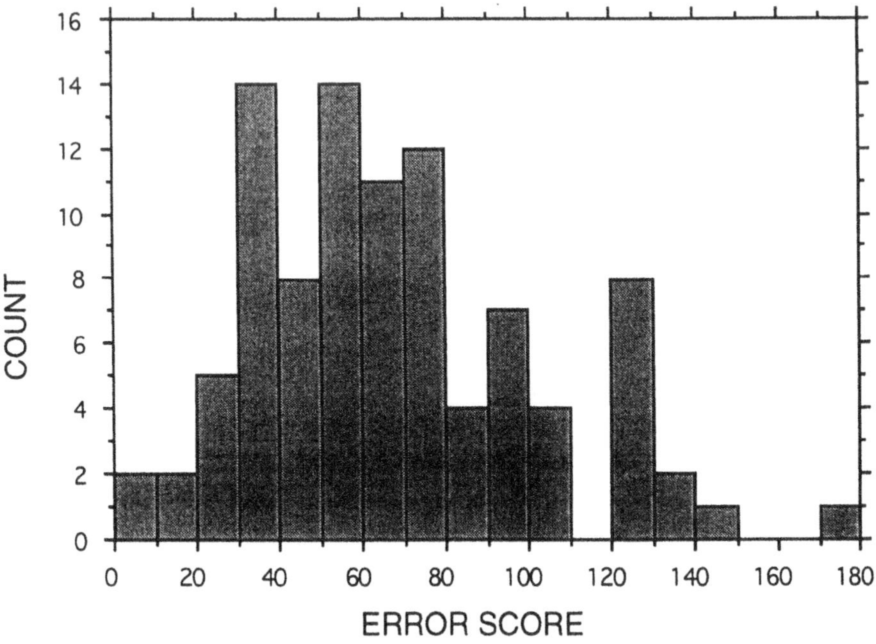

Fig. 1. Distribution of the error scores.

Table 1. Means, standard deviations, medians, minima and maxima of the error scores, by age group.

	Total	<20 years	20–39 years	40–59 years	>59 years
Count	95	28	56	9	2
Minimum	0	36	0	30	54
Maximum	174	138	144	174	66
Mean	66.4	75.8	57.9	91.3	60.0
SD	34.5	26.4	34.2	44.8	8.4
Median	60	75	54.	84	60
MAD	24	21	24	30	6

MAD = mean absolute deviation

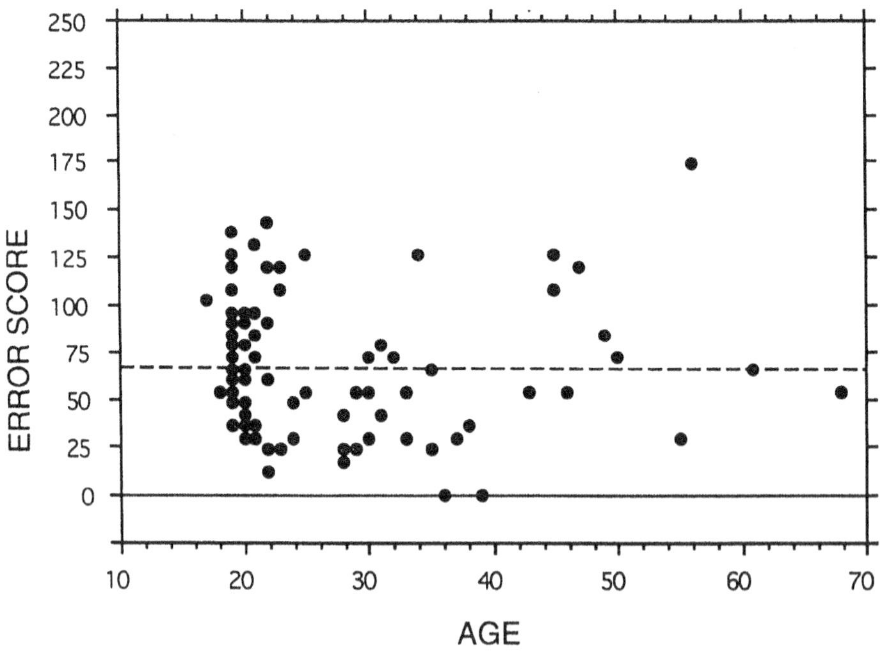

Fig. 2. Correlation between age and error scores (– – – = regression line).

Table 2. No differences were found between males and females.

Gender	Count	Mean	SD
M	71	66.085	31.448
F	24	67.500	43.058

Discussion

The colour test Roth 28 HUE desaturated has not previously been evaluated for normal values. It is a simple test, which can be performed more quickly than the FM-100, and is therefore less dependent on the patient's attention. Because of the simplicity of this test, it is suitable for children, as demonstrated by Roth (1968) in children with strabismic amblyopia. According to this study, this colour test is useful at the age of 5 years and above. Like Verriest *et al.* (1982), who used the FM-100 test, we found no difference between the results of females and males (Table 2). Additionally, there were no differences between the right and the left eye, which contradicts the results of Verriest *et al.* (1982), who found that the right eye had significantly higher values than the left eye. This could indicate a training effect, which is avoided by the Roth 28 HUE desaturated because of its simpler test administration. Therefore, the examiner or the patient can decide independently with which eye they start the test. This could be an advantage, if the patient wants to start with the preferred eye.

An important result of this study is the relatively stable error score with increasing age. This was unexpected, because in the FM-100 there is a higher error score in the younger subjects related to reduced attention, and a higher error score in the elder ones because of pre-retinal changes (Verriest *et al.*, 1982). As mentioned above, no correlation between age and error score was found with the Roth 28 HUE desaturated, for which there are various explanations. The Roth 28 HUE desaturated is much easier to administer than the FM-100 and does not required prolonged attention. This results in fewer errors due to reduced concentration, particularly in younger subjects.

The stability of the test over age may be based on the following: first, we only examined subjects who had a best-corrected visual acuity of 0.8 or higher. Therefore, we excluded persons with functional loss involving morphological changes such as a nuclear cataract or an intensive age-related macular degeneration to avoid colour disturbances resulting from these conditions (Moss *et al.*, 1995). Second, our subjects had no systemic diseases, especially no cardiovascular problems. Although this was mentioned in other studies, no detailed description was made, andwe consider this aspect to be very important: patients with coronary heart disease tend to have high error scores in spite of an otherwise normal ophthalmological status (Erb *et al.*, 1995). In patients with diabetes, acute changes in blood glucose alone can alter the short-wavelength

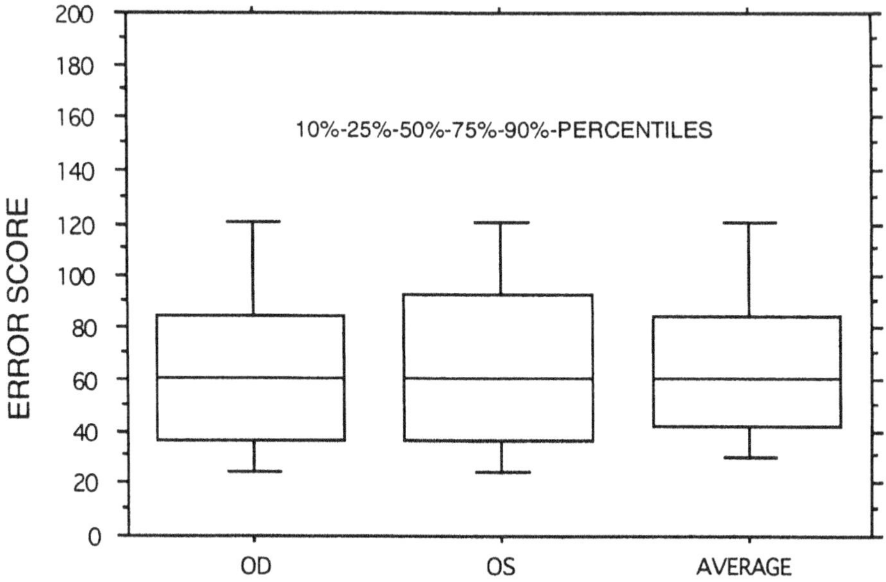

Fig. 3. Error scores in the right (OD) and left (OS) eye.

sensitivity (Schneck *et al.*, 1993). Additionally, in none of the evaluating studies (e.g. Verriest *et al.*, 1982) were smokers excluded, or no attention was given to drug therapy. But both factors have a well-known influence on colour vision and can change the error scores (Lyle, 1974a, 1974b; Pokorny *et al.*, 1979). However, these may be a reason for the increasing error scores of the FM-100 with age. The interaction between systemic diseases, drug use and smoking are unknown.

To conclude, we found a stable error score over age, but we need to examine more subjects over 40 years of age to obtain better evidence for this statement. If we can confirm our finding, the Roth 28 HUE desaturated will be a powerful tool in the follow-up of colour disturbances, caused, for example, by chronic drug use or chronic diseases. We also plan to create a catalogue of different influences on colour vision with this test to produce reliable information about the extent of colour vision disturbance caused by individual factors.

Note

We are aware of the fact that the Kruskal–Wallis test would be appropriate to test the different age classes against one another. In this study with preliminary results we have unequal numbers of subjects in the age groups. Therefore, a statistically correct comparison between the groups is not possible at present. We intend to do this later.

References

Erb, C., Voelker, W., Stübiger, N. and Thiel, H.-J. (1995). White-noise-campimetry abnormalities and colour-vision disturbances in patients with coronary heart disease. Perception 24 (suppl.): 53b.

Lyle, W.M. (1974a). Drugs and conditions which may affect color vision. Part I: Drugs and chemicals. J. Am. Optomet. Assoc. 45: 47–61.

Lyle, W.M. (1974b). Drugs and conditions which may affect color vision. Part II: Diseases and conditions. J. Am. Optomet. Assoc. 45: 173–182.

Moss, I.D., Wild, J.M. and Whitaker, D.J. (1995). The influence of age-related cataract on blue-on-yellow perimetry. Invest. Ophthalmol. Vis. Sci. 36: 764–773.

Pokorny, J., Smith, V.C., Verriest, G. and Pinckers, A.J.L.G. (1979). Congenital and Acquired Color Vision Defects. Grune & Stratton, New York.

Roth, A. (1966a). Test-28 hue de Roth selon Farnsworth-Munsell. (Manuel), Luneau, Paris.

Roth, A. (1966b). Le test 28-hue selon Farnsworth. Bull. Soc. Ophtal. Fr. 66: 231–238.

Roth, A. (1968). Le sens chromatique dans lamblyopie functionelle. Docum. Ophthal. 24: 113–200.

Schneck, M.E., Adams, A.J., Volbrecht, V.J. and Linfoot, J.L. (1993). Acute changes in blood glucose and their effects on colour vision function in diabetes. In: Drum, B. (ed), Colour Vision Deficiencies XI: 165–173. Kluwer, Dordrecht.

Verriest, G., Van Laethem, J. and Uvijls, A. (1982). A new assessment of the normal ranges of the Farnsworth-Munsell 100-HUE test scores. Am. J. Ophthalmol. 93: 635–642.

University Eye Clinic
Eberhard-Karls-University
Schleichstrasse 12
D-72076 Tübingen, Germany

36. Evaluation of Yu's Chinese pseudoisochromatic plates for acquired dyschromatopsia

J. LEID, J. POLETTI and Y. SOULEYMANOU

(Pau and Tarbes, France)

Abstract

The goal of this study was to validate clinically the Chinese pseudoisochromatic (PIC) plates of Yu Ziping and to evaluate their utility in detecting acquired dyschromatopsia, in particular that associated with diabetes. We tested 39 patients, 12 of whom are diabetic, with a battery of tests including the Ishihara test, Yu's test, Lanthony's desaturated D15 and the Farnsworth 100-hue. We analysed the results with the following goals in mind: (1) to establish the purpose of each plate from Yu's test, as a good translation of the instructions was lacking; (2) to determine the relative efficiency of Yu's test with respect to other tests; (3) to decide whether Yu's test is of value in the screening of dyschromatopsias arising from diabetes.

Overall, Yu's plates seemed particularly effective for detecting minor anomalies along the red–green axis, but were insufficient for finding anomalies along the blue–yellow axis. Use of the desaturated D15 in association with Yu's plates is possibly a better strategy for the screening of dyschromatopsias of diabetic origin.

Introduction

We initially learned about Yu's test at the 1985 IRGCVD Symposium in Avignon, where it was introduced by Dong (Dong *et al.*, 1987). Dr. J. Poletti brought a copy back on a subsequent trip to China. Since then, we have used it extensively on a daily basis for a large variety of diseases. After intuitively noticing its good quality, and intrigued by the sensitivity of certain plates, we decided to adopt a rigorous protocol in order to evaluate its clinical validity.

This study aimed to identify the advantages of each plate, since we lacked a complete translation of the instructions written in Chinese, to evaluate the relative effectiveness of this test relative to the tests which are better known in Europe and North America and to evaluate the efficacy of the test in the detection of diabetic retinopathy.

C.R. Cavonius (ed.), Colour Vision Deficiencies XIII, pp. 331–338.

Materials and methods

This study was conducted from October 1984 to January 1985 on patients examined at Tarbes Hospital.

Materials

Yu's test is PIC plates, manufactured in China, whose quality seems excellent. The test is packaged in the standard form of a book containing 50 plates and an accompanying 24-page booklet written in Chinese. The title page reads 'Diagnostic plates for dyschromatopsia, 4th edition, revised, written and illustrated by Ziping Yu'.

Publishing data show that the book was printed in Peking, and its first edition is dated May 1958. The latest (4th) edition is dated February 1982.

According to the preface, the book contains four groups of plates. The first (plates 1–13) contains simple digits for a rapid examination, the second group (plates 14–22) comprises geometric figures suitable for illiterate adults, the third group (plates 23–36) contains drawings for use in children, and the fourth group (plates 37–50) is made up of multiple digits for closer examination.

Further explanations suggest that the author has an incomplete knowledge of acquired defects. According to Yu, a patient with acquired dyschromatopsia suffers from both a red–green and a blue–yellow axis anomaly (at times, he actually calls these defects 'total'). These explanations are all the more surprising in that the numbers printed on the plates cited as examples contain not only some digits that cannot be read by an individual suffering from a dyschromatopsia with a red–green axis (i.e. a protanope, a deuteranope or an acquired dyschromate Verriest type I or II), but also others that cannot be read with a blue–yellow axis dyschromatopsia. Therefore, the test should logically allow for the detection of pure blue–yellow axis, seemingly unknown to the author. Hence, this test has a much wider scope than initially conceived.

At the end of the preface, there is an attempt to quantitatively classify congenital colour vision defects into two levels for dichromats and two levels for anomalous trichromats. The author suggests that his test actually allows this quantitative evaluation. This text is signed Ziping Yu, on behalf of the ophthalmology teaching and research team at the Nankin medical school, 1980. There follows a long instruction sheet for which we do not have a translation, except for a few explanations pertaining to the chapter entitled 'How to use the plates', rehashing the previously discussed issues concerning acquired defects. Charts indicate the purpose of each plate. The plates follow the pattern used in most pseudoisochromatic (PIC) tests, with demonstration plates, plates unreadable by dichromats only, others unreadable by those with red–green axis defects and transposition plates in which subjects with defective colour vision see characters different to those seen by normal individuals. Most of the plates are designed with coloured dots located on confusion axes. A few rare plates (used specifically to separate protans from deutans) display a design on a grey background, chosen in the protan neutral zone. For deutans, a green–brown confusion axis has primarily been chosen, which we feel gives this test its originality.

Patients

We examined 46 patients: a control group of 10 patients, young and free of systemic ocular pathology, 11 patients with known inherited defects of colour vision and no ophthalmological disease, and a group of 25 patients with acquired colour defects but no inherited defects. This latter group presented a diversity of diseases, including glaucoma, multiple sclerosis and hypertension, from which we selected 14 diabetics. Inclusion criteria included visual acuity $\geqslant 7/10$, age $\leqslant 65$ years and suspicion of a presumably isolated colour vision defect. Each patient underwent the Ishihara, Lanthony's desaturated D15 (D15D), Farnsworth's automated 100-hue ('Chromops') and Yu's tests, always in the same order Monocular examinations were performed by a single person, in daylight from a north-facing window with optimal untinted optical correction.

Results

The results were analysed individually and are shown in Tables 1–4. These do not show the entire set of results as we are primarily interested in the use of this test in diabetic patients. Yu's test proved to be very effective in detecting inherited anomalies, with a sensitivity close to that of the Ishihara test. Nevertheless, we remain relatively unconvinced of the test's true capacity to measure defects quantitatively, although because of the small number of patients with inherited colour defects tested, we cannot give a final opinion. It was possible to discriminate between abnormal dichromats and trichromats more often than with the Ishihara plates, but the Yu test was clearly far less effective than with the D15/desaturated D15 couple (we added Farnsworth's 15-hue test to the battery in patients with inherited defects of colour vision).

Controls

Eight of 10 controls were able to read all of the plates in Yu's test. One patient made an error on plate 8, which is green–brown, and this same patient made a reversal on the desaturated D15. That patient might have poor colour discrimination, although he performed the 100-hue test within normal limits (score of <168 with the Chromops, which is not too difficult). Moreover twice we found isolated inversions with Lanthony's test and once again a brown plate was misread with Yu's test.

Table 1. Controls – OD.

Patients	Ishihara	Desaturated D15	100-hue	Yu's test
1. ROL. M.	Normal	Normal	Normal	Normal
2. ESQ. X.	Normal	Normal	Normal	Normal
3. SOU. Y.	Normal	1 Reversal	Normal	Normal
4. ESQ. L.	Normal	Normal	Normal	2 Reversals Pl 2, 25
5. CAR. G.	Normal	Normal	Normal	Normal
6. LAB. S.	Normal	Normal	Normal	Normal
7. DED. C.	Normal	1 Reversal	Normal	1 Error Pl 8
8. ABA. S.	Normal	Normal	Normal	Normal
9. RAH. S.	Normal	1 Reversal	Normal	Normal
10. SAL. C.	Normal	Normal	Normal	Normal

Table 2. Controls – OS.

Patients	Ishihara	Desaturated D15	100-hue	Yu's test
1. ROL. M.	Normal	Normal	Normal	Normal
2. ESQ. X.	Normal	Normal	Normal	Normal
3. SOU. Y.	Normal	Normal	1 Reversal	Normal
4. ESQ. L.	Normal	Normal	1 Reversal	Normal
5. CAR. G.	Normal	Normal	Normal	1 Error Pl 22
6. LAB. S.	Normal	Normal	Normal	Normal
7. ED. C.	Normal	Normal	Normal	Normal
8. ABA. S.	Normal	Normal	Normal	Normal
9. RAH. S.	Normal	Normal	Normal	Normal
10. SAL. C.	Normal	Normal	Normal	Normal

Patients with inherited defects

Studying patients with inherited colour defects allowed us first to identify each plate's specific advantages. As expected, they were consistent with the booklet's indications (and with our expectations following the study of the plates) with the exception of the limited capacity to make quantitative measurements. We were nevertheless struck by the high sensitivity of the plates using a green–brown pair in detecting minor red–green axis anomalies. Likewise, plates 12 and 13 and 21–23, obviously designed to separate protans from deutans, proved to be more effective than similar plates in Ishihara's test.

Table 3. Diabetic patients – OD.

Patients	Ishihara	Desaturated D15	100-hue	Yu's test
1. MON. M.	Normal	Tetartan	Tritan	Blue Not Seen 24, 25
2. ABA. XL.	Normal	Tritan/Tetartan	Tritan	No Error
3. CAZ. A.	Normal	Deutan/Tritan	Tritan	2 Errors 38
4. COR. C.	Normal	Normal	No Axis	No Error
5. DEL. C.	Normal	Tritan/Tetartan	Mild Tetartan	No Error
6. DUP. J.	Normal	Tritan	Deutan/Tritan	Errors 40, 48
7. ZAN. M.	Normal	Mild Tetartan	No Axis	Errors 34, 40
8. CYN. A.	Normal	Normal	Deutan/Tritan	Errors 28, 34, 40
9. CAN. M.	Normal	2 Reversals	Tritan	Slow, No Error
10. GON. A.	Normal	Deutan/Tritan	Deutan/Tetartan	Slow, No Error
11. BAR. N.	Deutan	Normal	Deutan	Errors 38, 40, 48
12. BAR. D.	Normal	Normal	Mild Tritan	Errors 30, 32, 38, 40
13. CAZ. D.	Normal	Mild Tetartan	Normal	Errors 38, 40
14. ABA. LC.	Normal	Tritan	Mild Tritan	Errors 38, 40, 45

Table 4. Diabetic patients – OS.

Patients	Ishihara	Desaturated D15	100-hue	Yu's test
1. MON. M.	Normal	Tetartan	Tritan	No Error
2. ABA. XL.	Normal	Tritan/Tetartan	Tritan	No Error
3. CAZ. A.	Normal	Deutan/Tritan	Tritan	Errors 38, 42
4. COR. C.	Normal	Normal	No Axis	No Error
5. DEL. C.	Normal	Tritan	Deutan/Tritan	Errors 24, 26
6. DUP. J.	Normal	Normal	No axis	No Error
7. ZAN. M.	Normal	Tital/Tetartan	Deutan/Tritan	Errors 38, 43, 48
8. CYN. A.	Normal	Normal	Tritan	No Error
9. CAN. M.	Normal	2 Reversals	Deutan/Tritan	No Error
10. GON. A.	Normal	Deutan/Tritan	Deutan	No Error
11. BAR. N.	Deutan	Normal	Mild Tetartan	Errors 12, 22, 24, 38
12. BAR. D.	Normal	Normal	Normal	Errors 34, 40, 42
13. CAZ. D.	Normal	Normal	Normal	Errors 38, 40
14. ABA. LC.	Normal	Tritan	Tritan/Tetartan	Errors 40, 42, 45

Acquired defects

In order to have a homogeneous population, we decided to investigate only the cohort of diabetic patients (28 eyes). The conclusions drawn from the study of this group were perfectly applicable to the individual analysis of the other acquired defects tested. In this group of patients, results with Ishihara's test were always normal, except for one individual with a minor RG deficiency.

Among the 11 patients in whom the the D15D was normal (extremely mild defects), four were classified as normal by Yu's test, and once the 100-hue test showed only a slight score elevation without a clear axis. Nevertheless, under these conditions we found one mild deutan axis and two isolated mild tritan axes with the 100-hue when Yu's test remained normal. In another patient (2 eyes), a type II RG axis was found with the 100-hue test, both pseudo-isochromatic plates showing a deutan type axis. In one eye we found two errors with brown plates when the rest of the battery was normal (probable false positive). Finally, in one eye, the 100-hue test showed a deutan and tritan double axis and Yu's test confirmed this with errors, not only on brown plates but also on a violet (tritan) plate.

In the 17 cases where the D15D revealed an anomaly this was consistently a blue–yellow axis anomaly, associated in four cases with a red–green axis. In those four cases, Yu's test was normal twice (both eyes in the same patient), although in the other two cases it showed RG but not BY axis anomalies. In three cases, the D15D showed a BY axis and the 100-hue revealed an additional RG axis anomaly. In all three cases, the two anomalies were detected by Yu's test. On six occasions, Yu's test failed to recognize a BY axis revealed by an arrangement test, twice it was successful and twice the results were uncertain.

In all, Yu's test was normal eight times when the D15D showed an anomaly.

The 100-hue test revealed a RG axis anomaly in eight cases: this was detected by the D15D only twice, and by Yu's test on five occasions.

Finally only five of 19 BY axes found with the 100-hue were detected by Yu's test.

Discussion

In general, most patients were intrigued by the drawings presented. The animal figures are not for the most part common in Europe, and their representation differed from ours. Even the numbers have a slightly unusual shape. This may be due to the fact that the author wanted them to be readable in two orientations.

Identification of each plate's purpose

This was easy thanks to the examination of the inherited colour defective patients. Each plate's purpose appeared consistent with the directions provided, except for the quantification capabilities, which proved insufficient. The brown–green plates were particularly effective in detecting red–green defects. Plates that include blue or violet optotypes have been judiciously chosen (the colours are close to the AO HRR), but their effectiveness remains as limited as always in detecting acquired defects with PIC plates. Leid and Leid (1989) already emphasized the fact that a PIC plate cannot deliver fine detection of acquired colour vision defects. These defects are often not quantitatively significant, and their confusion zones are large and variable across subjects. For the most part, this renders the PIC charts ineffective as, by definition, they try to identify confusions along a well defined and narrow axis.

No matter how rigorous the conception and manufacturing of a PIC plate, it will only deliver what it is designed for. Consequently, although PIC plates are very useful in detecting inherited defects, they must be considered only as adjuncts in acquired defects.

Relative effectiveness in comparison with other tests

Yu's test is perfectly capable of detecting the RG axis anomalies seen by D15D, and is clearly more effective than Ishihara's test, which is ineffective in detecting small RG axis acquired defects. In this respect it almost matches, but cannot equal, the 100-hue. Yu's test is far inferior to the D15D and, of course, the 100-hue test in detecting small BY axes. Yu's test performs better than the Ishihara and D15D in detecting small RG axes. The axes that it detects are most often in agreement with the 100-hue test. Compared with other tests, it does not detect additional BY defects.

The advantage of Yu's test in detecting diabetic dyschromatopsia

Initially, diabetic colour vision deficiency usually presents with a mild BY axis. Later, it is accompanied by RG axes, especially according to its ophthalmoscopic nature. Consequently, Yu's test alone is certainly not suitable for that type of screening. On the other hand, paired with D15D it noticeably increases that test's already considerable effectiveness in this area, since it can detect small RG axis defects that the arrangement test may fail to detect.

Conclusions

Yu's Chinese PIC plates are interesting in several respects. They may be slightly better than the Ishihara for detecting most inherited colour vision deficiencies, but are not equal to the 15-hue test in protan–deutan differentiation. In the case of small RG axis acquired defects, they are capable of very fine discrimination: thsee plates are better than Lanthony's desaturated D15 test, and almost as good as the 100-hue. They can detect a few acquired BY axes (tritan and tetartan), but as in any PIC, this capability is globally weak, owing to the very principles of this type of test.

All of these features makes it a useful test in current practice, in view of its broad range of applications. We think that the simple and rapid combination of Yu's test and Lanthony's desaturated Panel D15 offers a quick and very high quality method for the detection of diabetic colour vision deficiencies in daily practice.

References

Dong, T.H., Xie, L.C. and Jin, W.Y. (1987). Introduction of Chinese designed colour vision testing charts and instrument. In: Verriest, G. (ed.), Colour Vision Deficiencies V: 299–302. Karger, Basel.
Leid, J. and Leid, V. (1989). Clinical study of the Lanthony's tritan album: initial results. In: Drum, B. and Verriest, G. (eds.), Colour Vision Deficiencies IX: 495–501. Kluwer, Dordrecht.
Souleymanou, Y. (1995). Intérêt du test des planches pseudo-isochromatiques chinoises du Yu dans les dyschromatopsies acquises. Mém. D.I.S. Ophtalmol. 47 pp.

Dr. Jean Leid
4, Place Royale
F-64000 Pau, France

J. Poletti, Y. Souleymanou
Department of Ophthalmology
Tarbes Hospital
F-65000 Tarbes, France

37. Screening colour vision with an LMS calibrated display

F. VIÉNOT, A. BEN M'BAREK and L. OTT

(Paris, France)

Abstract

We present a CRT colour-screening test that specifically addresses single classes of cones. The calibration of the colour monitor is maintained by using a photodiode attachment. Pairs of pseudoisochromatic plates with varying degrees of L, M or S cone signal difference have been designed. It successfully separates protan and deutan and agrees with Nagel anomaloscope settings. A set of three double plates for each type is sufficient to rate the severity of the defect. Testing for tritan defects is also possible. The variability of results originating from variations in individual observers' criteria is avoided by balancing S cone-based discrimination against L or M cone-based discrimination.

Introduction

Congenital colour deficits originate from a lack of, or an alteration of, one or more of the photopigments that are present in the normal eye. Protan, deutan and tritan defects are related respectively to the loss of, or change in, cone photopigments sensitive to long (L), middle (M) or short (S) wavelengths. They are associated with selective losses of discrimination along one direction of the colour space. No other visual function is usually impaired. About 8% of the population, mostly males, are affected to some degree.

Acquired colour deficits originate from various pathologies, and are often associated with other ophthalmic diseases. The severity of the impairment varies from subject to subject and may change with time. The losses of colour discrimination in acquired colour deficits are not as well defined as are congenital colour deficits. However, an accurate description of the cone inputs is necessary to investigate the loss of colour function.

Several sets of LMS fundamentals, i.e. the LMS corneal cone action spectra, have been proposed (Smith and Pokorny, 1975; Vos *et al.*, 1990; Stockman *et al.*, 1993), which offer valuable estimates of the cone inputs. We have developed a test on a video display that allows specific screening along fixed directions on the LMS colour space. Careful calibration of the video screen has been made and long-term calibration is ensured in order to guarantee a strict correspon-

C.R. Cavonius (ed.), Colour Vision Deficiencies XIII, pp. 339–349.

dence between the colour of each pixel and the LMS cone signals. Presenting the test on a video display offers free viewing, rapid screening and the possibility of future computer-controlled presentation.

Methods

Calibration

The tests are displayed on a 14″ Trinitron colour monitor. They are generated with a 3×8 bit videographics board and a Macintosh Quadra 700 computer using Colorys (Primera) software. Several steps in the calibration of the screen were necessary to ensure accurate and reproducible LMS-based colorimetry.

Absolute calibration relies on spectroradiometry of the three primaries of the display and on photometric measurement of the voltage-to-luminance transfer function of each primary. The spectroradiometric calibration was obtained using a computer-controlled Bentham spectroradiometer (bandwidth 2.4 nm, 2 nm step; 1 nm steps were used with the red primary for a control; Pefferkorn, 1992). The spectral response of the Bentham photomultiplier detector was calibrated with a luminance standard traceable to the National Physical Laboratory before and after the monitor measurement. Recording each spectrum took about 3 min; three records were averaged for each primary. Since a minor change in the high-voltage supply of the cathode ray tube (CRT) can result in a noticeable change in the luminance of the screen, an extra spectroradiometric calibration was performed on the nominal white of the monitor in order to obtain a reliable measure of the relative luminances of the three primaries (Travis, 1991). This yielded a comprehensive colorimetric calibration of the colour screen, including the chromaticities of the three primaries and of the neutral white.

The luminance-versus-voltage relationship (gamma function) of each primary was measured, using a Photo Research Pritchard luminance meter. Sixteen equally spaced video signal levels were chosen, and a non-linear function was fitted to the calibration data.

Long-term calibration was performed with a commercial Radius Precision Colour Calibrator attachment. This is a single photodiode detector, associated with software that drives the 256 available levels for each primary gun and readjusts the look-up tables of the videoboard. Choice of the gamma value and the colour temperature of the nominal white is given. The original absolute calibration – spectral distribution and luminance output of each primary – was made after running the calibration software. The nominal gamma value was set to 1.8 and the nominal white colour temperature was set at 9300 K. Several control calibrations were then performed and the monitor readjusted. Although the actual gamma value and the nominal white colour temperature were not exactly at the expected nominal value, the repeatability of the calibration was acceptable.

Colorimetry

Once the spectroradiometric and photometric calibrations have been performed, it is possible to specify any pixel in a LMS colour representation and to establish a strict correspondence between the L, M and S cone inputs and the I, J, K software settings associated respectively with the red, green and blue primaries.

The colour specification of a pixel can be obtained either by weighting the fundamental response function L(λ), M(λ) or S(λ) (Vos, 1978) by the spectral distribution $E_\lambda(\lambda)$ of the stimulus,

$$L_{red} = k\int E_{red,\lambda}(\lambda)\, L(\lambda)\, d\lambda$$
$$M_{red} = k\int E_{red,\lambda}(\lambda)\, M(\lambda)\, d\lambda$$
$$S_{red} = k\int E_{red,\lambda}(\lambda)\, S(\lambda)\, d\lambda$$

or by using a linear transformation between the colorimetric quantities related to the three primaries and the responses of the fundamental responses of the cones (Viénot and Brettel, 1991). We used the linear transformations between (LMS) and (XYZ) of Smith and Pokorny (1975). Given the linear transformation between (XYZ) and (RGB), the matrix and the inverse matrix for transforming the RGB tristimulus values into LMS cone signals can be computed.

$$\begin{pmatrix} L \\ M \\ S \end{pmatrix} = C \cdot \begin{pmatrix} R \\ G \\ B \end{pmatrix}; \quad \begin{pmatrix} R \\ G \\ B \end{pmatrix} = C^{-1} \cdot \begin{pmatrix} L \\ M \\ S \end{pmatrix}$$

For each primary, the software value is the output value of a non-linear function.

$$R = f_i(I),\ G = f_j(J),\ B = f_k(K)$$

while another non-linear function describes the converse relationships.

$$I = g_i(R),\ J = g_j(G),\ K = g_k(B)$$

Construction of the test

Several LMS-based colour plates were prepared (Viénot *et al.*, 1994). A reliable protocol was designed to specifically address one cone response, or a combination of cone responses.

Each plate is composed of four different colours forming a two-dimensional pattern in the LMS colour space (Fig. 1). Along one dimension, the colour

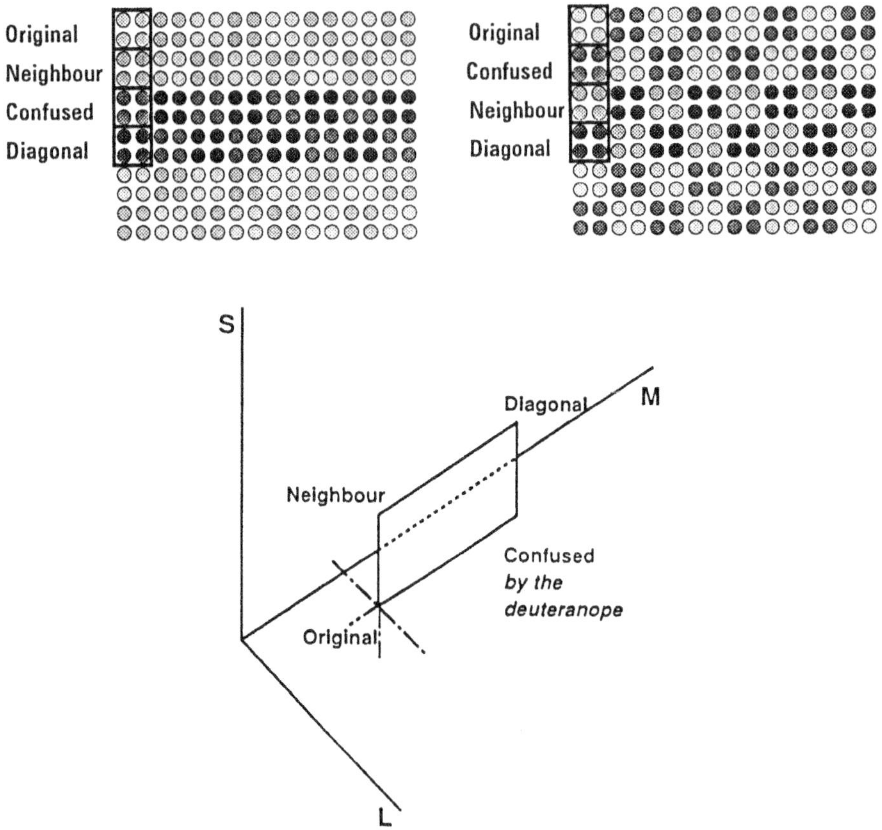

Fig. 1. Model of a pair of plates. In the left plate, two colours differing in either the L, M or S content are associated in order to make a pattern visible to the normal observer and invisible to the abnormal observer; the reverse configuration is achieved in the right companion plate. The observer is asked to point to the plate where the pattern is the most visible.

difference is related to the direction of the signal to be tested in the LMS colour space. In the other direction, the colour difference is related to the signal direction in the LMS colour space that serves as a control signal.

A plate comprises 6×8 squares of four discrete colour discs (diameter = 3.5 mm or 24 arcmin at the usual viewing distance of 50 cm) each embedded in a black background in order to avoid border effects. Rows and columns of the square spaces are interlaced to favour the salience of a pattern by proximal grouping of individual discs. The pattern consists of the joining of the two middle horizontal rows into a wide band of identical discs (15 mm width or 1.7 at 50 cm). Perceptual grouping occurs when colour differences are less within each horizontal band than between vertically adjacent squares.

Plates are paired in each presentation. In one plate, the pattern is composed of colours that differ by the colour signal that is being tested. It differs from the background discs by the control colour difference. The reverse organization is presented in the companion plate (see Fig.1). A comparison between the two plates allows an evaluation of test discrimination versus control discrimination.

For each plate, the locations of the four colours in LMS space are centred on a grey stimulus (not displayed in the plate) which is 0.3 times the luminance of the D65 white of the display. The differences along the L, M or (L–M) directions are sampled on a linear scale. The differences along the (L+M) or S directions are sampled on a log scale, since this should yield uniform discrimination steps if the Weber fraction Δ(L+M)/(L+M) or ΔS/S is constant (Fig. 2).

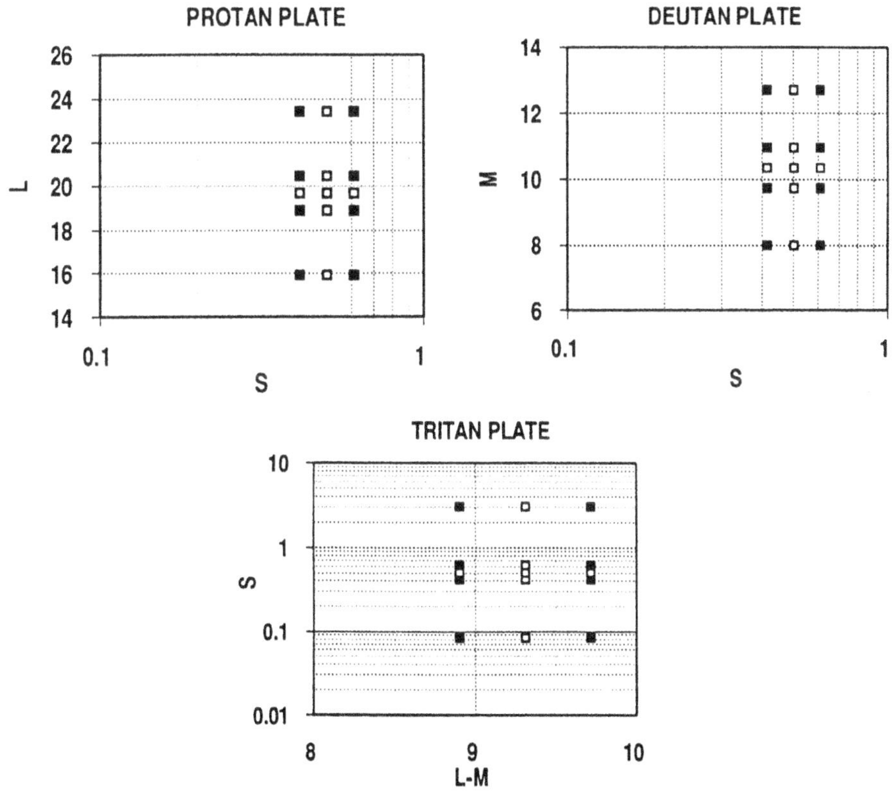

Fig. 2. Specification of the colours used in the construction of the screening test. Horizontally, the colour difference serves as a control signal. Vertically, the colour differences are related to the cone mechanism to be tested and correspond to a mild and to an extreme signal. The filled symbols represent the colours that are included in the paired plates. The central empty symbol represents the achromatic grey colour (D65, $Y=30$; $L=19.7$; $M=10.3$; $S=0.504$) around which the colour specification is constructed. L and M in relative luminance units; S in arbitrary units.

Differences are limited at the smallest values by the discrimination threshold and at the highest values by the limits of the colour gamut obtainable on the CRT display but the smallest difference is chosen to be several times the just noticeable difference, as the structure of the pattern elevates the difference threshold and only a comparison task is used. Several plates (5–8 depending upon the deficit to be screened) are prepared with increasing discrimination steps along the test direction.

Experimental protocol

Pairs of plates are presented. For screening tritan defects, a range in S difference is compared with a constant (L–M) difference and (L+M) remains constant. For screening protan defects, a range in L difference is compared with a constant S difference and M remains constant. For screening deutan defects, a range in M difference is compared with a constant S difference and L remains constant.

The observer is presented with a pair of plates and is asked to say on which plate the band is most visible. Short duration (about 2 sec), natural distance (1.5 times the display diagonal) and free viewing were used. In a rapid screening procedure, only three pairs of plates are presented for each type of colour vision defect: an extreme discrimination plate, in which the greatest difference obtainable on a CRT display along the test direction is presented, a moderate discrimination plate, in which the test difference is clearly larger than the reference difference for the normal trichromat but which is failed in cases of mild colour vision deficiency, and a faint discrimination plate, in which the test difference is only slightly more visible than the reference difference to a normal trichromat.

Results

First study

The CRT plate test was administered to colour-deficient subjects, classified on the basis of their performances on a battery of standard tests that included the Ishihara plates, the Panel D15, the desaturated D15, the FM 100-hue, and the Nagel anomaloscope. Testing on our LMS calibrated display was administered first, so as not to predjudice the tester.

Our test clearly discriminates betwen protan and deutan defects. For dichromats, the loss of discrimination is total in the direction corresponding to the deficit, compared to only a partial impairment of discrimination in the direction corresponding to other deficits. In the case of anomalies the impairment is more severe in the direction of the deficit than in the other (Fig. 3). No tritan observer has yet been screened.

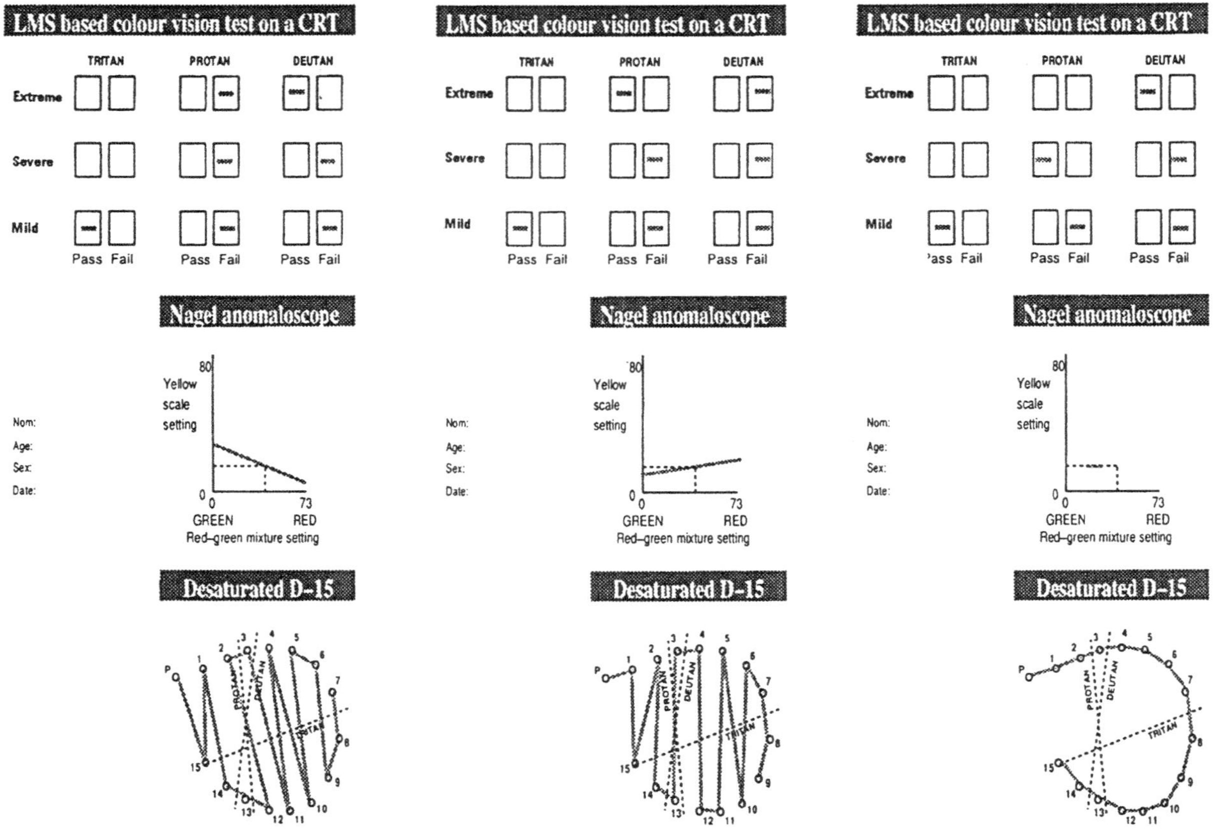

Fig. 3. Example of results achieved by observers on the LMS calibrated screening plates, on the Nagel anomaloscope and on the desaturated D15.

Second study

The CRT plate test was administered to student volunteers (aged less than 23 years) by inexperienced students of the same class. Among the 18 subjects classified as normal trichromats by their match on the Nagel anomaloscope or their performance on the desaturated D15 (one or no inversion), 12 passed the faint plates and six failed on the deutan faint plate only.

At the other extreme, the dichromats and the extreme anomalous trichromats failed on the extreme plate. One observer failed the faint deutan plate but passed the faint protan and the faint tritan plate. Questioning revealed that he had undergone retinal surgery. His Farnsworth D15 and desaturated D15 results show random errors with no clear confusion axis.

Discussion and conclusion

LMS-based construction

Several factors allow the LMS-based colour selection to optimize screening. In designing plates with discrete dots, we have followed successful examples such as the Ishihara plates or the tritan plates (Reffin *et al.*, 1991; Lanthony, 1987). Presenting discrete dots avoids artefacts due to edge effects or enhancement of luminous cues at borders. Grouping of several dots favours isolation of the chromatic versus achromatic visual response. It is known that the chromatic contrast sensitivity function is low-pass (Hilz and Cavonius, 1970; Mullen, 1985) and that chromatic sensitivity improves as the test area increases (King-Smith and Carden, 1975).

Secondly, no predetermination of the individual spectral luminous efficiency is required for our test. Substantial interobserver variability exists among normal trichromats. *A fortiori*, the standard photometric observer spectral luminous efficiency V(λ) cannot be used for dichromats or anomalous trichromats (Wright, 1946). Systematic luminous efficiency differences have also been shown among carriers. In all of these cases, an isoluminant match for the standard V(λ) is inappropriate. Consequently, any test based on V(λ) isoluminance is unsafe. In order to avoid an erroneous diagnosis based on brightness cues, earlier CRT tests used a number of stratagems: either a separate luminance match was required from the subject before starting the examination; or a luminance adjustment is allowed in order to compensate for any imbalance reported by the subject, as in the Nagel anomaloscope; or the problem of equating luminance is avoided by introducing random luminance noise (Reffin *et al.*, 1991). Our test makes no assumption about the subject's spectral sensitivity. By using three independent LMS signal variations we can ignore how the LMS signals combine to yield the luminance signal. Rather, as the functional cones of colour defectives form a subset or modification of the standard LMS cone population, the reference discrimination signals are as adequate for abnormal colour vision as for normal colour vision. In any case, the discrimination signal is always reduced for abnormals compared to the normals.

Finally, auto-scaling is introduced in the examination. By presenting paired plates for comparison, we avoid the individual variations of criterion that subjects use when they rate a colour difference. It is assumed that the psychometric function parameters show the same variations in both test and in reference directions. Screening is therefore valid with inaccurate subjects as well as with sensitive subjects.

Sensitivity of the test

Several factors are responsible for the enhanced sensitivity of our test. Compared with the threshold values either measured by other authors on CRT displays $\Delta l = 0.001$ (Krauskopf and Gegenfurtner, 1992) or $\Delta l = 0.005$ (Regan *et al.*, 1995), where $l = L/(L+M)$, or with MacAdam ellipses $\Delta L = 0.01$ (L in L-trolands; Nagy *et al.*, 1987), our lowest value of $\Delta l = 0.018$ or $\Delta L = 1.6$ (L in relative luminance units) is several times greater. Nevertheless, the sensitivity of our test is not determined by the size of the colour differences presented in one plate but by the difference between two colour differences which therefore benefit from being rather large.

The maximum difference is limited by the colour gamut of the CRT display. Given the geometry of the LMS specification of the colour gamut and the requirement that the three confusion lines cross as long a distance as possible in the physical colour space, centring the colours of the CRT plates on the grey that is 0.3 times the luminance of the nominal white of the display yields an equal range in any test direction.

In order to screen the S differences, we have adopted a geometrical spacing of S steps, unlike other authors (Krauskopf and Gegenfurtner, 1992; Regan *et al.*, 1995) who used linear scaling. From the results of Regan *et al.* it appears that linear scaling produces an increase of the S-axis of the ellipse as S increases, and that a geometrical scale would compensate for this non-uniformity, as has been suggested by Le Grand (1949) and Nagy *et al.* (1987).

Practical aspects

As far as the administration of the test is concerned, long-term calibration rapidity and free viewing make our test advantageous for mass screening. The long-term calibration permits us to rely on a single initial calibration, and allows comparison among individuals from day to day.

Examination is rapid, taking under 2 min when 9 pairs of plates are presented with the manual protocol. The instructions are easy to understand by a naive subject. Free viewing makes the observer comfortable. Extra plates for subtle gradations could be prepared whenever fine steps are necessary, but this would take time. Duration of the examination could be reduced in an automated procedure.

In the future, other directions could be probed, as the test can address any combination of cone inputs. The absence of luminance control, the auto-scaling property of the procedure, and the free viewing mode are advantageous for adapting our test for clinical purposes.

Acknowledgements

This research has been supported by the Cellule de préfiguration de la Grande Galerie du Muséum. Acknowledgements to Hans Brettel, Michel Chapron, Alain Chiron, Philippe Lanthony, Stéphane Pefferkorn, Richard Thibert and Michel Van Praët, to the Bureau National de Métrologie and National Physical Laboratory and to the reviewers.

References

King-Smith, P.E. and Carden, D. (1976). Luminance and opponent-colour contributions to visual detection and adaptation and to temporal and spatial integration. J. Opt. Soc. Am. 66: 709–717.

Krauskopf, J. and Gegenfurtner, K. (1992). Colour discrimination and adaptation. Vision Res. 32: 2165–2175.

Lanthony, P. (1991). A tritan album. In: Verriest, G. (ed.), Colour Vision Deficiencies VIII, Doc. Ophthalmol. Proc. Ser. 46: 141–143. Nijhoff/Junk, Dordrecht.

Le Grand, Y. (1949). Les seuils différentiels de couleurs dans la théorie de Young. Rev. Opt. 28: 261–278. (English translation, Colour Res. Appl. 19: 296–309, 1994.)

MacLeod, D.I.A. and Boynton, R.M. (1979). Chromaticity diagram showing cone excitation by stimuli of equal luminance. J. Opt. Soc. Am. 69: 1183–1186.

Mullen, K. T. (1985). The contrast sensitivity of human colour vision to red–green and blue–yellow chromatic gratings. J. Physiol. (Lond.) 359: 381–400.

Nagy, A.L., Eskew, R.T. and Boynton, R.M. (1987). Analysis of colour-matching ellipses in a cone-excitation space. J. Opt. Soc. Am. A, 4: 756–758.

Pefferkorn, S. (1992). Expertise d'un spectroradiomètre. Intégration dans une chaine de mesure. Diplôme d'Ingénieur, ESIP, Poitiers, France.

Reffin, J.P., Astell, S. and Mollon, J.D. (1991). Trials of a computer-controlled colour vision test that preserves the advantages of pseudoisochromatic plates. In: Drum, B., Moreland, J.D. and Serra, A. (eds.), Colour Vision Deficiencies X, Doc. Ophthalmol. Proc. Ser. 54: 69–76. Kluwer, Dordrecht.

Regan, B.C. and Mollon, J.D. (1995). Discrimination ellipses in the MacLeod-Boynton diagram: Results for normal and colour-deficient subjects obtained with a CRT display. In Drum B. (ed.), Colour Vision Deficiencies XII, Doc. Ophthalmol. Proc. Ser. 57: 445–451. Kluwer, Dordrecht.

Smith, V. C. and Pokorny, J. (1975). Spectral sensitivity of the foveal cone pigments between 400 and 500 nm. Vision Res. 15: 161–171.

Stockman, A., MacLeod, D.I.A. and Johnson, N.E. (1993). The spectral sensitivities of the human cones. J. Opt. Soc. Am. A 10: 2491–2521.

Travis, D. (1991). Effective Colour Displays. Theory and Practice. Academic, London.

Viénot, F. and Brettel, H. (1991). Vers une colourimétrie physiologique. L'onde électrique 71: 7–12.

Viénot, F., Ott, L. and Ben M'Barek, A. (1994). LMS based colour vision test on CRT. JERMOV Proc. 216.
Vos, J. J. (1978). Colourimetric and photometric properties of a 2 deg fundamental observer. Colour Res. Appl. 3: 125–128.
Vos, J.J., Estévez, O. and Walraven, P.L. (1990). Improved colour fundamentals offer a new view on photometric additivity. Vision Res. 30: 937–943.
Wright, W.D. (1946). Researches on Normal and Defective Colour Vision. Henry Kimpton, London.

Françoise Viénot
Laboratoire de Physique Appliquée
CNRS UPR 257
Muséum national d'Histoire naturelle
43 rue Cuvier
75005 Paris, France

A. Ben M'Barek, L. Ott
Ecole Nationale Supérieure de l'Electronique et de ses Applications
95014 Cergy-Pontoise, France

38. Coldef and tritan flicker test: PC-based evaluation of central and peripheral colour vision

Ch. NIMSGERN, H. LANG, G.U. AUFFART and H. KRASTEL

(Heidelberg and Darmstadt, Germany)

Abstract

Computer-supported evaluations of visual field and contrast sensitivity have become routine clinical tools in ophthalmology. The aim of this study was to develop and evaluate versatile computerized tests of central and peripheral colour sense for clinical application in congenital and acquired colour vision deficiencies. A personal computer with a calibrated monitor was used to perform central and peripheral colour tests which follow up aspects of procedures described earlier by Yu and by Korth. The colours displayed on the monitor are selected from the respective confusion axes of CIE-Lu'v' colour chart. Central colour testing Coldef (6°) was performed by presenting coloured optotypes on a coloured surround. Contrast between optotype and surround was varied in subsequent presentations, until the individual colour contrast threshold was determined by a staircase decision procedure. Coldef was performed along three confusion axes: protan, deutan and tritan confusion lines.

The peripheral tritan flicker test was performed along a tritan confusion axis. Prior to this test, luminance was adjusted to the observer's individual slope of luminosity. A 30° homogeneous blue stimulus was then presented on a congruent yellow background. Stimulus oscillation on the tritan axis amounted to 15 Hz. The colour contrast of stimulus and background was varied by a staircase procedure to determine the individual colour flicker threshold. Data were obtained from 20 normal observers, 22 patients with retinitis pigmentosa (RP), 15 with glaucoma and 7 with congenital colour vision deficiencies. In the central colour test RP and glaucoma patients showed impaired colour contrast sensitivity along the protan, deutan and tritan axis. Patients with congenital colour vision deficiencies (daltonians) showed reduced colour contrast sensitivity exclusively along the protan and deutan axes but not along the tritan axis. In addition, they showed no impairment of contrast thresholds in the tritan flicker test. Peripheral colour contrast sensitivity was also impaired in RP and glaucoma patients. The peripheral and central colour tests were shown to be rapid and sensitive tools in the quantitative classification of congenital and acquired colour vision deficiencies.

C.R. Cavonius (ed.), Colour Vision Deficiencies XIII, pp. 351–359.

Introduction

Conventional colour tests such as Ishihara plates are usually designed to detect congenital red–green colour deficiencies. Blue–yellow defects are not usually tested. Moreover, most of the traditional tests are qualitative or semiquantitative (e.g. panel D15, Lanthony desaturated panel D15, Ishihara plates). Computerized VDUs offer several advantages: colours can easily be generated and changed in hue, brightness and saturation. A computer test should offer the possibility of revealing congenital and acquired colour vision deficiencies and to allow quantitative classification.

Patients and methods

Sixty-four observers were studied: 20 normal subjects, 22 patients with retinitis pigmentosa (RP), 15 with glaucoma and 7 daltonians. The mean age of all observers was 43.25 years (SD 17.93, range 9–83). Observers were positioned on a chin rest, at 36 cm distance to the monitor. Colour tests are performed monocularly after best correction for refractive error. Test duration is at about 10 min per eye (both tests).

The principles of examination are partially based on Yu's VDU colour test (Yu *et al.*, 1991) and on Korth's flicker test (Korth *et al.*, 1992). A 486 IBM-compatible PC with a calibrated 14-inch colour screen served to display standardized colour stimuli. The primaries of the monitor defined the triangle shown in Fig. 1. The CIE coordinates of stimulus and surround were selected in agreement with the Smith–Pokorny–Verriest confusion lines (Smith and Pokorny, 1975; Pokorny *et al.*, 1979). Two different test strategies have been developed: the central and the peripheral colour vision test: Coldef and tritan flicker.

Central colour vision test: Coldef

Central colour vision was examined by defining the colour contrast threshold for coloured optotypes centred on a coloured surround. Each optotype was presented for 2 sec. The surround rectangle covered 6 × 6°, the optotype 3 × 4°.

A staircase procedure of decreasing and increasing colour contrast of optotype and surround was used to determine the individual colour contrast threshold. When correct recognition was achieved the colour contrast of the subsequent presentation was reduced. If the optotype was not identified correctly, colour contrast was increased. Colour contrast steps were varied along a linear scale on the respective confusion axis. Steps were based on ΔE values according to the CIE-Lu'v' 1978 colour chart (Lang, 1993). ΔE equals the minimal distance in the CIE-Lu'v' colour chart that a normal observer is able to discriminate. In our present test, colour-contrast units were defined for the protan, deutan and tritan axis: ΔProtan (ΔP) = 7.25 ΔE, ΔDeutan (ΔD) = 7.75 ΔE and ΔTritan (ΔT) = 2.7 ΔE. The contrast scale ranged from 0.4–9.0 ΔP and

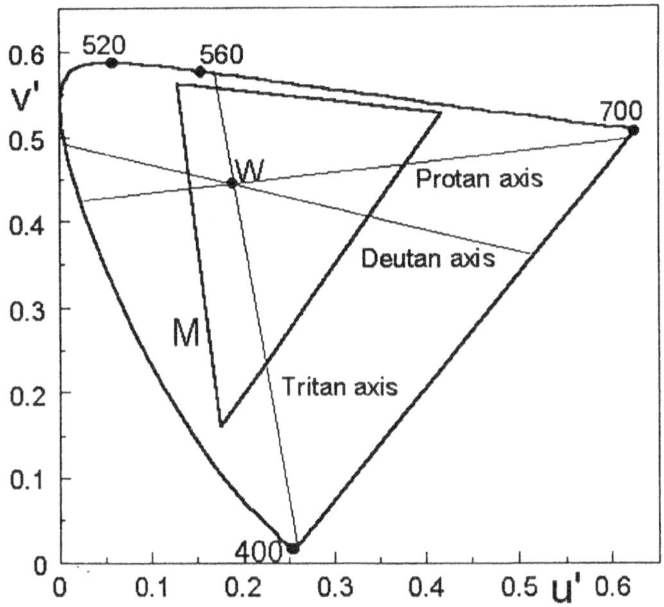

Fig. 1. CIE-Lu'v' 1978 colour chart with Smith–Pokorny–Verriest confusion axes, monitor primaries (M) and achromatic point of the screen (W).

0.4–9.8 ΔD respectively, divided into steps of 0.4. The minimum step on the tritan confusion axis was 1.6 (range 1.6–53.3 ΔT). The colour contrast threshold was defined by the lowest colour contrast at which optotypes were recognized correctly on three subsequent presentations.

Peripheral colour vision test: tritan flicker

This examination was performed using a flicker detection procedure. A pretest was performed to adjust stimulus and background brightness to the observer's individual slope of luminosity (luminance adjustment 1 and 2, Table 3) using a stimulus frequency of 15 Hz, as in the subsequent main test. Both parts of the test were presented on a 30° square field. After individual adjustment of luminance a 15 Hz blue flicker stimulus was presented for 2 sec on a coincident yellow background (dominant wavelength 562 nm). Colour loci of stimulus and background were selected from the tritan axis (Fig. 1). The stimulus alternates between two positions on the tritan confusion axis. Both positions were changed stepwise in distance until the flicker contrast threshold was determined, i.e. when a minimum of flicker is detected by the observer. Colour contrast steps are defined as ΔTritan values (T = 2.7 ΔE). The size of the minimum contrast steps was 4 ΔT. If no flicker was detected at 15 Hz and maximum contrast (88 ΔT), the stimulus repetition rate is reduced to 10 or 7.5 Hz. The pretest was then performed again at reduced frequency, because luminance perception depends on stimulus frequency (Krastel *et al.*, 1991).

Results

Colour contrast thresholds of the central colour test of all 20 normal observers are within the ellipses shown in Figures 2, 3 and 4. These do not exceed 0.8 ΔP/ΔD on the protan/deutan axis and 3.1 ΔT on the tritan axis. The thresholds of RP patients were on average 7.7 times higher along the tritan axis (18.13 ΔT) compared to normal observers. Results along the protan and deutan axes were also elevated, but less markedly (1.86 ΔP, 2.26 ΔD). Central tritan thresholds in glaucoma individuals exceeded the averaged normal values by a factor of 3.7 (Table 1).

Comparison of mean defect (MD) and loss variance (LV) in computerized perimetry (Octopus program 38) showed a significant positive correlation of MD with elevated tritan thresholds in glaucoma (+0.63) and a negative

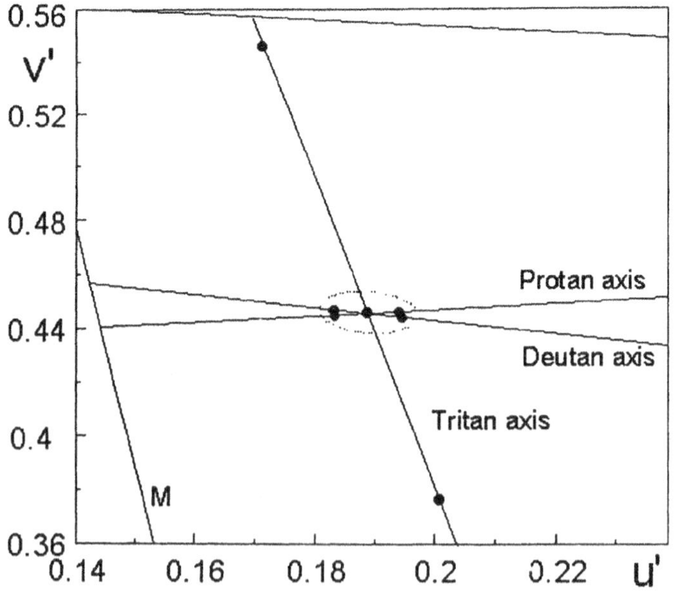

Fig. 2. Coldef: results of 47-year-old female RP simplex patient, dotted line = range of normal values, M = monitor triangle.

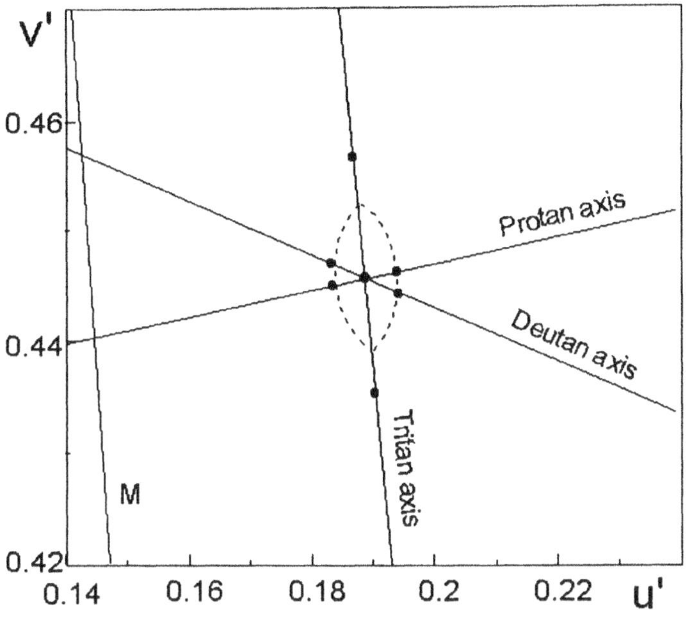

Fig. 3. Coldef: results of 36-year-old male open-angle glaucoma patient. M = monitor triangle, dotted line = range of normal values. Note that scaling of y-axis differs from Fig. 2.

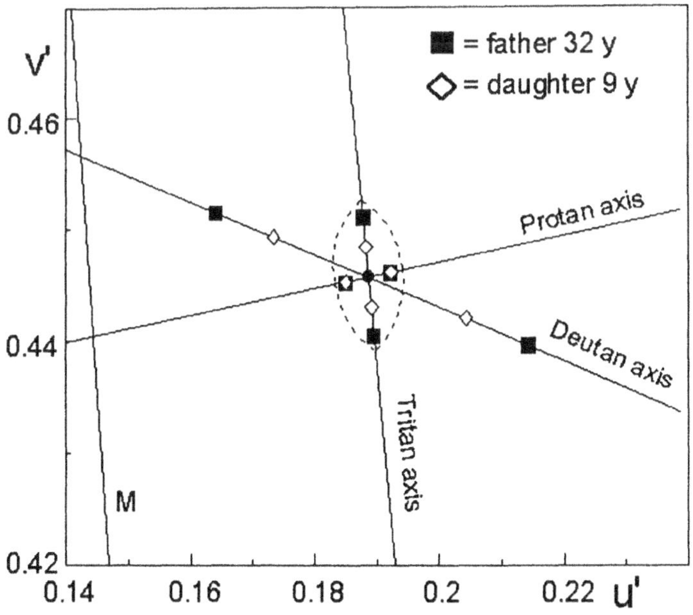

Fig. 4. Coldef: daltonian results. M = monitor triangle, ellipse indicates area of normal observers' range.

Table 1. Coldef: results of normals, RP and glaucoma, mean values and standard deviation.

Group	Visual acuity	Protan threshold (ΔP)	Deutan threshold (ΔD)	Tritan threshold (ΔT)
Normal ($n = 20$)	1.1 ± 0.14	0.5 ± 0.18	0.5 ± 0.18	2.35 ± 0.77
Retinitis pigmentosa ($n = 22$)	0.6 ± 0.28	1.86 ± 1.89	2.26 ± 2.2	18.13 ± 17.94
p (Normals vs. RP; u-test)		0.0001	0.0001	0.0001
Glaucoma ($n = 15$)	0.66 ± 0.36	1.24 ± 1.21	1.32 ± 0.97	8.87 ± 9.23
p (Normals vs. glaucoma, u-test)		0.0002	0.0001	0.0001

correlation with LV (–0.507, Spearman rank correlation).

Seven patients with congenital colour vision deficiencies underwent both colour vision tests. All results of the tritan examination were within the normal range, since congenital colour vision defects affect mostly red–green perception. This kind of colour deficiency is only examined exclusively by the central colour test, where testing is performed also along protan and deutan confusion axes. Results of Coldef testing are given in Table 2.

To detect peripheral blue flicker, normal observers need contrast thresholds of 4–20 ΔT at a maximum frequency of 15 Hz. RP patients need 3.1 times higher contrast levels than normal observers, while glaucoma individuals need 3.5 times greater contrast. Comparison of mean defect (MD) and loss variance (LV) in computerized perimetry (Octopus program 38) showed a positive correlation of MD with elevated tritan thresholds in glaucoma (+0.603) and a negative correlation with LV (–0.417, Spearman rank correlation).

Daltonians showed no statistically significant deterioration of tritan thresholds, as already shown in the central colour test. Blue flicker thresholds of all daltonians were within the normal range (4–16 ΔT). The luminance adjustment of the peripheral colour test was performed to compensate individual loss of perception in luminosity. Comparison of normal and RP results showed a significant difference ($p = 0.0001$, u-test). Results from glaucoma patients were also significantly different from those of normal observers ($p = 0.0002$, u-test). Results of peripheral colour testing are given in Table 3.

Case reports

A 47-year-old female RP patient (RP simplex) with visual acuity of 0.6 and no lens opacity showed a grossly elevated threshold along the tritan axis (50.3 ΔT, Fig. 2). Contrast thresholds along protan and deutan axes were within the

Table 2. Coldef: results of patients with congenital colour disturbances (+ = elevated thresholds).

				Right eye				Left eye			
ID	Age	Sex	Visual acuity	ΔP	ΔD	ΔT	Visual acuity	ΔP	ΔD	ΔT	Results of anomaloscopy
P1	12	m	1.25	0.8	3.1+	1.6	1.25	0.8	3.1+	1.6	Deuteranomal
P2	9	f	1.0	0.8	2.0+	1.6	1.0	0.8	3.1+	3.1	Deuteranomal
P3	32	m	1.0	1.2+	3.9+	3.1	1.0	0.8	5.1+	1.6	Deuteranope
P4	15	f	1.25	1.6+	2.7+	3.1	1.25	0.8	3.9+	3.1	Deuteranomal
P5	19	m	1.0	0.8	2.7+	3.1	1.0	0.8	3.1+	1.6	Deuteranomal
P6	20	m	0.8	7.8+	0.4	1.6	0.8	7.1+	0.8	3.1	Protanope
P7	47	m	1.0	9.0+	0.8	3.1	1.0	9.0+	0.8	1.6	Protanope

Table 3. Tritan flicker: results of normals, RP and glaucoma, mean values and standard deviation.

Group	Visual acuity	Luminance Adjustment 1	Luminance Adjustment 2	Tritan threshold (ΔT)
Normals ($n=20$)	1.1 ± 0.14	38.35 ± 0.81	31.95 ± 0.39	11.4 ± 4.55
Retinitis pigmentosa ($n=22$)	0.6 ± 0.28	36.46 ± 1.37	32.86 ± 0.83	35.44 ± 25.5
p (Normals vs. RP, u-test)		0.0001	0.0002	0.0001
Glaucoma ($n=15$)	0.66 ± 0.36	37.0 ± 1.31	32.8 ± 1.01	33.48 ± 30.86
p (Normals vs. glaucoma, u-test)		0.0011	0.002	0.0001

normal range (0.8 ΔP, 0.8 ΔD). The electroretinogram was diminished and her visual field was constricted to the central 12°. The fundus of the examined eye showed pigment clumping and narrow retinal arterioles.

A 36-year-old male with unilateral open angle glaucoma presented with visual acuity of 0.5, a cup–disc ratio of 0.9 and a nerve fibre bundle defect on computerized perimetry (Fig. 3). His tritan colour contrast threshold was elevated on the central colour test, while the protan and deutan thresholds were borderline (1.2 ΔP, 1.2 ΔD, 6.3 ΔT).

The specimen of daltonians is a special choice (Fig. 4). The father was a small-field deuteranope according to his 2° and 10° Rayleigh matches. On the central colour test, he appeared to have some functional peripheral M cones, as do nearly all deuteranopic observers. His daughter was a mixed hemizygote, because her maternal grandfather suffered from a congenital colour defect.

Phenotypically she behaved as simple deuteranomalous on the 2° Rayleigh match. Her central colour contrast threshold was less elevated than her father's one (father: 0.8 ΔP, 5.1 ΔD, 3.1 ΔT, daughter: 0.8 ΔP, 3.1 ΔD, 1.6 ΔT).

Discussion

Computerized psychophysical tests have been used in ophthalmology for several years. Static perimetry, contrast sensitivity and other visual functions can be evaluated by computerized test systems produced by several manufacturers. With the advance of microelectronics, data processing and better graphic devices, accurate colour presentation is possible, allowing for the development of standardized colour tests (e.g. King-Smith *et al.*, 1983, Arden *et al.*, 1988, Mollon and Reffin, 1989).

Our central and peripheral colour vision tests can detect colour deficiencies and offer a quantitative result. The Coldef test presents stimuli along three main colour confusion axes (protan, deutan and tritan) and offers the facility of quick screening for daltonism. Flammer and Drance (1984) described a positive correlation between mean defect and tritan colour disturbances of glaucoma patients. This is confirmed by our data, which also test peripheral colour vision.

The tritan flicker peripheral colour test works on a colour confusion line that is individually adjusted to the observer's luminosity perception. Tritan plates suffer the drawback of a fixed luminance ratio of stimulus and surround structures. Individual patients affected by tritan deficiencies of various degrees may exhibit slopes of luminosity that deviate from the prefixed luminance ratio of plate optotypes and background. Patients may pass such plate tests despite severe tritan defects. By adjusting for individual slopes of luminosity, the tritan flicker test offers a principal advantage in methodology. This facility has up to now offered only by spectral matching procedures, in particular the Moreland equation (Moreland and Kerr, 1979) which has found widespread application in acquired colour vision defects in ophthalmic disease evaluation (Krastel and Moreland, 1991). Our results indicate that individual adjustment of luminance makes sense: there is actually a significant deviation, as shown by the data in Table 3. Peripheral colour testing combines examination along individually adjusted luminance of the tritan confusion axis with presenting a flicker stimulus.

Conclusion

These computerized and standardized colour tests offer sensitive procedures to evaluate colour function in congenital as well as in acquired deficiencies. Results are obtained within few minutes and a quantitative evaluation of the defect is provided.

Acknowledgements

Supported by a grant from the DFG (Kr 584/2-2) and Oculus Inc., 35583 Wetzlar, Germany. HL is the only author with a commercial interest in these tests.

References

Arden, G.B., Gündüz, K. and Perry, S. (1988). Colour vision testing with a computer graphics system: preliminary results. Doc. Ophthalmol. 69: 167–174.

Flammer, J. and Drance, S.M. (1984). Correlation between color vision scores and quantitative perimetry in suspected glaucoma. Arch. Ophthalmol. 102: 38–39.

King-Smith, P.E., Chioran, G.M., Sellers, K.L. and Alvarez, S.L. (1983). Normal and deficient colour discrimination analyzed by colour television. In: Mollon, J.D. and Sharpe, L.T. (eds.), Colour Vision, Physiology and Psychophysics: 167–172. Academic Press, London.

Korth, M., Horn, F. and Martus, P. (1992). Einfacher zeitlicher Kontrastempfindlichkeits-Test in der Glaukom-Diagnostik. Klin. Monatsbl. Augenheilkd. 203: 99–103.

Krastel, H., Jaeger, W., Zimmerman, S., Heckmann, B. and Krystek, M. (1991). Systematics of human photopic spectral sensitivity. In Drum, B., Moreland, J.D. and Serra, A. (eds.), Colour Vision Deficiencies X. Doc. Ophthal. Proc. Ser. 54: 323–339. Kluwer, Dordrecht.

Krastel, H. and Moreland, J.D. (1991). Colour vision deficiencies in opthalmic diseases. In: Forster, D.H. (ed): Inherited and Acquired Colour Vision Deficiencies: 115–172. Macmillan, Basingstoke.

Lang, H. (1993). Farbmetrik, In: Bergmann and Schäfer (eds.), Lehrbuch der Experimentalphysik, Band 3, Optik: 665–754. De Gruyter, Berlin.

Mollon, J.D. and Reffin, J.P. (1989). A computer controlled colour vision test that combines the principles of Chibret and Stilling. J. Physiol. (Lond.) 414, 5P.

Moreland, J.D. and Kerr, J. (1979). Optimization of a Rayleigh-type equation for the detection of tritanomaly. Vision Res. 19: 1369–1375.

Pokorny, J., Smith V.C. and Verriest, G. (1979). Congenital color defects. In: Pokorny, J., Smith, V.C., Verriest, G. and Pinckers, A.J.L.G. (eds.), Congenital and Acquired Color Vision Defects. Grune and Stratton, New York: 183–241.

Smith, V.C. and Pokorny, J. (1975). Spectral sensitivity of the foveal cone photopigments between 400 and 500 nm. Vision Res. 15: 161–171.

Yu, T.C., Falcao-Reis, F., Spileers, W. and Arden, G.B. (1991). Peripheral colour contrast, a new screening test for preglaucomatous visual loss. Invest. Ophthalmol. Vis. Sci. 32: 2779–2789.

Ch. Nimsgern, G.U. Auffahrt, H. Krastel
Dept. of Ophthtalmology
Ruprecht-Karls-University, Heidelberg, INF 400
D-69120 Heidelberg, Germany

PART SIX

General considerations in testing colour vision

39. Effect of macular pigment on colour matching with field sizes in the 1° to 10° range

J.D. MORELAND and E.C. ALEXANDER

(Keele, UK)

Abstract

As part of the characterization of a new colorimetric standard observer it was necessary to define a continuously variable function for macular pigment absorbance for circular fields in the range 1°–10°. There were no published data from which this could be obtained directly. Using the mean retinal profile of macular pigment for the 15 subjects in the Moreland and Bhatt (1984) study together with a transform function measured for one subject, a set of equivalent macular pigment absorbances were found that were well described by an exponential function.

Introduction

Technical Committee TC1-36 of the CIE has the task of defining a continuous standard colorimetric observer for field sizes in the range 1°–10°. One of the problems encountered concerns the effect of macular pigment on colour matching for centred circular fields. A reasonable body of published data allows the calculation of a mean effective absorbance for a 2° field but not for larger field diameters. Colorimetric and other estimates of the retinal profile of macular pigment exist and have been reviewed by Moreland and Bhatt (1984), but the connection between this local profile and the equivalent absorbance for whole circular fields remains undetermined. In a first attempt, the average retinal absorbance profile of macular pigment, reported by Moreland and Bhatt (1984), was fit by an exponential function. This function was integrated for circular fields over the required range.

In colour matching, however, where macular pigment manifests visibly in large fields such as Maxwell's spot, that central spot region is ignored. In this instance, the match is determined outside the spot in a region of lower macular pigment concentration and so integration over the whole field could over-estimate the equivalent absorbance. Indeed, it has even been suggested, for flicker photometry of circular fields, that it is the macular pigment concentration at the very edge of the field which is important (Werner *et al.*, 1987). We addressed the problem by comparing colour matches made with annular and

C.R. Cavonius (ed.), Colour Vision Deficiencies XIII, pp. 363–368.

whole circular fields for a single subject to derive a suitable conversion function. This function was then used to convert from the exponential describing the mean retinal absorbance profile to whole field equivalents. The whole field equivalent could also be described by an exponential function.

Materials and methods

The Moreland and Bhatt (1984) colour matching method was used for assessing macular pigment. The apparatus has been described elsewhere (Moreland and Kerr, 1976). Bipartite fields were used, both annular and circular, which were divided vertically respectively by a wide gap and a fine line. Briefly, the matching stimuli employed were mixtures of 460 nm (B) with 570 nm (Y) in one half-field and of 490 nm (C) with 610 nm (O) in thc other. The end-point of matches made between the two half-fields was a desaturated colour, perceptually quite close to white. The fields of view were defined by masks. Annular fields were at eccentricities (mean of inner and outer radii) 1°, 1.5°, 2°, 2.5°, 3°, 4° and 5° and centred on a small dim red fixation spot. The annuli were 0.86° wide for eccentricities less than 2° and 1.72° for all others. The fields were exposed for 0.5 sec every 2 sec to reduce and control Troxler's fading and the subject maintained steady fixation whilst matching.

The circular fields had radii of 1°, 1.5°, 2°, 2.5°, 3°, 4° and 5°. These fields were exposed continuously and the subject fixated the centre of the border between the two half-fields whilst matching.

Five matches were made for each field in a session. Annular and circular fields were presented in separate sessions. There were four replications of each session followed by four more with the mixing stimuli exchanged between the left and right fields to reduce any effects of retinal asymmetry. Field eccentricity and size were presented in a counterbalanced random order and time of day (morning/afternoon) was counterbalanced between sessions to reduce any systematic ordering or temporal effects.

Results and discussion

Changes in macular pigmentation are reflected by changes in the colour matches. The change in macular pigment absorbance was evaluated by changes in log(B/Y) and in log(C/O) on the assumption that macular pigment absorbance is zero for the long wavelength Y and O stimuli. As Moreland and Bhatt (1984) noted, the two ratios obtained in the match provide simultaneous assessments of pigment absorbance for each half-field. The mean macular pigment absorbance profiles for the eight sessions are shown in Figure 1.

The similarity of the results for circles and annuli suggest that the subject made matches for circular fields that were determined near but not actually at the edge of the field. Using polynomial functions for smoothing and interpola-

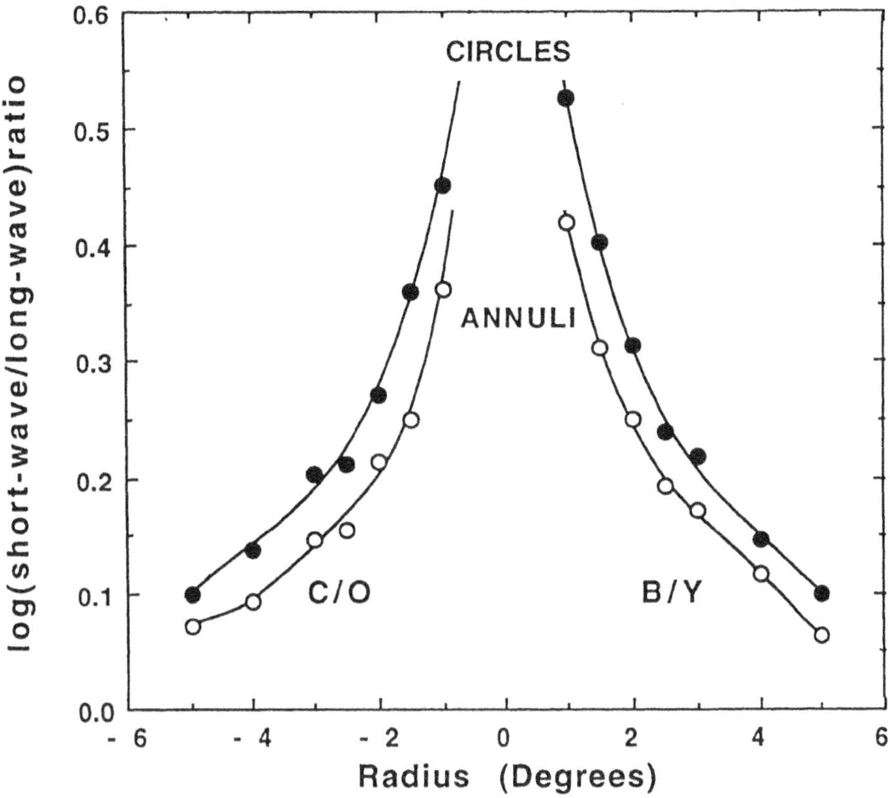

Fig. 1. Relative macular pigment absorbance profiles of subject HK as a function of eccentricity for annular fields and of field radius for circular fields. Radii for C/O are made negative for clarity of presentation. Ordinates for circles at $\pm 5°$ are set arbitrarily at 0.1 and all other data are relative. Points are the means of 40 matches. Full-line curves are best-fit polynomials (5th order for annuli, 4th order for circles). These polynomials were used to predict interpolated values of annular and circular field radii for equal macular pigment absorbance (see Fig. 2).

tion, the macular pigment equivalence function for annular and circular fields was determined. This function was very well described by a cubic constrained to pass through the origin, and is shown in Figure 2. It indicates that colour matching in a circular field is equivalent to matching for an annular zone at about 0.8 of the circular field radius.

The mean retinal absorbance profile of macular pigment for Moreland and Bhatt's (1984) 15 subjects is well described by an exponential (Fig. 3). This exponential was used to calculate the equivalent macular pigment profile for circular fields first by integration (via macular pigment transmittance) and

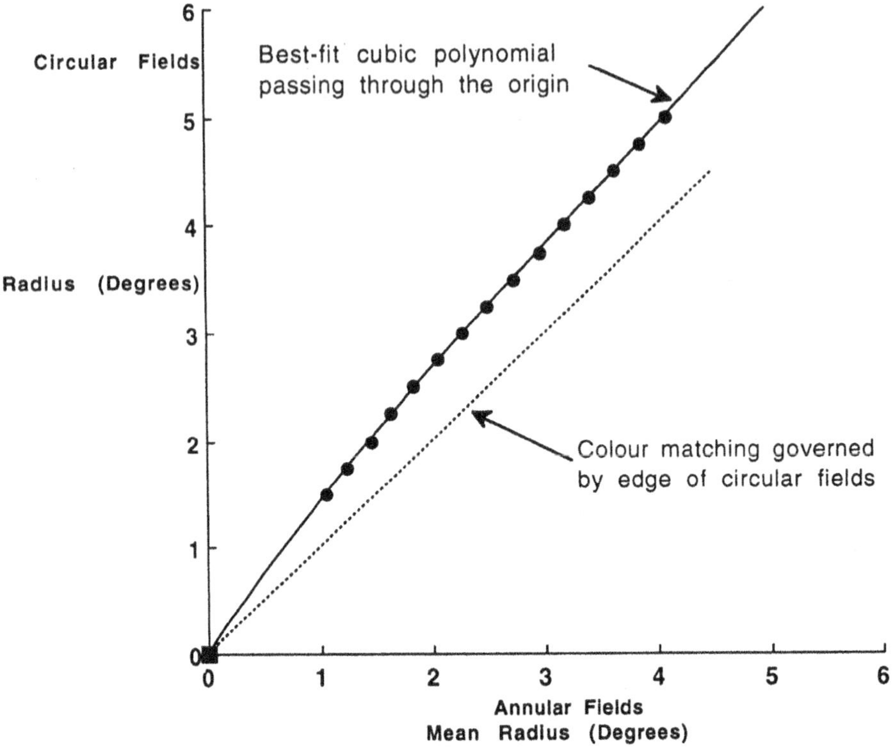

Fig. 2. Macular pigment equivalence for annular and circular fields. Points are derived by interpolation from Fig. 1 and are means for log(C/O) and log(B/Y). The data are described very well by a cubic passing through the origin. The broken line shows the function that should be obtained if colour matching were made at the very edge of the circular fields.

second by using the equivalent radii function in Figure 2. It is assumed that the latter function is not subject dependent.

These derived data are shown in Figure 4. As expected, the calculated equivalent absorbance functions agree for small field sizes and diverge increasingly with increasing field size. The function based upon equivalent radii is well described by the exponential

$$A = k.10^{(-D/7.96)}$$

where A = equivalent absorbance, D = field diameter in degrees and k = 0.723.

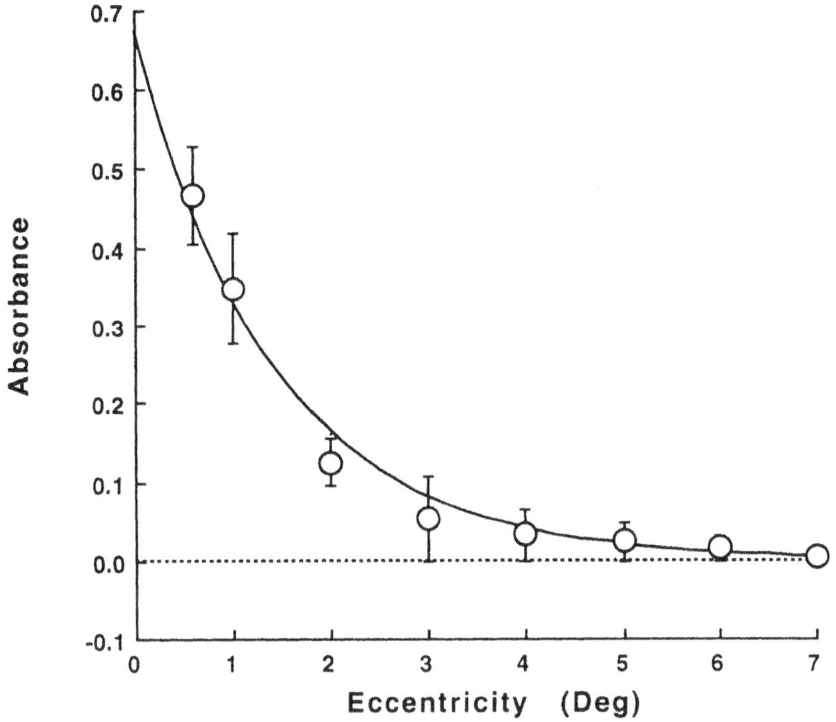

Fig. 3. Mean retinal absorbance profile for Moreland and Bhatt's (1984) 15 subjects. The datum at 0.4° corresponds to a foveally-centred field of 100 min arc diameter, which has been replotted as an annular equivalent using the equivalent radius from Fig. 2. The full line curve is a best-fit exponential. This exponential is used for calculating the equivalent macular pigment absorbance profile for circular fields.

Conclusions

Using the mean retinal profile of macular pigment from the Moreland and Bhatt (1984) study together with the transform function measured for one subject, a set of equivalent macular pigment absorbances were found for circular fields in the 1°–10° range which were well described by an exponential function. An alternative calculation by integration of the retinal absorbance profile over-estimates the equivalent macular pigment absorbance increasingly as field sizes increases.

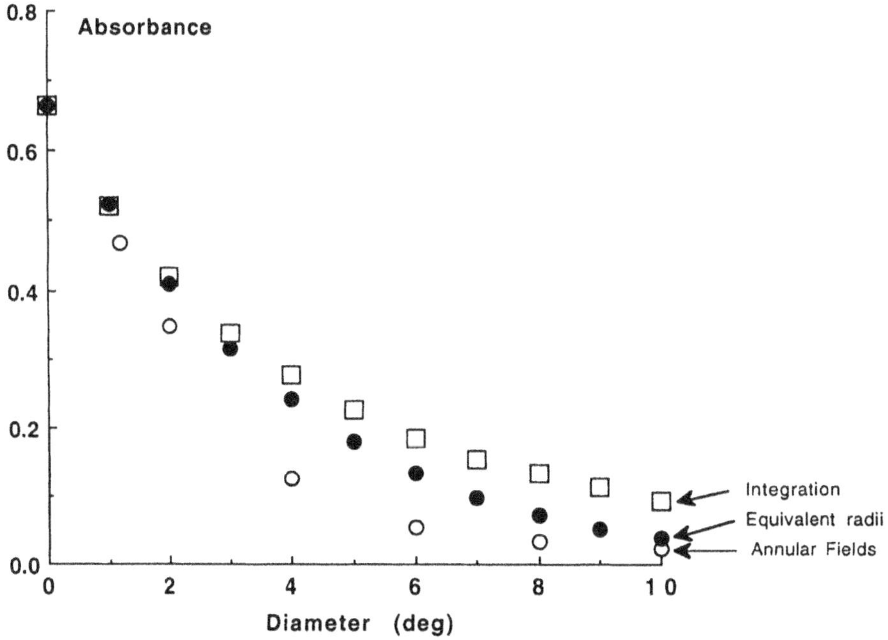

Fig. 4. Variation with field diameter of the equivalent macular pigment absorbance calculated by integration (squares) and calculated using equivalent radii (filled circles). The original annular field mean absorbance profile for 15 subjects is shown for comparison (open circles).

References

Moreland, J.D. and Bhatt, P. (1984). Retinal distribution of macular pigment. In: Verriest, G. (ed.), Colour Vision Deficiencies VII. Doc. Ophthalmol. Proc. Ser. 39: 127–132. W. Junk, The Hague.

Moreland, J.D. and Kerr, J. (1979). Optimization of a Rayleigh-type equation for the detection of tritanomaly. Vision Res. 19: 1369–1375.

Werner, J.S., Donelly, S.K. and Kliegl, R. (1987). Aging and human macular pigment density. Vision Res. 27: 257–268.

Department of Communication & Neuroscience
Keele University
Staffordshire, ST5 5BG, UK

40. Psychometric functions for chromatic discriminations

A. WATANABE, V.C. SMITH and J. POKORNY
(Chicago, IL, USA)

Abstract

Psychometric functions for equiluminant chromatic discriminations were measured by the method of constant stimuli. The functions were measured from three starting points on axes of a relative cone troland chromaticity space in a surround metameric to equal energy white. The data were fit by probability functions that yielded estimates of the slope of the psychometric curve and the chromaticity threshold. The slopes were similar for both axes with an average value near 2 (± 1). The thresholds were minimal near the adapting chromaticity and rose in a V-shape on either side, similar to data in the literature using other psychophysical protocols.

Introduction

Recent studies have suggested that chromatic discrimination is mediated by the spectral opponent pathways of the parvocellular pathway. In an equiluminant achromatic surround, chromatic discrimination is best when measured at the starting chromaticity of the surround. This measurement is a detection of chromatic change from the surround, comparable to a measurement of least colorimetric purity, as reviewed by Yeh *et al.* (1993). When the starting chromaticity differs from the surround, thresholds are higher. The data form a characteristic V-shape (Krauskopf and Gegenfurtner, 1992; Zaidi *et al.*, 1992; Miyahara *et al.*, 1993).

There have been few studies of the psychometric function for equiluminant chromatic detection or discrimination. The usual chromaticity threshold protocols use adaptive staircase procedures (Krauskopf and Gegenfurtner, 1992; Zaidi *et al.*, 1992; Miyahara *et al.*, 1993). A measurement of an equiluminant chromatic detection for red–green stimuli on a 3000 td yellow surround (Eskew *et al.*, 1991) gave a slope near 2.0. We know of no comparison of psychometric functions measured on cardinal axes of colour space nor of psychometric functions measured at starting chromaticities away from an adapting surround. Knowledge of the psychometric function may be important for modelling of threshold data and for evaluating the possibility of interaction of the cardinal

C.R. Cavonius (ed.), Colour Vision Deficiencies XIII, pp. 369–376.

axes in threshold paradigms (Smith and Pokorny, 1995). We therefore assessed psychometric functions for stimuli placed on the cardinal cone excitation axes in an equiluminant plane (MacLeod and Boynton, 1979; Boynton and Kambe, 1980; Derrington *et al.*, 1984). Adaptation was maintained by a 115 td surround near equal energy white. Discrimination was assessed both at the adapting chromaticity and at distant points.

Materials and methods

Equipment

Stimuli were generated by a Pixar 2 image-processor under computer control and were presented on a Nanao colour monitor. The instrument and calibration have been fully described (Miyahara *et al.*, 1993). The monitor was viewed with the natural pupil at a distance of 1 m. The monitor provided a $9.2^\circ \times 8.7^\circ$, 12 cd/m^2 achromatic field, metameric to equal energy white. On each trial a square array consisting of four $1^\circ \times 1^\circ$ squares separated by 0.05° gaps was presented using a 1.5: sec raised cosine temporal profile. The chromaticity of one square differed from the others and the location of the odd square was varied randomly from trial to trial.

Chromatic discriminations were measured along the axes of a relative cone troland chromaticity space (Boynton and Kambe, 1980). In this space the horizontal l-axis represents relative L trolands and the vertical s-axis represents relative S tds. We chose three starting chromaticities on each axis intersecting at equal energy white. On the s-axis the values of s/(l+m) of the starting chromaticities were 0.7, which appeared 'yellowish', 0.997, which appeared 'white' and 4.0, which appeared 'bluish'. On the l-axis, the values of l/(l+m) of the starting chromaticities were 0.62, which appeared 'greenish', 0.665, which appeared 'white' and 0.76, which appeared 'reddish'. For each axis the stimuli were varied from the starting chromaticities in both increment and decrement directions.

Observers

There were three observers, ranging in age from 24 to 30 years. All three were judged to have normal colour vision, according to examination with the Nagel anomaloscope and the Farnsworth-Munsell 100-hue test. Observers AC and AW completed the full set of protocols. Confirmatory data were obtained for observer NW.

Procedure

Psychometric functions were measured using the method of constant stimuli. There were 40 four-alternative forced-choice trials for each of nine chromaticities, giving a total of 360 trials. The choice of chromaticity on each trial was chosen at random by the computer. The observer used a mouse to move a cursor to the location of the square judged different from the others. The chromaticities included the starting chromaticity and eight test chromaticities. The range of test chromaticities was determined by an initial forced-choice staircase procedure, to give sufficient data in the slope region for fitting of a psychometric function. Once an appropriate range was determined, two replications were made. Observer NW was available only for a limited time and did not complete the full set of replications. Before each session the observer adapted to a white background for 2 min. The session lasted about 30 min.

Results and discussion

Figure 1 shows a typical result, the percentage correct for observer AW for a 'yellowish' starting chromaticity measured in the increment and decrement directions. The two replications are shown for each direction. To estimate the slopes of psychometric functions, we fit the data to the Quick (1974) psychometric function using a correction for guessing:

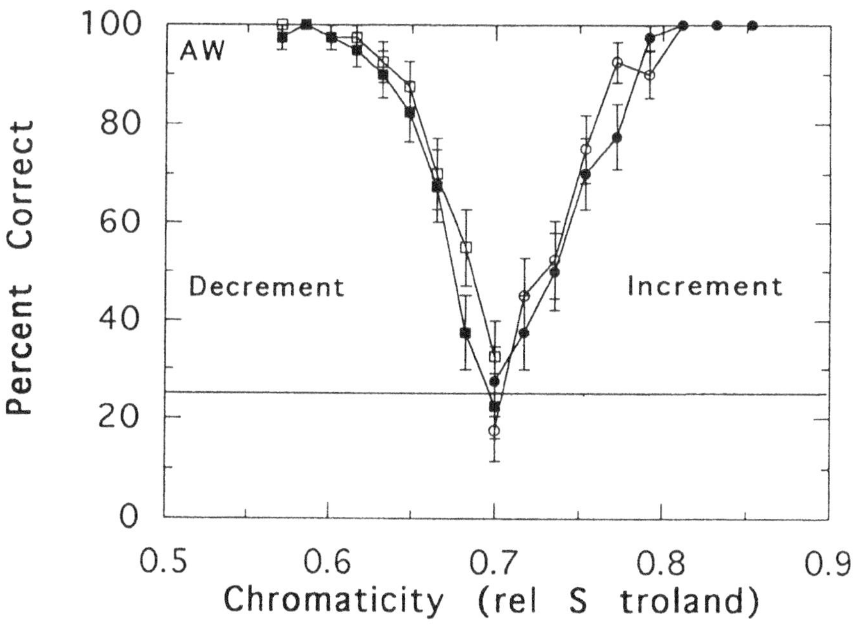

Fig. 1. Sample data for observer AW. The measurement is for the s-axis with a starting chromaticity of 0.7. The percent correct responses for both increment and decrement directions are plotted versus the test chromaticity. Both replications are shown.

$$\%\text{correct} = 100\{0.25 + 0.75(1-2^{-x^k})\} \quad (1)$$
$$x = c/c_{thr} \quad (2)$$

In this equation x is relative stimulus strength and the parameter k determines the slope of the function. The relative stimulus strength, x, is defined as the contrast of the test light, c divided by the threshold contrast, c_{thr}. For our experiment, the contrast is the difference between the chromaticity of the test light and the starting chromaticity. Thus c is defined as $(l-l_s)$ for the l-axis and $(s-s_s)$ for the s-axis. The cone trolands are obtained by multiplying the chromaticities (l,s) by 115. The data were fitted using a least squares criterion with the threshold, c_{thr} and k as free parameters. In these fits, data at the starting chromaticity, and data for which the proportion correct was 100%, were omitted.

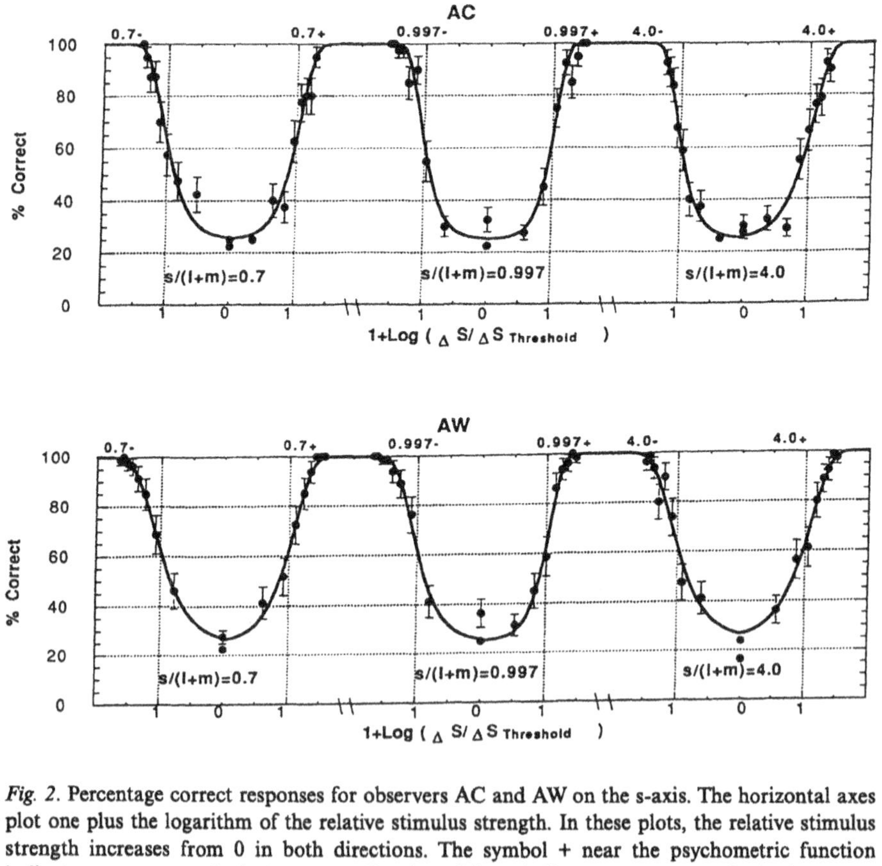

Fig. 2. Percentage correct responses for observers AC and AW on the s-axis. The horizontal axes plot one plus the logarithm of the relative stimulus strength. In these plots, the relative stimulus strength increases from 0 in both directions. The symbol + near the psychometric function indicates measurement in the increment direction while – indicates measurement in the decrement direction.

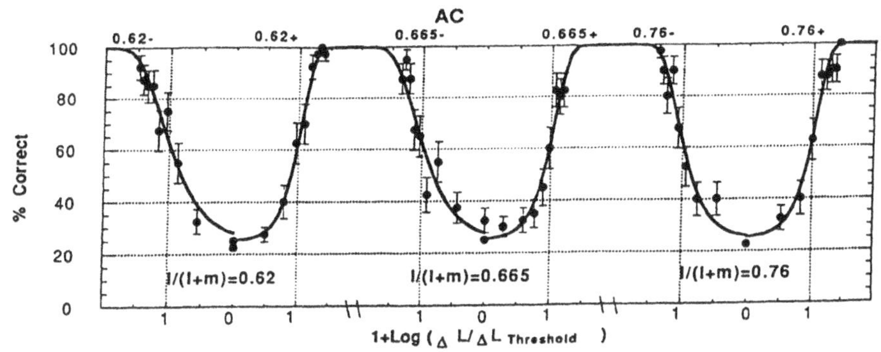

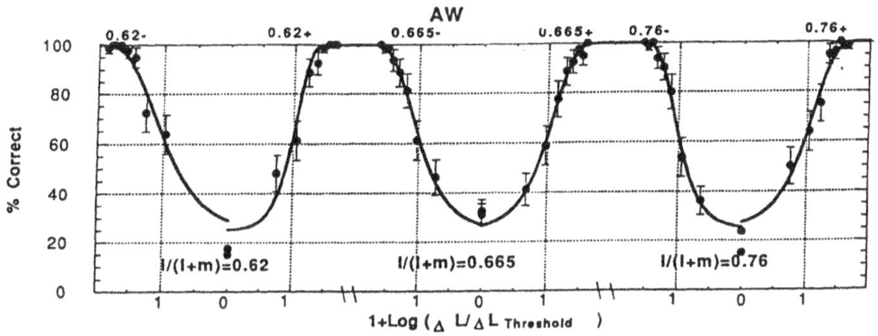

Fig. 3. Percentage correct responses for observers AC and AW on the l-axis. The plot format is as described for Fig. 2.

Table 1. The average values and range of the slope parameter, k, for observers AC and AW.

	Starting chromacity							
	s-axis				l-axis			
	0.7	0.997	4.0	Range	0.62	0.665	0.76	Range
AC	2.24	1.89	2.26	1.60–3.00	1.85	1.70	2.15	1.05–2.35
AW	1.49	1.97	1.48	1.00–2.49	1.62	1.53	1.62	0.94–2.35

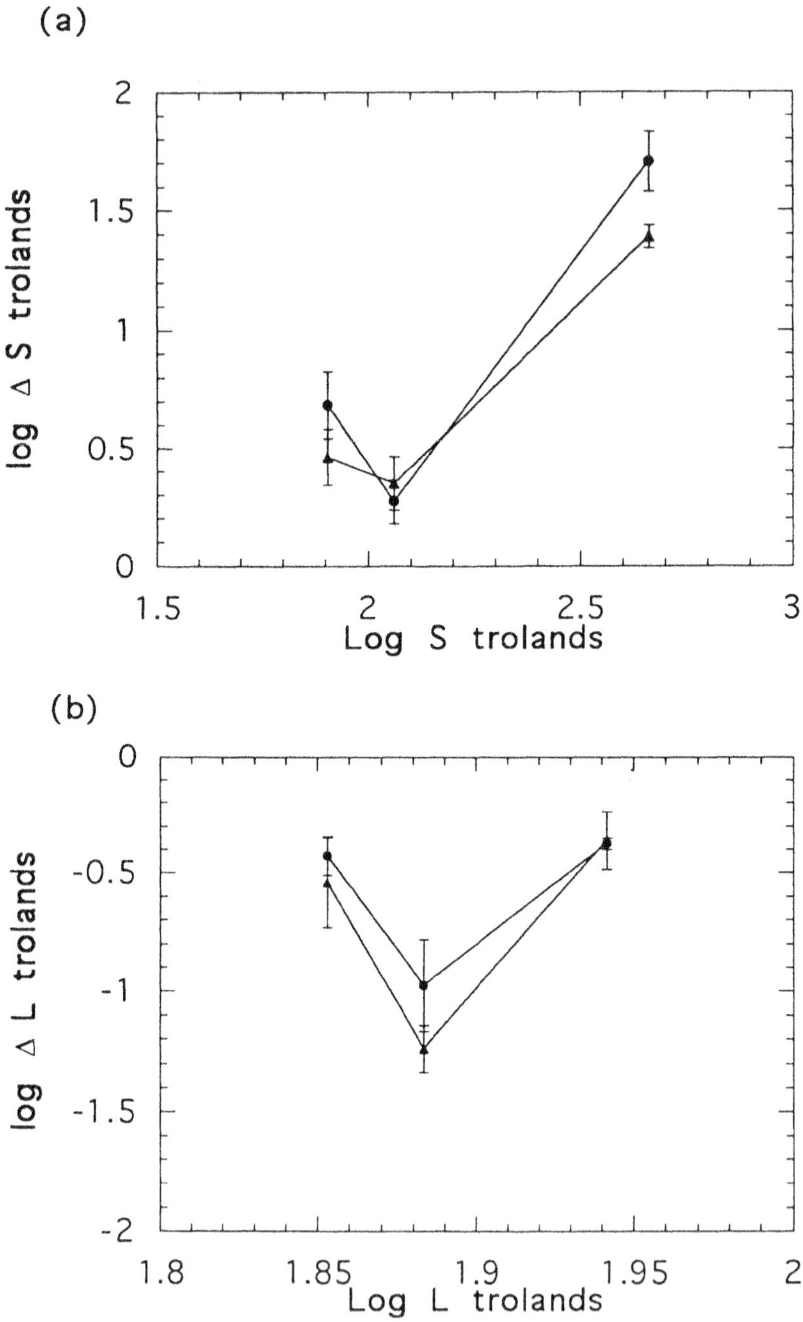

Fig. 4. The threshold chromaticities plotted as a function of the starting chromaticity in cone trolands for observers AC (circles) and AW (triangles). (a) Data for the s-axis, (b) data for the l-axis.

The results of one replication for observers AC and AW are shown for the s-axis in Figure 2 and for the l-axis in Figure 3. In these plots the horizontal axis shows one plus the logarithm of the relative stimulus strength, and the vertical axis shows the percentage of correct results. We chose this format in order to display the increment and decrement data in a single logarithmic plot. Thus relative stimulus strength increases from 0 in both directions and the threshold, c_{thr} plots at 50%. The fits describe the data reasonably well. The actual slope parameters vary considerably. Table 1 shows the values of the slope parameter, k. In this table we have averaged the estimates of both directions and replications. The slopes hover around 2 ± 1. This value is comparable to that reported by Eskew *et al.* (1991) for an l-chromaticity discrimination measured on a 3000 td yellow background and data reported by Vingrys and Mehta (this volume). On the s-axis there was a trend noted in all three observers for the slope to be higher at equal energy white. This effect was small, however, compared with the individual variability in the slopes.

Figure 4 shows the chromatic discrimination thresholds plotted as a function of the starting chromaticity in cone trolands. Both functions show a V-shape with best discrimination at the chromaticity of the surround. The chromatic discrimination threshold functions measured by the method of constant stimuli are similar to those measured using a staircase method (e.g. Zaidi *et al.*, 1992; Miyahara *et al.*, 1993) and control data in our equipment, indicating good quality of the data.

Discrimination thresholds measured at starting chromaticities distant from the adaptation point have been described by saturation functions (Zaidi *et al.*, 1992). We therefore asked whether the slope of the psychometric function might vary among the spectral opponent pathways. We conclude that although a saturation function may determine the arms of the V-shape, the function for chromatic contrast did not saturate within the available stimulus domain.

Acknowledgements

Supported in part by NEI grant EY07390, and a RPB Foreign Research Fellowship to AW. We thank Linda Glennie for assistance in programming the PIXAR color system.

References

Boynton, R.M. and Kambe, N. (1980). Chromatic difference steps of moderate size measured along theoretically critical axes. Col. Res. Appl. 5: 13–23.

Derrington, A.M., Krauskopf, J. and Lennie, P. (1984). Chromatic mechanisms in lateral geniculate nucleus of macaque. J. Physiol. (Lond.) 357: 241–265.

Eskew, R.T.J., Stromeyer, C.F.D., Picotte, C.J. and Kronauer, R.E. (1991). Detection uncertainty and the facilitation of chromatic detection by luminance contours. J. Opt. Soc. Am. A8: 394–403.

Krauskopf, J. and Gegenfurtner, K. (1992). Color discrimination and adaptation. Vision Res. 32: 2165–2175.
MacLeod, D.I.A. and Boynton, R.M. (1979). Chromaticity diagram showing cone excitation by stimuli of equal luminance. J. Opt. Soc. Am. 69: 1183–1185.
Miyahara, E., Smith, V.C. and Pokorny, J. (1993). How surrounds affect chromaticity discrimination. J. Opt. Soc. Am. A 10: 545–553.
Quick, R.F. (1974). A vector-magnitude model of contrast detection. Kybernetik 16: 65–67.
Smith, V.C. and Pokorny, J. (1995). Chromatic discrimination axes, CRT phosphor spectra and individual variation in color vision. J. Opt. Soc. Am. A 12: 27–35.
Yeh, T., Smith, V.C., and Pokorny, J. (1993). Colorimetric purity discrimination: Data and theory. Vision Res. 33: 1847–1857.
Zaidi, Q., Shapiro, A. and Hood, D. (1992). The effect of adaptation on the differential sensitivity of the S-cone color system. Vision Res. 32: 1297–1318.

Visual Sciences Center
University of Chicago
939 East 57th Street
Chicago, IL 60637, USA

41. Psychometric functions in the central visual field

A.J. VINGRYS and A.B. METHA

(Melbourne, Australia)

Abstract

We consider the nature of the luminance (L) and chromatic (RG or BY) psychometric functions in the central visual field. Stimuli were generated on a colour TV monitor with spatiotemporal characteristics that minimized transients (3° Gabor – 1 c/deg sine wave, 1 sec raised-cosine duration). Testing was performed along the vertical meridian in 3° steps between 0° and 12°. Thresholds were expressed as cone contrasts (%) for probes from a white background (CIE 1931 $x = 0.31$, $y = 0.32$; $L = 33$ cd/m^2). Two observers performed a method of constant stimuli, with data fit by a Weibull function (base 2). Foveal RG thresholds were significantly more sensitive than others (RG = 0.29%; L = 1.03%; BY = 10.02%: $p < 0.05$). The average eccentricity-related fall off in sensitivity (deci-Log/10°: $L = 2.81 \pm 0.08$; $RG = 6.13 \pm 0.07$; $BY = 5.96 \pm 0.64$) was significantly steeper for both colour mechanisms than for luminance. The average slope (beta) for $RG = 3.61 \pm 0.33$ was significantly greater than $Lum = 2.97 \pm 0.19$ or $BY = 2.63 \pm 0.20$ ($p < 0.05$). Both colour processes showed a complex eccentricity-related trend for beta with low values in the fovea. The opponent RG process showed low variability and the highest sensitivity up to 6°. Beyond 6°, the non-opponent luminance mechanism has the highest sensitivity, but high variability.

Introduction

There are many reasons for understanding the nature of the mechanisms subserving the detection of luminous or coloured targets. However, regardless of such need, the psychometric functions for detection mechanisms have been poorly defined, often with conflicting results: Maloney (1990) demonstrated the difficulties associated in establishing robust estimations for such relationships. His data show that while there were trends for shallower slopes for the short wavelength-detecting mechanism, there were no statistical differences between the slopes of any of the colour detection mechanisms, due to the uncertainties in their estimates. However, it is unclear whether the test paradigm used by Maloney isolated the opponent or non-opponent post-receptoral processes (see Figs. 9 and 10 in Kalloniatis and Harwerth, 1990).

C.R. Cavonius (ed.), Colour Vision Deficiencies XIII, pp. 377–384.

Slopes for foveal detection mechanisms (beta values) were also determined by Cole *et al.* (1994) by contour-fitting to RG, L, and BY detection thresholds. These authors reported that the average slopes of the BY (3.0) and L (3.0) post-receptoral mechanisms were similar, and greater than that of the RG (2.2) opponent process. Stromeyer and colleagues (1992) approximated the slopes for the RG and L processes by fitting a Weibull to data obtained from their Quest staircases (Watson and Pelli, 1983). Generally, these slopes were found to increase with eccentricity and, at any given eccentricity, the slope of the luminance mechanism was always steeper than that of the colour-opponent process.

None of these studies measured the slope directly. Stromeyer *et al.* (1992) and Cole *et al.* (1994) both approximated their slopes from fits to staircase data or threshold contours. Cole *et al.* noted that estimating slopes from detection contours can give large errors due to the influence of one or a few strategically located points. Cole *et al.* reported substantially steeper slopes from contour fits than those determined from a frequency of seeing curve (Cole *et al.*, 1994, p. 1272). A similar argument can be made for staircase data. It is possible that some of the differences in this previous literature may have arisen from such methodological discrepancies: we feel that slopes derived from frequency of seeing curves will be more robust to the artefacts described by Cole *et al.* (1994).

Apart from the theoretical issues involved, it has been proposed that modern static perimetry should be adapted to test colour, as well as luminance mechanisms (King-Smith *et al.*, 1987; Vingrys *et al.*, 1988, 1989; Hart *et al.*, 1990; de Jong *et al.*, 1990; Sample and Weinreb, 1992; Johnson *et al.*, 1993; Wild *et al.*, 1995). Moreover, it has been shown that colour perimetry is more sensitive for the early detection of disease than is white-on-white perimetry (Vingrys *et al.*, 1988, 1989; Johnson *et al.*, 1993). There is a clear need to ensure that optimal testing algorithms are available for such purposes in order to maximize the early detection of disease. The development of optimal algorithms requires a knowledge of the underlying frequency of seeing curves for each mechanism mediating detection. Moreover, the shape of the psychometric function is related to the internal noise within a mechanism (Wilson, 1980) and there is evidence that the noise might itself be affected by the disease processes (Chauhan *et al.*, 1993). In order to make sense of any differences in the slopes derived from patients, it is important that these are not confused with differences inherent in psychophysical tasks (e.g. colour or luminous stimuli) or eye movements (see Maloney, 1990, p.132).

Here we report the nature of the frequency of seeing curves for stimuli selected to isolate the RG, BY and L post-receptoral mechanisms over the central 12° of visual field. Such information is fundamental for a better understanding of visual processing, in optimizing staircase test procedures, and in comprehending perimetric test results that report increases in fluctuation (noise) or that make use of different-coloured stimuli.

Materials and methods

Stimuli were generated by a VSG2/1 graphics device and presented on a calibrated and gamma-corrected Barco CDCT 6551 CRT (Metha *et al.*, 1993). Their spatio-temporal limits minimized transients; 3° (diameter) Gaussian spatial windowed sinusoidal (1 c/deg) grating, shown in a 1.0 second raised cosine (Hann) temporal window. The CRT was surrounded by a large (55° × 55°) screen that approximated the luminance and chromaticity of the monitor (CIE 1931 x = 0.31, y = 0.32, L = 33 cd/m^2).

Testing was performed at a viewing distance of 175 cm along the inferior vertical midline by having observers fixate small (6 arcmin) fixation marks at 3° steps, out to 12°. The two authors, who have normal acuity and colour vision, acted as observers. Monocular testing was performed with the fellow eye occluded by a Ganzfeld occluder (Fuhr *et al.*, 1990).

Stimuli were modulated either in the L and M cone, or S cone planes. These axes cross the white point on the dichromatic confusion loci (1931 CIE; L cone, x = 0.75, y = 0.25; M cone, x = 1.4, y = –0.4; S cone, x = 0.26, y = 0.23). The cardinal directions that isolate the RG (colour opponent) and Lum (additive) post-receptoral mechanisms were chosen after analysing the detection contours in the L/M cone plane; these stimuli did not modulate the S cones (for details see Metha *et al.*, 1994). BY isolation was achieved by modulating S cones only. Thresholds are expressed in terms of the RMS cone contrasts based on the Smith and Pokorny (1975) cone fundamentals. No attempt was made to allow for peripheral changes in cone fundamentals: such changes produce minor effects compared with the noise in the estimates (Metha *et al.*, 1994).

A method of constant stimuli (MOCS) was used to generate frequency of seeing curves at all eccentricities for the three mechanisms (RG, Lum, BY). The MOCS comprised 40 trials at 6–14 contrast levels within the limits of our monitor that spanned the expected detection threshold; threshold estimates were made by a staircase procedure just prior to testing. Contrast levels were randomly interleaved during a single 1.5-hour test session with all data collection completed over a period of 2 weeks. Stimulus colour and eccentricity were fixed during any single session.

The psychophysical task was the detection of the stimulus using a two-interval forced-choice paradigm with a 500 msec interstimulus delay. Observers identified, by button press, the interval that contained the stimulus. An audible tone marked the duration of each interval, and feedback was supplied following response.

Results and discussion

Curve fitting procedures

Representative data for three eccentricities (0°, 6° and 12°) are given for one

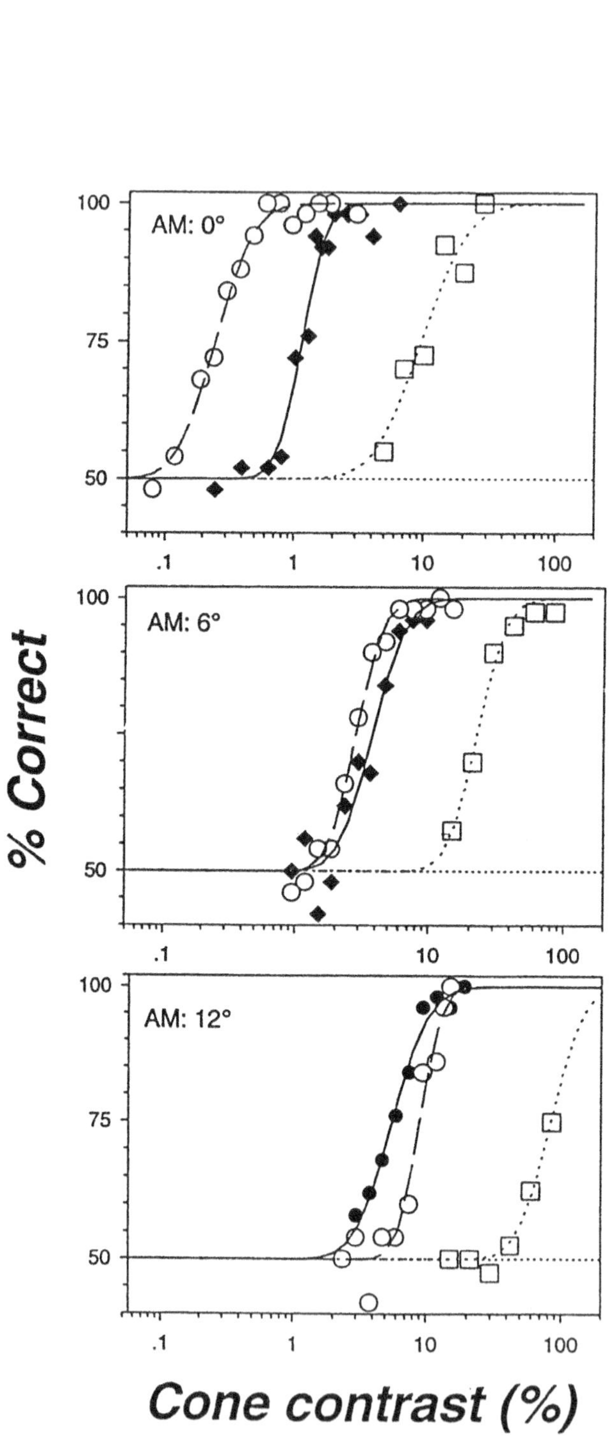

Fig. 1. Representative data obtained at three eccentricities for one observer (AM). Unfilled symbols are for colour processes (circles = RG; squares = BY); filled for the luminance mechanism.

observer (AM) in Figure 1. Best fitting curves are shown for each data set. In Figure 1 the unfilled symbols show thresholds mediated by colour-opponent processes (circles = RG; squares = BY) while the filled diamonds show luminance detection thresholds. The horizontal dotted line shows chance performance. Our analysis is performed after fitting a base-2 Weibull of the form:

$$Pc(C) = 1-(1-\gamma).2^{-(c/\alpha)\beta} \qquad \text{(Equation 1)}$$

where c is the contrast of the stimulus, Pc is the proportion correct when stimulus contrast is c, γ is the guess factor (0.5 for 2-IFC), α is the threshold contrast at the 75% performance level and β is an index of the slope of the psychometric function.

An iterative Marquard–Levenberg least χ^2 curve-fitting procedure was used to fit all data sets by adjusting parameters α and β. Figure 2 shows the results returned from the fitting procedure for both observers as a function of eccentricity. In Figure 2 the symbols are as defined previously (Figure 1). The threshold data of Figure 2a show a linear trend in semi-log axes which suggests the following relationship:

$$\text{Threshold} = 10 \hat{} (n \times \text{Eccentricity}+k) \qquad \text{(Equation 2)}$$

where k is a sensitivity constant and n is the slope of the relationship.

Alpha or the threshold parameter

The thresholds of Figures 1 and 2a show that the RG opponent process is the most sensitive (lowest threshold) foveal mechanism under the test conditions. The average foveal RG threshold is significantly less than any of the others: RG = 0.29%; Lum = 1.03%; BY = 10.02% ($p < 0.05$ for all comparisons). The BY process has – by an order of magnitude – the highest threshold at all eccentricities.

The eccentricity-related change in sensitivity varies among mechanisms. If the raw data are fitted by a linear trend, then the average eccentricity-related change (dLog/10°) is greatest for the colour processes: Lum = 2.81 ± 0.08; RG = 6.13 ± 0.07 and BY = 5.96 ± 0.64 ($p < 0.05$). However, Equation 2 produces a much better fit to our data over the central 12°, confirming that all processes have a non-linear relationship with eccentricity. The RG process has the steepest fall-off (n = 0.11) followed by BY (n = 0.08) and Luminance (n = 0.04). Our trends indicate that the RG mechanism is most sensitive within the first 8° and extrapolation shows that RG and BY thresholds should become similar at about 28–30°. The BY mechanism always remains a magnitude less sensitive than does the Luminance mechanism.

Although we acknowledge that extrapolation should be viewed cautiously, our data are consistent with those reported by others (Dain and King-Smith, 1989; Mullen, 1991; Stromeyer *et al.*, 1992; Metha *et al.*, 1994). In terms of RMS

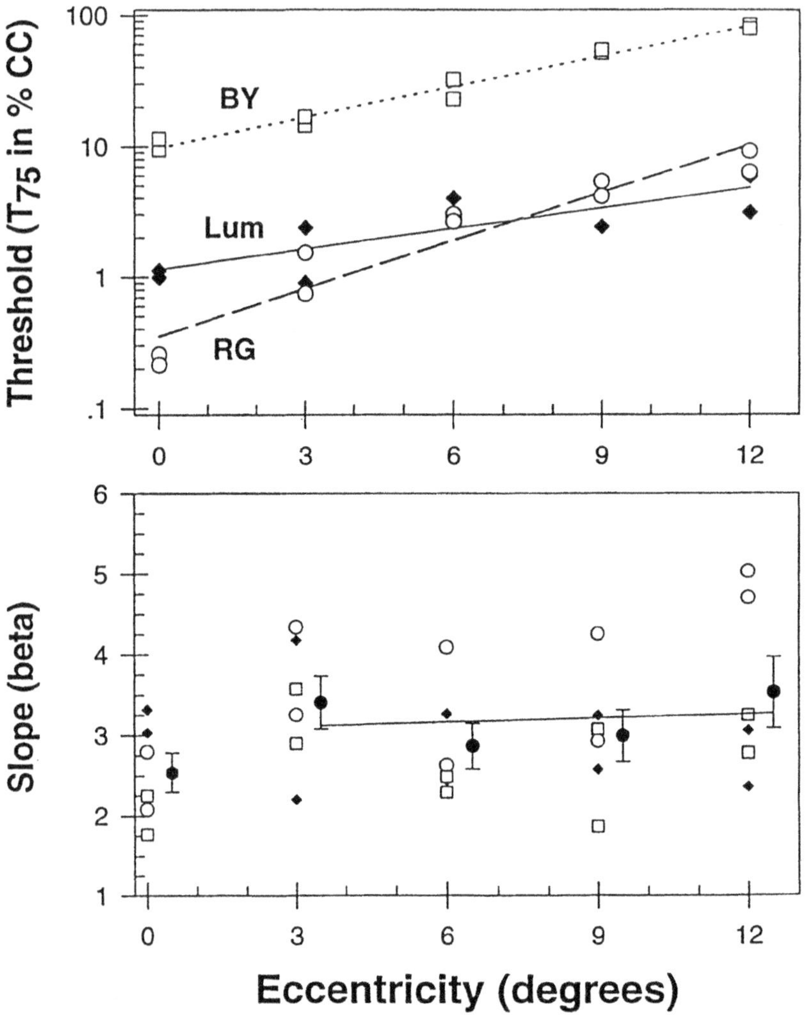

Fig. 2. The alpha (upper; T_{75}) and beta (lower) values of both subjects returned from the Weibull fit to the data: symbols are as per Figure 1. The trends shown by the lines give the power function relationship between eccentricity and threshold; details are given in the text.

cone contrast, RG is by far the most sensitive mechanism at the fovea, and this sensitivity advantage is quickly lost at greater eccentricities. As Mullen (1991) points out, these results suggest that colour vision is indeed a specialization of the central visual field. If disease were to affect such regions in a non-selective manner then testing for colour vision losses, and especially RG losses, might provide the greatest yields.

Beta or the slope parameter

Contrary to the results of Stromeyer *et al.* (1992), no consistent eccentricity-related dependence emerged from our data for the luminance process ($\beta = 2.97 \pm 0.19$; Fig. 2b). However, several interesting observations can be made regarding the β values of the colour mechanisms: Firstly, there is a tendency for the shallowest slopes (lowest β) to be found in the fovea (mean 2.58) for colour mechanisms. When these foveal data are extracted from the averaging process, the eccentricity-related trend (3.06) is not significantly removed from zero (solid line in Fig. 2b). Nevertheless, trend analysis for all data shows that the average colour slopes (RG and BY) do have a significant cubic component ($p < 0.05$): i.e. a tendency to kick up and down at either end. This is consistent with the suggestion of Stromeyer *et al.* (1992) that RG slopes steepen with eccentricity after about 10°.

At any given eccentricity, the average β for RG is greater than for BY. When β is pooled across observers and eccentricities, a single-factor ANOVA shows that the β value for RG (3.61 ± 0.33) is significantly greater than that for BY (2.63 ± 0.20; $p < 0.05$). This result is consistent with that of Maloney (1990), who proposed that the blue mechanism tends to have shallower slopes.

Comment

Our data indicate that psychometric slopes do not vary in a predictable manner within the central 12° for any of the mechanisms tested. This means that assuming a slope of ~3.0 will be adequate for most applications, although it may overestimate foveal slopes of colour mechanisms, especially the RG process. However, a more interesting and practical question arises from our findings. We have found that the RG mechanism has the lowest threshold and least variability (steepest slope) in the fovea. On the other hand the BY mechanism shows the opposite trend: it has both the highest thresholds and variability. If colour mechanisms were to be affected in a non-selective manner by disease, as is evident by recent work in patients with glaucoma (Yu *et al.*, 1991), then this finding would suggest that testing RG detection thresholds might hold substantial benefits for the early detection of dysfunction.

Acknowledgments

L.E. Mahon performed some analysis of this data, D.R. Badcock provided useful comments on an early draft.

References

Chauhan, B., Tompkins, J., LeBlanc, R. and McCormick, T. (1993). Characteristics of the frequency-of-seeing curves in normal subjects, patients with suspected glaucoma, and patients with glaucoma. Invest. Ophthalmol. Vis. Sci., 34: 3534–3540.

Cole, G.R., Hine, T. J. and McIlhagga, W. (1994). Estimation of linear detection mechanisms for stimuli of medium spatial frequency. Vision Res. 34: 1267–1278.

Dain, S.J. and King-Smith, P.E. (1989). Thresholds for iso-luminous colors across the visual field. In: Drum, B. and Verriest, G. (eds.), Colour Vision Deficiencies IX, 561–572. Kluwer Academic, Dordrecht.

de Jong, L.A.M.S., Snepvangers, C.E.J., van den Berg, T.J.T.P. and Langerhorst, C.T. (1990). Blue–yellow perimetry in the detection of early glaucomatous damage. Doc. Ophthalmol. 75: 303–314.

Fuhr, P.S., Hershner, T.A. and Daum, K.M. (1990). Ganzfeld blackout occurs in bowl perimetry and is eliminated by translucent occlusion. Arch. Ophthalmol. 108: 983–988.

Hart, W.M. Jr., Silverman, S.E., Trick, G.L., Nesher, R. and Gordon, M.O. (1990). Luminance and colour contrasts in glaucoma. Invest. Ophthalmol. Vis. Sci. 31: 359–367.

Johnson, C.A., Adams, A.J., Casson, E.J. and Brandt, J.D. (1993). Blue-on-yellow perimetry can predict the development of glaucomatous field loss. Arch. Ophthalmol. 111: 645–650.

Kalloniatis, M. and Herwerth, R. S. (1990). Spectral sensitivity and adaptation characteristics of cone mechanisms under white light adaptation. J. Opt. Soc. Am. A7: 1912–1928.

King-Smith, P.E., Vingrys, A.J. and Benes, S.C. (1987). Visual thresholds measured with color video monitors. Color Res. Appl. 12: 73–80.

Maloney, L.T. (1990). The slope of the psychometric function at different wavelengths. Vision Res. 30: 129–136.

Metha, A.B., Vingrys, A.J. and Badcock, D.R. (1993). Calibration of a color monitor for visual psychophysics. Behav. Res. Meth. Inst. Comp. 25: 371–383.

Metha, A.B., Vingrys, A.J. and Badcock, D.R. (1994). Detection and discrimination of moving stimuli: the effects of color, luminance and eccentricity. J. Opt. Soc. Am. A11: 1697–1709.

Mullen, K.T. (1991). Colour vision as a post-receptoral specialization of the central visual field. Vision Res. 31: 119–130.

Sample, P.A. and Weinreb, R.N. (1992). Progressive color visual field loss in glaucoma. Invest. Ophthalmol. Vis. Sci. 33: 2068–2071.

Smith, V.C. and Pokorny, J. (1975) Spectral sensitivity of the foveal cone photopigments between 400 and 500 nm. Vision Res. 15: 161–171.

Stromeyer III, C.F., Lee, J. and Eskew, R.T. Jr. (1992). Peripheral chromatic sensitivity for flashes: a post-receptoral red-green asymmetry. Vision Res. 32: 1865–1873.

Vingrys, A.J., King-Smith, P.E., Benes, S.C. and Weber, P.A. (1988). The value of color perimetry for detecting glaucomatous field losses. Opt. Soc. Am. Tech. Digest 3: 180–183.

Vingrys, A.J., King-Smith, P.E. and Benes, S.C. (1989). Color perimetry can be more sensitive than achromatic perimetry. Clin. Vision Sci. 4: 197–209.

Watson, A.B. and Pelli, D.G. (1983). QUEST: A Bayesian adaptive psychometric method. Percept. Psychophys. 33: 113–120.

Wild, J.M., Moss, I.D., Whitaker, D. and O'Neill, E.C. (1995). The statistical interpretation of blue-on-yellow visual field loss. Invest. Ophthalmol. Vis. Sci. 36: 1398–1410.

Wilson, H. (1980). A transducer function for threshold and suprathreshold human vision. Biol. Cybernet. 38: 171–178.

Yu, T.C., Falcao-Reis, F., Speilers, W. and Arden, G.B. (1991). Peripheral color contrast. Invest. Ophthalmol. Vis. Sci. 32: 2779–2789.

Department of Optometry and Vision Sciences
University of Melbourne
Parkville, Victoria, Australia 3052

42. Sons and mothers: classification of colour-deficient and heterozygous subjects by counterphase modulation photometry

G. JORDAN and J.D. MOLLON
(Cambridge, UK)

Abstract

In the OSCAR test of Estévez *et al.* (1983) red and green lights are modulated in counterphase and the subject is asked to adjust their relative depths of modulation so as to minimize flicker. In a population consisting of normal mothers and carriers of colour deficiency (classified by their sons' performance on the Nagel anomaloscope), the OSCAR settings of the mothers were strongly correlated with those of their sons. Protan and deutan carriers formed discrete populations; and many individual carriers of protan deficiencies could be distinguished from normals with confidence. Protan and deutan sons were distinguished from each other with complete reliability, but some deutan sons, and most deutan carriers, fell within the distribution of normal settings.

Introduction

Can carriers of colour deficiency be identified psychophysically? Since 1920 it has been suspected that red–green colour deficiencies are not always completely recessive and that some carriers may reveal themselves by abnormalities of colour vision (Fleischer, 1920; Schiötz, 1920; Waaler, 1927). In fact, abnormalities of spectral matching or of colour discrimination are rather rare in heterozygotes (Jordan and Mollon, 1993). More diagnostic have been tests of increment threshold or spectral sensitivity. Thus protan carriers exhibit depressed sensitivity to long wavelengths (a condition known as Schmidt's sign; Schmidt, 1934, 1955; Walls and Mathews, 1952; Crone, 1959; Ikeda *et al.*, 1972; Swanson, 1991). An analogous loss of middle wavelength sensitivity has been reported in deutan carriers (de Vries, 1948; Crone, 1959; Adam, 1969) and has been named de Vries' sign, although this phenomenon has been more difficult to demonstrate than Schmidt's sign.

Such manifestations of the carrier state have been explained by the phenomenon of X-chromosome inactivation: although a woman inherits two X-chromosomes, a random process determines which of the two is expressed in

C.R. Cavonius (ed.), Colour Vision Deficiencies XIII, pp. 385–392.

any individual cell of her body (Lyon, 1972). The retinal mosaic of the heterozygote is thought to contain a subset of cones that express the abnormal chromosome which her colour-deficient son inherits.

Estévez *et al.* (1983) introduced a rapid measure of relative luminosity, called the OSCAR test (**O**bjective **S**creening of **C**olour **A**nomalies and **R**eductions). Since earlier reports suggest that carriers of colour deficiency reveal themselves rather readily in measurements of this kind, we have been prompted to ask whether the OSCAR test is clinically useful for identifying heterozygotes.

The test presents to the observer a mixture of two flickering lights, one a nearly monochromatic red light of 650 nm and one a relatively broad-band green light of dominant wavelength 560 nm. The two sources are light-emitting diodes and their outputs are mixed within a solid perspex rod. One end of the rod is slightly roughened to give a diffusing surface; and this end is viewed by the subject. The two component lights are alternated in counterphase at approximately 16 Hz (Fig. 1). A single control knob allows the subject to increase the depth of modulation of one component while concurrently decreasing the depth of modulation of the other component. The subject is asked to adjust the knob until the flicker is minimally visible. The advantage of this counterphase modulation photometry over flicker photometry is that the time-averaged luminance and chromaticity of the stimulus remain constant as the adjustment is made.

Materials and methods

Subjects

The 144 subjects comprised 43 carriers of X-linked colour deficiencies and their colour-deficient sons, and 25 colour normal mothers and their sons. In some cases, more than one son was tested from a given family. Most carriers had already participated in an earlier study of the colour vision of heterozygotes (Jordan and Mollon, 1993). All subjects were naive as to the aim of the experiment and were paid a small sum for their participation.

OSCAR test

The subject held the OSCAR test in two hands while operating the knob with his or her right hand. The stimulus field subtended approximately 0.7° at the eye. The dominant wavelength was approximately 594 nm and the luminance was 15 cd/m^2. Between successive settings the control knob was offset by the experimenter, alternately to one or other end of its range. Subjects made five settings binocularly. The OSCAR test is known to be sensitive to the lighting conditions (Birch, 1993): all the present subjects were tested in the presence of a low level of Illuminant C (Macbeth easel lamp) giving approximately 3 lux in an otherwise dark room. The version of the test we used was manufactured commercially by Medilog (The Netherlands).

Nagel anomaloscope

The Rayleigh match was measured with the Nagel anomaloscope (model I) for the dominant eye of each subject. The procedure was as follows: the experimenter set the red–green mixture and the subject adjusted the luminance of the monochromatic yellow to equate the fields in brightness. After this was done, the subject looked at a white piece of paper illuminated by Illuminant C to neutralize his or her adaptation and then rated the goodness of the match. The procedure was repeated until both ends of the matching range were securely determined.

Results

The main panel of Figure 2 plots the OSCAR settings of sons against the settings of their mothers. The ordinates represent the arbitrary scale units of the OSCAR test, normalized so that the mean setting of our population of normal mothers and sons is zero. Projected to the left of the main panel is the histogram of settings of individual sons; and projected below the main panel is the

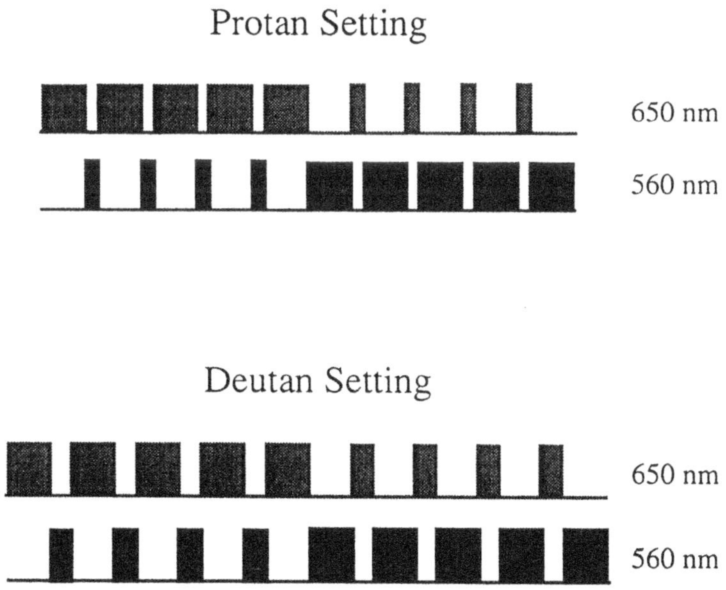

Fig. 1. In the OSCAR test the red and the green components of the stimulus are modulated in counterphase at 16 Hz. The depth of modulation is controlled not by varying the LED current (which would change the spectral output of an individual LED) but by turning the lamps fully on and off at a much higher frequency and varying the proportion of time the lamp is on during the two phases of each cycle. Thus the diagram shows schematically the high-frequency modulation during the opposite phases of just one period of the 16 Hz stimulus.

histogram of settings of mothers.

From the histogram for sons, it is seen that counterphase modulation photometry achieves a complete separation of subjects defined as protan and deutan by the Nagel anomaloscope. Moreover, all the protan subjects lie outside the distribution of colour-normal subjects. However, there is significant overlap between the normal and the deutan sons. In judging this overlap, and the similar overlap in Figure 1 of Cavonius and Kammann (1984), it is important to remember that normal individuals are about 15 times more common in the population than are deutans and thus, to obtain likelihood ratios for clinical use,

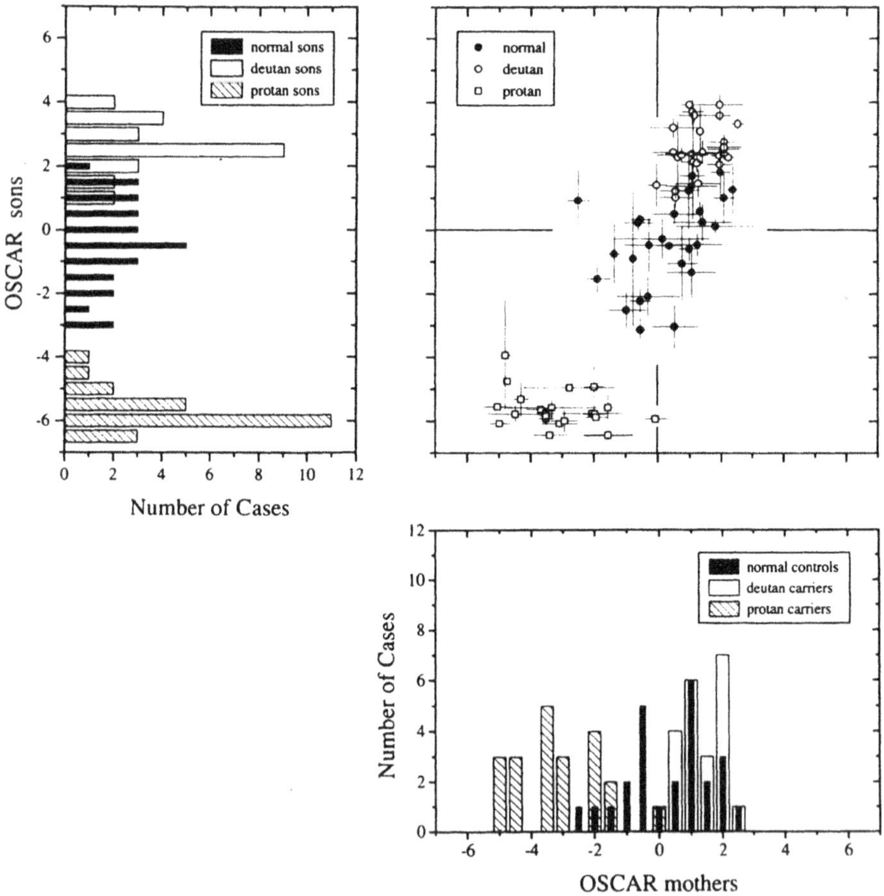

Fig. 2. The upper right panel shows the relationship between the settings of mothers and sons on the OSCAR test. Note the clear difference between settings of deutan and protan carriers. The distributions of settings for sons and those for the mothers are projected on to the two subsidiary histograms. The cell centred on 0.0 in the histogram for mothers represents one normal, one protan heterozygote and one deutan heterozygote.

we must increase the relative height of the colour-normal distribution so that areas of the normal and deutan distributions are in the ratio 15:1; 'colour-normal' thus becomes the more likely diagnosis over a significant fraction of the range occupied in the histogram by deutan settings.

The main panel of Figure 2 exhibits a strong overall correlation ($R = 0.85$, $p < 0.001$) between the minimum-flicker settings of mothers and sons. As can be seen in the lower histogram, this rapid clinical test achieves an almost complete separation of protan and deutan carriers. There is overlap between the control mothers and the two types of carrier, especially the deutan carriers. Nevertheless, many protan carriers exhibit Schmidt's sign, in that they are less sensitive to long wavelengths than are any of the normals. None of our population of protan carriers exhibited a setting as extreme as the most extreme protan sons, but some carriers are as extreme as some of the protanopic sons.

The OSCAR test did not distinguish between dichromatic and anomalous sons, nor between the corresponding subsets of mothers. A moderate but significant correlation between OSCAR and anomaloscope settings was found for normal sons ($R = -0.62$; $p = 0.003$), but not for normal mothers ($R = -0.15$; $p = 0.44$).

Discussion

Counterphase modulation photometry as a test for colour blindness

We confirm the judgement of others (Cavonius and Kammann, 1984; Zisman *et al.*, 1987; Verriest and Uvijls, 1987) that the OSCAR test is not ideal as a screening test for colour blindness. Its failure to distinguish some deutans from normal individuals reflects a difficulty that has classically faced tests that depend on the measurement of spectral luminosity (Crone, 1959; Adam, 1969); Nevertheless, the OSCAR test has a role in clinical diagnosis. We found, as did Cavonius and Kammann (1984) and Verriest and Uvijls (1987), that the test was completely reliable in distinguishing deutans from protans, a discrimination that is only poorly made by pseudoisochromatic plates (Walls, 1959; Frey, 1958). If a patient has failed a plate test, the OSCAR test offers an economical means to classify him or her as protan or deutan. But to distinguish protanomalous subjects from protanopes, and deuteranomalous from deuteranopes, it will still be necessary to turn to the anomaloscope.

Counterphase modulation photometry and the heterozygote

The OSCAR test has a role to play in the examination of carriers of colour deficiency. As groups, the protan and deutan heterozygotes are statistically quite distinct from each other and from the putatively homozygous women. A subgroup of heterozygotes can be individually identified with some confidence using the OSCAR test: these are the protan carriers who make a very negative

Table 1. The relationship between OSCAR settings and the probability that a woman is a protan carrier rather than a normal.

OSCAR	–6.0	–5.5	–5.0	–4.5	–4.0	–3.5	–3.0	–2.5	–2.0
$p_{(protan)}$	1.0	1.0	0.99	0.97	0.91	0.77	0.53	0.28	0.14

setting. If we take Verriest's (1972) estimates of the relative frequencies of normal mothers and protan carriers (84/3.7) and if we fit normal distributions to our OSCAR settings for normal mothers and for protan carriers, then we can calculate that 'protan heterozygote' becomes the more likely diagnosis when the OSCAR reading is less than –3.0 and becomes almost certain for readings of –4 or less. Table 1 shows the calculated probabilities over a range of OSCAR settings. Our estimates are conservative because statistically our sample of 'homozygous' mothers ought to contain one or more heterozygotes who have not yet revealed themselves by bearing a colour-deficient boy. It would be desirable to have data for a larger sample of colour-normal women, each of whom had several sons, none of whom are colour blind.

Theory of counterphase modulation photometry

To understand why deutan and protan heterozygotes perform differently on the OSCAR test, we must consider what is happening when a subject sets the instrument to minimum flicker. Only in the case of the dichromat is there a straightforward basis for the setting. Such a subject can find a setting that completely abolishes flicker. In the case of, say, a protanope, this is thought to be the setting at which an increment in the 650 nm light gives an increase in the quantum catch of the middle wavelength cones that is exactly balanced by a decrement in the quantum catch from the 560 nm light (see Estévez *et al.*, 1983). This setting corresponds closely to the setting that Rushton called the 'isolept' when discussing the exchange-threshold method of isolating individual classes of receptor (Rushton *et al.*, 1973).

We do not know the exact basis on which a trichromat finds a setting of minimum flicker. A setting that corresponds to the isolept for his long wavelength cones will not correspond to the isolept for his middle wavelength cones, and vice versa. Perhaps he has independent access to the two cone signals and compromises between their isolepts, finding an intermediate setting where the counterphase modulations of the two separate cone signals are both equally small (Mollon, 1987); or perhaps, as more commonly assumed, there is a post-receptoral channel (the 'luminance' channel) that combines signals from the long and middle wavelength cones, so that decrements in one signal cancel increments in the other (Ingling and Tsou, 1985). The first possibility might be called the compromise hypothesis and the second the cancellation hypothesis. It is empirically difficult to distinguish psychophysically between compromise and addition; for both hypotheses require a setting intermediate between the two isolepts. In either case, we should expect the OSCAR setting to be easier and more precise for an anomalous trichromat than for a normal observer. In anomalous trichromacy the spectral sensitivities of the long and middle wavelength cones are thought to be closer together than normal; the two isolept settings would, therefore, also be close and an intermediate setting would not modulate strongly the signal of either class of cone.

How are we to explain the different settings of protan and deutan heterozygotes? Consider a carrier for protanopia. Owing to X-chromosome inactivation, we may suppose that a subset of her cones expresses the X-chromosome that her protan sons inherit. In consequence, an increased proportion of cones in her retina will contain a middle wavelength-sensitive pigment. Since the total strength of the signal from a given class of cones is likely to depend on their numerosity, our protan carrier should require more modulation of the red LED to compensate a given modulation of the green LED. In terms of the compromise hypothesis, her abnormal setting is necessary to keep the absolute modulation of the middle wavelength cone signal small and equal to that of the long wavelength signal. In terms of the cancellation hypothesis, the shift is necessary because a larger decrement in the signal of individual long wavelength cones is needed to cancel a given increment in the middle wavelength signal. In the case of a carrier of protanomaly, the argument must be somewhat less certain since the nature of anomalous trichromacy remains unsettled. However, most accounts of protanomaly suppose that the protanomalous trichromat depends for his residual red–green discrimination either on two normal middle wavelength pigments with different spectral sensitivities or on the combination of a normal middle wavelength pigment and an anomalous pigment that occupies a spectral position close to that of the middle wavelength pigment(s). Whichever hypothesis one prefers, a heterozygote for protanomaly will effectively end up with an excess of cones that contain pigment(s) with their peak sensitivity in the middle wavelength region; and the signal of these cones will dominate flicker photometric settings. *Mutatis mutandis*, the same arguments can be applied to carriers of deuteranopia and deuteranomaly.

Acknowledgements

Supported by MRC grant G9113630N and by Wellcome Trust Fellowship to GJ. We are grateful to O. Estévez for the gift of the OSCAR test.

References

Adam, A. (1969). Foveal red–green ratios of normals, colourblinds and heterozygotes. Proc Tel-Hashomer Hosp. Israel 8: 2–6.

Birch, J. (1993). Diagnosis of Defective Colour Vision. Oxford University Press, Oxford.

Cavonius, C.R. and Kammann, J. (1984). A clinical evaluation of the 'OSCAR' color vision set. In: Verriest, G. (ed.), Colour Vision Deficiencies VII: 275–279. W. Junk, The Hague.
Crone, R.A. (1959). Spectral sensitivity in color-defective subjects and heterozygote carriers. Am. J. Ophthalmol. 48: 231–238.
de Vries, H.L. (1948). The fundamental response curves of normal and abnormal dichromatic and trichromatic eyes. Physica 14: 367–380.
Estévez, O., Spekreijse, H., Van Dalen, J.T.W. and Verduyn Lunel, H.F.E. (1983). The Oscar color vision test: Theory and evaluation (Objective Screening of Color Anomalies and Reductions). Am. J. Optom. Physiol. Opt. 60: 892–901.
Fleischer, R. (1920). Die Vererbung geschlechtsgebundener Krankheiten. 42. Vers. Deut. Ophthalmol. Ges. 4–14.
Frey, R.G. (1958). Welche pseudoisochromatischen Tafeln sind für die Praxis am besten geeignet? Graef. Arch. Ophthalmol. 160: 301–320.
Ikeda, M., Hukami, K. and Urakubo, M. (1972). Flicker photometry with chromatic adaptation and defective color vision. Am. J. Ophthalmol. 73: 270–277.
Ingling, C.R. and Tsou, B.H.-P. (1985). Flicker photometry and achromatic-channel structure. J. Opt. Soc. Am. A2: 1375–1378.
Jordan, G. and Mollon, J.D. (1993). A study of women heterozygous for colour vision deficiencies. Vision Res. 33: 1495–1508.
Lyon, M.F. (1972). X-chromosome inactivation and developmental patterns in mammals. Biol. Rev. 47: 1–35.
Mollon, J.D. (1987). On the nature of models of colour vision. Die Farbe 34: 29–46.
Rushton, W.A.H., Powell, D. and Wright, K.D. (1973). Exchange thresholds in dichromats. Vision Res. 13: 1993–2002.
Schiötz, I. (1920). Rotgrünblindheit als Erbeigenschaft. Klin. Monatsbl. Augenheilkd. 68: 498–526.
Schmidt, I. (1934). Über manifeste Heterozygotie bei Konduktorinnen für Farbsinnstörungen. Klin. Monatsbl. Augenheilkd. 92: 456–467.
Schmidt, I. (1955). A sign of manifest heterozygosity in carriers of color deficiency. Am. J. Optom. 32: 404–408.
Swanson, W.H. (1991). Heterochromatic modulation photometry in heterozygous carriers of congenital color defects. In: Drum, B., Moreland, J.D. and Serra, A. (eds), Colour Vision Deficiencies X, Doc. Ophthal. Proc. Ser. 54: 457–471. Kluwer, Dordrecht.
Verriest, G. (1972). Chromaticity discrimination in protan and deutan heterozygotes. Die Farbe 21: 7–16
Verriest, G. and Uvijls, A. (1987). About the OSCAR test. In: Verriest, G. (ed), Colour Vision Deficiencies VIII, Doc. Ophthal. Proc. Ser. 46: 177–179. Martinus Nijhoff, W. Junk, Dordrecht.
Waaler, G.H.M. (1927). Über die Erblichkeitsverhältnisse der verschiedenen Arten von angeborener Rotgrünblindheit. Z. Indukt. Abstam. Vererbungslehre 45: 279–333.
Walls, G.L. (1959). How good is the H-R-R test for color blindness? Am. J. Optom. Arch. Am. Acad. Optom. Monogr.249: 1–25
Walls, G.L. and Matthews, R.W. (1952). New Means of Studying Color Blindness and Normal Foveal Color Vision. University of California Publications in Psychology, No. 7.
Zisman, F., Seger, K.R. and Adams, A.J. (1987). Specificity evaluation of the OSCAR colour vision test. In: Verriest, G. (ed.), Colour Vision Deficiencies VIII, Doc. Ophthal. Proc. Ser. 46: 173–176. Martinus Nijhoff, W. Junk, Dordrecht.

Department of Experimental Psychology
University of Cambridge
Downing St.
Cambridge, CB2 3EB, UK

43. Influence of lightness (Munsell value) on the desaturated Panel D15

P. LANTHONY and M. GOBERVILLE
(Paris, France)

Abstract

Comparison of Panels D15 at values 4, 6 and 8 indicates that the Panel D15 8/2 is more sensitive to anomalous trichromacy.

Introduction

The first so-called 'desaturated test' was the Farnsworth B-20, which had Munsell value 5 and chroma 2 (Farnsworth, 1943). Several other desaturated tests were then devised, retaining the same value 5 and chroma 2 (Lanthony, 1990). Our desaturated test was different, since it had Munsell value 8 (Lanthony and Dubois-Poulsen, 1973). It was therefore important to determine whether this high value had influenced the efficiency of the test.

In a previous study of a population of normal subjects, we found 16% false-positive results using D15 8/2, versus 3% with D15 6/2 and D15 4/2 (Lanthony, 1995). The purpose of the present study, made with a population of definitely anomalous subjects (anomalous trichromats), was to determine which of these tests is most effective in detecting anomalous trichromacy.

Materials and methods

Three desaturated Panel D15 were used: the commercially-available D15 8/2, and two experimental prototypes with values 6 and 4, retaining the same chroma 2 and the same hues; i.e., 10 B (reference), 5 B (1), 10 BG (2), 5 BG (3), 10 G (4), 5 G (5), 10 GY (6), 5 GY (7), 5 Y (8), 10 YR (9), 2,5 YR (10), 7,5 R (11), 2,5 R (12), 5 RP (13), 10 P (14), 5 P (15). The three D15 panels were presented to each subject in a random order, with a 1000 lux illuminant C.

The population studied included 22 subjects with anomalous trichromacy (5 protanomalous and 17 deuteranomalous subjects), assessed by their anomalous quotients on the Nagel anomaloscope. The results are presented using the scoring method described earlier (Lanthony, 1986).

C.R. Cavonius (ed.), Colour Vision Deficiencies XIII, pp. 393–395.

Results

The results for individual subjects are presented in Table 1. According to our method of quantification, the scores were abnormal in 16 of 22 subjects with D15 8/2, in four subjects with D15 6/2, and in three subjects with D15 4/2.

Table 1. Scores of the subjects on the three D15 tests.

		Scores		
Subject/age	Quotient	4/2	6/2	8/2
01 S.F. 24	0.30	74	82	244
02 P.J. 28	0.26	74	128	182
03 R.J. 27	0.28	62	70	128
04 A.G. 16	0.44	18	8	128
05 L.A. 27	0.19	30	8	56
06 A.D. 21	4.70	114	200	212
07 M.B. 23	3.44	0	0	16
08 G.E. 17	4.40	76	166	372
09 D.E. 14	3.21	90	66	90
10 M.A. 16	3.20	52	24	44
11 G.M. 25	3.20	44	230	294
12 F.C. 20	3.44	8	8	152
13 B.T. 13	3.20	44	8	126
14 M.L. 20	3.45	58	38	198
15 S.J. 13	2.63	0	0	166
16 V.E. 25	3.40	0	0	170
17 W.J. 20	4.70	30	0	220
18 M.E. 14	3.41	76	16	132
19 T.J. 18	3.20	160	8	276
20 Z.J. 17	2.33	56	24	88
21 P.B. 16	3.44	94	0	132
22 R.P. 29	3.20	82	34	124

Discussion

These results suggest that the desaturated Panel D15 at value 8 was more sensitive than Panel D15 at value 6 and value 4, since there were only six false-negative results among 22 subjects, versus 18 false-negative at value 6, and 19 at value 4. Conversely, our previous study of normal subjects showed 16% false-positive at value 2, but only 3% at values 6 and 4. Hence, the Panel D15 8/2 is more sensitive than D15 6/2 or D15 4/2, but at the expense of specificity. The explanation of these results is given by comparing the location of the three Panel D15 in the CIE triangle: Panel D15 at 8/2 is closer to the reference white of the triangle than is either Panel D15 6/2 or 4/2 (see Wyszecki and Stiles, 1982, pp. 482, 484, 486).

Conclusion

The high level of lightness in the Panel D15 8/2 influences the results of this test. It allows a higher sensitivity than in Panel D15 6/2 and 4/2, (i.e. fewer false-negative results among colour-deficient subjects), but this is at the expense of lower specificity (more false-positive results in normal subjects), which is a common problem in sensory examination.

References

Farnsworth, D. (1943). The Farnsworth-Munsell 100-Hue and dichotomous tests for color vision. J. Opt. Soc. Am. 33: 568–578.

Lanthony, P. (1986). Evaluation du Panel D-15 désaturé. I: methode de quantification et scores normaux. J. Fr. Ophtalmol. 9: 843–847.

Lanthony, P. (1990). Clinical examination of the chromatic saturation. Neuro-Ophthalmol. 10: 119–127.

Lanthony, P. (1995). Evaluation du Panel D-15 désaturé. IV: effet de la répétition du Panel D-15 désaturé. J. Fr. Ophtalmol. 18: 578–583

Lanthony, P. and Dubois-Poulsen, A. (1973). Le Farnsworth 15 désaturé. Bull. Soc. Ophtalmol. Fr. 73: 861–866.

Wyszecki, G. and Stiles, W.S. (1982). Color Science. Wiley, New York.

Laboratoire de la vision des couleurs
Centre Hospitalier National d'Ophtalmologie des Quinze-Vingts
28 rue de Charenton
F-75012 Paris, France

44. Temperature and Rayleigh matching on Moreland's anomaloscope

J.D. MORELAND and E.C. ALEXANDER
(Keele, U.K.)

Abstract

The Moreland anomaloscope is subject to substantial self-warming which is due primarily to radiant heat from its own light source. The spectral transmission of the three bandpass interference filters, used to define the stimuli of the Rayleigh equation, was measured spectrophotometrically at two controlled temperatures representing a 'cold-start' and the end of the self-warming phase. Small changes in bandpass transmission characteristics were recorded for all three filters. In a separate experiment, Rayleigh matches were made continuously over the whole self-warming phase. The observed change in anomaly quotient was less than 0.01 and has no practical importance.

Introduction

Jordan and Mollon (1993) confirmed Köllner's long overlooked note (1915) that Nagel's anomaloscope is sensitive to ambient temperature and they report a rate of change in the anomaly quotient of $0.01°C^{-1}$. The source of variation was identified as the compound dispersing prism of the instrument: temperature-dependent changes in refractive index producing wavelength changes in the instrument's stimuli.

The Moreland anomaloscope employs a 500 W light source and 3-cavity bandpass interference filters to define its stimuli (Moreland and Kerr, 1979). After switching the instrument on, self-warming occurs which is largely due to radiant heat from the lamp. Filter temperature increases with time and stabilizes at about 35°C above the ambient room temperature. Calculations, made at the design stage and based on the filter manufacturer's catalogued temperature coefficients, indicated maximum changes in centre wavelength of about 0.5 nm. These were judged to be unimportant at the time, but the work of Jordan and Mollon (1993) suggested a need for a more careful control.

C.R. Cavonius (ed.), Colour Vision Deficiencies XIII, pp. 397–401.

Materials and methods

Experiment 1

The bandpass characteristics of the three interference filters (G, 546 nm; Y, 589 nm; R, 671 nm) used for the Rayleigh equation in the Moreland anomaloscope were measured on a recording spectrophotometer (Varian DMS100) at two representative temperatures, 20°C ('cold-start' ambient room temperature) and 55°C (end of self-warming phase). Temperature stabilization was achieved by mounting the filters, in turn, in a purpose-built thermally controlled jacket within the measuring chamber of the spectrophotometer. The jacket temperature was measured by a calibrated thermocouple.

Experiment 2

The anomaloscope was switched on at room temperature and Rayleigh matches (approximately 100) were made continuously during the self-warming phase from about 20°C to 55°C. Temperature was recorded close to the green (546 nm) interference filter by means of a thermocouple in contact with the metal filter holder.

Results and discussion

In Experiment 1, the centre wavelengths of the interference filters increased with temperature by amounts greater than those calculated from the manufacturer's data. These changes were accompanied by small reductions in bandwidth and in peak transmission (see Fig. 1 and Table 1). Blocking performance, outside the filter bandpass, was unaffected.

In experiment 2, a linear regression of log(G/R) on filter temperature accounted for only 2% of the variance and yielded a slope of $-0.0001°C^{-1}$, equivalent to a rate of change in the anomaly quotient of $-0.0002°C^{-1}$ This temperature coefficient is much smaller than, and opposite in sign to, that reported by Jordan and Mollon (1993) for the Nagel anomaloscope. However, the short-term temperature changes resulting from self-warming in Moreland's anomaloscope are larger than those of ambient seasonal origin and so this comparison lacks relevance.

The change in anomaly quotient, corresponding to the self-warming temperature rise of 35°C, is –0.007. Ambient temperature variations could add to the effect. However, even if we make the pessimistic assumption that an ambient temperature variation of 15°C is added cumulatively, then the largest total variation in anomaly quotient would be –0.01. This is 15 times smaller than the corresponding change due to seasonal variations in ambient temperature which has been reported for Nagel's anomaloscope (Jordan and Mollon, 1993) and is too small to have practical significance.

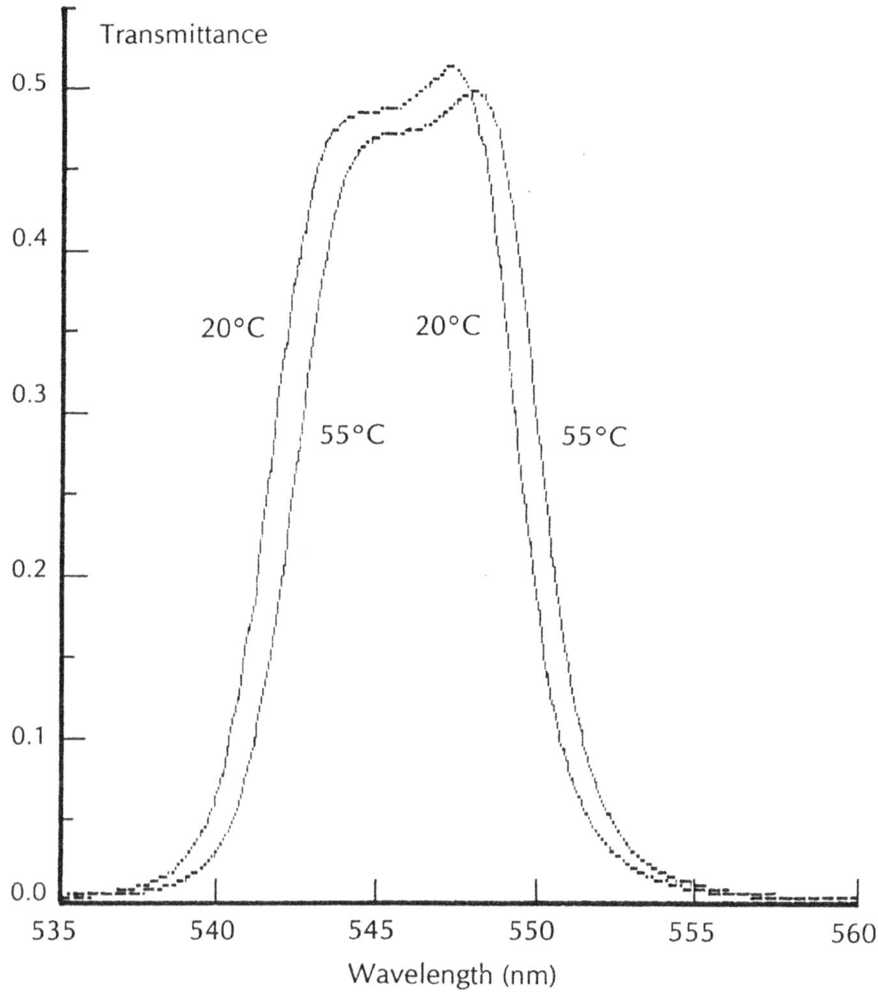

Fig. 1. Bandpass transmittance spectra of the 546 nm interference filter at two temperatures representing the 'cold-start' and the end of the self-warming phase of the Moreland anomaloscope. Recordings were made on a Varian DMS100 spectrophotometer using a probe bandwidth of 1 nm at a speed of 100 nm/min.

Our normal practice, when striving after calibrational or observational accuracy, is to switch the anomaloscope on some 30 min or more before making measurements. In such instances the variation in temperature and, consequently in anomaly quotient, would be even smaller.

Table 1. Interference filter characteristics at 20°C with changes, shown in parentheses, for a temperature increase of 35°C.

Stimulus	Green (G)	Yellow (Y)	Red (R)
Nominal wavelength nm	546	589	671
Centre wavelength nm	545.6 (+0.7)	588.4 (+1.0)	669.9 (+0.6)
Half-height bandwidth nm	7.9 (–0.1)	9.8 (–0.3)	11.9 (–0.0)
Transmittance per cent	51.3 (–1.6)	63.9 (–2.4)	58.1 (–1.7)

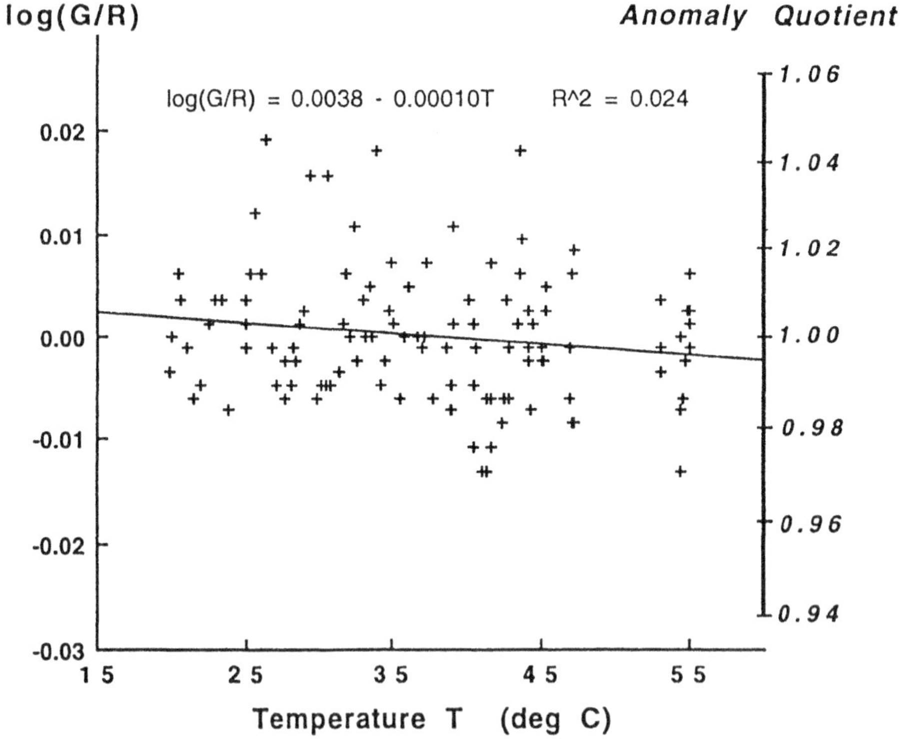

Fig. 2. Variation of Rayleigh matches with temperature for the Moreland anomaloscope. Crosses denote single colour matches made by a colour normal subject during the self-warming phase of the anomaloscope. The origin of the left ordinate is arbitrary, corresponding to an anomaly quotient of 1.

Conclusions

Finite temperature-related changes are observed in the bandpass spectra of interference filters used in the Moreland anomaloscope after it is switched on. These physical changes have no significant effect on colour matching with the Rayleigh equation.

References

Jordan, G. and Mollon, J.D. (1993). The Nagel anomaloscope and seasonal variation of colour vision. Nature 363: 546–549.

Köllner, H. (1915). Über Übergänge zwischen normalen Farbensinn und angeborener Rotgrünblindheit und über die Möglichkeit ihrer quantitativen Bestimmung. Arch. Augenheilkd. 78: 302–335.

Moreland, J.D. and Kerr, J. (1979). Optimization of a Rayleigh-type equation for the detection of tritanomaly. Vision Res. 19: 1369–1375.

Department of Communication and Neuroscience
Keele University
Staffordshire, ST5 5BG, UK

45. Effect of longitudinal chromatic aberration on photometric matches using a heterochromatic square-wave grating

S. PEFFERKORN, A. CHIRON and F. VIÉNOT
(Paris, France)

Abstract

We have developed a heterochromatic spatial photometry technique (HSP) based on minimizing the contrast between the odd and the even bars of a two-colour square-wave grating, which is illuminated by an adjustable monochromatic test light and a fixed white reference light, respectively. We have conducted a systematic investigation of the effect of chromatic aberration on photometric matches, driving the accommodation of the observer from under-corrected to over-corrected values by ophthalmic lenses in front of the eye. Heterochromatic spatial matches depended upon the ophthalmic lens power. The test/reference illuminance ratio varies with spatial frequency and with the ophthalmic correction. This could be due to the contrast of the retinal image of the grating, which depends upon the difference in focus between the white and coloured bars. When the correction is intermediate, between the values that optimize the monochromatic test and the white reference, it minimizes the standard deviation and seems to yield a test/reference illuminance ratio that is approximately invariant for spatial frequencies below 4 cpd.

Introduction

The everyday visual scene is usually composed of contiguous small patches. Although photopic photometry using a bipartite field has been explored for many years (e.g. Kaiser, 1971), some difficulties remain concerning brightness matches obtained with an heterochromatic isoluminant square-wave grating (Hilz and Cavonius, 1970; Mullen, 1985). These difficulties originate from the difference in focus of the differently coloured bars of the grating. In this study, we address two questions: (1) what is the influence of defocusing on heterochromatic brightness matches? (2) Can we accurately control the effect of longitudinal chromatic aberration of the eye on heterochromatic brightness matches?

We conducted a systematic investigation into the effect of the chromatic

C.R. Cavonius (ed.), Colour Vision Deficiencies XIII, pp. 403–407.

aberration (Bedford and Wyszecki, 1957) on photometric matches. The accommodation of the observer was forced from under-corrected to over-corrected values by ophthalmic lenses placed in front of the eye, while correcting for the chromatic aberration of the optical system.

Materials and methods

In our two-channel system, the test and the reference channels come together at the bars of a mirrored square-wave grating. The common exit channel includes a variable focus lens which allows the spatial frequency of the heterochromatic grating to be adjusted without changing field luminance. In situ calibration and independent controls of the reference and test lights are provided.

The observer chooses the ophthalmic lens that gives the best image of the white bars (100% luminous contrast) with no blur or colour fringes, and then repeats the procedure for the coloured bars. Five ophthalmic lenses based on these choices are then tested: the two corrections, an intermediate value between the two choices, and 0.5 diopters beyond the two choices (e.g. away from the mid-point). If, for example, the two choices were 0.75 D and 1.75 D, the series would be 0.25, 0.75, 1.25, 1.75 and 2.25 D (see Fig. 1).

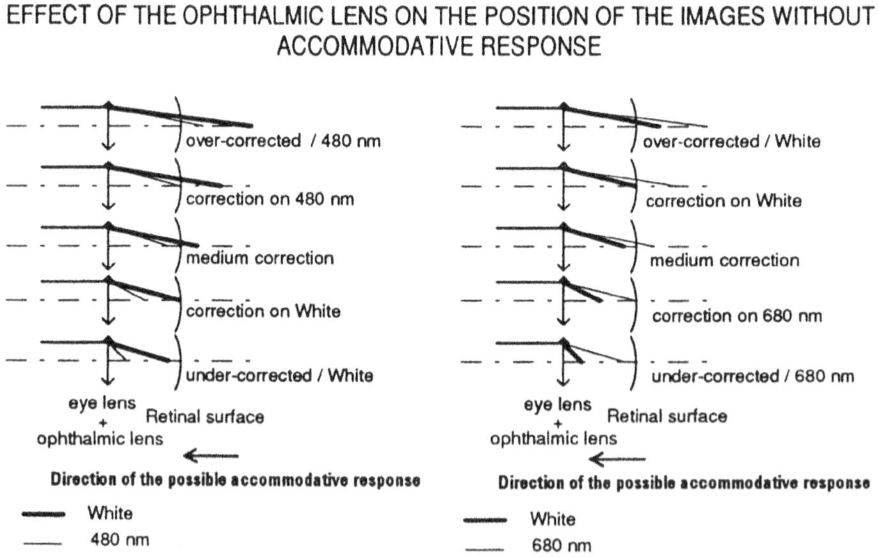

Fig. 1. Relative positions of the image of the white and monochromatic bars (480 and 680 nm), according to five suitably chosen ophthalmic lenses. In this simplified case, accommodation is ignored. Only the longitudinal chromatic aberration plays a significant role.

During the experiment, the observer adjusted the radiance of the test light to a fixed white reference light in order to minimize the apparent contrast between odd and even bars. (The subject's task differs from judging the distinctness of the border between the two central bars as it would be with the minimum distinct border criterion of Kaiser, 1971.) Not all the conditions cause the spatial pattern to disappear, but a minimum brightness contrast can always be found. Using their individual corrections, three colour-normal observers made matches on a 10° field, at 100 Td, for 11 spatial frequencies (0.1–7 c/deg) and two test wavelengths (480 and 680 nm). For every condition, 10 matches were recorded, using a computer-controlled procedure.

Results

Brightness matches depend on the ophthalmic lens that controls the focusing of the grating. For high spatial frequency photometric matches, the monochromatic test/white reference illuminance ratio increases with spatial frequency using the ophthalmic correction for the white reference. It decreases with spatial frequency when the ophthalmic correction is made to the monochromatic test. The test/reference illuminance ratio changes at middle spatial frequencies. The direct comparison brightness match also depends upon the ophthalmic correction (Fig. 2).

Discussion and conclusion

The longitudinal chromatic aberration of the eye modifies the contrast of the retinal image, thus defocusing or focusing coloured bars on the retina (Levi and Austing, 1968). The luminous flux of the defocused bars spreads on the retina, adding further to the flux of the focused bars. This results in a decrease of the retinal illuminance produced by the defocused bars and a variation of the average retinal contrast, together with a variation of the local luminance contrast at the edge of the bars. For high spatial frequencies, the contrast of the defocused bars may be much lower than the contrast of the focused bars. Moreover, the observer is not able to concentrate on an edge while making the match, but integrates the whole energy of the illuminated bars and compensates for the lack of energy in the lower contrast bars. However, for lower spatial frequencies, the contrast of the focused and defocused bars is nearly the same, but an artefact of luminance raises the local luminance contrast on the border of a bar (Bradley *et al.*, 1992). The observer is able to fixate an edge and make a match, so this luminance artefact may change the criterion adopted by the observer to assess the luminous equivalence of the coloured bars, and may explain why brightness matches depend on the quality of the retinal image.

The difference in focus between the test and the reference bars also depends upon the steady-state accommodative response of the observer (Kruger *et al.*,

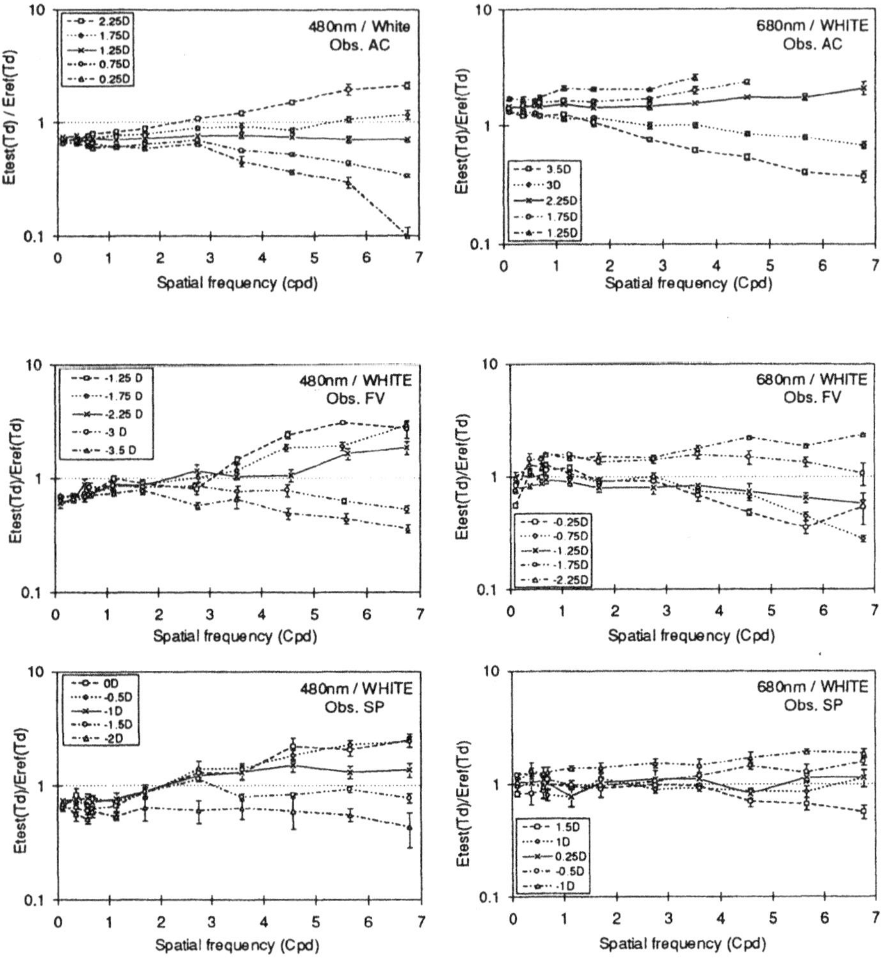

Fig. 2. Influence of the ophthalmic lens on heterochromatic brightness matches. Ratio of the luminance of the monochromatic test field (480 nm, 680 nm) to the luminance of the white reference field, versus the spatial frequency (cpd) of the grating. The error bars represent the standard errors calculated from 30 matches recorded during three sessions. For some data, the error bars are smaller that the size of symbols. The experimental data usually do not fall on 1.0 because the individual brightness judgements may differ from CIE luminance. Three observers: AC, FV, SP.

1993), which is a much more difficult parameter to control. Consider the Fourier transform of a square-wave grating as the sum of fundamental and harmonic components. With low spatial frequency square-wave gratings, the contrast of the fundamental and the contrast of the harmonics might counteract one another, so the accommodative response to an isoluminant grating could be ambiguous. The fact that observers SP and FV are younger than AC and have relatively more accommodation might explain their greater variability.

It seems possible to control the longitudinal chromatic aberration and to keep a test/reference illuminance ratio approximately steady as spatial frequency increases, provided that the spatial frequency is limited to 4 cpd and that an ophthalmic correction is chosen that is intermediate between those that are optimal for the wavelength distributions of the two sets of grating bars.

Acknowledgements

Research supported by BNM. Acknowledgements to NPL.

References

Bedford, R.E. and Wyszecki, G. (1957). Axial chromatic aberration of the human eye. J. Opt. Soc. Am. 47: 564–565.

Bradley, A., Zhang, X. and Thibos, L.N. (1992). Failures of isoluminance caused by ocular chromatic aberrations. Appl. Optics 31: 3657–3667.

Hilz, R. and Cavonius, C.R. (1970). Wavelength discrimination measured with square-wave gratings. J. Opt. Soc. Am. 60: 273–277.

Kaiser, P.K. (1971). Minimally distinct border as a preferred psychophysical criterion in visual heterochromatic photometry. J. Opt. Soc. Am. 61: 966–971.

Kruger, P.B., Mathews, S., Aggarwala, K.R. and Sanchez, N. (1993). Chromatic aberration and ocular focus: Fincham revisited. Vision Res. 33: 1397–1411.

Levi, L. and Austing, R.H. (1968). Tables of the modulation transfer function of a defocused perfect lens. Applied Optics 7: 967–974.

Mullen, K.T. (1985). The contrast sensitivity of human colour vision to red–green and blue–yellow chromatic gratings. J. Physiol. (Lond.) 359: 381–400.

CNRS UPR 257
Muséum national d'Histoire naturelle
Laboratoire de Physique appliquée
43 rue Cuvier
75005 Paris, France

46. Detecting blue-cone monochromats by electroretinography

M.S. ROY, P. GOURAS and C. MACKAY
(Newark, NJ, and New York, NY, USA)

Abstract

The aim of this study was to specify the electroretinogram (ERG) criteria that allow identification of the X-chromosome linked condition, blue-cone monochromatism. Ganzfeld spectral flashes superimposed on strong rod-saturating backgrounds (10 000–20 000 photopic trolands) were used to examine the ERG in one child with X-linked blue-cone monochromatism. Computer-averaged corneal ERGs to blue (450 nm), green (534 nm) and red (633 nm) flashes at 5.1 Hz were examined. A characteristic ERG response was obtained to blue flashes that is completely absent with green and red flashes of any intensity. This S cone response has a conspicuous b-wave with a maximum amplitude of 5–10 mV and an implicit time of 35–45 msec. Rod monochromats and patients with selective cone degeneration have no detectable ERG under these conditions. Using a rod-saturating background, the presence of an S cone ERG to blue flashes and the absence of an L/M cone ERG can, therefore, objectively and rapidly identify blue cone monochromats.

Introduction

Blue cone monochromatism is an X-linked ocular disorder in which affected males have normal blue cone and rod function but lack red and green cone function. Patients usually present with decreased visual acuity in the range 20/100–20/200, myopia, nystagmus and minimal fundus findings except for some granularity of the macular area, which may become more marked with advancing age (Fleishman and O'Donnell, 1981). Their colour vision is abnormal. Patients show numerous crossing errors along the protan–deutan axis on the Farnsworth Panel D15 test (François *et al.*, 1966) and name blues correctly on the Sloan achromatopsia test while calling other colours yellow–green (Smith *et al.*, 1980). Other colour vision tests such as AO-HRR and Berson colour test (Berson *et al.*, 1983) may be helpful but are not universally available. Colour vision tests, however, are subjective and may be difficult for young patients in whom the diagnosis of blue cone monochromatism is most often made. The simpler Ishihara colour vision test does not differentiate blue

C.R. Cavonius (ed.), Colour Vision Deficiencies XIII, pp. 409–413.

cone monochromats from rod monochromats.

In contrast to colour vision tests, the electroretinogram (ERG) provides an objective measure, and in blue cone monochromats typically demonstrates marked reduction in amplitude of the cone ERG to high frequency flicker, with normal rod responses (Berson *et al.*, 1983). We previously reported the case of a 21-year-old blue cone monochromat identified by electroretinography using spectrally selective stimuli, i.e. blue, green and yellow full-field test lights on a full-field bright background (Gouras *et al.*, 1989). We now report a second case of blue cone monochromatism identified by spectral ERG.

Case report

The patient is a 9-year-old black male who presented with bilateral decreased vision and nystagmus since an early age. At age 3 he was given prescription glasses (OD −0.25 and OS +0.50) which did not significantly improve vision. At the time of presentation best corrected vision was 20/100 in each eye for distance, and J5 and J3 for near and for right and left eyes, respectively. The anterior segment was normal; the lens was clear. The fundus showed normal optic nerve colour and normal retinal vessels. The foveal reflex was absent. There was no abnormality of either the macula or fundus pigmentation and no intraretinal pigment spicule deposition. The patient was unable to identify any colour plate on the Ishihara test and could not perform the Farnsworth D15 test.

The ERG was recorded with a Nicolet Spirit (Madison, WI) signal averager coupled to a Ganzfeld stimulator providing a relatively strong adapting field (20 000 td) at maximum brightness. Superimposed on this steady white-light field were flashes obtained from a Grass strobe (P52, Grass instruments, Quincy, MA).

The hue of the flash could be modified by inserting filters into the strobe housing. The following Kodak Wratten filters (Eastman Kodak, Rochester, NY) were used: 98 (450 nm), 48 (471 nm), 75 (488 nm), 61 (534 nm), 21 (593 nm) and 29 (633 nm). The number in parentheses indicates the dominant wavelength for maximum transmission of each filter at the approximate colour temperature of the strobe.

The flash rate was 5.1 Hz for cone ERG recording and 1.1 Hz for rod ERG. For the blue cone ERG, 1000 responses were averaged, allowing detection of signals as low as 0.1 μV. The a- and b-wave amplitude and implicit time were measured in a conventional manner by determining the amplitude of the negative a-wave from baseline to trough, and of the subsequent b-wave from the trough of the a-wave to the peak of the b-wave. The normal blue cone ERG appears as a separate b-wave riding on an earlier b-wave of the long (L) and middle (M) cone ERG. Details regarding ERG recording have been published previously (Gouras and MacKay, 1990).

Figure 1 shows ERG responses obtained with white, blue, green, red and green flashes. Compared with normal individuals, the cone ERG to white

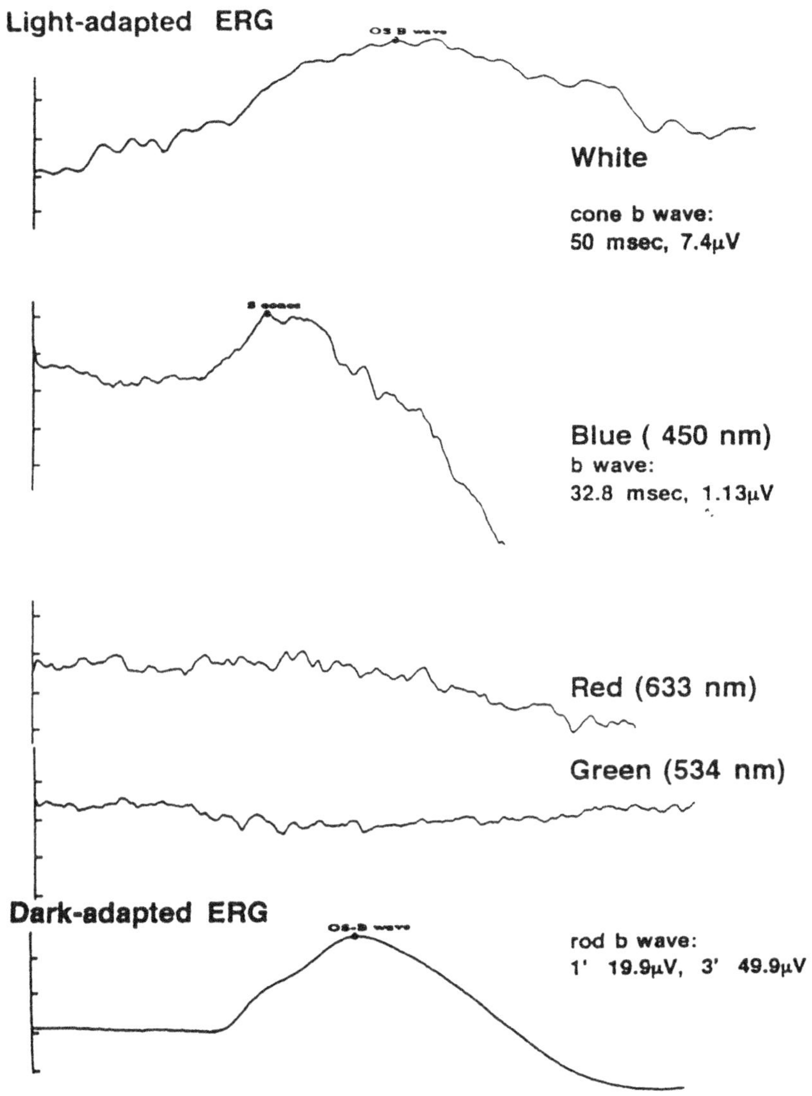

Fig. 1. Light-adapted cone responses to white, blue, red and green flashes and dark-adapted rod responses to blue flash in the left eye of a 9-year-old child. Normal ERG blue cone and rod responses and absent L and M cone responses identify blue cone monochromatism.

flashes was reduced in amplitude to 7.4 μV and delayed to about 50 msec. There were no light-adapted cone ERG responses to either red or green flashes, while a 1.13 μV ERG blue cone response was obtained. Rod responses were normal, being 19.9 μV after 1 min of dark adaptation and 49.9 μV at 3 min. These values are within normal limits according to the standard protocol followed in our laboratory.

Discussion

The electroretinographic response of the human short-wavelength (S) cone system can be distinguished from the longer wavelength (L and M) cone system by using Ganzfeld short-wavelength stimulation at high levels of retinal adaptation (Gouras *et al.*, 1990). The short wavelength responses have a larger, later positive b-wave that is absent with long-wavelength flashes. This later b-wave has its largest amplitude at 450 nm and is undetectable at wavelengths longer than 500 nm, while the earlier b-wave is elicited by all spectral lights but has its largest amplitude at 540 nm, a pattern indicative of an L or M cone response. It is possible to examine both of these cone systems simultaneously in the same ERG as long as the retina is sufficiently light adapted. This light adaptation saturates the rod response and speeds the response of the L and M cones more than that of the S cone ERG so that the two distinct responses are separated in time. The implicit time of the S cone ERG b-wave is 38.2 msec (SD 2.5). Its amplitude varies with age (Gouras *et al.*, 1993). Other investigators have suggested using the silent-substitution method for isolating the blue cone ERG at suprathreshold levels (Sawusch *et al.*, 1987).

In our patient, the blue cone ERG recording was obtained with 20 000 td background illumination and computer averaging of 1000 responses. The implicit time of the blue cone b-wave response was somewhat shorter than previously reported, at 32.8 msec, and its amplitude was within normal limits for the patient's age, at 1.1 μV (Gouras *et al.*, 1993). No earlier (L and M) b-wave was observed. With either red or green flashes the ERG tracing was flat, while the dark-adapted rod response to blue flashes was normal. These results are comparable to those obtained in our previous blue cone monochromat patient (Gouras *et al.*, 1989). The cone ERG response to white flashes was low in amplitude (7.4 μV) and delayed in implicit time (50 msec), supporting the fact that the white flash used might have been a somewhat weak stimulus for the blue cones. Because of the age of the patient, it was difficult to examine the cone ERG response in response to increasing intensities of the white and blue flashes. However, it is worth noting that despite the young age of the patient and the presence of nystagmus, ERG recordings were easily obtained, in contrast to the observation of other investigators that separation of the blue cone ERG in young patients with photophobia and nystagmus is difficult to obtain.

In summary, we present the case of a 9-year-old male in whom bilateral decreased vision normal blue cone and rod ERG with absent red and green cone

ERG led to the diagnosis of blue cone monochromatism. We conclude that spectral ERG is an objective, short, and reliable method to identify blue cone monchromatism, particularly in young children.

References

Berson, E.L., Sandberg, M.A., Rosner, B. and Sullivan, P.L. (1983). Color plates to help identify patients with blue cone monochromatism. Am. J. Ophthalmol. 95: 741–747.

Fleishman, J.A. and O'Donnell, F.E. (1981). Congenital X-linked incomplete achromatopsia. Evidence for slow progression, carrier fundus findings, and possible genetic linkage with glucose-6-phosphate dehydrogenase locus. Arch. Ophthalmol. 99: 468–472.

François, J., Verriest, G., Matton-Van Leuven, M.Th., De Rouck, A. and Manavian, D. (1966). Atypical achromatopsia of sex-linked recessive inheritance. Am. J. Ophthalmol. 61: 1101–1108.

Gouras, P., Mackay, C.J. and Lewis A.L. (1989). The blue cone electroretinogram isolated in a sex-linked achromat. In: Drum, B. and Verriest, G. (eds.), Colour Vision Deficiencies IX, 89–93. Kluwer, Dordrecht.

Gouras, P. and MacKay, C.J. (1990). Electroretinographic responses of the short-wavelength-sensitive cones. Invest. Ophthalmol. Vis. Sci. 31: 1203–1209.

Gouras, P., MacKay, C.J. and Yamamoto, S. T. (1993). The human S-cone electroretinogram and its variation among subjects with and without L and M cone function. Invest. Ophthalmol. Vis. Sci. 34: 2437–2442.

Sawusch, M., Pokorny, J. and Smith, V. (1987). Clinical electroretinography for short wavelength sensitive cones. Invest. Ophthalmol. Vis. Sci. 28: 966–974.

Smith, V.C., Pokorny, J. (1980). Cone dysfunction syndromes defined by colour vision. In: Verriest, G. (ed.), Colour vision deficiencies, vol. 5: 69–82. Bristol, Adam Hilger.

M. Roy
Department of Ophthalmology, UMD
90 Bergen St., 6th Floor
Newark, NJ 07103-2499, USA

P. Gouras, C.J. MacKay
Department of Ophthalmology, Columbia University
635 West 135th St.
New York, NY 10032, USA

47. Fast determination of the spectral modulation sensitivity function: a comparison between trichromats and deuteranopes

V. BONNARDEL, D.L. RUDERMAN and H.B. BARLOW
(Cambridge, UK)

Abstract

Using drifting comb-filtered spectra, rapid determinations of the spectral modulation sensitivity function (SMSF) were performed in two normal trichromats and two deuteranopes. Detection thresholds were measured for comb-frequencies (f) varying from 0.5 to 3.5 cycles/300 nm with a constant temporal frequency (1 Hz). The dichromatic SMSF shows a corner frequency at near 1.2 cycles/300 nm, whereas the corner frequency of the trichromatic SMSF is near 1.7 cycles/300 nm. The impairment of sensitivity of deuteranopes at high frequencies is explained by the loss of the red–green opponent-colour channel with its triphasic characteristic on the wavelength axis.

Introduction

In 1982, Barlow proposed that the modulation transfer function (MTF) of the chromatic system could be estimated by measuring threshold sensitivity to comb-filtered spectra. Such stimuli are produced by a sinusoidal modulation of the radiant energy over the visible spectrum (400–700 nm) and are defined by their frequency (cycles/300 nm), phase (degrees) and amplitude (Fig. 1).

In a previous study, discrimination thresholds for comb-filtered spectra were measured for one normal trichromat for various frequencies over the full range of phases (Bonnardel *et al.*, 1996). As was suggested by early studies using different kinds of periodic spectral modulation (Barlow *et al.*, 1983; Bonnardel and Varela, 1991), or more recently by computations from classical discrimination data (Romero *et al.*, 1995a), the chromatic system shows a large phase dependence. The spectral modulation sensitivity function (SMSF) was defined as sensitivity measured for the optimal phase (i.e. the phase which reveals the highest sensitivity). In the present study, drifting comb-filtered spectra provide faster determination of the SMSF, since for each frequency only one measurement is obtained, which necessarily corresponds to the optimal phase. Measurements of the SMSF in two normal trichromats and two deuteranopes are presented.

C.R. Cavonius (ed.), Colour Vision Deficiencies XIII, pp. 415–424.

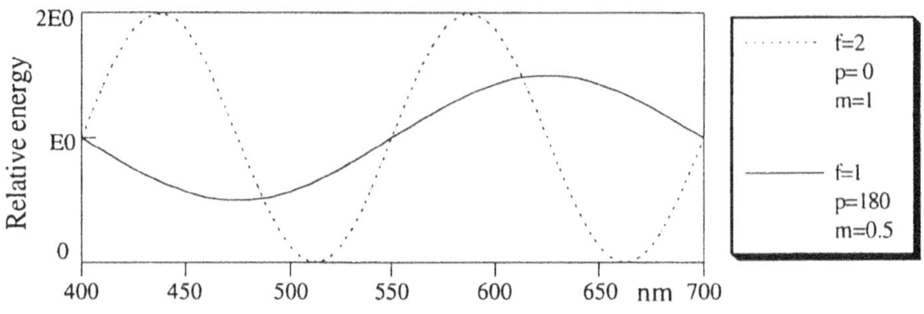

Fig. 1. Comb-filtered spectra are illuminants whose spectral power distributions are sinusoidal modulations over wavelength, expressed by:

$$E(\lambda) = E_0(\lambda)[1 + m\sin(fp(\lambda) + p_0)] \qquad 400 < \lambda < 700 \qquad [1]$$

where $E_0(\lambda)$ is the DC level of energy; m is the amplitude of the modulation varying from 0 to 1; and f is the frequency expressed in cycles/300 nm, hereinafter denoted c. p_0 is the initial phase (degree). $p(\lambda) = (1.2\lambda - 480)$ transforms the 300 nm interval into 360°.

Materials and methods

The stimulator

Comb-filtered spectra were produced by a stimulator based on an interference wedge and a liquid crystal display (LCD) (Fig. 2). The LCD screen is of double super-twisted nematic type with a contrast ratio of 15:1. The screen is a passive-matrix addressing system of 640 × 480 pixels on which sinusoids are drawn in a rectangle whose size fits that of the continuous spectrum formed by the interference wedge. The software application, developed for Macintosh computers, allows us to display sinusoidal profiles of any frequency and phase, the level of modulation being set by the number of pixels corresponding to the width of the interference wedge at each wavelength.

It is worth noting that temporal modulation of the three primaries of a CRT display would provide relative absorption rates in the three types of cones equivalent to most of the drifting comb-filtered spectra by an appropriate setting of the relative intensity of the three phosphors. However, the advantage of comb-filtered spectra is that a direct measurement of the visual system sensitivity is obtained without a prior knowledge of the fundamentals or even of their number.

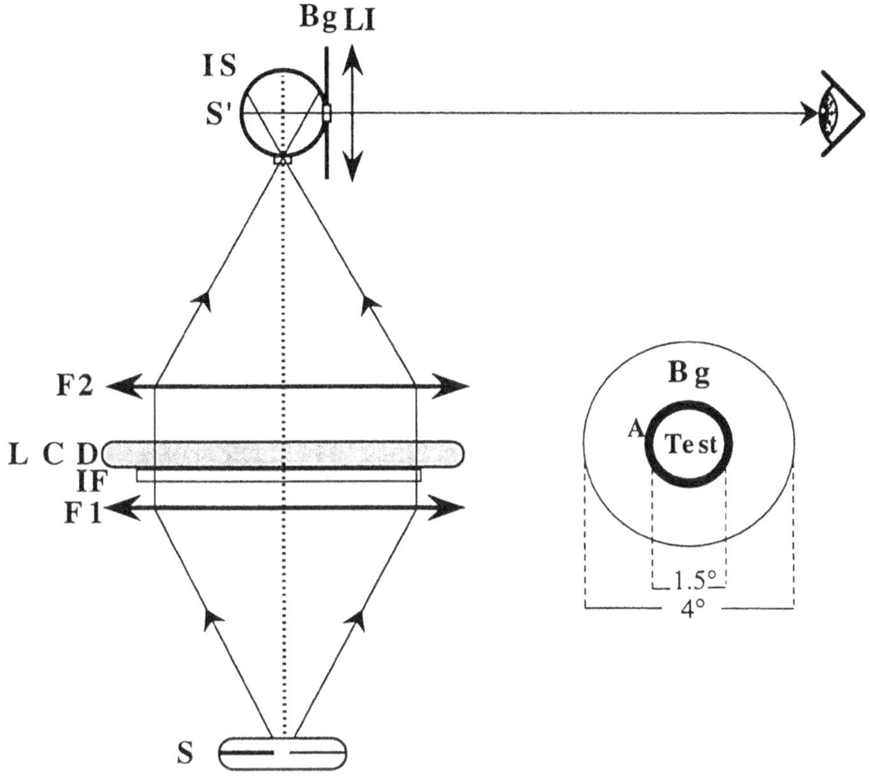

Fig. 2. The light from the source (**S**, short-arc xenon bulb equipped with UV and IR filters) located at the focal point of a Fresnel lens (**F1**) is focused by a second lens (**F2**) to form a real image **S'** of **S** at the aperture of the integrating sphere (**IS**). The liquid crystal display (**LCD**) is mounted in the collimated beam directly after the interference wedge (**IF**) which gives a continuous linear spectrum from 400 to 700 nm. The output of the integrating sphere (**IS**), viewed though a lens (**L1**), produces a homogeneous field 1.5° in diameter. A narrow dark annulus (**A**) delimits the test spot from a 4° background (**Bg**) of the same luminance (5 cd/m^2). In the absence of modulation ($E(\lambda) = E_0(\lambda)$), the chromaticity coordinates of the test spot are $x_0 = 0.275$ $y_0 = 0.358$. The chromaticity coordinates of the background are $x_0 = 0.355$, $y_0 = 0.427$.

Psychophysical procedure

The observer sat in a darkened enclosure and was adapted to the luminance of the test spot and uniform surround. His head was supported on a chin rest while he viewed monocularly with his dominant eye a 1.5° test spot of the required spectral composition on a 4° luminous background adjusted to match the unmodulated test spot in luminance and to approximate it in colour. In a

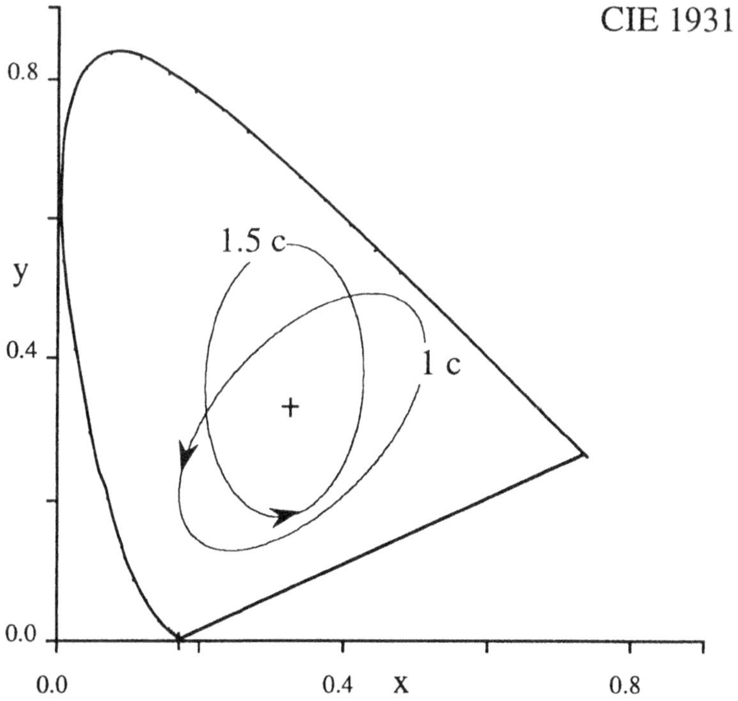

Fig. 3. The chromaticity coordinates of fully-modulated comb-filtered spectra ($m = 1$) were computed for two frequencies (1 and 1.5 cycles) for a continuous variation of phase (arrows indicate the starting phase p_0 and the direction of variation). When plotted in the CIE (1931) diagram they describe an ellipse. Variation in depth of modulation corresponds to movement along a line that joins the unmodulated spectrum (cross) to an elliptical contour in a given direction, i.e. phase.

detection task, the observer indicated whether or not a change occurred during the 2 sec test presentation of a drifting comb-filtered spectrum. A trial was initiated after the observer had given his response and began with a 1 sec presentation of a static comb-filtered spectrum. The test interval during which the comb-filtered spectrum scrolled was signalled by beeps. Between test presentations the unmodulated spectrum was displayed. Twelve comb frequencies from 0.5 to 3.5 cycles with a constant temporal frequency (1 Hz) were tested in three runs of four independent and interleaved staircase procedures. The four last measures of six reversals were used to compute the thresholds. The experimental data points correspond to the mean of three experimental sessions of about 45 min each.

Drifting comb-filtered spectra produce luminance modulations whose amplitude is greater at low comb frequencies. The size of the spot, the duration of the test presentation and the temporal frequency were, therefore, chosen to favour detection on the basis of chromatic cues, and a dark annulus around the test spot was added to mask the achromatic component (Mollon, 1982; Kulikowski and Walsh, 1991).

Observers

A vision scientist (ST) and one of the authors (VB) participated in the experiment as normal trichromat observers. Their colour vision was previously tested with the usual clinical tests. The dichromatic observers are two male adults of the same age as the trichromat observers and were classified as deuteranopes according to Ishihara, Farnworth-Munsell 100-hue, and the Nagel anomaloscope.

Results and discussion

Calibrations

The spectral power distribution of the test field was measured at the position of the observer's eye with a spectroradiometer (model PR-650, Photo Research). The parameters of the sinusoidal spectral modulation represented by the fraction $E(\lambda)/E_0(\lambda)$ (see equation [1] in legend Fig. 1) were estimated by mean of a fitting procedure (Kaleidagraph-Macintosh software). For 72 best-fitting sinusoids the mean correlation was 0.98. The linear relationship between the amplitude of the best-fitted sinusoid and the number of pixels in the width of the electronic mask was determined from these 72 fits and served to compute the amplitude for the rest of the threshold measures. The reciprocal of the amplitude was then plotted as a function of frequency.

Spectral modulation sensitivity functions

The mean SMSF of the two trichromat observers and the mean SMSF of the two dichromat observers are shown in Figure 4. The trichromatic SMSF shows a corner frequency near 1.7 cycles whereas the dichromatic SMSF starts to decrease at frequencies above 1.2 cycles. These results are consistent with previous measurements performed with a Michelson interferometer by Barlow *et al.* (1983) and Gemperlein *et al.* (1990), where the maximum sensitivity of a deuteranope observer is shifted toward low frequencies.

In addition to the relative loss of sensitivity at high frequencies, the deuteranopes show a general impairment of sensitivity compared to normal observers: deuteranopes needed illuminants of higher purity in order to detect a change in the test. Without drawing any conclusion from four observers, it is interesting to note that in their study Regan *et al.* (1994) reported that dichromats, as a population, had higher thresholds than normal trichromats in the tritan direction. The test consisted of computer-controlled displays of

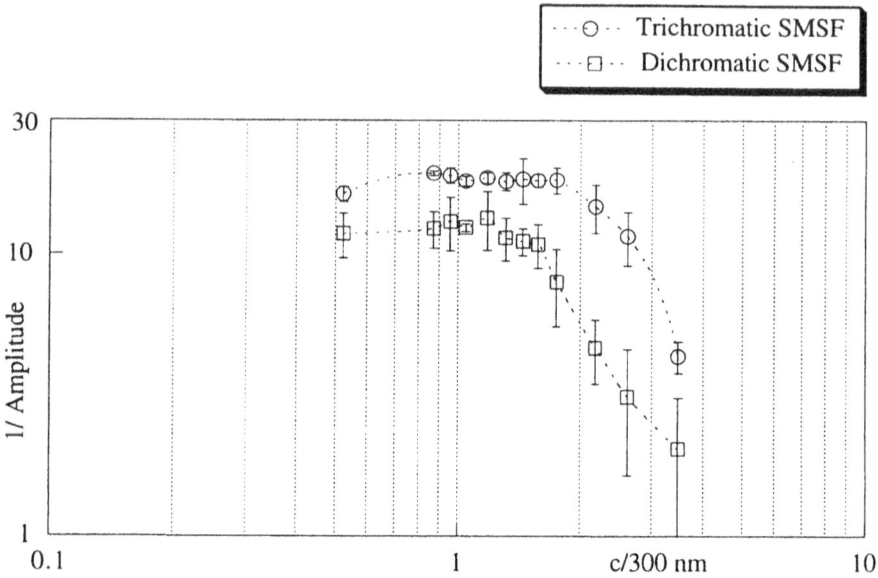

Fig. 4. **Mean SMSF of the two trichromat observers (circles) compared with the mean SMSF of the two dichromat observers (squares). For each curve, the y-bars represent the difference between the threshold measurements of the two observers.**

pseudoisochromatic plate patterns whose chromaticities vary discretely along different directions in the colour space. When the sensitivity was probed from the achromatic point, the measurements, as in our test, corresponded to saturation threshold.

Finally, it is not clear whether the high frequency cut-off, here estimated as being near 4 cycles for the two curves, occurs at lower frequencies in the case of dichromat observers.

Theory

Benzschawel *et al.* (1986) and more recently Romero *et al.* (1995b) used current opponent-colours models to predict the sensitivity of the whole visual system to comb-filtered spectra. In these models, the spectral opponency that is achieved by neural interaction between the three classes of receptors is expressed as the result of a linear combination of a set of cone fundamentals. In the Ingling and Tsou model (1977), Smith and Pokorny fundamentals (1975) are used to derive one achromatic and two opponent-colour mechanisms (Fig. 5).

The achromatic channel (A) results from the positive combination of L and M cones. The r–g channel which receives its inputs from a positive combination of L and S cones opposed to M cones exhibits a triphasic spectral profile with

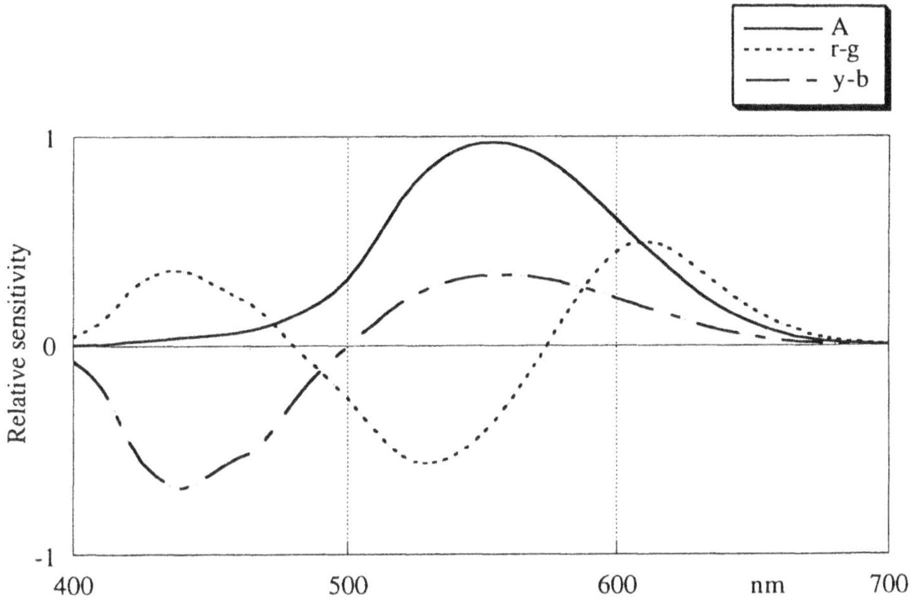

Fig. 5. Spectral sensitivity curves, A(λ) (solid line), r–g(λ) (dotted line) and y–b(λ) (dashed line), of the three mechanisms postulated in the Ingling and Tsou model (1977).

two zero-crossings (480 and 570 nm). The y–b channel corresponds to a positive combination of L and M cones opposed to S cones and shows a diphasic profile with a zero-crossing at 500 nm.

The individual response for a given frequency and a given phase, $S_A(f, p)$, $S_{r\text{-}g}(f, p)$ and $S_{y\text{-}b}(f, p)$, are then computed for a maximal modulation (m = 1) with frequencies varying from 0.1 to 5.9 cycles in steps of 0.1 cycles and p varying from 0 to 330° in steps of 30°.

Following Benzschawel *et al.* (1986):

$$S_A(f, p) = [(A_{f,\,p})^2 - (A_0)^2]^{1/2} \qquad [2]$$

with $A_{f,\,p} = \Sigma E(\lambda) \times A(\lambda)$ and $A_0 = \Sigma A(\lambda)$ for $400 < \lambda < 700$ and similarly for $S_{r\text{-}g}$ and $S_{y\text{-}b}$.

The maximal response computed over the 12 phases (i.e. response to the optimal phase, p_{opt}), is then plotted as a function of frequency (Fig. 6). The response of the achromatic channel is maximal for the lowest frequency (0.1 cycles) and decreases by 50% at 1.3 cycles, where it is equal to that of the two opponent-colour channels. The maximal response function of the two opponent-colour channels shows bandpass profiles with similar bandwidth and a centre frequency of 0.9 and 1.6 cycles, respectively, for the y–b and r–g channels.

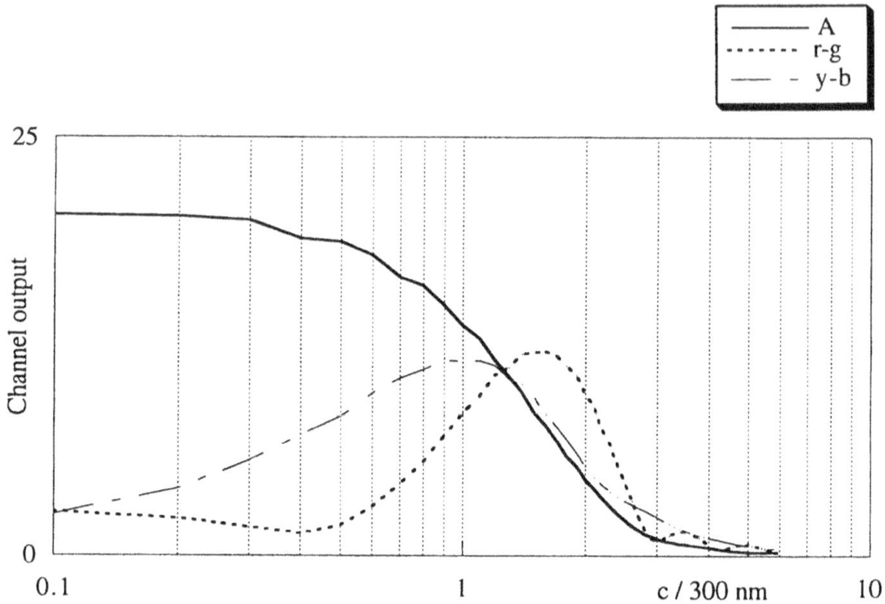

Fig. 6. The response of the three channels (A(λ), solid line; r–g(λ), dotted line; y–b(λ), dashed line) were computed for comb-filtered spectra with frequency varying from 0.1 to 5.9 cycles. Maximal response functions plotted on this graph correspond to the response of each channel to the optimal phase; S_A (f, p_{opt}), $S_{r\text{-}g}$(f, p_{opt}), $S_{y\text{-}b}$(f, p_{opt}).

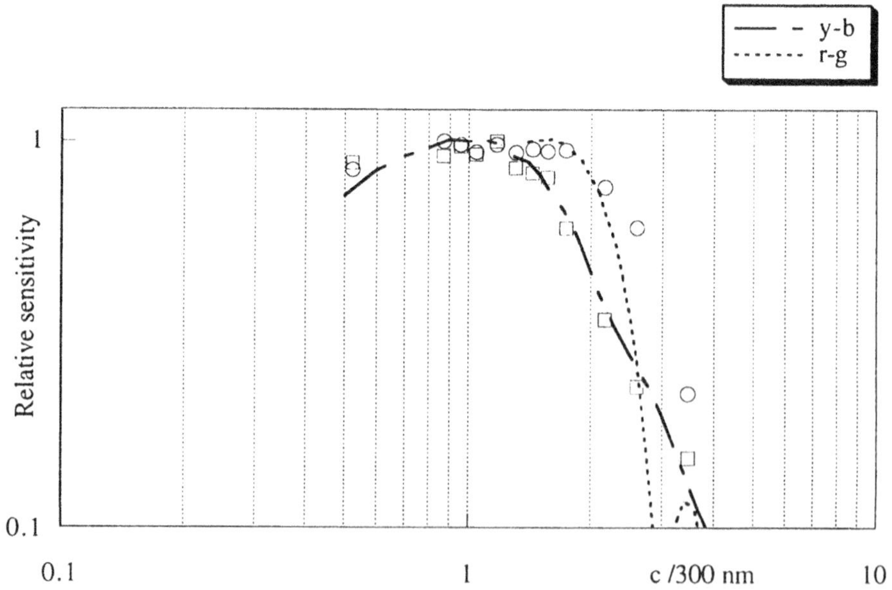

Fig. 7. The data points of the trichromatic (circles) and of the dichromatic (squares) SMSF are plotted with the maximal response function of r–g and y–b mechanisms in normalized ordinates.

Although there are noticeable differences in their response decay, after 5 cycles the response of the three channels is residual as shown in the studies of Benzchawel *et al.* (1986) and Romero *et al.* (1995b).

For comparison, the experimental data points of Figure 4 are plotted on the same graph as the maximal response functions of r–g and y–b channels on a linear normalized Y-axis (Fig. 7). Except for the measurement obtained at the lowest frequency (0.5 cycles), where the contribution of the luminance component is the strongest, the maximal response function of the y–b channel can account for both the span of the maximum of sensitivity and the descending slope of the deuteranopic SMSF. The response of the r–g channel accounts for the extension of the maximum of sensitivity to 1.7 cycles observed in the trichromat SMSF, but poorly predicts its descending slope, which is shallower. A full account of the trichromatic SMSF will need a rule of combination of the three channels.

Conclusion

This first determination of the SMSF in two deuteranope observers suggest that the colour system of red–green dichromats based on one diphasic opponent-colour channel suffers a loss of modulation sensitivity for high frequencies compared with a trichromatic system.

Acknowledgements

We thank Dr. J.D. Mollon for lending the equipment and for comments, Mr B.C. Regan for providing the tests of the dichromat observers and Dr. S. Tripathy for acting as an observer. This work was carried out while VB was supported by a Fellowship from the European Neuroscience Program and while DLR was supported by a Fellowship of NSF NATO grant RCD-9353730.

References

Barlow, H.B. (1982). What causes trichromacy? A theoretical analysis using comb-filtered spectra. Vision Res. 22: 635-643.

Barlow, H.B., Gemperlein, R., Paul, R. and Steiner, A. (1983). Human contrast sensitivity for comb-filtered spectra. J. Physiol. 340: 50 P.

Benzschawel, T., Brill, M.H. and Cohn, T.E. (1986). Analysis of human color mechanisms using sinusoidal spectral power distributions. J. Opt. Soc. Am. A9: 1713–1725.

Bonnardel, V. and Varela, F.J. (1991). A frequency view of color: measuring the human sensitivity to square-wave spectral power distributions. Proc. Roy. Soc. Lond. B. 245: 165–171.

Bonnardel, V., Bellemare, H. and Mollon, J.D. (1996). Measurements of human sensitivity to comb-filtered spectra. Vision Res. 36: 2713–2720.

Buchsbaum, G. and Gottschalk, A. (1984). Chromaticity coordinates of frequency-limited functions. J. Opt. Soc. Am. A1: 885–887.

Gemperlein R., Paul, R. and Steiner, A. (1990). Fourier interferometric stimulation (FIS): the method and its applications. In: Blakemore, C. (ed.), Vision: Coding and Efficiency: 42–150. Cambridge University Press, Cambridge.

Ingling, C.R. and Tsou, B.H.P. (1977). Orthogonal combination of the three visual channels. Vision Res. 17: 1075–1082.

Kulikowski, J.J. and Walsh, V. (1991). On the limits of colour detection and discrimination. In: Kulikowski, J.J., Walsh, V. and Murray, I.J. (eds.), Limits of Vision: Vision and Visual Dysfunction Vol 5: 16: 202–220. Macmillan, London.

Mollon, J.D. (1982). Color vision. Annu. Rev. Psychol. 33: 41–85.

Regan, B.C., Reffin, J.P. and Mollon, J.D. (1994). Luminance noise and the rapid determination of discrimination ellipses in colour deficiency. Vision Res. 34: 1279–1299.

Romero, J., Nieves, J.L. and García-Beltrán, A. (1995a). Human processing of colour information in the chromatic-frequency domain. Vision Res. 35: 867–871.

Romero, J., Nieves, J.L., García-Beltrán, A. and Hita, E. (1995b). Analysis of colour-vision mechanisms in the chromatic frequency domain. J. Optics (Paris) 26: 9–15.

Smith, V.C. and Pokorny, J. (1975). Spectral sensitivity of the foveal cone pigments between 400 and 500 nm. Vision Res. 15: 161–171.

Valérie Bonnardel and Horace B. Barlow
Physiological Laboratory, University of Cambridge
Downing Street
Cambridge CB2 3EG, UK

Daniel L. Ruderman
Current address: Department of Biomedical Engineering
University of Southern California
Los Angeles, CA 90089-1451, USA

48. Colour discrimination in a monoptic and dichoptic task

L. RÜTTIGER, A. KURTENBACH and E. ZRENNER
(Tübingen, Germany)

Abstract

Wavelength discrimination thresholds, Δλ, were measured under monoptic and dichoptic conditions to obtain information about the site and the characteristics of the stages involved in the processing of colour information. In the dichoptic condition, fusion of the visual field was facilitated by presenting a coincident white background to both eyes and only one half-field of the chromatic stimulus component to each eye. In the monoptic condition, both halves were presented to the same eye. Dichoptic Δλ values were in general 3.5 times higher than monoptic values. Shortening the presentation time led to an increase of Δλ in the monoptic task (2.8 times higher) but only to a marginal increase in the dichoptic task (1.3). Processing of colour information seems to be determined at a peripheral (retinal) stage, dependent on the presentation time of the stimulus, and a central binocular stage mainly independent from the duration of the stimulus because of long-term memory functions.

Introduction

Neurones in the retina and the lateral geniculate nucleus (LGN) are anatomically and functionally monocular and thus represent only the visual information of one eye (Grüsser and Saur, 1960). It is not until the level of the primary visual cortex (V1) that the information from both eyes, including colour information, is combined or fused at a binocular stage (de Weert and Levelt, 1976). Stimuli presented to both eyes separately (dichoptic presentation) have to be consistent in their dimensions of brightness, size and surround to be fused (Thomas *et al.*, 1961; Ikeda and Sagawa, 1979).

Only a few experiments have been run to elucidate the contribution of the binocular stages of colour processing. Dichoptic mixing experiments done by Crook and Feehrer (1969) as well as by de Weert and Levelt (1976) did not address the question of whether colour-opponent signals from retinal ganglion cells are combined first for binocularity or whether they are transformed to 'higher order' chromatic values before being fused at the binocular stage. Obviously the processing of colour in central and peripheral stages is different:

C.R. Cavonius (ed.), Colour Vision Deficiencies XIII, pp. 425–430.

red and green light superimposed on the same side of the retina always leads to the sensation of yellow – never does the sensation change to green or red as it is the case in dichoptic presentation because of interocular rivalry.

Dichoptic wavelength discrimination functions ($\Delta\lambda$) should be determined mainly by the characteristics of the binocular stage. A difference in the $\Delta\lambda$ values between monoptic and dichoptic presentations should therefore be an indication of the importance of retinal representation of the stimulus and retinal colour opponent interactions. In the present experiment, monoptic and dichoptic $\Delta\lambda$ functions were measured with presentation times of 0.5 and 0.05 sec to bring out the effect of temporal integration of colour information within the complete visual system, peripheral and central stages. To obtain more information about the dynamic characteristics of the binocular stage, a dichoptic memory task was introduced in which stimuli were delayed for 3 sec or more. According to colour discrimination being only determined by the central stage, no differences in the performance on monoptic and dichoptic presentation had to be expected.

Methods

Experiments were run on a conventional 4 channel Maxwellian view system, controlled and driven by computer, laboratory interface (CED 1401) and stepping motor devices (Elbag). For details see Reitner *et al.*, 1992.

First, heterochromatic brightness matches were run for wavelengths from 420 to 680 nm in steps of 10 nm to obtain individual spectral brightness functions. These luminance values were used in the wavelength discrimination experiments ($\Delta\lambda$) to avoid luminance errors introduced by the $\Delta\lambda$ steps. The stimulus was a circular bipartite field (1°, separated by 5 min) with a dark surround. Brightness matches were made in one of the half-fields against a standard light in the other half-field (540 nm, 8.9 log quanta/s/deg^2, 625 td), with presentation times of 0.5 and 0.05 sec using an ascending and descending procedure. All stimuli were presented on an achromatic white, coincident background (10.9 log quanta/s/deg^2) with a dark surround (see also Fig. 1). $\Delta\lambda$ thresholds were determined in the direction of the longer wavelengths in an ascending and descending 2-AFC task. Subjects answered by means of a press button whether the changed wavelength had been seen in the left or the right half-field.

Monoptic thresholds were determined in a side-by-side stimulus (both half-fields of the bipartite field presented to one eye). For the dichoptic measure, the stimulus was presented as shown in Figure 1: the achromatic background was presented to both eyes to help achieve a fused visual image. Each chromatic half-field was presented to one eye only. The resulting cyclopean impression was the same as that presented to both eyes. In a dichoptic memory task, only one of the half-fields was presented to the subject, randomly in the right or the left eye, in the left or the right half-field and as the unchanged or the longer wavelength.

Thresholds were calculated from the single answers by probit analysis as well

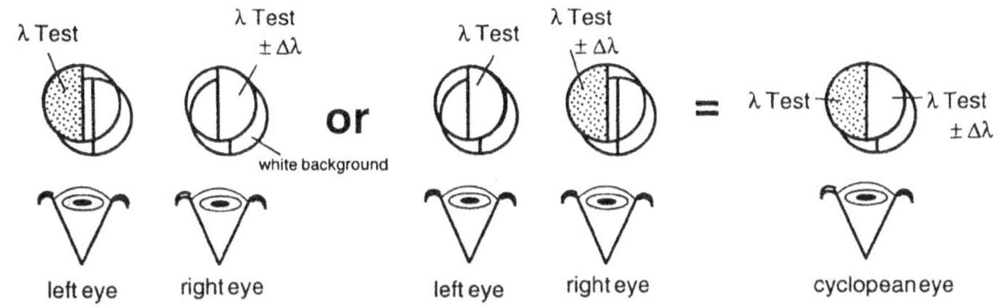

Fig. 1. Stimulus configuration for presentation in the dichoptic task. The chromatic components from both eyes (λ and λ±Δλ) are fused at a binocular stage and lead to the impression of a complete circular field within both eyes (cyclopean eye). Changed wavelengths were distributed randomly.

as from the last presented Δλ value, which was a measure for the threshold. If both values differed, measurements for this wavelength were repeated and the mean was taken as threshold. Two subjects were tested in this experiment. Both were experienced observers in psychophysical experiments and had normal colour vision as tested by the FM-100 and Ishihara plates. LR, male, one of the authors, was 31 years old, and the other, female, subject was 33 years old.

If the images from both eyes can be fused but they are very different, interocular rivalry can result in the alternating perception of the two single images. Differences in form, brightness and colour of an object can evoke rivalry and therefore deteriorate dichoptic examinations. We chose stimulus conditions to prevent interocular rivalry. Receding tests confirmed the stability of Δλ functions against purity and brightness variations of as much as 0.6 log units (see also Aulhorn *et al.*, 1969).

Results

Figure 2 shows the Δλ values for one of the subjects (LR). For both presentation times dichoptic thresholds were higher than the monoptic thresholds throughout the spectral range.

Figure 3a–b shows the Δλ values of two subjects for five wavelengths. For comparison, values are also shown as means, absolute and relative to the Δλ values for 0.5 sec and monoptic presentation (Fig. 3c–d).

Vertical broken lines are the mean of all wavelengths to clarify the extent of Δλ value elevation. Δλ values for the dichoptic conditions (filled symbols) are generally higher than for the monoptic measured 0.5 sec presentation (open circles).

With dichoptic conditions, Δλ values changed only a little with decreased presentation time (factor 1.3) whereas with monoptic conditions, Δλ values

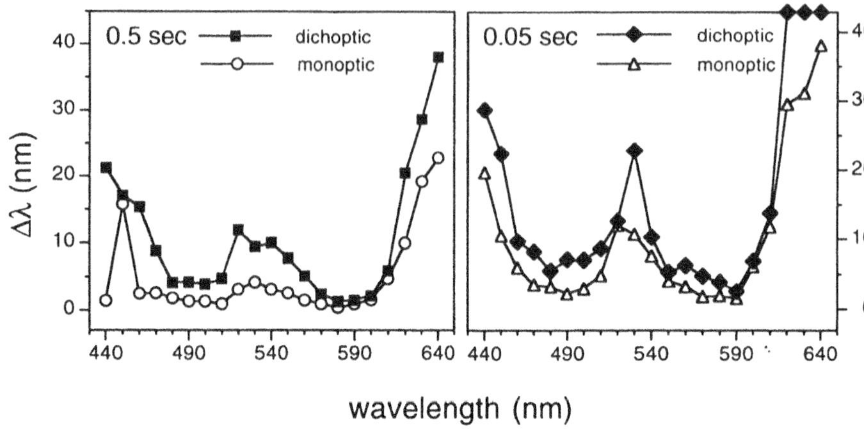

Fig. 2. **Wavelength discrimination functions for one subject (LR) for exposure duration of 0.5 sec and 0.05 sec for monoptic and dichoptic conditions. The ordinate gives the just noticable difference in nm to the longer wavelengths as determined by a 2-AFC task and probit analyses (see text).**

increased by 2.8.

In the midspectral range near 540 nm Δλ values for dichoptic and monoptic conditions differed only marginally (Fig. 3b). At the ends of the spectral range measured, differences were factor 4.5 at 620 nm and 7.15 at 440 nm.

Discussion

If all proximal colour opponent mechanisms follow similar time characteristics, as proposed by several authors (see e.g. Smith *et al.*, 1984), central visual stages should determine the time course of integration. So, differences in Δλ values between monoptic and dichoptic presentation should indicate the importance of peripheral colour coding for colour discrimination (see e.g. de Weert and Levelt, 1976).

The shape of the Δλ curves found in this study are similar for dichoptic and monoptic presentation and fit well with threshold measurements for dichoptic colour fusion limits (Ikeda and Sagawa, 1979; Ikeda and Nakashima, 1980) and show the same typical maxima and minima as Δλ curves known from the literature (Wright and Pitt, 1934). But Δλ values are slightly higher for the dichoptic task, indicating that a common representation of the comparable stimuli on a monocular stage is necessary to perform optimal colour discrimination. Because the visual pathways contain no binocular neurones up to the LGN, the exact site of this first stage of colour processing remains unclear. However, the missing short wave pessimum in the dichoptic presentation (Fig.

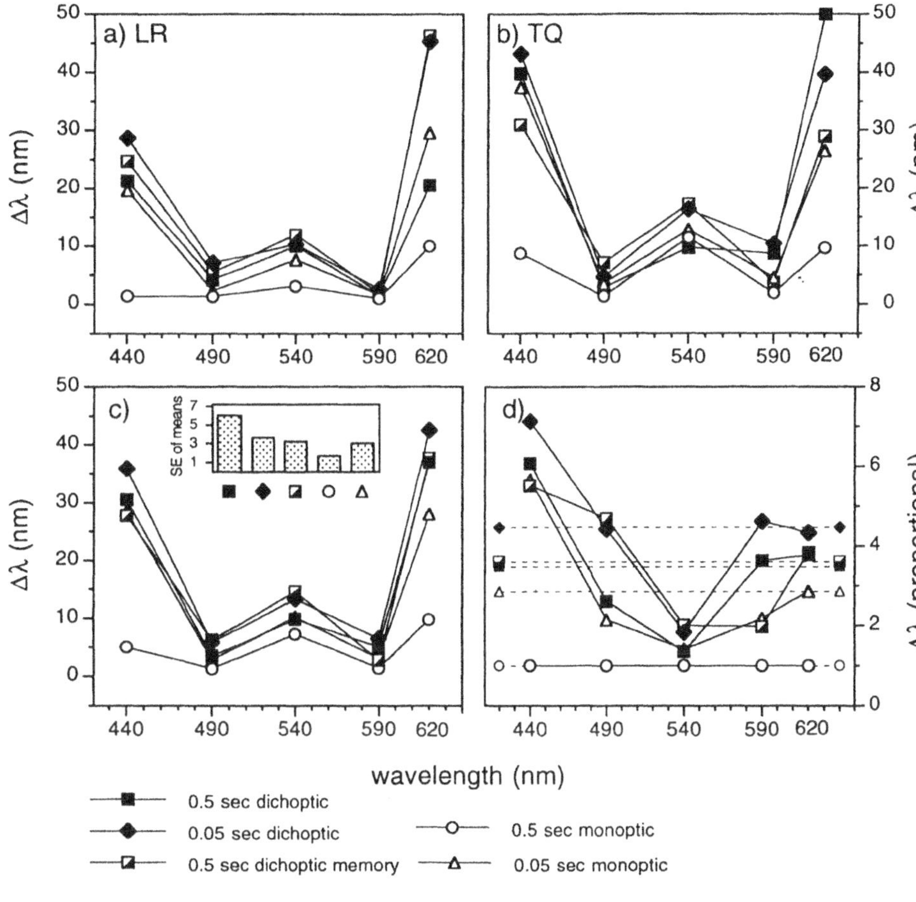

Fig. 3. Wavelength discrimination functions between 440 and 620 nm for three dichoptic and two monoptic conditions. Shown are the results for monoptic and dichoptic presentation with exposure durations of 0.5 sec and 0.05 sec. In the dichoptic memory task, only one half-field of the bipartite field was presented randomly to the left or the right eye. (a) and (b) Single subjects. (c) Means and standard error of the absolute Δλ values. (d) Means of the Δλ values normalized to the monoptic 0.5 sec condition.

2, 0.5 sec) makes it likely that this stage is located in the retina: this short wave pessimum is thought to be derived from opponent interactions of colour sensitive retinal neurones (Mollon and Estévez, 1988).

The binocular stage was less affected by the presentation time of the stimulus: shortening led to only marginally enlarged Δλ values (factor 1.3) in contrast to monoptic measurements (factor 2.8). The involvement of a memory task had no further effect (factor 1.15), although the exact colour value has to be kept in mind for more than 3 sec. This confirmes the results of Sachtler and Zaidi (1992), who found that chromatic signals are more stable in memory than achromatic.

Discrimination of colour seems to occur at different peripheral and central stages. For optimal discrimination ability coloured stimuli have to be present at a monocular stage for contrast processing and sharpening of the opponent signal. This stage is dependent on the luminance level, the size and the presentation time of the stimulus.

Processing of colour information at the cortical, binocular, stage was found to be independent of duration and intensity of the stimulus.

References

Aulhorn, E., Eisert, S. and Harms, H. (1969). Die Beeinflussung der binocularen Helligkeits-Empfindung durch die sensorische Prävalenz. Doc. Ophthalmol. 26: 230–239.

Crook, M.N. and Feehrer, C.E. (1969). Binocular color-vision asymmetries. Vision Res. 9: 717–718.

de Weert, Ch.M.M. and Levelt, W.J.M. (1976). Comparison of normal and dichoptic color mixing. Vision Res. 16: 59–70.

Grüsser, O.-J. and Saur, G. (1960). Monoculare und binoculare Lichtreizung einzelner Neurone im Geniculatum laterale der Katze. Pflügers Arch. 271: 595–612.

Ikeda, M. and Sagawa, K. (1979). Binocular color fusion limit. J. Opt. Soc. Am. 69: 316–321.

Ikeda, M. and Nakashima, Y. (1980). Wavelength difference limit for binocular color fusion. Vision Res. 20: 693–697.

Mollon, J.D. and Estévez, O. (1988). Tyndall's paradox of hue discrimination. J. Opt. Soc. Am. A 5: 151–159.

Reitner, A., Sharpe, L.T. and Zrenner, E. (1992). Wavelength discrimination as a function of field intensity, duration and size. Vision Res. 32: 179–185.

Sachtler, W.L. and Zaidi, Q. (1992). Chromatic and luminance signals in visual memory. J. Opt. Soc. Am. A 9: 877–894.

Smith, V.C., Bowen, R.W. and Pokorny, J. (1984). Threshold temporal integration of chromatic stimuli. Vision Res. 24: 653–660.

Thomas, F.H., Dimmick, F.L. and Luria, S.M. (1961). A study of binocular color mixture. Vision Res. 1: 108–120.

Wright, W.D. and Pitt, F.H.G. (1934). Hue discrimination in normal colour-vision. Proc. Phys. Soc. (Lond.) 46: 459–473.

University Eye Hospital
Department of Pathophysiology and Neuro-Ophthalmology
Schleichstr. 12
D-72076 Tübingen, Germany

49. Optimal step sizes for colour and luminance staircasing

A.J. VINGRYS and L.E. MAHON
(Melbourne, Australia)

Abstract

We investigated the best step size for measuring luminous and isoluminant coloured (red or blue/tritan) detection thresholds. Stimuli were generated on a colour TV monitor as 1°, 600 msec spots at one of four eccentric (1.6°) locations. Thresholds were expressed as contrasts (%) in cardinal colour space (RG, L or BY) for incremental probes from a white background (CIE 1931 $x = 0.30$, $y = 0.31$; $L = 33\ cd/m^2$). One observer (aged 28) was extensively tested, and trends were confirmed on three others (aged 25–33). Staircase step sizes varied from 0.2 to 4.8 deci-log units (dLog) with starting points chosen to lie within 2–3 steps of threshold. A logarithmic staircase comprising 13 reversals was performed at each step size, with threshold being taken as the geometric mean of the last 12 reversals of four randomly interleaved staircases. Both threshold and variance increased with larger step sizes. Thresholds show a V-shaped relationship with step size; minimum thresholds were found with 1.25 dLog steps. Variance showed a broken-line relationship with step size, increasing for step sizes > 1.0 dLog. No differences were found among the three coloured stimuli. We conclude that small steps (0.4–1.0 dLog) enhance precision at the expense of efficiency (time), whereas large steps (1.5–3 dLog) improve efficiency but reduce precision. In a clinical setting, 2–3 dLog step sizes should be adequate for all colours.

Introduction

Staircase procedures are efficient methods for the estimation of thresholds especially when no *a priori* knowledge of the endpoint is available (King-Smith *et al.*, 1994). A staircase typically comprises an up/down method regulated by certain rules to determine the endpoint (Cornsweet, 1962; Wetherill and Levitt, 1965). The critical parameters that affect staircase efficiency, accuracy and precision are the starting point, the step size and the number of reversals (Cornsweet, 1962). Of these, step size is one of the most important, and is the issue of this study.

If the step size is inappropriate the efficiency of the staircase can be reduced.

C.R. Cavonius (ed.), Colour Vision Deficiencies XIII, pp. 431–437.

Cornsweet observed that large steps can make the experiment insensitive whereas small steps can render it inefficient (Cornsweet, 1962). Johnson *et al.* (1992) confirmed the trade-off between efficiency and accuracy in a clinical setting. Using perimetry, they found that larger step sizes yielded shorter test durations but less accurate thresholds. They concluded that a 3 dLog step size optimized performance. Dixon (1965) suggested that if efficiency is not an issue, the greatest accuracy can be achieved with a step size that is 1.5 times the standard deviation of the endpoint. Cornsweet (1962) found, by empirical study, that a staircase has the greatest accuracy when the step approximates the standard deviation of the threshold.

This study addresses the specific question of how the accuracy and precision of a threshold are affected by the step size of a staircase procedure. We consider this issue for both the colour (R and B/T) and luminance mechanisms, since the optimal step size for measuring thresholds of these mechanisms might vary due to differences in their underlying response functions.

Materials and methods

Detection thresholds were measured using a four-alternative spatial forced-choice paradigm (4AFC). The staircase was a modified up/down procedure that required three correct or one incorrect responses for reversal (Wetherill and Levitt, 1965). For a 2AFC this procedure would converge onto the 79% correct level. However, with a 4AFC the staircase converges to the 63% level (Wetherill and Levitt, 1965).

Stimuli were generated using a VSG2/1 video board (Cambridge Research Systems) after gamma-correcting the colour TV monitor (Barco CDCT 6551) via three look-up tables. The 1°, 600 msec spot could be located at one of four eccentric (1.6°) locations. The subject's task was to nominate, by button press, the location of the spot. Auditory tones warned of impending presentations with a different tone used to provide feedback on the correctness of the response. Foveation was maintained by a small fixation target.

Thresholds are expressed as contrasts (%) in cardinal colour space (Krauskopf *et al.*, 1982) for isoluminant red and blue (tritan axis, indicated as B/T), and white increments seen on a white background (CIE 1931 $x = 0.30$, $y = 0.31$; $L = 33\ cd/m^2$). This cardinal colour space was adopted since it is quantified in terms of channel operation and should be suited for comparisons of colour and luminance processes (Krauskopf *et al.*, 1982).

One observer (aged 28) was extensively tested, with trends confirmed on three others (aged 25–33) using a limited set of five step sizes. Step size could vary from 0.2 to 4.8 dLog units with starting points chosen to lie within 2–3 steps of threshold. Four randomly interleaved logarithmic staircases comprising 13 reversals were performed at each step size: threshold was taken as the geometric mean of the last 12 reversals. Staircasing commenced with a step size twice the final value; this was halved after the first reversal and held constant for the final

12 reversals. For the single observer who was extensively tested, the mean and variance were calculated from eight staircases whereas for the other three subjects four staircases were used. The order of presentation of the step sizes was randomly varied between subjects and between sessions, although during any single session four thresholds were obtained at a fixed step size. All other staircase parameters and testing conditions were kept constant between sessions.

All subjects had 6/6 visual acuity or better in the tested eye and wore their habitual refractive correction as needed. They also had normal colour vision as tested by the Ishihara plates and Nagel anomaloscope. Observations were made monocularly using the eye with best acuity or the preferred eye when acuity was similar in both eyes. In most cases, all experiments were performed within a week but in the case of the subject extensively tested, the sessions were completed after a one-month break; no significant changes were found after this break.

Results

The results obtained from our single extensively tested observer are shown in Figure 1. Thresholds are given on the ordinate of Figure 1a in terms of the percentage contrast for the three mechanisms (RG, BY and Luminance) at detection threshold. It is apparent from this Figure that the red isoluminant thresholds (circles) were substantially lower than were the thresholds for the other colours (blue/tritan, squares: luminance, diamonds). It is also apparent that the underlying trends in the thresholds are parabolic with respect to step size and that a minimum threshold (greatest sensitivity) is found using a step of about 1–2 dLog. While this parabolic trend is evident for all three colours it appears to be most pronounced for the luminous spots (diamonds) and least for red probes (circles).

Variance is plotted in Figure 1b for the same observer as a function of step size. Here the trend is for a marked increase in variance after about 1–3 dLog depending upon the colour of the probe. Luminous spots show the earliest and greatest increase whereas red probes have the latest and least effect. In fact, Figure 1 suggests that the coloured stimuli are relatively robust to step size changes. In order to appreciate the relative changes better, Figure 2a shows thresholds normalized to the best performance of each observer. The unfilled symbols are the data of our single observer shown in Figure 1, whereas the filled symbols are averages ($\pm$SEM) for our three other subjects. The data for the single observer have been fitted by the two linear trends shown (Bailey *et al.*, 1991). Figure 2b shows similar data for the coefficient of variation: the symbols and lines are as described previously. It is evident in Figure 2 that the relative trends are similar for all colours and subjects. The trends predict that the greatest accuracy (lowest threshold) is achieved with a step size of 1.25 dLog whereas the greatest precision (minimum variance) is obtained with steps

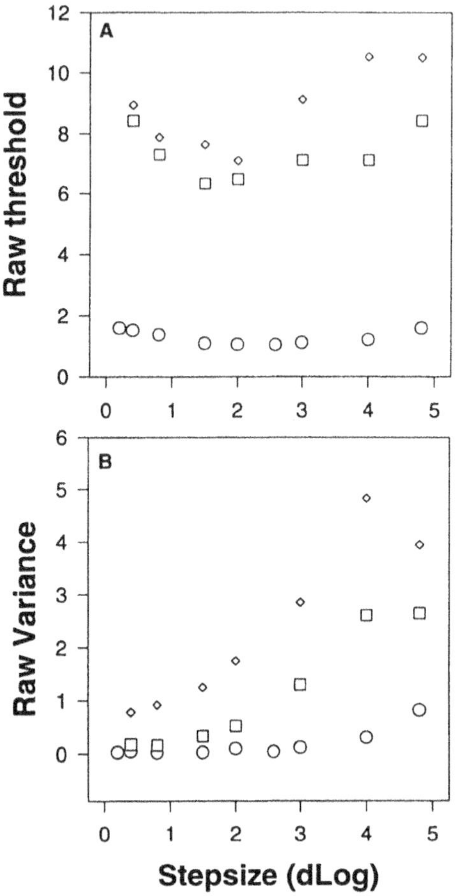

Fig. 1. The effect of staircase step size on (A) threshold (% contrast) and (B) variance for an isoluminant red (circles), isoluminant blue/tritan (squares) and luminous (diamonds) increments for one observer. Errors (SEM) are smaller than the symbols.

<1 dLog. Interestingly, decreasing step size improved precision but had a more deleterious effect on threshold (slope = –0.46) than did increasing step size (slope = 0.12).

Discussion

In choosing the optimal step size our results show that there is a trade-off between precision (low variance) and accuracy (low threshold). A small step size will yield precise outcomes but may result in higher threshold values whereas a slightly larger step size will give a lower threshold at the expense of precision

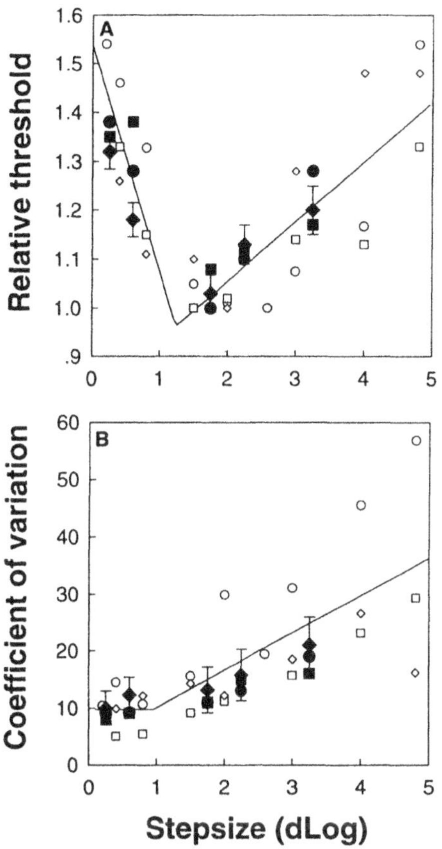

Fig. 2. The effect of step size on: (A) relative threshold and (B) the coefficient of variation. The open symbol and the lightly-filled diamonds are data from one extensively tested observer; large filled symbols show means (±SEM) for 3 other subjects. Trends have been fitted to the data of the first observer. Other details are as in Figure 1 or given in the text.

(Figs 1 and 2). Small step sizes also suffer the disadvantage of substantially increasing test duration. This effect is shown for one observer in Figure 3.

Consistent with the literature we find a V-like function for the relationship between threshold and step size (Dixon, 1965; Levitt, 1970). Levitt (1970) has shown that a step size twice the optimal value will reduce accuracy by 25% whereas a step size half the optimal value produces a doubling (100%) in threshold (ratio = 4). The slopes of our trends (0.12 vs –0.46) give an asymmetry consistent with this literature (3.8 vs 4.0). Dixon concluded that the greatest accuracy can be achieved with stepsizes up to 1.8 dLog (Dixon, 1965). Cornsweet (1962) found that a staircase has the greatest accuracy when the step approximates the standard deviation of the threshold (about 1 dLog). Our data are consistent with this literature and show that a step size of 1.25 dLog gives

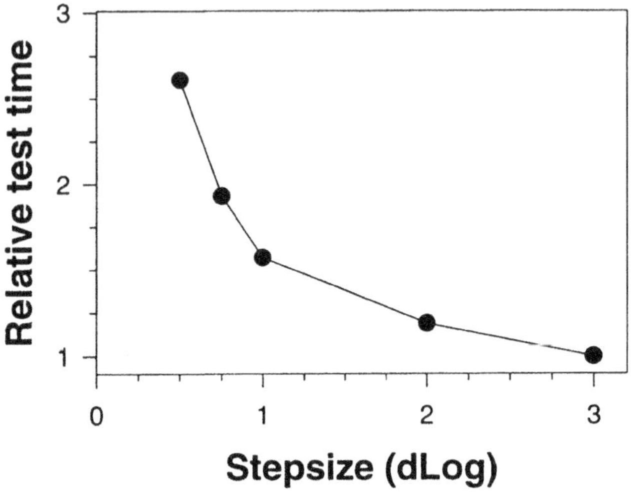

Fig. 3. The effect of step size on relative test duration.

the greatest accuracy and < 1.0 dLog the greatest precision; the precision in the measurement does not appear to have been considered by these investigators. The applicability of our findings to other spatio-temporal configurations is not readily apparent. Since our data were obtained with a fixed (1°, 600 msec spot) spatio-temporal profile, they should be interpreted in this light. However, the consistency of our data across the two colour and luminance mechanisms, when plotted on a relative scale, and the similarity of our findings with the previous literature, where various stimulus configurations were used, suggests that similar trends might be expected from different profiles.

In designing a psychophysical study, the investigator needs to consider the above mentioned trade-offs among precision, accuracy and test duration. For clinical trials, test duration and accuracy are likely to be the two most important issues so step sizes of 1–2 dLog should be used to enhance these factors. If eccentric retinal regions are to be tested (e.g. perimetry), then the step size could be slightly larger (2–3 dLog) since the fluctuation in peripheral regions is often greater and the optimal step size value is related to threshold variability (Cornsweet, 1962; Dixon, 1965). However, if experiments are being conducted to ellucidate aspects of basic visual processing where precision is important, then the use of small step sizes (0.6–1 dLog) will provide the most precise outcomes. Such a step size, however, will significantly prolong the test duration and the investigator should consider allowing appropriately timed breaks in order to minimize the effects of fatigue and boredom.

References

Bailey, I.L., Bullimore, M.A., Raasch, T.W. and Taylor H.R. (1991). Clinical grading and the effects of scaling. Invest. Ophthalmol. Vis. Sci. 32: 422-432.

Cornsweet, T.N. (1962). The staircase method in psychophysics. Am. J. Psych. 75: 485–491.

Dixon, W.J. (1965). The up-and-down method for small samples. J. Am. Stat. Assoc. 60: 967–978.

Johnson, C.A., Balwantray, C. and Shapiro, L.R. (1992). Properties of staircase procedures for estimating thresholds in automated perimetry. Invest. Ophthalmol. Vis. Sci. 33: 2966–2974.

King-Smith, P.E., Grigsby, S.S., Vingrys, A.J., Benes, S.C. and Supowit, A. (1994). Efficient and unbiased modifications of the QUEST threshold method: Theory, simulations, experimental evaluation and practical implication. Vision Res. 34: 885–912.

Krauskopf, J., Williams, D.R. and Heeley, D.W. (1982). Cardinal directions of colour space. Vision Res. 22: 1123–1131.

Levitt, H. (1970). Transformed up-down methods in psychoacoustics. J. Acoustic. Soc. Am. 49: 467–477.

Wetherill, G.B. and Levitt, H. (1965). Sequential estimation of points on a psychometric function. Br. J. Math. Stat. Psychol. 18: 1–10.

Department of Optometry and Vision Sciences
University of Melbourne
Parkville, Victoria, Australia 3052

50. Light scattered in the eye and its effect on the measurement of the colour constancy index

J.L. BARBUR, J.A. HARLOW and C. WILLIAMS
(London, UK)

Abstract

It is well established that at least some of the neural processes that mediate colour constancy effects are very rapid and practically instantaneous with the change of illuminant. This aspect of colour constancy has been labelled instantaneous or immediate in order to distinguish it from other factors, such as adaptation, which also contribute to colour constancy and have a longer time course. Instantaneous colour constancy has been studied using either simultaneous or dichoptic presentation of a test and a comparison field surrounded by a Mondrian pattern or a uniform background. In order to avoid local adaptation to each of the two illuminants, we developed a dynamic colour matching technique, based on sequential changes of illuminant which affects only the surrounding background and not the test target. This technique makes it easy to measure conveniently an index of constancy on a scale of zero to one, and is based on the change in the target chromaticity needed to keep the perceived colour of the test patch invariant under each of the two background illuminants. Using this dynamic technique, we were also able to investigate the spatial properties of immediate constancy effects using a dark annulus of variable width which separates the test target from the surrounding background. The results show that the effect of the immediate, adjacent surround is very important, although a significant C_{index} value can also be elicited with distant surrounds when the light scattered in the dioptrics of the eye from the surrounding background onto the test target is very small. The computations involved in such measurements are normally carried out on changes of chromaticity as measured on the visual display. The forward scatter of light caused by the dioptrics of the eye can change significantly the chromaticity of the test target as imaged on the retina even when the test chromaticity measured on the screen remains unchanged. In this study we have investigated how scattered light in the eye affects the index of colour constancy. The scatter function which describes the amount and the angular distribution of light scattered in the eye has been measured for each subject. C_{index} values have been computed with and without correction for the changes in test target chromaticity caused by scattered light. The results show that scattered light in the eye cannot explain immediate constancy, and that on average its effect is to reduce significantly the estimated colour constancy index.

C.R. Cavonius (ed.), Colour Vision Deficiencies XIII, pp. 439–448.

Introduction

Land's elegant demonstrations of colour constancy have shown that the perceived colour of objects can be largely independent of the spectral power distribution of the illuminant and that the neural processes that mediate such constancy effects can be very rapid, practically instantaneous with the change of illuminant (Land, 1974). This aspect of colour constancy has been labelled instantaneous (Arend and Reeves, 1986) or immediate (Foster *et al.*, 1992) to distinguish it from other factors and mechanisms (Lennie and D'Zmura, 1988), such as adaptation, which also contribute to colour constancy and have a longer time course (Fairchild and Lennie, 1992).

Immediate constancy has been studied using either simultaneous (Arend and Reeves, 1986; Arend *et al.*, 1991) or dichoptic presentation of a test and a comparison field surrounded by a Mondrian pattern or a uniform background (Valberg and Lange-Malecki, 1990). In order to avoid local adaptation to one or the other of the two illuminants, we developed a dynamic colour matching technique, based on a sequential change of illuminant that affects only the surrounding background and not the test target (Barbur *et al.*, 1989). This technique makes it easy to measure an index of constancy, C_{index}, on a scale of zero to one, and is based on the change in test-target chromaticity needed to keep the perceived colour of the test patch invariant under sequential changes of background illuminant. Using this method, we were also able to investigate the spatial properties of immediate constancy effects using a dark annulus of variable width, which separates the test target from the surrounding background. The results show that the effect of the immediate, adjacent surround is very important, although a statistically significant C_{index} value can also be produced with distant surrounds when the light scattered in the dioptrics of the eye from the surrounding background onto the test target is very small. The computations involved are carried out on the changes of target chromaticity as measured on the visual display (see Fig. 1A). The forward scatter of light caused by the dioptrics of the eye can change significantly the chromaticity of the test target as imaged on the retina even when the test chromaticity measured on the screen remains unchanged (Walraven, 1973). In this study we aim to investigate how scattered light in the eye affects the index of colour constancy. The scatter function that describes the amount and the angular distribution of light scattered in the eye has been measured for each subject. C_{index} values have been computed with and without correction for changes in test target chromaticity caused by scattered light. The results show that scattered light in the eye cannot explain immediate constancy and that on average its effect is to reduce significantly the estimated colour constancy index. More generally, we computed how the change in target retinal chromaticity caused by forward scatter in the eye varies with the difference in chromaticity between target and background. This approach is relevant to understanding how scattered light affects threshold chromatic discrimination measurements. The results suggest that in subjects with normal levels of scatter, chromatic thresholds are not affected significantly by scattered light. This is not the case when the test involves the use of saturated colours, as in measurements of colour constancy, or when subjects have unusually large amounts of forward scatter in the eye.

Materials and methods

Measurement of the C_{index}

Figure 1A shows a typical shift in test target chromaticity, which the subject needs to make in order to perceive no change in the colour of the test patch when the illuminant changes from daylight to tungsten: i.e. a change from D_{65} to A, respectively. The change of illuminant is only applied to the surrounding background field and not to the test target. The initial chromaticity of the test target corresponds to that computed under the D_{65} illuminant. The subject's task is to alter the chromaticity of the test target so that the change of background illuminant does not affect the perceived colour of the test target. Experimentally one finds that a shift in the chromaticity of the test patch towards its expected chromaticity under the new illuminant is required, if the test patch is to remain colour-invariant under the two illuminants. A colour match that requires a shift of chromaticity from T_D to T_A implies perfect constancy and results in a C_{index} of 1. Zero constancy requires no change in target chromaticity under the two background illuminants and results in an index of 0. In other words, the perceived colour of the test target is entirely determined by its wavelength radiance distribution on the retina, and changes of surround illuminant have no effect.

Measurement of scattered light

In order to be able to account for the changes in test target chromaticity caused by scattered light in the presence of daylight or tungsten backgrounds, we need to know how much the luminance of the test target changes when scattered light of chromaticity equal to that of the background field is added to it. The scatter function of the human eye can be described adequately by an empirical equation of the form $L_s = kE\theta^{-n}$ (Stiles and Crawford, 1937). Since we cannot measure directly the amount of scattered light and its angular distribution, we measure instead the luminance of an external target that cancels the retinal illuminance produced by scattered light, using a method similar to the flicker compensation method (van den Berg, 1986; van den Berg and Spekreijse, 1987). The measurement procedure makes use of extended annuli of different eccentricities, which are similar to those employed in our colour constancy measurements (Barbur *et al.*, 1993). The luminance of the test target on the screen is modulated sinusoidally in counterphase with the scatter source to cancel the modulation of retinal illuminance over the test target caused by scattered light from the annulus (see inset to Fig. 3). At threshold, the two components are approxi-

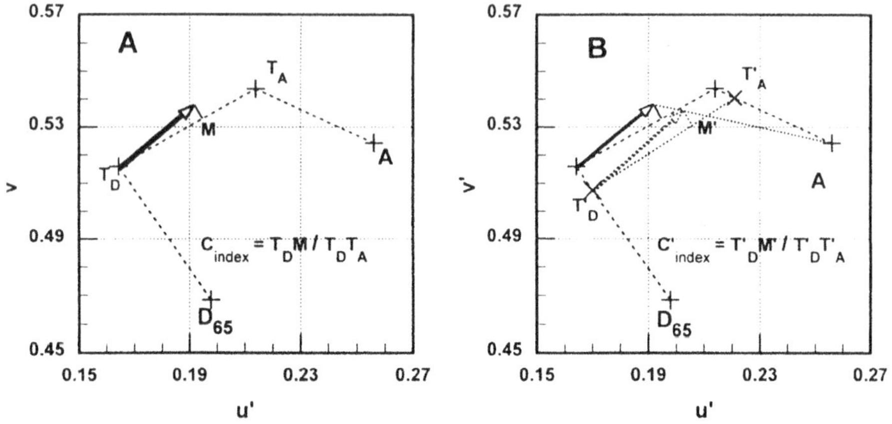

Fig. 1. Definition of C_{index} for a typical test target patch using chromaticity changes measured in the CIE (u′, v′) diagram. The stimulus was either a disc surrounded by an annulus or a square test patch surrounded by other patches (similar to the Mondrian display employed by McCann *et al.*, 1976). The visual stimuli were generated on a 20" high resolution, colour display (Hewlett Packard, Model D1187A), which was calibrated for both luminance and chromaticity co-ordinates using a Gamma Scientific Model DR-2 telespectroradiometer. The calibration data make it possible to generate any specified luminance and chromaticity (within the limits imposed by the phosphors of the display) using standard colorimetric transformations (Wyszecki and Stiles, 1982). Initially, the whole display is as if illuminated by D_{65}. The illuminant then changes from D_{65} to A every 0.8 sec, but the change of illuminant affects only the surrounding background, the test target being unaffected. (A) Location of CIE illuminants A and D_{65}, and the corresponding locations of the same Munsell chip (i.e. the test stimulus) when illuminated with each of the two illuminants. The arrow shows a typical shift in the chromaticity of the test target that the subject makes to ensure that the perceived colour of the test patch remains invariant when the background illuminant changes from daylight to tungsten (i.e. D_{65} and A, respectively). A shift in the chromaticity of the test patch towards its expected chromaticity under the new illuminant is required if the test patch is to remain colour-invariant under the two illuminants. A colour match that requires a shift of chromaticity from T_D to T_A implies perfect constancy and results in a C_{index} of 1. This condition corresponds to complete invariance of perceived colour under changes of illuminant that affect the entire scene. Zero constancy, on the other hand, requires no change in target chromaticity under the two background illuminants and results in an index of 0 (i.e. the perceived colour of the test patch is determined entirely by its spectral radiance distribution on the retina). The chromaticity changes are as measured on the display monitor without correction for forward scatter in the eye. Test target chromaticity changes caused by forward scatter in the eye have been computed in the case of a disc target surrounded by a large annulus. (B) The diagram shows how correction for forward scatter in the eye can cause significant changes in the measured C_{index} value. The computation of C_{index} is identical to that shown in (A) but the chromaticities involved (i.e. T'_D, T'_A and M′) have been corrected for the effect of scattered light.

mately equal and as a result no flicker is detected. An iterative procedure is then used to compute the scatter function of the eye from these data. We start with a scatter index, n, and compute the effective eccentricity of each extended annulus, which is a function of n. This is defined as the eccentricity of a narrow annulus that would contribute the same amount of scattered light over the test stimulus for the same illuminance level at the pupil. Error-weighted regression analysis on the measured data and the set of computed effective eccentricities is carried out, which yields the scatter function parameters, n and k, for a given eye. The iteration then continues until the scatter index, n, predicted by the regression analysis equals that used for the computation of effective eccentricities. The use of large, extended annuli reduces the need for high luminance stimuli and yields accurate and repeatable results (Barbur *et al.*, 1993). For the annular geometry employed in this study, a knowledge of the scatter function of the eye makes it possible to compute the change in the luminance of the central test target caused by light that is scattered onto it from the surrounding annulus. Changes in the chromaticity of the test patch on the retina under the two different illuminants that scattered light causes can thus be computed. This in turn leads to the computation of a new index, C_{index}, that is corrected for test target chromaticity changes caused by scattered light. This is illustrated graphically in Figure 1B.

Results and conclusions

Figure 2 shows the effect of the width of a dark border surrounding the central test target (see inset) on the measurement of C_{index}. The different Munsell chip reflectances used as test targets are labelled on the graph. The data show the importance of the adjacent surround on such measurements, although distant backgrounds can also contribute significantly to immediate constancy. The decrease in C_{index} with increasing width of the dark border of the separation annulus was similar for all Munsell chips tested, but its absolute value varied greatly with test-target chromaticity. A similar variation in immediate constancy across chromaticity space has been observed in other studies and has been attributed to the location of colours within perceptual categories (Kulikowski and Walsh, 1993).

Measurements of the scatter function of the eye using the flicker cancellation technique are shown in Figure 3 for an older observer (subject PF) and two young observers (CW and JB). The mean luminance of the scatter source annulus was 50 cd/m^2. The results show that in the case of the nearest scatter source, the increase in retinal illuminance over the test target caused by scattered light is equivalent to a screen luminance of at least 3 cd/m^2.

Figure 4 shows C_{index} measurements, averaged over the same Munsell patches shown in Figure 2, as a function of the width of the dark separation annulus, with and without correction for forward light scatter in the eye (see Fig. 1). Although the amount of correction involved depends on several parameters, including the chromatic saturation of the test target and its location on the

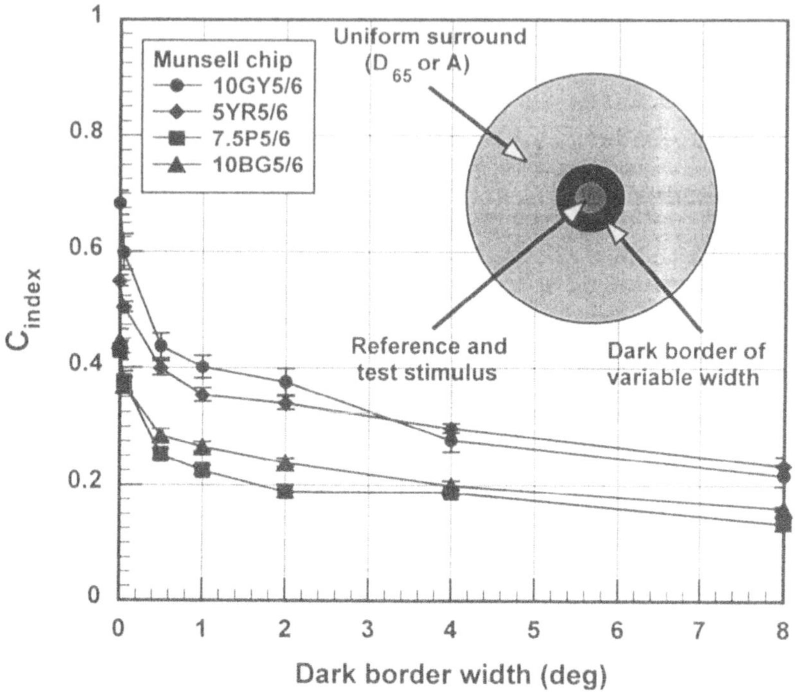

Fig 2. Measurement of C_{index} for four Munsell chips plotted as a function of the width of a dark annulus separating the test target from the surrounding background field. The inset shows a schematic representation of the test target and the uniform surround.

chromaticity diagram in relation to the chromaticity of the illuminants involved, the results show that on average scattered light in the eye can reduce significantly the estimated C_{index}. Since the stray light parameter, k, shows large variation in normal subjects and a steady increase with age, correction for scattered light may reduce some of the inter-subject variability.

The results show that immediate constancy cannot be attributed to, nor explained by, scattered light. Correction for changes of test target chromaticity caused by scattered light tends to increase the estimated C_{index} values. So far we have considered only chromaticity changes caused by scattered light in measurements of the colour constancy index when saturated colours are involved. It is also of interest to establish whether increased scatter in older subjects is sufficient to affect their measurements of chromatic displacement thresholds. In order to determine whether this is the case we carried out computations of target chromaticity changes caused by scatter in the case of a subject with normal levels of scattered light and in an older subject that showed an almost four-fold increase in scatter (Fig. 5, in which the stimulus parameters are given).

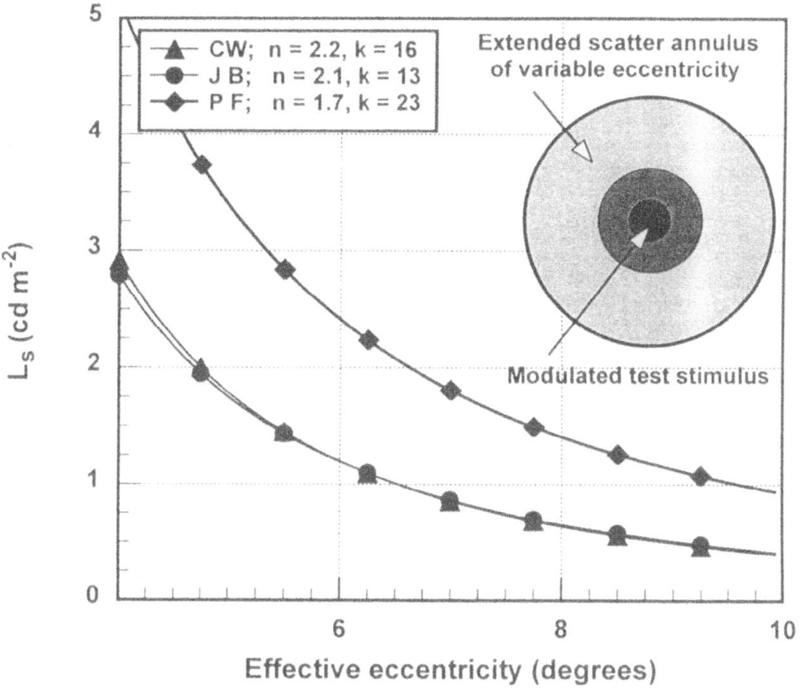

Fig. 3. Measurements of the scatter function of the eye using the flicker cancellation technique. Data are shown for three normal subjects with no signs of ocular opacities. The parameters n and k that describe the scatter function of the eye (i.e. $L_s = kE\theta^{-n}$) are as given in the legend. The mean luminance of the scatter source was 50 cd/m^2.

The change in test-target chromaticity caused by forward scatter in the eye was computed as a function of the chromatic displacement of the test target away from background chromaticity. This displacement was restricted to the colour confusion lines of dichromats (Wright, 1946), largely because the same directions were used in another study that investigated the effects of age on chromatic discrimination sensitivity (Knoblauch *et al.*, 1995). The corresponding change in test-target chromaticity caused by scattered light is shown in Figure 5 for increasing chromatic saturation of the target. Thresholds for chromatic displacement measured under similar conditions are described as elliptical contours in the CIE x, y diagram with typical minor and major semi-axes of 0.004 and 0.01, respectively (Barbur *et al.*, 1994). The results shown in Figure 5 suggest that normal chromatic displacement thresholds will be largely unaffected by scattered light, since even for a target chromatic displacement that is four times larger than the major semi-axis of the chromatic discrimination ellipse (i.e. a value of 0.04), the change in target chromaticity caused by scattered light is well below detection threshold.

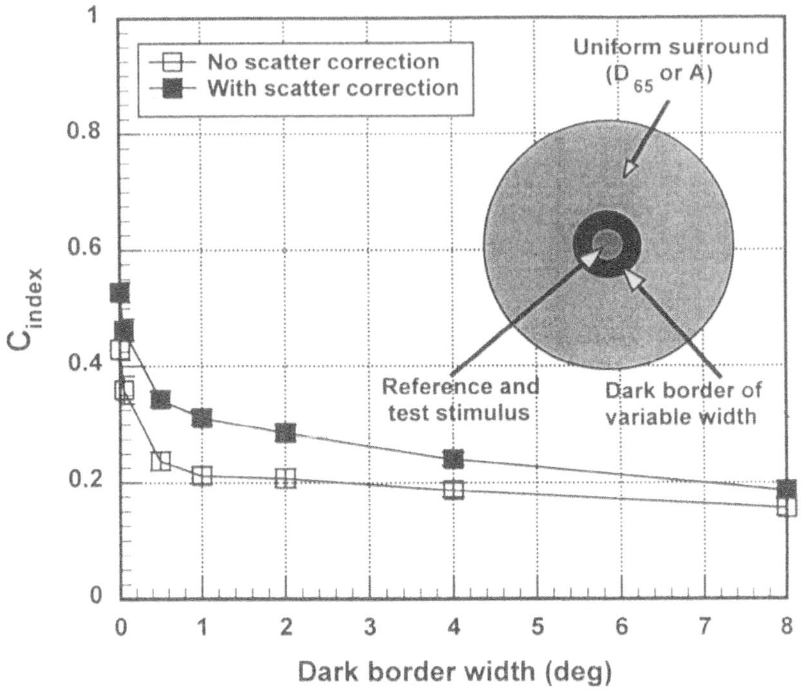

Fig. 4. Comparison of C_{index} measurements, averaged for the four Munsell chips shown in Figure 2, with and without correction for forward scatter in the eye. The results show that correction for chromaticity changes caused by scattered light from the surrounding background onto the central test patch can increase the computed C_{index} value. The exact increase depends on several parameters, including the chromatic saturation of the test stimulus with respect to the surrounding background (see Fig. 5) and the amount and angular distribution of scattered light in the eye. The results above show that for an older subject with clear ocular media (subject PF, Fig. 3), the increase in C_{index} value when correction for scattered light is applied can be large (>25% on average).

This is also the case for the older subject, who exhibits increased light scatter, although his change in target chromaticity is much greater. In summary, the results suggest that thresholds for chromatic discrimination are not affected significantly by forward light scatter in the eye even in subjects with above average levels of scattered light. On the other hand, when the experiment involves the use of saturated coloured targets (i.e. when the difference in chromaticity between test target and surrounding background is large), as is often the case in measurements of colour constancy, increased scatter in the eye can cause large changes in target chromaticity. The C_{index} parameter is frequently used to quantify simultaneous colour constancy. The results suggest that by not taking into account the changes in test target chromaticity caused by forward scatter in the eye, the C_{index} can be underestimated often by as much as 25 %.

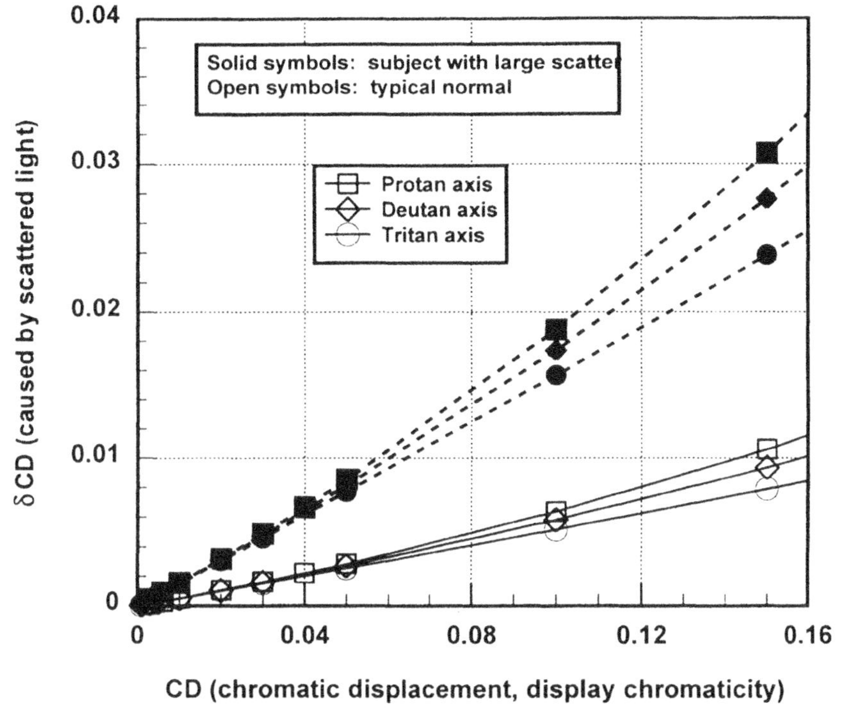

Fig. 5. Computed change in target chromaticity caused by forward light scatter in the eye as a function of the difference in chromaticity between the test target and the surrounding background field. Two light scatter functions were selected for this computation: a normal subject ($n = 2.09$, $k = 13$, solid lines) and an older subject with much more scatter (i.e. $n = 2.15$, $k = 46$, broken lines). The test target was a disc of 1.5° angular subtense and was surrounded by a uniform background of 15° diameter. The background field was neutral of x, y chromaticity 0.305, 0.323 and luminance 34 cd/m^2. The chromatic displacement (CD) of the test target was always taken along the colour confusion lines of dichromats and represents the distance on the CIE x, y diagram between the chromaticity of the test target and that of the surrounding background.

References

Arend, L. and Reeves, A. (1986). Simultaneous colour constancy. J. Opt. Soc. Am., 3: 1743–1751.

Arend, L., Reeves, A., Schirillo, J. and Goldstein, R. (1991). Simultaneous colour constancy: papers with diverse Munsell values. J. Opt. Soc. Am., 3: 1743–1751.

Barbur, J.L., Pinney, H.D. and Saunders, J.E. (1989). A method for quantifying changes in colour appearance. Perception 18: 530.

Barbur, J.L., DeCunha, D., Harlow, A. and Woodward, E.G. (1993). Methods for the measurement and analysis of scattered light in the human eye. In: Noninvasive Assessment of the Visual System Technical Digest, 3: 170–173. Opt. Soc. Amer., Washington, D.C.

Barbur, J.L., Harlow, A.J. and Plant G. (1994). Insights into the different exploits of colour in the visual cortex. Proc. Roy. Soc. Lond. B 258: 327–334.

Fairchild, M.D. and Lennie, P. (1992). Chromatic adaptation to natural and incandescent illuminants. Vision Res. 32: 2191–2193.
Foster, D.H., Craven, B.J. and Sale, E.R.H. (1992). Immediate colour constancy, Ophthal. Physiol. Opt. 12: 157–160.
Knoblauch, J.L., Barbur, J. and Vital-Durand, F. (1995). Development and ageing of chromatic sensitivity. Invest. Ophthal. Vis. Sci. 36: 910.
Kulikowski, J.J. and Walsh, V. (1993). Colour constancy for categorical centroids and border colours, Perception 22: 13.
Land, E.H. (1974). The retinex theory of colour vision. Proc. Roy. Inst. Gt. Br. 47: 23–58.
Lennie, P. and D' Zmura, M. (1988). Mechanisms of colour vision. CRC Crit. Rev. Neurobiol. 3: 330–400.
McCann, J., McKee, S.P. and Taylor, T.H. (1976). Quantitative studies in retinex theory. A comparison between theoretical predictions and observer responses to the 'Color Mondrian' experiments. Vision Res. 16: 445–458.
Stiles, W.S. and Crawford, B.H. (1937). The effect of a glaring light source on extra foveal vision. Proc. Roy. Soc. Lond. B. 122: 255–280.
Valberg, A. and Lange-Malecki, B. (1990). 'Colour constancy' in Mondrian patterns: a partial cancellation of physical chromaticity shifts by simultaneous contrast. Vision Res. 30: 371–380.
van den Berg, T.J.T.P. (1986). Importance of pathological intraocular scatter for visual disability. Doc. Ophthalmol. 61: 327–333.
van den Berg, T.J.T.P. and Spekreijse, H.(1987). Measurement of the straylight function of the eye in cataract and other optical media disturbances by means of a direct compensation method. Invest. Ophthal. Vis. Sci. (Suppl.) 28: 397.
Walraven, J. (1973). Spatial characteristics of chromatic induction; the segregation of lateral effects from straylight artefacts, Vision Res. 13: 1739–1753.
Wright, W.D. (1946). Researches on Normal and Defective Colour Vision. Henry Kimpton, London.
Wyszecki, G. and Stiles, W.S. (1982). Color Science: Concepts and Methods, Quantitative Data and Formulae. Wiley, New York.

J.L. Barbur, J.A. Harlow and C. Williams*
Applied Vision Research Centre
City University
Northampton Square
London EC1V OHB, UK

*Present address: McGill Vision Research Centre
Department of Ophthalmology
Montreal, Canada

PART SEVEN

Interactions among systems and applied problems

51. Do rod signals add with S cone signals in increment detection?

S.L. BUCK, E. SANOCKI and R. KNIGHT
(Seattle, WA, USA)

Abstract

Test additivity experiments reveal the combination rules for increment detection by rods and an S cone-dominated mechanism isolated by means of chromatic adaptation. The measured additivity was corrected (reduced) to compensate for the partial detection by each mechanism of both test wavelengths in the combined stimuli. The small degree of additivity found between rod and S cone signals under these conditions is consistent with either a small rod–cone interaction or with probability summation between independent mechanisms. This contrasts with the higher degree of additivity previously found between rod signals and M and L cone signals in increment detection.

Introduction

Signals originating in rod and cone photoreceptors can interact when they are processed simultaneously by the visual system. Past studies have examined how rod and cone signals combine to jointly influence the detection of threshold stimuli (Benimoff *et al.*, 1982; Drum, 1982; Frumkes *et al.*, 1973; Ikeda and Urakubo, 1969). These studies have generally shown that the effects of rod and cone signals are at least partially additive.

More recently, attempts have been made to determine whether rod signals combine to the same degree with signals from each of the three types of cone. For detection of increments, Buck and Knight (1989, 1994) found that the degree of additivity with rods is high and is similar for both M cones and L cones: less than the ideal prediction of linear addition; but greater than that predicted by orthogonal vector addition. The similarity of rod additivity with M and L cones suggests the simple hypothesis that rods may be equally additive with S cones as well. Such a hypothesis might be expected if rod and cone signals merge non-differentially very early in the visual pathways, say by means of gap junctions between photoreceptors.

In the present study (portions presented by Buck *et al.*, 1991) we determined the degree of additivity of rod and S cone signals for detection of increments. The chromatic adaptation technique used by Stiles (1949, 1953) and many

C.R. Cavonius (ed.), Colour Vision Deficiencies XIII, pp. 451–458.

others allows simultaneous isolation of rods and a psychophysical mechanism whose primary mode (peak around 440 nm) is determined by S cones (Pugh, 1976; Pugh and Mollon, 1979). The degree of additivity between rod and S cone signals can then be determined by means of the same type of test-additivity procedure we used to assess the additivity of rod and M or L cone signals (Buck and Knight, 1994). This type of procedure has long been used to examine the combination rules for separate contributions to psychophysical detection threshold (e.g. Guth, 1965; Ikeda, 1963) or to identify chromatic opponency (Thornton and Pugh, 1983).

Materials and methods

All observations were made with a conventional Maxwellian-view apparatus. (See Buck and Knight, 1994, for details of apparatus and spatial, temporal, and irradiance calibrations.) Two channels produced 2° diameter test stimuli that were viewed 7° to the right of a dim fixation cross. The output of either test channel could be used alone to create a single wavelength test, or both could be superimposed to create a mixed wavelength test. The test stimulus was presented once every 2.5 sec for a duration of 100 or 200 msec, depending on the condition. The third channel produced a 10° diameter background stimulus that was continuously presented concentric to the test stimulus. The spectral composition of the background was determined by a long wavelength, broad-band Wratten #29, which passes wavelengths above 600 nm. For one condition for observer ES, the test diameter was increased to 6°, the background diameter was increased to 12.2°, and the eccentricity was increased to 9.5°. Throughout these procedures, an observer's test eye was adapted to the dark for 30 min before each experimental session and was adapted to the background for 2 min before each condition.

Four experienced observers, each having normal colour vision, were tested. Observers SB, RK, and DG used their right eyes; observer ES used her left eye. This resulted in presentation of stimuli to the nasal and temporal retinas, respectively.

Pilot testing revealed background wavelength conditions for simultaneous isolation of rods and S cones. The long-wavelength Wratten #29 background was presented at 3.68 (SB) or 3.46 (ES, RK, and DG) log photopic trolands. Using this background, complete test-spectral-sensitivity (TSS) functions were measured for each observer (see Fig. 1). Rod ($V'\lambda$) and S cone ($\pi 1$) spectral templates were fitted by eye to appropriate portions of the TSS functions. For S cones, Stiles' foveal $\pi 1$ template was converted from quanta to energy and from wavenumber to wavelength and corrected for the average reduction of macular pigment density found at 7° (Wyszecki and Stiles, 1982). Spectral sensitivity and threshold versus background illuminance measurements confirmed that $V'\lambda$ and $\pi 1$ were the best-fitting spectral templates for the data of the present observers.

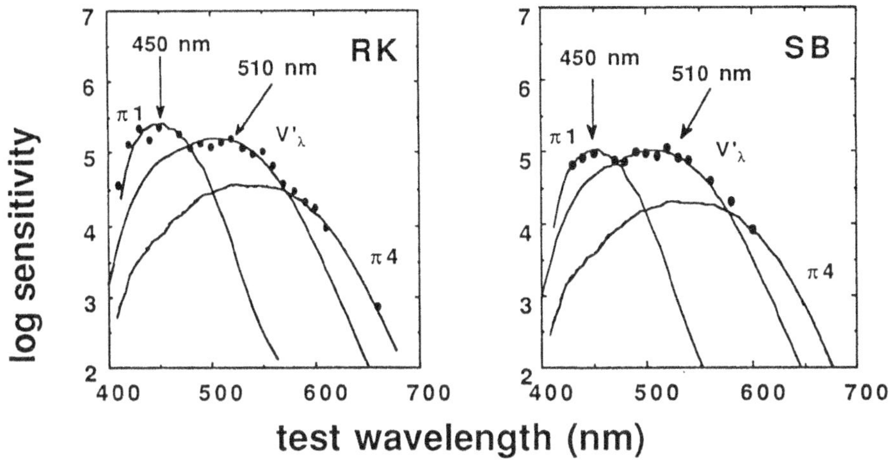

Fig. 1. Sample test spectral sensitivity functions obtained on long wavelength backgrounds used to isolate rods (V'λ) and S cones (π1) simultaneously. Test wavelengths used for the additivity studies are indicated.

Two test wavelengths, 450 nm for S cones and 510 nm for rods, were selected that maximized the sensitivity of each detection mechanism relative to the other, and to additional cone mechanisms (such as π4 shown in Fig. 1) on the background. Thresholds were measured for each wavelength, singly and in combination, by means of the method of adjustment. In a session, the background composition and duration of the test stimulus remained constant. Only the spectral composition of the test stimulus changed. The observer first set five thresholds for test stimulus 1, then five thresholds for test stimulus 2, then one or more sets of five thresholds for each combination of the two test stimuli, then five thresholds for test stimulus 2, and finally five thresholds for test stimulus 1. Each session was performed at least three times.

In the combination conditions, the ratio of the irradiances of the two test wavelengths was proportional to their thresholds when each wavelength was presented alone during preliminary sessions. For ES, additional combination conditions were tested in which the ratios of the two test stimuli were also 1:3 and 3:1. An important aspect of the procedure was that, in a given condition, the observer adjusted the total amount of the combined stimulus while the ratio of irradiances of the components remained constant.

The basic data for all subsequent analyses were the ratios determined by the energy of each test stimulus at threshold in the combined stimulus divided by the energy of that same test stimulus at threshold when it was presented alone. Thus, each additivity test produced two ratios, one each for 450 nm and 510 nm test stimuli or, alternatively, for S cones and rods.

Table 1. Test addtivity data: cone and rod correction factors, corrected additivity coordinates, and *k* values.

	Correction factors		Corrected additivities		
Observer	S cone	Rod	S cone	Rod	*k* values
100 msec test duration					
RK	0.0478	0.1138	0.5690	0.8439	2.13
	0.0310	0.1848	0.7419	0.7917	2.62
Mean (s.e.)					2.38 (0.24)
ES	0.0366	0.1497	0.6929	0.8663	2.93
	0.0540	0.1238	0.6302	0.9143	3.09
	0.0509	0.1160	0.6782	0.8714	2.88
Mean (e.s.)					2.97 (0.06)
200 msec test duration					
ES	0.1125	0.0313	0.4547	0.9296	2.35
	0.0341	0.0906	1.0231	0.3631	–[a]
	0.1266	0.0199	0.3434	1.0305	–[a]
	0.0915	0.0609	0.8017	0.6801	2.35
	0.1293	0.0414	0.6116	0.9158	2.99
	0.0556	0.0787	0.9733	0.4549	3.18
	0.0365	0.1519	0.6884	0.8229	2.53
Mean (s.e.)					2.68 (0.17)
SB	0.0361	0.1504	0.7189	0.8031	2.57
	0.0500	0.1234	0.7390	0.7954	2.63
Mean (s.e.)					2.60 (0.03)
DG	0.0313	0.1597	0.5692	0.9202	2.80
	0.0310	0.1779	0.6947	0.8102	2.48
Mean (s.e.)					2.64 (0.16

[a]Observer's *k* values undefined. See text for explanation.

An additional transformation was applied to the data to correct for the necessarily incomplete specificity of the two test wavelengths, both of which actually stimulate each detection mechanism to some extent. (See Buck and Knight (1994) for detailed discussion and validation tests.) This non-specificity correction simply adds together the separate stimulations of a mechanism due to the two test wavelengths in the mixture, where each is expressed as a proportion of the threshold for that mechanism. The principal assumption behind the correction factor is that each mechanism is completely test-additive.

Results

Table 1 shows the corrected additivity coordinates (corrected additivities) as well as the calculated non-specificity corrections (correction factors) for each observer and test duration. In general, the correction factors were larger for rod detection than for S cone detection, reflecting a greater difference in rod and S cone test sensitivities at 510 nm than at 450 nm under these conditions.

The additivity results are presented graphically on a summation square in Figure 2. On the x- and y-axes are plotted, respectively, the two ratios representing the proportion of S cone- and rod-threshold stimulation that

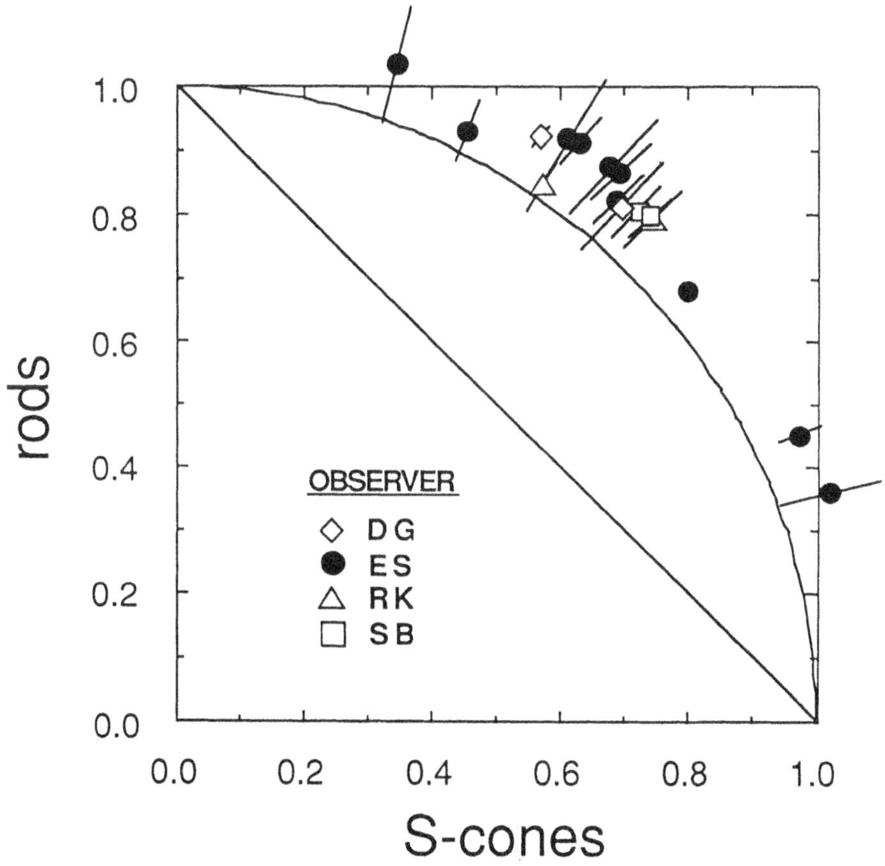

Fig. 2. Corrected additivity results for all observers and conditions. The degree of corrected additivity (overall $k = 2.65$) was less than that predicted by orthogonal vector addition (curved line, $k = 2.0$). This small amount of additivity may reflect a weak interaction of rod and S cone signals or only probability summation between independent mechanisms. Error bars represent ± 1 s.e. of the means of the combined test–stimulus conditions for at least three sessions.

existed at threshold for each combined test–stimulus condition. In such a plot, different regions indicate different degrees of additivity, and therefore different combination rules for the two mechanisms. Complete additivity would result in data falling along the diagonal of slope –1.0 between 1.0 on the ordinate and the abscissa. Complete independence would result in data falling along the upper and right axis lines (ignoring effects of probability summation). Partial additivity is shown by data falling between these two regions.

The degree of partial additivity can be characterized by a k-value metric that describes a best-fitting isoadditivity function (Stiles, 1949; Graham, 1989). Thus, data for the mixture of different proportions of two stimuli that display the same degree of additivity will fall along an arc that can be described by a single k value. Complete additivity, orthogonal vector addition, and complete independence are associated with k values of 1, 2, and ∞, respectively.

Figure 2 shows that the corrected additivity measurements fall between the unit axes and the orthogonal vector addition line. Individual observers' means were all >2.0, as shown in Table 1, with an overall mean of 2.65 across all observers and conditions. This is a degree of partial additivity that may be associated either with probability summation of independent mechanisms or with a small degree of additivity caused by interaction between the mechanisms. There were no significant overall differences between the 100 msec and 200 msec test durations ($k = 2.67 \pm 0.30$ and 2.64 ± 0.02, respectively).

For two conditions for ES, the corrected additivities fell just beyond the unit axes, resulting in indeterminate k values. The test stimuli for these two conditions were composed of extreme ratios of the two wavelengths (1:3 and 3:1). It seems likely that detection was mediated by a single mechanism in each case. This is supported by the overlap of the error bars with the unit axes. Modification of stimulus parameters for ES (see Materials and methods) yielded an additivity within the bounds of those obtained in the other condition (condition with asterisk in Table 1).

Discussion

Our results are consistent with two possible relationships between rod and S cone signals: either there is a small but real amount of additive mixing of the signals, or there is independent detection with enhancement of sensitivity caused only by probability summation.

Whatever the underlying cause, the weak additivity of rod and S cone signals contrasts with the stronger additivity of rod signals with either M or L cone signals ($k = 1.63$ and 1.55, respectively) shown previously by Buck and Knight (1989, 1994). The conditions of measurement were made as similar as possible among these studies.

In the present study, because the calculated non-specificity correction for S cones was small, the validity of the correction does not depend on assumptions of complete test additivity or exact spectral sensitivity (e.g. $\pi 1$, $\pi 3$) of the S cone mechanism. Instead, the validity of the non-specificity correction depends primarily on the univariance and known spectral sensitivity of rods. Although there is a substantial literature concerning the adaptation of rod signals by cone signals (e.g. Buck *et al.*, 1979), no challenge has ever been made to the univariance of test detection by rods. If we apply the non-specificity correction only to rod detection, there is only a slight reduction of the overall mean k value, from 2.65 ($\pm$0.09 s.e.) to 2.41 ($\pm$0.08).

The present results argue against indiscriminate mixture of rod signals with all three types of cone signals early in retinal processing. A general possibility is that rod signals could enter M and L cone pathways but not S cone pathways. If rod and M and L cone signals mix additively by means of gap junctions between photoreceptors or in subsequent neural pathways, then the pathways determining increment threshold appear to lack similar types of interaction between rod and S cone signals.

There is, however, considerable evidence in colour vision for interactions between rod and S cone signals. Numerous investigators have argued that scotopic stimulation appears blue or adds blueness to colour mixtures (e.g. Trezona, 1970, 1974; Hunt, 1952; Willmer, 1950; Richards and Luria, 1964; Ambler, 1974). Several studies have shown that rod signals influence S cone colour pathways that are involved in scotopic colour contrast (Buck, 1995; Buck and Ayers, this volume; Buck *et al.*, 1994; Buck and Brandt, 1993, 1995). Reitner *et al.* (1991) have shown that S cones and rods can work together to mediate wavelength discrimination for blue cone monochromats, who apparently have only rods and S cones. Haegerstrom-Portnoy *et al.* (1993) and Blackwell and Blackwell (1961) have shown cancellation between rod and S cone signals for other tasks. It is not yet clear how to model this diverse set of rod/S cone interactions for both colour vision and increment detection.

Acknowledgements

This work was supported by USPHS/NIH grants EY03221 and EY07031.

References

Ambler, B.A. (1974). Hue discrimination in peripheral vision under conditions of dark and light adaptation. Percept. Psychophys. 15: 586–590.

Benimoff, N.I., Schneider, S. and Hood, D.C. (1982). Interactions between rod and cone channels above threshold: a test of various models. Vision Res. 22: 1133–1140.

Blackwell, H. and Blackwell, O. (1961). Rod and cone receptor mechanisms in typical and atypical congenital achromatopsia. Vision Res. 1: 62–107.

Buck, S. (1995). Influence of rod signals on the perceptual opponent color channels: evidence from successive scotopic contrast. Invest. Ophthalmol. Vis. Sci. Suppl. 36: 960.

Buck, S.L. and Brandt, J. (1993). Scotopic contrast colors show preferential involvement of rod signals with S cone color pathways. Invest. Ophthalmol. Vis. Sci. Suppl. 34: 764.

Buck, S.L. and Brandt, J. (1995). The range of scotopic contrast colors matches that associated with S-cone activation. In: Drum, B. (ed.), Colour Vision Deficiencies XII, Doc. Ophthalmol. Proc. Ser. 57: 309–316. Kluwer, Dordrecht.

Buck, S.L. and Knight, R. (1989). Rods combine subadditively with π4 and π5 in increment detection. Invest. Ophthalmol. Vis. Sci. Suppl. 30: 455.

Buck, S.L. and Knight, R. (1994). Partial additivity of rod signals with M- and L-cone signals in increment detection. Vision Res. 34: 2537–2545.

Buck, S.L., Peeples, D. and Makous, W. (1979). Spatial patterns of rod–cone interaction. Vision Res. 19: 775–782.

Buck, S.L., Sanocki, E. and Knight, R. (1991). Rod signals show less additivity with S-cone than with M- and L- cone signals. Invest. Ophthalmol. Vis. Sci. Suppl. 32: 1212.

Buck, S.L., Ayers, M. and Brandt, J. (1994). Successive and simultaneous scotopic contrast compared. Invest. Ophthalmol. Vis. Sci. Suppl. 35: 1657.

Drum, B. (1982). Summation of rod and cone responses at absolute threshold. Vision Res. 22: 823–826.

Frumkes, T.E., Sekuler, M.D., Barris, M.C., Reiss, E.H. and Chalupa, L.M. (1973). Rod–cone interaction in human scotopic vision – I. Temporal analysis. Vision Res. 13: 1269–1282.

Graham, N. (1989). Visual Pattern Analyzers. Oxford, New York.

Guth, S.L. (1965). Luminance addition: general considerations and some results at foveal threshold. J. Opt. Soc. Am. 55: 718–722.

Haegerstrom-Portnoy, G., Schneck, M., Fisher, S. and Hewlett, S. (1993). Achromats' saturating rods prohibit cone flicker detection. Invest. Ophthalmol. Vis. Sci. Suppl. 34: 743.

Hunt, R. (1952). Light and dark adaptation and the perception of color. J. Opt. Soc. Am. 42: 190–199.

Ikeda, M. (1963). Study of interrelations between mechanisms at threshold. J. Opt. Soc. Am. 53: 1305–1313.

Ikeda, M. and Urakubo, M. (1969). Rod–cone interrelation. J. Opt. Soc. Am. 59: 217–222.

Pugh, E.N. (1976). The nature of the π1 mechanism of W. S. Stiles. J. Physiol. 257: 713–714.

Pugh, E.N. and Mollon, J.D. (1979). A theory of the π1 and π3 color mechanisms of Stiles. Vision Res. 19: 293–312.

Reitner, A., Sharpe, L.T. and Zrenner, E. (1991). Is colour vision possible with only rods and blue-sensitive cones? Nature 352: 798–800.

Richards, W. and Luria, S. (1964). Color mixture functions at low luminance levels. Vision Res. 4: 281–313.

Stiles, W.S. (1949). Investigations of the scotopic and trichromatic mechanisms of vision by the two colour threshold technique. Rev. d'Opt.: 28: 215–237. (Reprinted in part in Stiles, W.S. (1978): 139–144.)

Stiles, W. S. (1953). Further studies of visual mechanisms by the two-color threshold technique. Coloq. Probl. Opt. Vis. 1: 65–103. (Reprinted in Stiles, W.S. (1978): 183–121.)

Stiles, W.S. (1978). Mechanisms of Colour Vision. Academic Press, New York.

Thornton, J. and Pugh, E. (1983). Red/green color opponency at detection threshold. Science 219: 191–193.

Trezona, P. (1970). Rod participation in the 'blue' mechanism and its effect on color matching. Vision Res. 10: 317–332.

Trezona, P. (1974). Additivity in the tetrachromatic color matching system. Vision Res. 14: 1291–1303.

Willmer, E.N. (1950). Low threshold rods and the perception of blue. J. Physiol. 111: 17P.

Wyszecki G. and Stiles W. S. (1982). Color Science. 2nd edn. Wiley, New York.

University of Washington
Department of Psychology
Box 351525
Seattle, WA 98195-1525, USA

52. Distant colour induction on the Fechner–Benham top

J. LE ROHELLEC and F. VIÉNOT
(Paris, France)

Abstract

Subjective colours can be obtained by rotating a Fechner–Benham (FB) top. We describe a modification of colour occurring when several arcs are simultaneously present. We presented a series of rotating discs in which one arc (the stimulus) was flanked by two other arcs (the inducers) separated from the stimulus by 2°. Seven stimuli differing by their start angle from the black half of the disc, and 4 inducers were tested. The disc rotated at 7 cycles/sec and was illuminated for 1, 2, 4, 8 or 16 cycles of rotation. The responses of four observers were collected using a hue-naming procedure. We examined the proportion of perceived hue for each arc, subject, preadaptation condition and exposure. Complementary pairings of inducers were compared with respect to their effect on each unique hue.

We found a distant colour-induction effect. In general, inducers located 2° from the stimulus modify the Fechner–Benham subjective colours, complementary contexts induced opposite changes and the colour perceived on the stimulus is shifted away from the perceived hue of the inducers. This strong distant induction effect is discussed in terms of colour constancy.

Introduction

Rotating a Fechner–Benham (FB) top produces colours on each ring, which are called pattern-induced flicker colours (PIFCs). The colours are determined mainly by the phase relationship between the temporal luminous pattern of the background signal and that of the arc (Fig. 1). Under steady white illumination, patterns with start angle 0°, 45°, 90° and 135° produce reddish, yellowish, greenish and bluish colours, respectively.

PIFCs are assumed to originate from retinal sites in the post-receptoral pathways (Campenhausen, 1968; Zrenner, 1983). It has been shown in previous studies that a change occurs in the appearance of the colours with exposure time, suggesting that more than one mechanism is involved in producing PIFCs (Le Rohellec and Viénot, 1995) and that this change in colour depends upon the state of adaptation of the observer, either steady (light preadaptation) or

C.R. Cavonius (ed.), Colour Vision Deficiencies XIII, pp. 459–464.

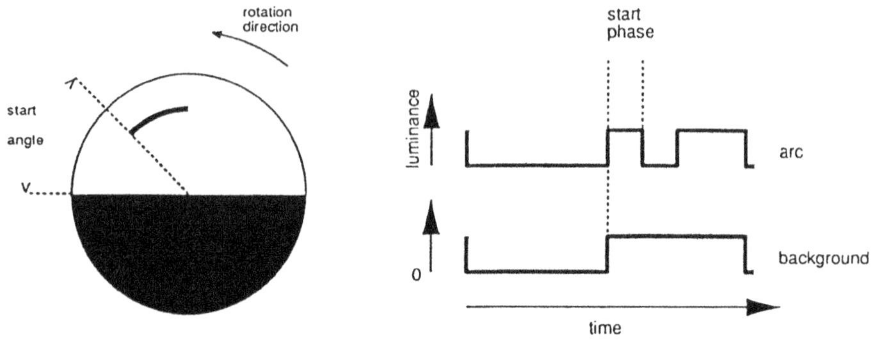

Fig. 1. A Fechner–Benham top that produces coloured rings when it is rotated. Left: start angle specifying the location of the arc on the white half disc and identifying the ring. Right: temporal pattern of luminance created by the background and by the arc of a rotating Fechner–Benham top.

transient (dark preadaptation) (Le Rohellec and Viénot, 1994). PIFCs also depend upon luminance adaptation (Kozac *et al.*, 1989). A reversal in the sequence of FB colours has been observed with a change in the wavelength of illumination (Viénot and Le Rohellec, 1992). Considerable inter-observer variability may be expected in the colours reported on the FB top. In the present study, we demonstrate that modification of colour occurs when several arcs are simultaneously present.

Materials and methods

Forty-two FB discs were prepared with three single stripes: the stimulus arc, flanked by two inducers, 2° on either side of the stimulus. The stimulus was an arc covering a 45° sector chosen among seven arcs differing in their start angle (the angle from the preceding black half of the disc: 0°, 22.5°, 45°, 67.5°, 90°, 112.5° or 135°; see Fig. 1). The inducers were two different arcs, each covering a 45° sector, chosen from four arcs differing in their start angle (0°, 45°, 90°, 135°). Thus, six combinations of two inducers, or contexts, were available, with external and internal start angles of 0° and 45°, 90° and 135°, 135° and 0°, 45° and 90°, 90° and 0°, 135° and 45°.

Each disc was rotated at 7 cycles/sec and was presented to the observer for 1, 2, 4, 8 or 16 rotation cycles. All discs were presented during a single session in a pseudorandom order, so that the observer could not anticipate the next stimulus. Each combination of stimulus and inducers was equally probable. Average luminance was 22 cd/m^2. Measurements were repeated during five sessions. For one series of five sessions, the presentation of the disc followed a dark period (transient preadaptation). For another series of five sessions, the presentation of the disc followed the presentation of the white uniform 22 cd/m^2 field (steady preadaptation).

Responses were collected using a hue-naming procedure. After each exposure, the observer gave a verbal description of the foveally-fixated stimulus arc, in terms of unique hues (red, yellow, green and blue). If the observer needed two hues, 2/3 of the weight was given to the first hue, and 1/3 to the second.

Four colour-normal observers participated in the experiment: two naive (AL, female, 28 years, and SP, male, 24 years) and two experienced (FV, female, 47 years, and JLR, male, 30 years).

Results

We computed the proportion of perceived hue for every arc stimulus and for every class of exposure (short: 1 or 2 cycles exposure; long: 8 or 16 cycles exposure), preadaptation, context and subject, by averaging 10 data (5 sessions, 2 exposures). Hence, the database consisted of 2688 average proportions of unique hue (672 for each unique hue). We compared the effect of context on the average proportion of unique hue, keeping constant all other variables – subject, preadaptation, exposure, and stimulus arc.

The results showed that some contexts facilitated certain unique hues and others did not. Consequently, we decided to focus the comparison within pairs of contexts that were complementary with respect to the selection of inducers. Complementary associations of inducers were compared with respect to their effect on each unique hue. The following associations were therefore compared: 0° and 45° inducers with 90° and 135° inducers (forward and backward contexts); 135° and 0° inducers with 45° and 90° inducers (edge and edgeless contexts); 90° and 0° inducers with 135° and 45° inducers (mixed contexts) (Fig. 2).

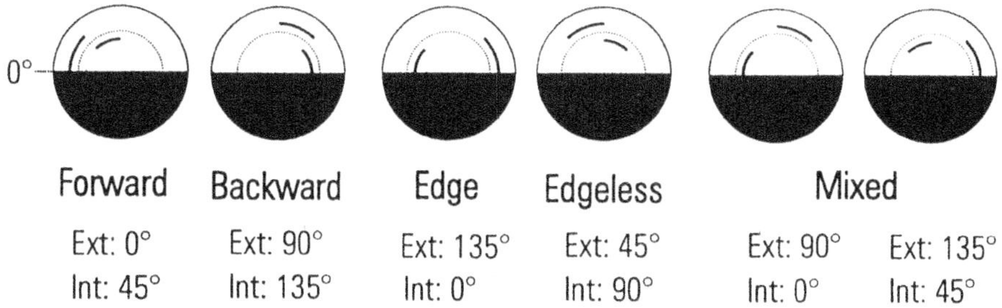

Fig. 2. Six combinations of inducers are shown and grouped in complementary presentations: external and internal start angle for forward and backward contexts: 0°–45° and 90°–135°; edge and edgeless contexts: 135°–0° and 45°–90°; mixed contexts: 90°–0° and 135°–45°. The dotted ring shows the location of the stimulus arc. 0° indicates the origin of the start angle.

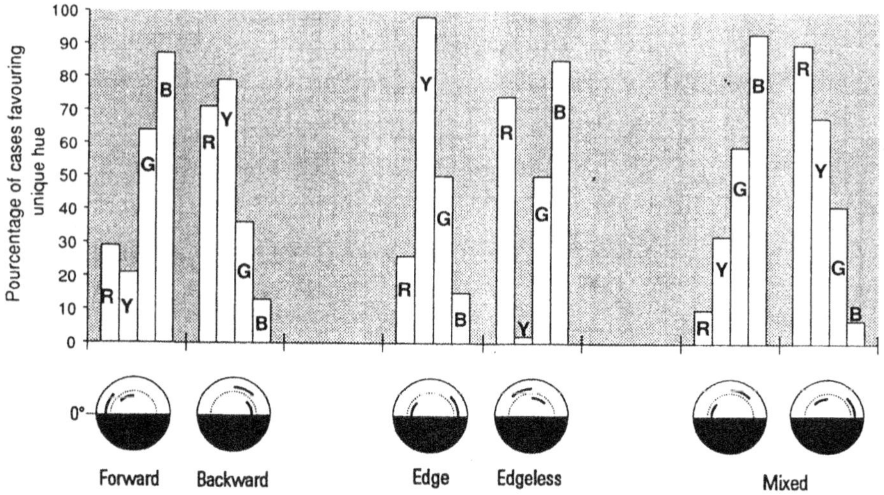

Fig. 3. Effect of context on the average proportion of unique hue, maintaining all other variables – subject, preadaptation, exposure, arc stimulus – constant. Ordinates: percentages of cases for which a given association of inducers favours each unique hue (R: red; Y: yellow; G: green; B: blue), compared with the complementary association of inducers (as shown in the bottom line). Left: comparison of the association of 0°–45° inducers with the association of 90°–135° inducers (forward and backward contexts). Middle: comparison of the association of 135°–0° inducers with the association of 90°–0° inducers (edge and edgeless contexts). Right: comparison of the association of 90°–0° inducers with the association of 135°–45° inducers (mixed contexts).

In order to eliminate noise due to under-represented hue for some arcs, we restricted the database to pairs in which one average proportion of unique hue was higher than 0.3 (577 out of 1344 pairs). The 0° and 135° arc stimuli are usually seen as red or blue, respectively, and the 45° and 90° arc stimuli are often seen as yellow or green, respectively. This yields very low proportions of other unique hues and justifies such a restriction. Further, in order to eliminate noise due to slight changes in the average proportion of unique hue between two complementary contexts, we only compared data in cases in which the difference between proportions was higher than 0.1 (395 out of 577 pairs; i.e. 68% of the pertinent average proportions). This selection restricted the data analysed to 29% of the total database. Then, given the arc, subject, preadaptation and exposure, we counted how many times each context favoured a given hue, compared with the complementary context.

Figure 3 shows the group-mean comparisons for 4 observers, 2 preadaptations, 2 exposures and 7 arc stimuli for each unique hue. In general, it appears that the inducers located at 2° from the stimulus modify the PIFCs, that complementary contexts induce opposite changes, and that the perceived colour is shifted away from the hue elicited by the inducers.

Forward and backward contexts

Short starting angle inducers (0° and 45°) tended to elicit blue and green rather than red and yellow responses. Conversely, long starting angle inducers (90° and 135°) tended to elicit red and yellow naming.

Edge and edgeless contexts

Edgeless context (45° and 90°) tended to elicit red and blue. Conversely, edge context (135° and 0°) highly favoured yellow, but we could not establish a change on green.

Mixed context

The opposite changes in relative proportions of unique hue induced by complementary mixed contexts resembled the changes induced by forward and backward contexts and were even more noticeable.

Controls of the reproducibility of the results have shown that the distant induction effect is robust. The results relative to each exposure time or to each preadaptation (not shown here) show the same shifts for all adaptations and exposure times.

Discussion and conclusion

A distant colour induction effect has been shown, because spatial context selectively changes the colours of PIFCs. Although our approach was qualitative, and our method of removing noisy data was arbitrary, clear examples of the distant colour induction effect were found. The change was always a shift of the PIFCs away from the colours induced by the neighbouring arcs. This may explain the variability of PIFCs reported in the literature and why the most effective configuration of the FB top consists of four arcs, equally distributed in terms of start angle.

The colour induction effect on PIFCs thus acts in the same way as normal colour contrast. This means that the PIFCs, once they are created, are subject to the same influences as colours produced by conventional means. It may be asked whether PIFCs are more fragile than spectral colours. For instance, the fact that yellow noticeably emerges when boundary inducers are simultaneously present suggests a very sensitive mechanism.

Further, several facts suggest that spatial induction could be located central to the retina. We would like to emphasize that inducers act at a distance (2° visual angle). Let us suppose that the subject could not maintain fixation, so that different rings could be successively imaged on the same retinal area. In this case, we would expect distant induction to occur only during long exposures, as eye movements would not be possible during short exposures. However, we have shown as strong a distant induction effect with short exposures as with long ones. Therefore, we think that the distant induction occurs at a visual site where signals originating from 2 distant retinal areas interact. Other authors (e.g. Supply, quoted by von Campenhausen and Schramme, 1995) have reported similar interactions over long distances between PIFCs and have demonstrated that dependence upon neighbouring colours is maintained in dichoptic presentations.

This strong distant induction effect, together with the changes of elicited colours with luminous preadaptation or with changes of illumination wavelength, suggests that PIFCs may not follow the laws of colour constancy.

Acknowledgements

Acknowledgements to the observers and to Hans Brettel and the anonymous referees for helpful comments on the manuscript.

References

von Campenhausen, C. (1968). Über die Farben der Benhamschen Scheibe. Z. vergl. Physiol. 60: 351–374.

von Campenhausen, C. and Schramme, J. (1995). 100 years of Benham's top in colour science. Perception 24: 695–717.

Kozak, W.M., Reitboeck, H.J. and Meno, F. (1989). Subjective color sensations elicited by moving patterns: effect of luminance. In: Kulikowski, J.J., Dickinson, C.M. and Murray, I.J. (eds.), Seeing Contour and Colour: 294–310. Pergamon, Oxford.

Le Rohellec, J. and Viénot, F. (1995). Contribution of two colour opponent mechanisms to Fechner-Benham subjective colours. In: Drum, B. (ed.), Colour Vision Deficiencies XII. Doc. Ophthalmol. Proc. Ser. 57 : 251–258. Kluwer, Dordrecht.

Le Rohellec, J. and Viénot, F. (1994). Dependence of Fechner–Benham colours upon luminance adaptation. Perception 23: 91.

Viénot, F. and Le Rohellec, J. (1992). Reversal in the sequence of Benham colours with change in wavelength illumination. Vision Res. 32: 2369–2374.

Zrenner, E. (1983). Neurophysiological Aspects of Color Vision in Primates: 106–109. Springer, Berlin.

CNRS UPR 257
Laboratoire de Physique Appliquée
Muséum National d'Histoire Naturelle
43, Rue Cuvier
F-75005 Paris, France

53. Influence of rod adaptation upon chromatic and achromatic cone-vision

T.E. FRUMKES, E. LEMBESSIS, J. VOLLARO, D. MOSHE and T. EYSTEINSSON

(Flushing, NY and Reykjavik, Iceland)

Abstract

We compared the influence of monocular and interocular rod-adaptation upon monocular cone-mediated vision. Dark-adapted rods exerted no detectable influence upon cone-mediated detection thresholds, but suppressed all other aspects of cone-mediated vision examined. The suppressive rod influence upon sensitivity to rapid achromatic flicker was monocular, involving a distal retinal mechanism, whereas influences upon sensitivity to high spatial frequency chromatic and achromatic gratings were completely binocular, suggesting a cerebral origin. Although essentially monocular, the rod influence upon specific thresholds was influenced by the adapted state of the non-viewing eye; this influence appears to be mediated at site(s) at which chromatic processing is segregated into different colour pathways.

Introduction

We have examined the influence of rod light and dark adaptation upon different aspects of chromatic and achromatic cone-related vision. Dark-adapted rods suppress all aspects of cone-mediated vision examined, with the exception of detection thresholds, but they must do so by means of several different neural mechanisms.

Materials and methods

The test display, which had an overall spatial extent of 1°–3° and was presented 4°–7° parafoveally, was presented monocularly to normal observers in Maxwellian view (left and middle columns of Fig. 1), or on a cathode ray display (right column of Fig. 1); head position was fixed by means of a full-mouth bite bar. To collect the data presented in Figure 1, the adapted state of the test eye, the contralateral eye, or in some cases both eyes was manipulated by an adapting field > 15° in diameter, which was presented prior to stimulation as a 50 000 td

C.R. Cavonius (ed.), Colour Vision Deficiencies XIII, pp. 465–469.

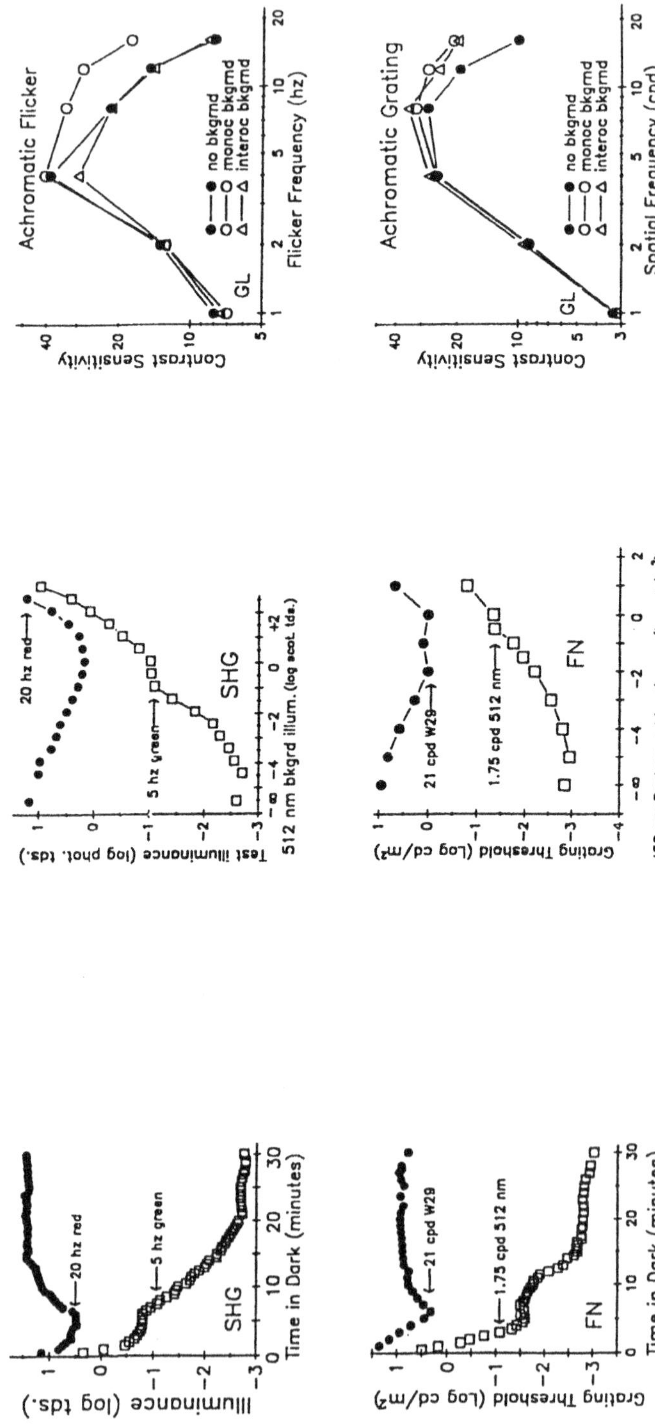
Achromatic Flicker
Contrast Sensitivity
Flicker Frequency (hz)
GL
no bkgrnd
monoc bkgrnd
interoc bkgrnd
Achromatic Grating
Spatial Frequency (cpd)
Test Illuminance (log phot. tds.)
512 nm bkgrd illum. (log scot. tds.)
20 hz red
5 hz green
SHG
Grating Threshold (Log cd/m²)
480 nm Background Luminance (log cd/m²)
21 cpd W29
1.75 cpd 512 nm
FN
Illuminance (log tds.)
Time in Dark (minutes)

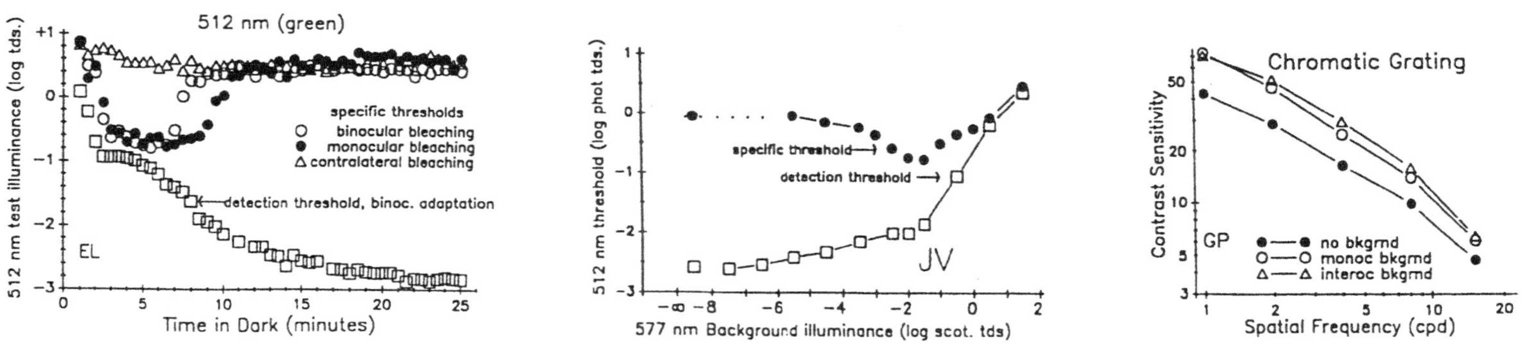

Fig. 1. Composite results from six different normal observers showing the influence of rod-adaptation upon achromatic flicker sensitivity (upper row), achromatic grating sensitivity (middle row), specific threshold (bottom row, left two functions), or chromatic grating sensitivity (bottom right). From left to right, the three different columns show dark adaptation, light adaptation (increment threshold), or contrast sensitivity functions. See text for further explanation.

'white' bleaching stimulus for 30 sec (left column), a variable luminance background (middle column), or as a fixed field of 0.03 cd/m^2 (top 2 rows, right column) or 0.3 cd/m^2 luminance (bottom right). 2.5 mm artificial pupils or smaller exit pupils ruled out any role for pupillary dilatation.

Results and discussion

The upper row of Fig. 1 summarizes older achromatic flicker data (Goldberg *et al.*, 1983; Frumkes *et al.*, 1995). The upper left plot shows the minimal illuminance just necessary to detect flicker in an intermittent stimulus during the recovery period following photopigment bleaching. When the subject's task is to adjust the luminance so as just to detect flicker of a 'green' LED that is modulated sinusoidally at 5 Hz, a typical dark adaptation function is obtained, with distinct cone and rod recovery stages. However, for 20 Hz 'red' (LED) flicker, which is only detectable by cones, sensitivity decreases during the rod recovery stage of dark adaptation. These and other similar data indicate that dark-adapted rods suppress cone-mediated flicker (for a review, see Frumkes and Eysteinsson, 1988). The middle plot shows corresponding light adaptation data. The 5 Hz green function resembles a typical, Stiles-like increment threshold function with distinct rod and cone limbs, but sensitivity to 20 Hz red flicker increases as background illuminance increases up to about 1 scotopic td. Following the usual convention, the upper-right plot shows modulation sensitivity (the reciprocal of contrast at threshold) for a stimulus of 10 cd/m^2 average luminance as a function of flicker frequency in the parafoveal retina. Presentation of a large, steady, background of 0.03 cd/m^2 to the viewing eye increases sensitivity to higher flicker frequencies alone, a result consistent with the data shown to the left. However, contralateral eye adaptation has little influence upon flicker sensitivity: this latter finding is consistent with electrophysiological data indicating that the rod adaptation flicker effect is mediated in retinal outer plexiform layer (Frumkes and Eysteinsson, 1988; Pflug *et al.*, 1990).

In a similar fashion, the second row of data in Figure 1 summarize older achromatic grating data (Naarendorp *et al.*, 1988; Frumkes *et al.*, 1995). The left and middle plots show the minimal luminance necessary to detect the orientation of a grating which was 1.75 cpd detectable by both rods and cones, or 21 cpd only detectable by cones. The 21 cpd data obtained during the 'rod recovery stage' of adaptation (left data) or with backgrounds dimmer than –1 log cd/m^2 luminance (middle data) suggest that dark-adapted rods suppress cone-mediated sensitivity to higher spatial frequencies. Thus, in the middle panel, even though presentation of the background field reduces slightly the contrast of the grating, it enhances grating detectability. The contrast sensitivity data on the right (obtained with a grating of 10 cd/m^2 average luminance) show that interocular as well as monocular presentation of a 0.03 cd/m^2 background enhances sensitivity to higher spatial frequencies. Since this effect is truly binocular, it is likely to be mediated more proximally than the analogous flicker effect. The data shown at the bottom right of Figure 1 show sensitivity to equiluminant red/green chromatic gratings of 5 cd/m^2 average luminance. A dim background field (0.3 cd/m^2) presented to the viewing eye or contralateral eye enhances sensitivity to all examined spatial frequencies. Detection of high spatial frequency chromatic gratings probably involves neurones with double-opponent receptive fields. Since such units are unknown subcortically in primates (e.g. Livingston and Hubel, 1987), the chromatic grating effect probably involves an adaptation influence upon striate or extrastriate visual cortex. Although not illustrated, these dim backgrounds similarly enhance sensitivity to blue/yellow chromatic gratings.

Finally, the data presented in the bottom left of Figure 1 show the influence of light and dark adaptation upon both the detection threshold and specific (hue identification) threshold of a 512 nm stimulus. Consider first the dark adaptation data shown on the left, a replication and extension of findings reported by Lie (1963). Following a monocular bleach, sensitivity to hue increases during the cone recovery stage of adaptation, but then rapidly decreases. A large number of control experiments by Lie (1963), as well as corresponding increment threshold data (bottom row, middle column), indicate that dark-adapted rods suppress hue sensitivity; in the case illustrated, to green. There are several reasons for believing this 'Lie effect' is mediated by a different mechanism to those responsible for the upper two rows of data. First, although the adaptation state of the non-viewing eye influences the Lie effect (compare the influence of binocular, monocular, and interocular bleaching upon the time-course and magnitude of the specific-threshold curves; also see Prestude *et al.*, 1978), this effect cannot be classified as either purely monocular or binocular. Second, unillustrated data show that the magnitude and time course of the Lie effect depend upon the wavelength or perceived hue of the test probe. In particular, the suppressive rod influence upon blue and yellow hues is similar in magnitude and time course, while the influence upon green and red is greater in magnitude, and follows different time courses. This suggests that the Lie effect is mediated at a site or sites where colour vision is segregated into several different pathways.

References

Frumkes, T.E. and Eysteinsson, T. (1988). The cellular basis for suppressive rod–cone interaction. Visual Neurosci. 1: 263–273.

Frumkes, T.E., Lange, G., Naarendorp, F. and Eysteinsson, T. (1995). Suppressive rod–cone interactions: underlying mechanisms and practical application. Doc. Ophthalmol. Proc. Ser. 57: 330–334.

Goldberg, S.H., Frumkes, T.E. and Nygaard, R.W. (1983). Inhibitory influence of unstimulated rods in human retina: evidence provided by examining cone flicker. Science 221: 180–182.

Lie, I. (1963). Dark adaptation and the photochromatic interval. Doc. Ophthalmol. 17: 411–510.

Livingston, M.S. and Hubel, D.H. (1987). Psychophysical evidence for separate channels for the perception of form, color, movement, and depth. J. Neurosci. 7: 3418–3468.
Naarendorp, F., Denny, N. and Frumkes, T.E. (1988). Rod light and dark adaptation influence cone-mediated spatial acuity. Vision Res. 28: 67–74.
Pflug, R., Nelson, R. and Ahnelt, P.K. (1990). Background-induced flicker-enhancement in cat retinal horizontal cells. I: Temporal and spectral properties. J. Neurophysiol. 64: 313–325.
Prestude, A.M., Watkins, L., and Watkins, J. (1978). Interocular light adaptation effect on the 'Lie' specific threshold effect. Vision Res. 18: 855–857.

Thomas E. Frumkes*, E.Lebessis, J. Vollaro, D. Moshe
Dept. of Psychology
Queens College of CUNY
Flushing, New York 11367, USA

T. Eysteinsson
Department of Physiology
University of Ireland
Reykjavik 101, Iceland

*Corresponding author

54. Defective colour vision is a risk factor in driving

B.L. COLE AND J.D. MADDOCKS
(Melbourne, Australia)

Abstract

Verriest *et al.* (1980) claimed 'definite proof that colour-defective drivers do not have more accidents than people with normal colour vision' by showing that colour-defective drivers were not over-represented in a sample of 2058 male drivers who had caused a traffic accident. In the same paper, however, it is shown that protans have significantly more rear-end collisions and other accidents caused by overlooking signal lights. Deutans had more accidents at traffic lights. The differing conclusions can only be reconciled by assuming that colour-defective drivers have fewer other accidents to compensate for their greater frequency of accidents involving signal lights, and this is inherently improbable.

Verriest *et al.* were incorrect to conclude, from their data, that defective colour vision is not a risk factor in driving, because their sample size was too small and the study lacked a control group. Even so, they found that 8.41% of the accident-causing sample had defective colour vision, somewhat more than the usually-assumed 8%.

The odds ratio for defective colour vision as a risk factor, calculated from comparison of the Verriest *et al.* data with those reported by European investigators, was 1.11. This is not significant, but the confidence interval raises the statistical possibility that the odds ratio may be as high as 1.31.

It is to be expected that drivers with defective colour vision will have difficulty locating and recognizing traffic signals because of their reduced ability to differentiate colours and, in the case of protans, their reduced ability to see red signals. We present data that shows that about 20% of anomalous trichromats and nearly 60% of dichromats admit to difficulty recognizing colours while 10–15 % of protans admit to difficulty seeing red signal lights. The accident data of Verriest *et al.* (1980) and also those of Hager (1963), taken with other evidence, lead to the conclusion that defective colour vision is a risk factor for driving. However, it is a risk that bears only on the 8% of drivers who have defective colour vision and the number of accidents due to defective colour vision will therefore be relatively small, probably no more than 4 in 1000 accidents. Nevertheless it would be in the public interest to apply a colour vision standard at least for drivers of commercial and public transport vehicles, since they drive greater distances than private drivers and the social cost of accidents involving commercial and public transport vehicles is higher.

C.R. Cavonius (ed.), Colour Vision Deficiencies XIII, pp. 471–481.

Introduction

There has been a long and somewhat inconclusive debate as to whether or not defective colour vision is a risk factor in driving motor vehicles. Persons with defective colour vision do make errors recognizing signal colours and are slower in their response to them (Nathan *et al.*, 1964) and protans have a reduced visual range to red signals (Verriest *et al.*, 1980; Cole and Vingrys, 1983). Of course, it does not follow that they will have accidents as a result: they may successfully use other cues such as the relative brightness of signals, signal light position, and the movement of other traffic, to help them recognize signal lights; or they may drive more cautiously.

Norman (1960), Gramberg-Danielsen (1961) and Zehnder (1971) did not find that colour-defective drivers had more accidents than colour-normal drivers, and Sachsenweger and Nothaass (1961) did not find defective colour vision to be over-represented in a group of drivers involved in an accident. These findings lend credence to the view that colour-defective drivers compensate in some way for their reduced ability to see and recognize signal lights, so that abnormal colour vision is not a risk factor.

The Verriest *et al.* study (1980)

Verriest *et al.* reported that 8.41% of a sample of 2058 male drivers who had been responsible for a traffic accident had defective colour vision. This is almost exactly the same prevalence of abnormal colour vision as is recorded among non-selected male Caucasian populations and they concluded 'we now have definitive proof that colour-defective drivers, either protan or deutan, do not have more traffic accidents than people with normal colour vision'.

There are a considerable number of difficulties with this conclusion. The first is that the inductive inference of science is such that it is not possible to prove that something does not happen: scientific inference requires the observation of differences and effects as a consequence of the operation of an independent variable for a conclusion to be reached. If no difference or effect is observed, there always remains the possibility that a replication of the investigation or a better-designed investigation may reveal an effect. The second difficulty is that this investigation was not a case control study: there was no control group of drivers not involved in an accident to enable proper comparison of the prevalence of defective colour vision (DCV) in accident-causing and accident-free drivers. The third difficulty is that the sample number of 2058 drivers was not large enough. Sample size for both the accident-causing and the accident-free control groups needs to be between 4000 and 800 000, depending on the level of risk of accident due to abnormal colour vision (Fig. 1).

Despite these difficulties the data of Verriest *et al.* suggest that DCV could be a risk factor, contrary to the conclusion so emphatically reached by the authors. The prevalence of defective colour vision found in the accident-causing male

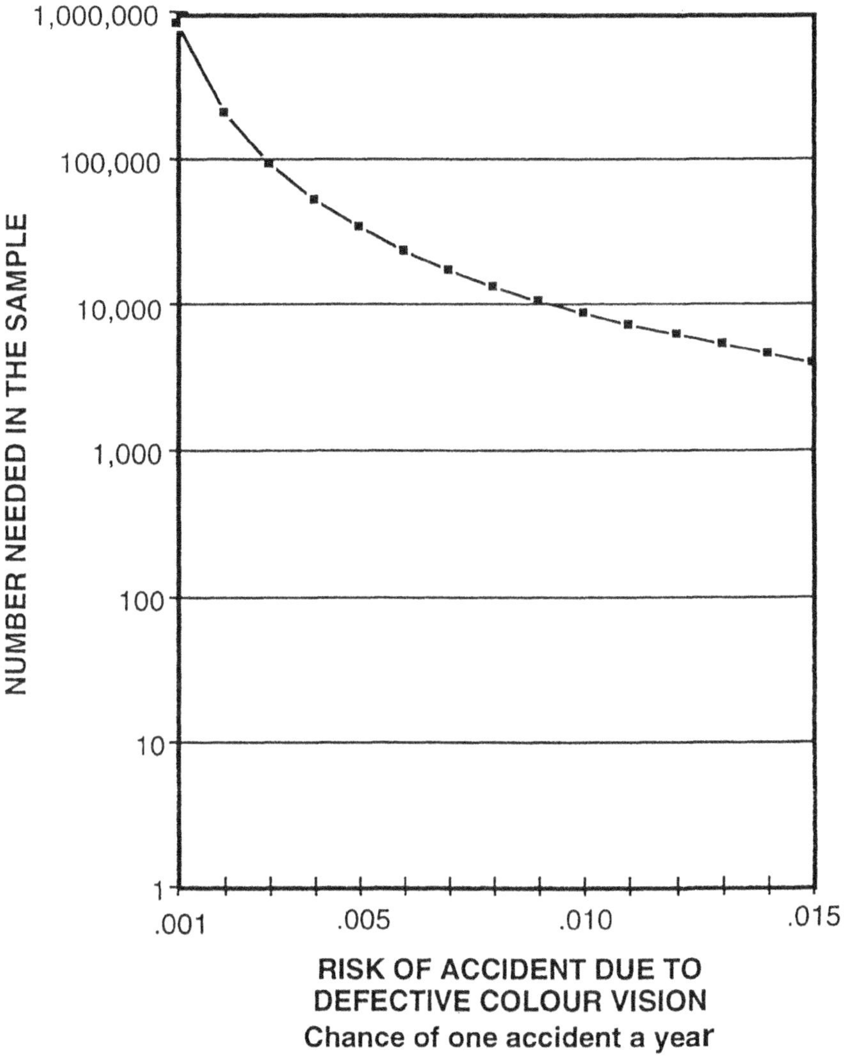

Fig. 1. The number of subjects needed in both an accident-causing and accident-free group for a case-control study to test whether defective colour vision is over-represented in the accident-causing group, for various assumed levels of risk of accident due to defective colour vision. The sample sizes needed are for a power of 80% for a one-tailed test with a level of significance of 0.05.

Table 1. Prevalence of abnormal colour vision amongst males.

Reference (Year)	Sample size	Prevalence
European studies		
Waaler (1927)	9 049	8.01
von Planta (1928)	2 000	7.95
Schmidt (1936)	6 863	7.75
Nelson (1938)	1 338	8.82
Vernon and Straker[a] (1943)	123 414	7.49
Francois *et al.* (1957)	1 243	8.37
Kherumian *et al.* (1959)	6 635	8.95
Crone (1968)	3 167	7.95
Koliopoulos *et al.* (1976)	21 231	7.95
Weighted average		7.67
Australian studies		
Mann and Turner (1956)	558	7.30
Robbins and Bailey (1975)	1 184	5.50
Steward and Cole (1989a)	3 040	6.32
Weighted average		6.23
Weighted average of all studies		7.63

[a] Vernon and Straker give the prevalence for their whole sample as 7.49% in their original paper, but do not state their sample size. Crone (1968), citing the Vernon and Straker report, lists their sample size as 123 414, the value we have used in this table. He quotes 7.25% as the prevalence for DCV. We have used a prevalence of 7.49% as it was the value given in Vernon and Straker's original report.

drivers was 8.41%, when the usually quoted prevalence among males is 8%. This numerical difference suggests a higher prevalence in the accident-causing group. Table 1 shows the reported prevalences of defective colour vision found in 12 studies of males, the weighted average of which is 7.67%: this widens the numerical difference.

Statistical comparison of the prevalences reported in the nine European investigations of male colour vision, listed in Table 1, showed that they were significantly different when the French study by Kherumian and Pickford (1959) was included. When this outlier study was excluded there was no longer any significant difference.

The prevalence of defective colour vision in the Verriest *et al.* sample of accident-causing drivers was compared with the European studies listed in Table 1. With the Kherumian and Pickford study included in the calculations and then excluded, an odds ratio of 1.11 was derived in each computation, with 95% confidence intervals (CIs) of (0.94, 1.30) and (0.95, 1.31) respectively. The odds ratio is not significantly different from 1.0 in each case. However, the difference

Table 2. Rear-end accidents and those involving overlooking red rear, stop and warning lights for protans and colour vision normal drivers (data from Verriest *et al.*, 1980).

	No. of accident-causing drivers	No. of accidents involving not seeing red lights	% accidents
Protans	41	18	43.9
Colour vision normal	1885	490	26.0
Total	1926	508	

$\chi^2 = 5.98$; df = 1; $p = 0.015$

is not far from statistical significance and the CIs suggest that the odds ratio might be as high as 1.31.

The three Australian population studies listed in Table 1 were not statistically different from each other. If the prevalence of 8.41% in Verriest's accident-causing group of drivers is compared with the lower estimates of prevalence reported by the three Australian studies then the difference is significant (95% CI (1.13, 1.68)).

It is worth noting that Hager (1963) found a significantly higher prevalence of defective colour vision among 342 drivers responsible for accidents (8.5%) compared with the prevalence of DCV among 2236 drivers presenting to an eye clinic (4.6%), although his paper fails to say if the samples were exclusively male or, if not, what was the proportion of females.

Protan (and possibly deutan) colour vision defects are a risk factor

Verriest *et al.* also reported the number of accidents involving signal lights for normal and DCV drivers in the sample of 2058 accident-causing drivers. These are reproduced in Tables 2 and 3. Protans have significantly more accidents involving red signal lights than colour vision normal (CVN) drivers while deutans have twice as many accidents at traffic-light controlled intersections, although the latter difference does not reach statistical significance. Hager (1963) reported data (Table 4) that also show that protan defects of colour vision are a risk factor, and his data suggest that deutan defects might likewise be associated with a higher risk of accident.

It is difficult to reconcile the emphatic conclusion reached in the Verriest *et al.* paper that colour-defective drivers do not have more accidents on the basis of a lack of over-representation of DCV in accident-causing drivers, when the same paper reports a significantly greater number of signal-related accidents for protans and a similar trend for deutans. In fairness, it must be said that the

Table 3. Accidents at traffic-light-controlled intersections reported for deutans and colour vision normal drivers (Data from Verriest *et al.*, 1980).

	No. of accidents	No. of accidents	% accidents
Deutans	132	6	4.6
Colour vision normal	1885	41	2.2
Total	2017	47	

$\chi^2 = 2.44$; df = 1; n.s. ($p = 0.12$)

Table 4. Number and proportion of protan and deutan drivers in a group of 342 drivers who had been responsible for an accident compared with the number and proportion in a group of 2236 drivers attending an eye clinic (data from Hager, 1963).

	Attending an eye clinic		Responsible for an accident		
	No.	%	No.	%	
Protans	25	1.1	12	3.5	$p<0.001$
Deutans	78	3.5	17	5.0	n.s ($p = 0.15$)

several authors of the Verriest *et al.* paper had different views about the evidence relating defective colour vision to risk of accident. Verriest was of the view that 'colour defective drivers are not responsible for more traffic accidents (and) it is useless to exclude them, even for professional driving'. His co-author Marré is reported in the paper as concluding that protans should be excluded from professional driving while Neubauer concludes that protanopes, deuteranopes and achromats should be excluded from professional conveyance of passengers and of dangerous goods.

There is also a need to reconcile the findings of Verriest *et al.* (1980) and Hager (1963) with those that find no differences in the accident rates for colour vision normal and abnormal drivers (Norman, 1960; Gramberg-Danielsen, 1961; Sachsenweger and Nothaass, 1961; Zehnder, 1971). Norman's study was well designed and controlled. It compared the accident rates of 149 colour-normal bus drivers to the same number of colour-defective bus drivers. The subjects were matched for age and duration of service and the drivers all drove the same route. Although the sample sizes were relatively small, the accumulated number of accidents recorded was of reasonable size (712 and 681 accidents for the two groups respectively). However, if the risk of accident due to defective colour vision is small, this sample size might still not be sufficient. In any event, bus drivers drive a familiar route, drive at low speeds and are subject to disciplined driving by the rules of the authority and it may be unwise to generalize this result to drivers of long-distance buses and trucks and private drivers.

If the odds ratio is about 1.1, the sample numbers for the comparison of accident rates of CVN and DCV drivers in the Gramberg-Danielsen (1961) and Zehnder (1971) studies are too small: there were 507 in each of the DCV and CVN groups in the Gramberg-Danielsen study and 169 in each group of the Zehnder study. Sachsenweger and Nothaass (1961) sought to show that DCV drivers were over-represented in an accident-involved group but only had 196 in their sample, which is far too few to hope to show a difference if the odds ratio is of the order of 1.10.

What do people with defective colour vision say?

Steward and Cole (1989b) asked 102 persons with defective colour vision about their everyday experiences with colour. Some of the questions related to their difficulty with driving. Part of that data is reproduced in Figures 2 and 3, together with some additional data not included in the original report .

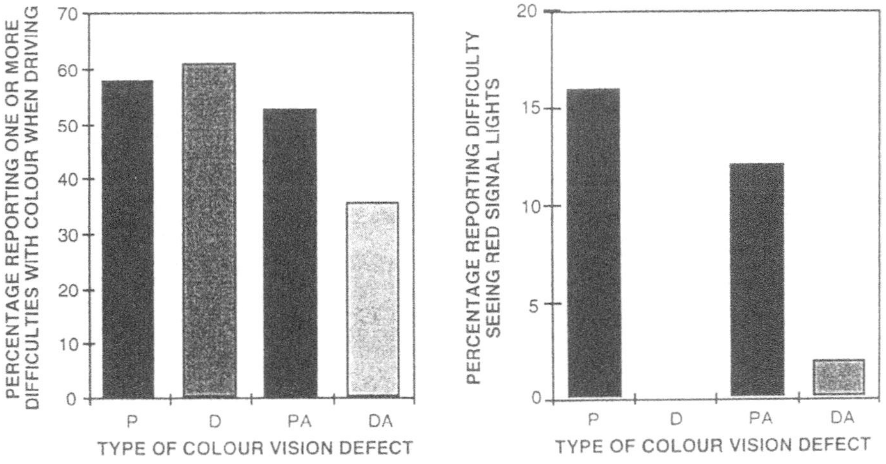

Fig. 2. The percentage of subjects with defective colour vision reporting (a) one or more difficulties with colour when driving and (b) difficulty seeing red signals. Data from Steward and Cole (1989b), in which 102 persons with defective colour vision consecutively presenting to an optometric practice for a reason other than their colour vision were given a comprehensive questionnaire about their problems with colour. A control group of 102 persons with normal colour vision reported no difficulties in answer to the same questions. P = protanope; D = deuteranope; PA = protanomal; DA = deuteranomal

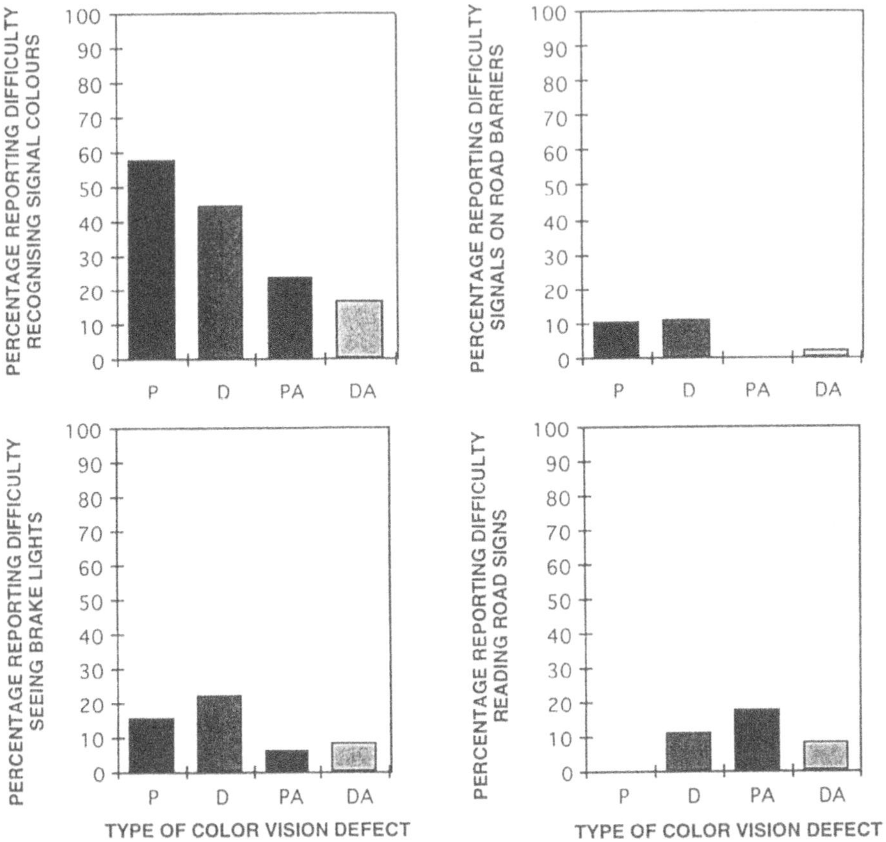

Fig. 3. The percentage of 102 subjects with defective colour vision reporting difficulty with (a) recognizing signal colours, (b) seeing signals on road barriers, (c) seeing brake lights and (d) reading road signs. Data from Steward and Cole (1989b). A control group of 102 persons with normal colour vision reported no difficulties in answer to the same questions.

Figure 2a shows that some 60% of dichromats and 35% to 55% of anomalous trichromats have one or more difficulties with colour when driving. Figure 2b shows that 12% of protanomals and 16% of protanopes report difficulty seeing red signals, a difficulty rarely reported by deutans. About half of the dichromats and one in five anomalous trichromats report problems recognizing signal colours (Fig. 3a). A few DCVs reported difficulties with lights on road barriers, seeing brake lights and reading road signs (Figure 3b–d). One in four protanopes reported difficulty seeing dashboard warning lights.

It should be noted that the questionnaire given to DCV subjects asked for an affirmative response to the questions about difficulty with colour only when the difficulty occurred often and was of some practical significance to them. A trivial or rarely occurring problem did not qualify for an affirmative answer. An equal-sized control group of subjects with normal colour vision who were given the same questionnaire did not reply in the affirmative to any of the questions.

Defective colour vision is a risk factor in driving

The convergence of evidence leads to the conclusion that DCV is a risk factor in driving. Laboratory experiments and field trials show that colour-defective persons do make errors recognizing signal colours and that protans have a reduced visual range for red signal lights. The results of the questionnaire of Steward and Cole (1989b) about difficulties with colour show that over half of those with DCV admit to problems with signals when driving. The data of Verriest *et al.* (1980) and that of Hager (1963) suggest that DCV drivers may be over-represented among drivers who have accidents and, less equivocally, it seems that protans, and perhaps deutans as well, have more accidents associated with signal lights than drivers with normal colour vision.

Cost-benefits of a standard for colour vision for driving

In a developed country of 20 million people there will be about 700 000 accidents a year (Bureau of Transport and Communication Economics, 1992) of which about 400 000 will involve male drivers. If there is no exclusion for colour vision, approximately 8% of those male drivers will have defective colour vision and they will be involved in about 8% of the accidents, some 32 000 accidents a year. If the odds ratio for DCV as a risk factor is 1.1 and the probability of an accident for all other causes is 0.07, then, ignoring the possibility of drivers having more than one accident in a year, DCV will be a contributing cause in 2700 of those accidents. The social cost of an accident ranges widely from those that involve property damage only (about $US 4000 per accident) to those that cause death (about $US 475 000). Weighted by the relative frequency of the various kinds of accidents, the average cost is about $US 8000 (These costings are from Australian Government statistics and were originally in Australian dollars converted here to $US; Bureau of Transport and Communication Economics, 1992.) The 2700 accidents related to defective colour vision can therefore be ascribed an annual social cost of about $US 21 million for a country of 20 million people. However, it is unlikely that there would be community acceptance of any proposal to exclude persons with defective colour vision from private driving. There could be community support for a colour vision standard for commercial drivers, especially those involved in driving public transport vehicles, large trucks and those conveying hazardous goods. About 20% of vehicles are commercial and are responsible for more than 30% of the distance travelled. Exclusion of protans and those with more severe loss of colour discrimination has the potential to reduce road accidents by some 1 in 1000 accidents.

Conclusion

It needs to be recognized that people with DCV do have problems with traffic signals and that DCV is probably a risk factor in driving. The first response to this should be to take care in the design of roads and traffic signals to minimize the problems experienced by drivers with defective colour vision. This can be done by careful choice of signal colours to minimize uncertainty in recognizing the colours and by ensuring that all red signal lights have an intensity sufficient to enable protans to see the signal at a safe distance. The Commission Internationale de l'Éclairage (CIE) provides good guidance on these two matters (CIE, 1980). Increasing the discriminability of signals by making the colours different sizes should also be cosidered.

There are also social and economic reasons to adopt a colour vision requirement for commercial drivers to exclude protans, deuteranopes and the more severely affected deuteranomals.

Acknowledgement

We are grateful for the assistance of Dr Ken Sharpe, Department of Statistics, University of Melbourne, Parkville, Australia, who performed the statistical analyses.

References

Bureau of Transport and Communication Economics (1992). Social cost of transport accidents in Australia. Report 79. Aust. Govt. Pub. Service., Canberra.

CIE (1980). Light signals for road traffic control. CIE Publication 48, CIE Central Bureau, Paris.

Cole, B.L. and Vingrys, A.J. (1983). Do protanomals have difficulty seeing red lights? Proc. CIE 20th Session, Amsterdam, 1983. CIE: 1–3. CIE Central Bureau, Paris.

Crone, R.A. (1968). Incidence of known and unknown colour vision defects. Ophthalmologica. 155: 37–55.

Francois, J., Verriest, G., Mortier, V. and Vanderdonck, R. (1957). De la fréquence des dyschromatopsies congénitales chez l'homme. Ann. Oculist., Paris. 190: 5–16. Cited in Crone (1968).

Gramberg-Danielsen, B. (1961). Untersuchungen über die Unfallhäufigkeit von Farbenuntüchigen im Strassenverkehr. Klin. Mbl. Augenheilk. 139: 677–681.

Hager, G. (1963). Das Sehorgan und das Unfallgeschehen im Strassenverkehr. Klin. Mbl. Augenheilk. 142: 427–433 .

Kherumian, R. and Pickford, R.W. (1959). Hérédité et fréquence des dyschromatopsies: p. 109. Vigot Frères, Paris. Cited in Crone (1968).

Koliopoulos, J., Iordanides, P., Palimeris, G. and Chimonidou, E. (1976). Data concerning colour vision deficiencies amongst 29,985 young Greeks. Mod. Probl. Ophthal. 17: 161–164.

Mann, I. and Turner, C. (1956). Color vision in native races in Australasia. Am. J. Ophthal. 41: 797–800.
Nathan, J., Henry, G.H. and Cole, B.L. (1964). Recognition of colored road traffic light signals by normal and color-vision-defective observers. J. Opt. Soc. Am. 54: 1041–1045.
Nelson, J.H. (1938). Anomalous trichromatism and its relation to normal trichromatism. Proc. Phys. Soc. (Lond.) 50: 661–702.
Norman, L.G. (1960). Medical aspects of road safety. Lancet i: 1039–1045.
Robbins, H.G. and Bailey, I.L. (1975). The organisation and findings of a school vision screening service. Aust. J. Optom. 58: 392.
Sachsenweger, R. and Nothaass, E. (1961). Eine Analyse von 4011 Verkehrsunfällen aus augenärztlicher Sicht. Dts.Gesundh. Wed. 12: 567–872.
Schmidt, I. (1936). Ergebnis einer Massenuntersuchung des Farbensinns mit dem Anomaloscop. Z. Babnärzte. 2: 1–10. Cited in Crone (1968).
Steward, J.M. and Cole, B.L. (1989a). Incidence of congenital colour vision defects in an Australian optometric population. In: Drum, B. and Verriest, G. (eds), Colour Vision Deficiencies IX: 109–111. Kluwer, Dordrecht.
Steward, J.M. and Cole, B.L. (1989b). What do color vision defectives say about everyday tasks? Optom. Vis. Sci. 66: 288–295.
Vernon, P and Straker, A. (1943). Distribution of colour blind men in Great Britain. Nature, Lond. 152: 690.
Verriest, G., Neubauer, O., Marré, M. and Uvijls, A. (1980). New investigations concerning the relationships between congenital colour vision defects and road traffic security. Int. Ophthal. 2: 87–99.
von Planta, P. (1928). Die Häufigheit der angeborenen Farbensinnstörungen bei Knaben und Mädchen und ihre Feststellung durch die üblichen klinischen Proben. Graefes Arch. Ophthal. 120: 253–281. Cited in Crone (1968).
Waaler, G.H.M. (1927). Über die Erblichkeitsverhältnisse der verschiedenen Arten von angeborener Rot-Grün-Blindheit. Acta Ophthal., Kbh. 5: 309–347. Cited in Crone (1968).
Zehnder, E. (1971). Die Bewahrung farbensinngestorter Motorfahrzeuglenker im Verkehr. Schweiz. Med. Wochenshr. 101: 530–537

The University of Melbourne
Department of Optometry and Vision Sciences
CNR Keppel and Cardigan Streets
Carlton, Victoria 3053, Australia

55. Doctors with inherited colour vision deficiency: their difficulties in clinical work

J.A.B. SPALDING
(Weybridge, UK)

Abstract

The range of difficulties in clinical practice of doctors – mainly GPs – with inherited colour vision deficiency was determined by questionnaire and related to results of colour vision testing for type and severity. Many difficulties were found, more amongst those with moderate or severe defects, and doctors' attitudes to their own colour deficiencies varied. Vocational screening and counselling for prospective doctors should be considered.

Introduction

Studies relating the clinical skills of doctors to the results of tests of colour discrimination are rare (Voke, 1980; Vakil *et al.*, 1991; Rigby *et al.*, 1991; Koningsberger *et al.*, 1994) and only three doctors seem to have published articles on their own difficulties in clinical work due to their inherited defect of colour vision (Logan, 1977; Spalding, 1992; Currier, 1994). Colour plays a part in many observations in medicine, however, and colour deficiencies are not uncommon. There have been no studies of its prevalence among doctors but such information as there is suggests that about 2000 general practitioners and 3000 hospital doctors practising in the UK in 1992 had a colour vision defect (Voke, 1980; Rigby *et al.*, 1991; Koningsberger *et al.*, 1994; MacLean, personal communication). Medical students in most countries are not screened for colour deficiencies. School-children in the UK are screened but not tested for severity.

This study aimed to discover the range of difficulties of general practitioners and to relate these to the type and severity of their defect. General practitioners have many ways to compensate for their defect, but certain features of their work may still be adversely affected.

Materials and methods

Doctors completed a questionnaire in response to letters in medical journals. They were given a battery of colour vision tests either in university departments

C.R. Cavonius (ed.), Colour Vision Deficiencies XIII, pp. 483–489.

– mainly of optometry and visual science – or by being visited. They were told the results of their tests and counselled.

The tests and numbers taking them were: Ishihara (39), City University (33), Farnsworth D15 (22), Farnsworth-Munsell 100-hue (22), Nagel Anomaloscope (18); Pickford-Nicholson Anomaloscope (2), American Optical HRR (6).

The word 'difficulty' in the questionnaire was intended to have its dictionary meaning, but because one doctor was uncertain of this, it was clarified in the last 15 questionnaires in the following way: 'A difficulty is any problem believed to be caused by the defect whether or not it is overcome'.

Results and discussion

Of the 42 doctors who completed the questionnaire, 40 were tested. The sample comprised 38 males and 2 females, with a mean age of 48.3 years; all but five were general practitioners; nine were retired.

Some verbatim answers

(D = deuteranope (dichromat); P = protanope (dichromat); DA = deuteranomalous trichromat; PA = protanomalous trichromat)

DA (moderate) Difficulties are overcome by awareness, self-training and effort. Nevertheless, red/green blind doctors would be better advised to avoid careers in haematology, histology and bacteriology.

DA (severe) I was aware of my colour blindness before going to medical school, but I only just realized (after being qualified 6 years) why I was useless at certain things, e.g. histology, clinical examination. My loss of confidence might have been sympathetically handled by medical school staff had they been on the lookout for this problem.

D (i.e. severe) Nowadays I am a trainer for young physicians in vocational training and advertise my own handicap in order to teach the young people to do the same, and then we consciously develop methods to overcome the problem.

D (i.e. severe) I did a year in pathology and had not realized that stains showed different tissue – no one had told me. My eyes pick up very fine detailed physical features of rashes etc. I feel this and body language are major advantages I have developed. I frequently ask my colleagues for advice, especially over babies with rashes and fevers etc. and the chance of a red ear or throat. I feel very vulnerable at the end of a busy surgery. I believe there are times when patients describe red rashes and I cannot see them and nurses point to the invisible dots.

I cannot believe that I could be guided to two careers which depended heavily on colour, e.g. pathology and dermatology. The profession offers no guidance and I feel vulnerable at times.

P (i.e. severe) In my ophthalmological training I will probably choose a subspeciality that does not concentrate on fundal examination, e.g. oculoplastic, strabismus, etc., as opposed to medical retina, vitreo-retinal surgery.

DA (severe) I once diagnosed a haematemesis as bile. The patient was lucky to survive. I once very nearly missed a haemorrhage into the anterior chamber in an Indian boy – but eventually saw the line of the blood–fluid level.

DA (moderate) I worry about missing the so-called pink ear.

DA (moderate) [For pallor]. In bedside work assessment by skin and conjunctival colour is not so easy as for the normal.

DA (moderate) I believe I am a safer anaesthetist because I insist on the appropriate monitors, especially pulse oximetry which can detect a drop in oxygen rather earlier than any non-colour-blind clinician.

Results of colour vision tests

All doctors had previously received a screening test (Table 1). One doctor first learnt that he had a defect (moderate) 6 years after qualification. Only 6 could state whether they were protan or deutan.

There were 33 deutans (6 slight, 10 moderate and 17 severe, of which 13 were dichromats) and 7 protans (1 slight, 1 moderate and 5 severe, of which 4 were dichromats).

What difficulties and advantages did you have in everyday life?

For difficulties, 28 mentioned dress sense, 15 decor, 13 traffic lights and signals. Others mentioned included sport, navigation, aesthetic sense and maps. Those with moderate and severe defects tended to report more difficulties. For advantages, 26 reported none, 5 the detection of camouflage and 4 enhanced perception in ways not involving colour.

A comparison can be made with the rather similar findings of Steward and Cole (1989) and Cole and Steward (1996).

The protan/deutan ratio is about the same as for the general population but the number of dichromats is higher. For anomalous trichromats the proportions of mild, moderate, and severe gradings is not known for the general population.

What difficulties did you have in medical practice and as a student?

A total of 197 difficulties due to colour vision defects was reported (Table 2).

Table 1. Colour vision tests performed and doctors' knowledge of their own defect prior to this study.

Screening test.	Ishihara, 38; Coloured wool, 1; Lantern, 1
Other tests done	D15, 2; 100-Hue, 3; City University, 2; Anomaloscope, 2; Bacon, 1
Tests performed by:	School medical service: 18; At college: 11; Others: 11
Advice reported to be given on screening	None: 27; About professions: 12; Misinformed: 1
Knowledge of:	
Severity:	Unknown or incorrect, 28
Type:	Partly known (i.e. red/green): 30
Doctors who learned late that they had a defect	6 years after qualification: 1 Within 2 years following qualification: 2

The verbatim answers given earlier help in appreciating the significance of these results. Doctors were aided by the following headings in the questionnaire: Paediatrics, General Medicine, Surgery, Dermatology, Infectious Diseases, Otoscopy, Ophthalmology, Mouth and Throat, Reading Test Results (e.g. test-tapes), and Charts.

If you have difficulties in medical practice how do you overcome them?

Seventeen answered that they did so by close observation or cross-checking: they specified looking, touching, doing special investigations and attention to lighting; seven asked for help from others; four gave more attention to the history. One reported that the rash of scarlet fever felt like sandpaper and that he felt for the rash of typhus. Some reported that when reading test tapes they relied on 'shade' or 'tone' rather than colour and only one reported using a meter.

Difficulties in medical practice and as a student related to severity and type of defect

There appeared to be no difference between moderate and severe defects for numbers of difficulties reported. Some doctors with moderate or severe defects reported no or few difficulties and make a striking contrast to those who reported many difficulties. One doctor with a mild defect reported as many as eight difficulties, but a further interview with him suggested that he was misinterpreting the cause of some difficulties. No differences were shown between protans and deutans for type or number of difficulties.

Table 2. Difficulties reported in medical practice and as students.

Subject	*n*
Widespread body colour changes: pallor (12), cyanosis (9), jaundice (3), cherry-red (2)	26
Dermatology/rashes/erythema of skin (6)	25
Charts (13), slides (5), prints (2), codes (4)	24
Test-strips for blood and urine	22
Ophthalmology: disc pallor (3), diabetic changes (2), haemorrhage versus pigment (1), glaucoma (1), haemorrhage in anterior chamber (1), Kayser-Fleischer rings (1), others (9)	18
Body products: blood or bile in urine, faeces, sputum,vomit	18
Otoscopy: the inflamed drum (8), wax versus blood (1), others (5)	14
Microscopy (students: 11)	13
Mouth and throat conditions	9
Ishihara test giving	8
Chemistry end-points (students)	7
Colour naming	5
Tissue identification (surgery: 3)	4
What am I missing?	3

Table 3. Significance given to difficulties related to their number, and to type and severity of defect.

Response	Mean no. of difficulties	Severity			Type of defect	
		Slight	Moderate	Severe	Deutan	Protan
No or very little significance	2.1	5	6	2	11	2
Some significance	6.8	0	4	11	11	4
Major significance	9.5	1	2	5	7	1
No difficulties	0.0	1	0	3	4	0

Do your think your defect has given you any advantages in medicine?

Twenty-six answered no; five thought that it enhanced their ability in other types of observation; three felt that they were able to observe jaundice at an earlier stage than other doctors and one had confirmed this by bilirubin blood levels. If reliable this may be a new finding.

The significance doctors gave to difficulties related to colour vision defects

The majority thought that overall they were functioning effectively. Some thought their defect of little or no significance, particularly those with a mild defect, and others that it was of some significance but did not cause a major problem. Eight doctors thought the defect was of major significance by virtue of the fact that they believed that they should not practise in certain specialities, or, in the case of one doctor, because he thought he had put a patient's life at risk due to his defect, which resulted in him concluding that bile rather than blood was present in vomit. The specialities mentioned were: histology, haematology, bacteriology, surgery, pathology, dermatology, anaesthetics and retinal work in ophthalmology. They may or may not have been correct in coming to these conclusions, but the fact that they did indicates that they took their response seriously. However, no doctor stated that he or she should not be working in general practice. The 11 doctors with moderate or severe defects who noticed no difficulties or gave them no or little significance may have been failing to notice observational errors.

The results of this study point towards the need for pre-vocational screening for colour vision deficiency and counselling about its implications in the medical profession.

Acknowledgements

I thank J. Birch (City University, London) for much valuable advice and for testing subjects; the following for their part in performing tests: C. Dickinson (Manchester), H. Yorke (Aston), G. Heron (Glasgow), J.D. Mollon (Cambridge) and A. Hill (Oxford); and all the doctors who took part in the study.

References

Cole, B.L. and Steward, J.M. (1996). Some (but only a few) colour vision defectives have no difficulty with colour. In: Dickinson, C., Murray, I. and Carden, D. (eds.), John Dalton's Colour Vision Legacy: 239–245. Taylor and Francis, London.

Currier, R.D. (1994). A two-and-a-half coloured rainbow. Arch. Neurol. 51: 1090–1092.

Koningsberger, J.C., van Norren, D., van Niel, J.C. and Dekker, W. (1994). Does colour vision in the endoscopist influence the accuracy of endoscopic diagnosis? Endoscopy 26: 549–553.

Logan, J.S. (1977). The disability in so-called red–green blindness. An account based on many years of self-observation. Ulster Med. J. 46: 41–45.
Rigby, H.S., Warren, H.F., Diamond, J., Carter, C. and Bradfield, J.W.B. (1991). Colour perception in pathologists: The Farnsworth-Munsell 100-hue test. J. Clin. Pathol. 44: 745–48.
Spalding, J.A.B. (1993). A doctor with an inherited defect of colour vision – effect on clinical skills. Br. J. Gen. Pract. 43: 32–33.
Steward, J.M. and Cole B.C. (1989). What do colour vision defectives say about everyday tasks. Optom. Vis. Sci. 66: 288–295.
Vakil, N., Knyrim, K. and Everbach, E.C. (1991). The appreciation of colour in endoscopy. Bailliére's Clin. Gastroenterol. 5: 183–194
Voke, J.E. (1980). Colour Vision Testing in Specific Industries and Professions. 13–15. Keeler, London.

5 Curzon Road
Weybridge
Surrey KT13 8UW, UK

56. How much light reaches the retina?

J. POKORNY and V.C. SMITH
(Chicago, IL, USA)

Abstract

There are three major prereceptoral factors governing the amount of light reaching the retina: pupil area, lens optical density and the optical density of the macular pigment. Standardized tabulations are available for each of these factors. These standard tabulations represent well the characteristics of an average observer and thus have application in modelling the behavior of a representative observer. In recent years significant advances have been made in the understanding of receptoral and postreceptoral retinal mechanisms, and it is now feasible to characterize aspects of these processes in individual observers from a relatively limited amount of psychophysical data. This communication addresses the question: are average tabulations for pupil, lens and macular pigment applicable to individual observers or is individual variability so great that independent assessment of these factors need be obtained?

Not all light impinging on the cornea reaches the photoreceptor layer of the retina. The incident light is controlled by the size of the pupil and may be spectrally selectively attenuated by the lens and macular pigment. There are characterizations of the effect of each of these prereceptoral factors for an average observer in the literature. Here we ask how applicable these 'Standard Observer' tabulations are in assessing visual function of individual observers.

Lens

Density spectrum for an average young observer

For the lens, Wyszecki and Stiles (1967; 1982) proposed a lens optical density function representative of a 32-year-old observer. Van Norren and Vos (1974) suggested a lens density function which agrees with the Wyszecki and Stiles function for wavelengths above 420 nm but shows higher densities in the 400–420 nm range. Recent analyses favor the van Norren and Vos density spectrum (Stockman *et al.*, 1993; van den Berg, 1993; van den Berg and Felius, 1995).

C.R. Cavonius (ed.), Colour Vision Deficiencies XIII, pp. 491–511.

Effect of age on lens density

Variation in lens density with adult age has been modelled as a scalar multiplier of the lens density function of a young observer (van Norren and Vos, 1974; Savage *et al.*, 1993) or with a two-component model (Pokorny *et al.*, 1987). The two-component model is characterized by two density spectra, one of which is age invariant while the second varies with age. Several recent studies (van den Berg, 1993; van den Berg and Felius, 1995) found that the two-component model better describes the changes in lens density with age. van den Berg and Felius (1995) have associated the two components with two classes of chromophores present in the human lens. The component which does not vary with age, T_{L2}, has been associated with the kynurenines (Cooper and Robson, 1969; van Heyningen, 1971; Bando *et al.*, 1981) while T_{L1}, the age-dependent component, has been associated with another class of proteins, the aged proteins (Spector *et al.*, 1975; Lerman and Borkman, 1976). There is a minor attenuation of short wavelength light by the cornea (Boettner and Wolter, 1962; van den Berg and Tan, 1994); in the two-component model this is included in the density spectrum of T_{L2}.

Contemporary with the Pokorny *et al.* publication was an analysis by Weale (1988), based upon measurements of spectra of excised lenses. Weale developed an exponential model to account for the effect of age on lens density. Delori and Burns (1996) show that the Weale exponential model may also be characterized as a two-component model with age unvarying and varying components.

Does lens yellowing accelerate at older ages?

With unselected normal populations the decrease in lens transmission with age is gradual and approximately linear between the ages of 20 and 60 years. After age 60 there is evidence that the rate of density increase rises (literature surveyed by Pokorny *et al.*, 1987). The rate after age 60 is dependent on the population under study, with a less dramatic increase in lens ageing in populations screened to eliminate observers with incipient cataracts (Moreland *et al.*, 1991). Accelerated lens ageing occurs earlier in Type I diabetics but, once begun, progresses at a similar rate as occurs in normals over the age of 60 (Lutze and Bresnick, 1991).

The two-component model

Pokorny *et al.* (1987) derived the age-dependent component T_{L1} from differences in lens density between populations of different ages. T_{L1} is independent of the age-invariant lens-density component (T_{L2}). The differences between the Wyszecki and Stiles and van Norren and Vos lens density functions are thus all in the age invariant component T_{L2}. Pokorny *et al.* published T_{L2} spectra for both the Wyszecki and Stiles and van Norren and Vos lens density functions. Table 1 gives the Pokorny *et al.* (1987) model using the van Norren and Vos lens

Table 1. Tabulation of the optical density of the total lens transmission function T_L for an average 32-year-old observer and separation of T_L into components. T_{L1} represents portion affected by ageing after age 20, and T_{L2} represents portion stable after age 20 (After Pokorny *et al.*, 1987).

Wavelength	T_L	T_{L1}	T_{L2}
400	1.933	0.600	1.333
410	1.280	0.510	0.770
420	0.787	0.433	0.354
430	0.493	0.377	0.116
440	0.360	0.327	0.033
450	0.300	0.295	0.005
460	0.267	0.267	–
470	0.233	0.233	–
480	0.207	0.207	–
490	0.187	0.187	–
500	0.167	0.167	–
510	0.147	0.147	–
520	0.133	0.133	–
530	0.120	0.120	–
540	0.107	0.107	–
550	0.093	0.093	–
560	0.080	0.080	–
570	0.067	0.067	–
580	0.053	0.053	–
590	0.040	0.040	–
600	0.033	0.033	–
610	0.027	0.027	–
620	0.020	0.020	–
630	0.013	0.013	–
640	0.007	0.007	–
650	0.000	0.000	–

The optical density of the lens of an average observer between the ages of 20 and 60 years may be estimated by: $T_L = T_{L1}\,[1+0.02(A-32)] + T_{L2}$. For an average observer over the age of 60 $T_L = T_{L1}\,[1.56+0.0667(A-60)] + T_{L2}$, where A is the observer's age.

T_L is the van Norren and Vos (1974) tabulation of lens density scaled by 1.333 to represent a 32-year-old observer (the average age of the Stiles and Burch observers) with a small pupil (<3 mm). To estimate the lens density function for a completely open pupil (>7 mm), multiply the tabulated values by 0.86.

For the Wyszecki and Stiles (1982) lens density function, the following values may be substituted:

	T_L	T_{L1}	T_{L2}
400	1.600	0.600	1.000
410	1.093	0.510	0.583
420	0.733	0.433	0.300

density function as representative of a 32-year-old observer. Though Table 1 lists a density of 0.00 at 650 nm, the literature (van Norren and Vos, 1974) suggests a density value of about 0.15 at long wavelengths.

Figure 1 shows model estimates of average lens density spectra for a 20- and a 70-year-old observer. The young observer's lens is relatively transparent throughout the visible spectrum, with significant density only at wavelengths below about 450 nm. The older observer's lens shows a more gradual increase in lens density as wavelength decreases; relative to long wavelengths, significant density is seen in the middle wavelength region of the spectrum. Figure 2 shows model estimates of the change in lens density as a function of age for wavelengths across the visible spectrum. The graph shows the acceleration in lens ageing beyond the age of 60 and emphasizes that the effect of lens ageing on visual function is highly wavelength dependent.

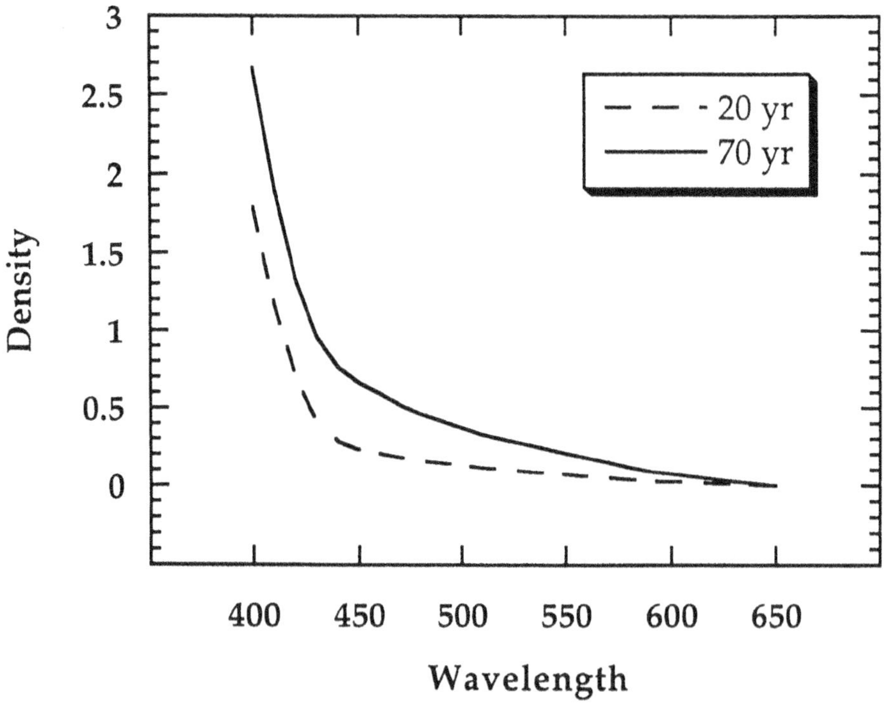

Fig. 1. Spectral lens density for an average 20-year-old and an average 70-year-old individual. Data are based on the van Norren and Vos (1974) spectral lens density function and the Pokorny *et al.* (1987) two-factor lens equation.

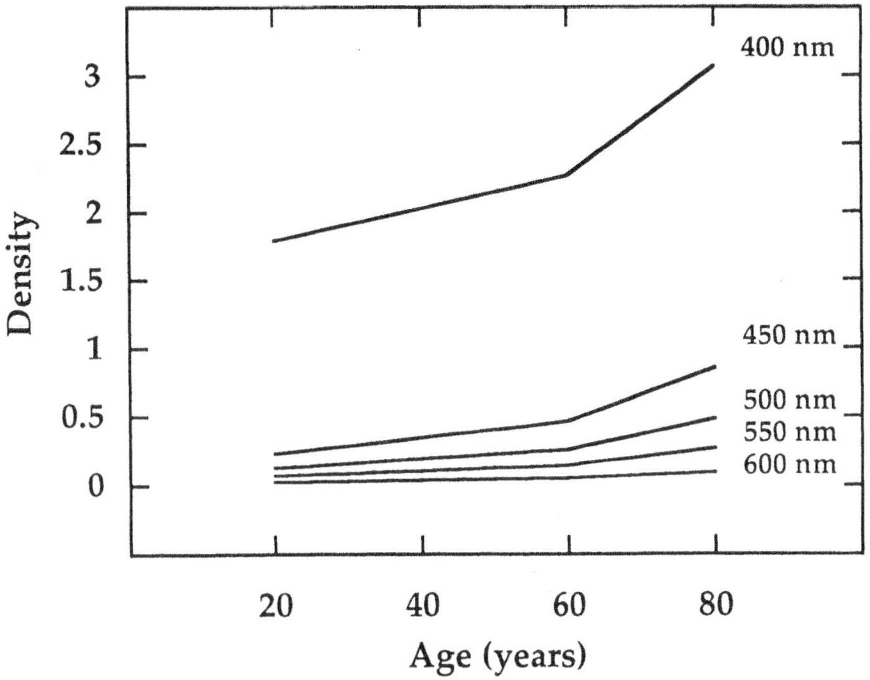

Fig. 2. The lens density as a function of age for five spectral wavelengths. Data sources as in Figure 1.

Intra-individual variation in lens density

Van Norren and Vos (1974) calculated from Crawford's (1949) scotopic spectral sensitivity data (50 observers, 17–30 years old) that about 95% of the individual variation in ocular absorption at short wavelengths is encompassed by about ±25% of the mean spectral density function for a young eye. In association with Jun Xu, a graduate student in our laboratory, we have undertaken a study of lens density to assess the generality of the two-component model in characterizing the lens density spectra of individual observers (Xu *et al.*, in press). We estimated lens density with two techniques, a reflectometric method in which light reflected off the optic disc is measured and a psychophysical method in which a flicker photometric match is used to estimate lens density. In each case we calculate an estimated lens age from the measurements which may be compared with the chronological ages of the observers.

Several reflectometric techniques have been used to measure the transmission spectrum of the human lens (or lens and ocular media). Said and Weale (1959) determined the ratio of radiances for the third and fourth Purkinje images, reflections from the front and rear surfaces of the lens. Johnson ét al. (1993) described an improved version of this method. Delori and Burns (1996) measured reflected light from the fundus at a parafoveal location.

In this study reflectometric measurements from optic disc were accomplished using a Utrecht fundus-reflection densitometer (van Norren and van der Kraats, 1989). The observer's head was stabilized using a bite bar and temple pads. The measuring beam was centered on the pupil and was, with direct visualization of the well-illuminated fundus through the densitometer eyepiece, directed to a region of the disc with a minimum amount of blood vessels. The stimulus field was 2.6° in diameter, and the measurement field (fundus area viewed by the photomultiplier) was 2.1°. Reflectance was measured at 13 wavelengths between 430 and 740 nm. A measurement took 140 sec. We modelled the measured spectrum with the following assumptions: (1) the reflectance spectrum of the optic disc is spectrally non-selective (we have made reflectometric measurements on pseudophakes to confirm this), (2) the hemoglobin reflectance is characterized by 0.05 Hb + 0.95 HbO_2 (spectra from van Assendelft, 1970; oxygenation level from Delori and Pflibsen, 1989), and (3) two-way passage of light through the lens. The model was fit using four parameters: (1) a vertical scaling (overall reflectance) factor, (2) a hemoglobin density, (3) a lens age parameter $2\{(k\,T_{L1}) + T_{L2}\}$, and (4) a factor accounting for superficial reflected light within the measurement field. The model solution for k, the T_{L1} multiplier, was used to calculate the estimated age of the lens. Figure 3 shows examples of reflectance spectra and model fits from the optic discs of 26- and 65-year-old observers. Estimated lens age from the model solutions are shown in parentheses, 23.2 years and 60.1 years. The density spectrum for the younger observer shows a pronounced mid-spectrum hump due to a high amount of absorption by hemoglobin on the optic disk.

In a second procedure we used a psychophysical method similar to that described by Delori *et al.* (1994) and Delori and Burns (1996). A heterochromatic flicker photometric match was made between a 430 nm standard and a 495 nm variable light. The rationale for the chosen wavelengths include: (1) a similar macular pigment density for the two wavelengths, (2) a similar L/M cone ratio (tritan pair) for the two wavelengths, and (3) sufficient wavelength separation to have a sizeable difference in lens absorption. The flicker photometric match was performed for a 10° stimulus field. This large field size was used to decrease any residual effect of macular pigment and to improve match precision. This procedure has advantages over the more common psychophysical procedure for estimating lens density of establishing scotopic equivalence between pairs of wavelengths (Sample *et al.*, 1988; Johnson *et al.*, 1989), because measurements are made with foveal view and no extended period of dark adaptation is needed.

Figure 4 shows estimated lens age as a function of chronological age as derived from the reflectometric and psychophysical data. The regression lines fit to the data both have slopes near 1. On average, the reflectometric regression line underestimates chronological age by about 4 years, the psychophysical

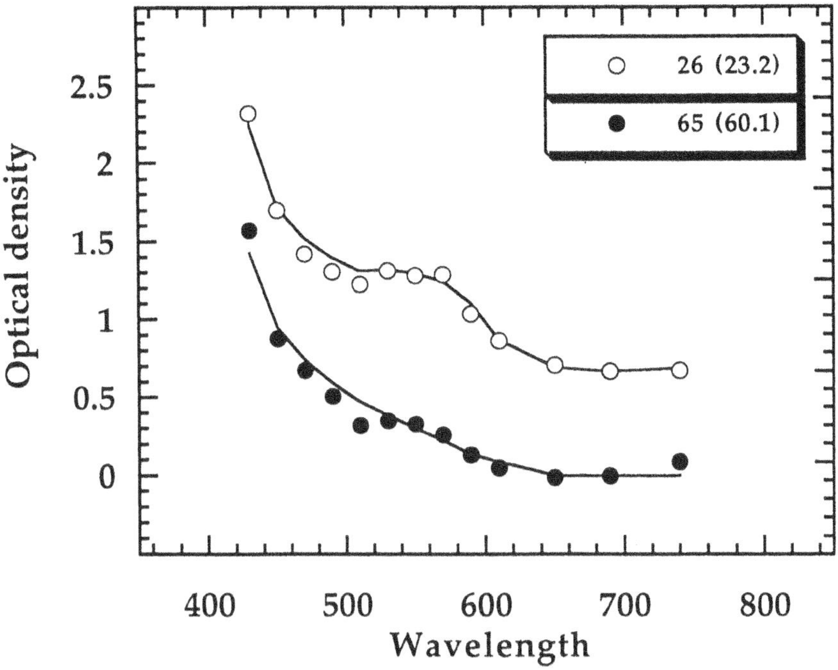

Fig. 3. Optical density as a function of wavelength estimated from retinal reflectometry at the optic disc. Open circles show data for a 26-year-old, closed circles show data for a 65-year-old. The solid lines are estimates of a model incorporating hemoglobin and the lens factors shown in Table 1. For clarity the data and model fit for the 26-year-old observer are shifted upward by 1.0 log unit. The best fit of the two-factor lens equation predicts a lens age of 23.3 for the 26-year-old and of 60.1 for the 65-year-old. The added density at mid-spectrum for the 26 year old demonstrates a high hemoglobin factor. Data and model fits of Xu *et al.* (in press).

estimate is closer to chronological age. The standard deviations of the data from the regression lines are 20% and 18% for the reflectometric and psychophysical measurements. For wavelengths greater than 450 nm the lens density function T_L is comprised solely of T_{L1}. With a 20% standard deviation the 95% confidence limits (2.06 standard deviations for 27 observers) in total lens density are $\pm 41\%$. Given that T_{L1} represents 31% of the density at 400 nm, the 95% confidence limits at 400 nm due to a 20% standard deviation in T_{L1} are $\pm 12.7\%$. The $\pm 25\%$ estimated by van Norren and Vos (1974) is intermediate between what is expected from the two-component model at intermediate and short wavelengths. An additional source of population variation at short wavelengths may be individual differences in the density of T_{L2} (van den Berg and Felius, 1995).

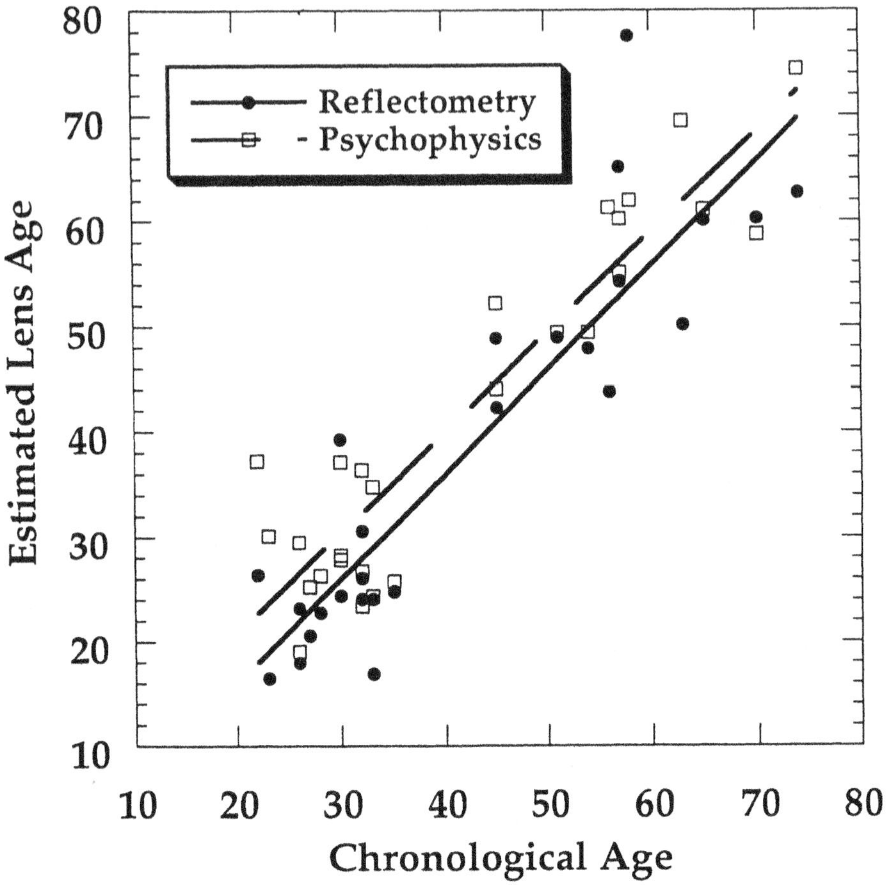

Fig. 4. A scatter plot of the estimated lens age vs the chronological age of 27 observers. Solid circles are estimates based retinal reflectometry at the optic disc. Open squares are estimates based on a psychophysical test. Lines are the correlations. For reflectometry (solid line) $r = 0.90$; for psychophysics (dashed line) $r = 0.93$. The estimates from reflectometry consistently underestimate chronological age by 4 years. Data source as in Figure 3.

Pupil

In a young observer, pupil diameter may vary between about 2 and 8 mm with variation in light level. This results in a 16-fold change in pupil area. Changes in pupil size produce a number of changes in retinal stimulation, which are relevant to visual function. There are changes in retinal illuminance, the ratio of rod/cone stimulation, spectral sensitivity and spatial resolution. This section

Table 2. Factors affecting pupil size.

Stimulus variables
Light level
Spectral composition
Spatial configuration
Field size
Spatial structure of field
Monocular/binocular view
Accommodative state
Non-visual stimuli
Pain
Noise
Observer variables
Individual differences
Age
Day-to-day within observer variance
Biomechanical factors
Respiration
Heart beat
Cognitive factors
Arousal, attention, fright
Workload
Hedonistic content

first reviews the major determinants of pupil size and indicates the limitations of the use of models in estimating the size of the natural pupil in order to calculate retinal illuminance. Then there is a description of the ways in which pupil size can be known including techniques to measure pupil diameter and the use of an artificial pupil.

Determinants of pupil size

Pupil size is affected by diverse extrinsic and intrinsic factors. Table 2 lists some of the major factors which cause alteration in pupil size. The pupil is a dynamic system which changes size on a continual basis. A comprehensive compilation of facts and literature on the pupil may be found in Loewenfeld (1993).

Standard representations of average pupil size

There have been a number of characterizations of how average pupil diameter varies with stimulus luminance, notably those of LeGrand (1948, 1968) and de Groot and Gebhard (1952). The literature studies used in developing these models were all from studies which employed binocular view of large stimulus fields, a stimulus situation that is relatively infrequently encountered in vision research. Trezona (1983) developed a more complex formulation which takes into consideration field diameter.

LeGrand (1948, 1968) proposed a method for calculating 'Effective Trolands', an estimate of retinal illuminance from knowledge of the luminance of a large viewing area. The technique is based upon estimation of pupil size from stimulus luminance and taking account of the loss of photopic efficacy of light entering the eccentric portions of the pupil (the Stiles–Crawford effect). LeGrand used empirical relations between pupil diameter and luminance (Crawford, 1936) and of directional efficiency of light entering the pupil (see also Moon and Spenser, 1944). To estimate pupil diameter from field luminance:

$$d = 5 - 3 \tanh(0.4 \log L) \qquad (1)$$

where d is pupil diameter in mm and L is luminance in cd/m^2.

To estimate retinal illuminance:

$$\text{Effective Trolands} = (d^2/4)\{(1-0.085(d^2/8) + 0.002(d^4/48)\} \qquad (2)$$

Because of the large dynamic range of the human visual system, luminance and retinal illuminance are frequently expressed as the logarithms of the respective units. We derived the following equation to describe the relation between the logarithm of luminance and the logarithm of effective retinal illuminance for the photopic luminance range of the LeGrand tabulation (0.1–500,000 cd/m^2; –1.0 to 5.7 log cd/m^2):

$$\log \text{Effective Trolands} = 1.147 + 0.80738 (\log L) + 0.013181 (\log L)^2 \qquad (3)$$

Spectral determinants of tonic pupil size

Photopic and scotopic contributions vary with field size and light level. For fields $>5°$, pupil size is predominantly determined by the Scotopic Luminous Efficiency Function $V'(\lambda)$ for light levels at which the rods are active (Laurens, 1923; Bouma, 1965; Berman *et al.*, 1992). For fields $<2°$, pupil size is predominantly determined by the Photopic Luminous Efficiency Function $V(\lambda)$ (Alpern and Campbell, 1962; Alexandridis, 1985).

Monocular/binocular view

Literature data indicate that pupil diameter is smaller with binocular view but reports differ on the magnitude of the phenomenon. Reeves (1918) data show the difference in pupil area between binocular and monocular view with the other eye in darkness to be a factor of about 2 while ten Doesschate and Alpern (1967) present data indicating considerable individual differences, with some observers showing factors of about 10.

Age

The pupil gradually becomes smaller with increasing age (age-related miosis). Figure 5 shows pupil diameter as a function of age from a representative study which included 91 observers (Winn *et al.*, 1994). Pupil diameter was found to decrease linearly as a function of age at all luminance levels studied, ranging from 9 to 4400 cd/m^2. Pupil diameter decreased by 0.043 mm per year at the lowest luminance level and by 0.015 mm per year at the highest.

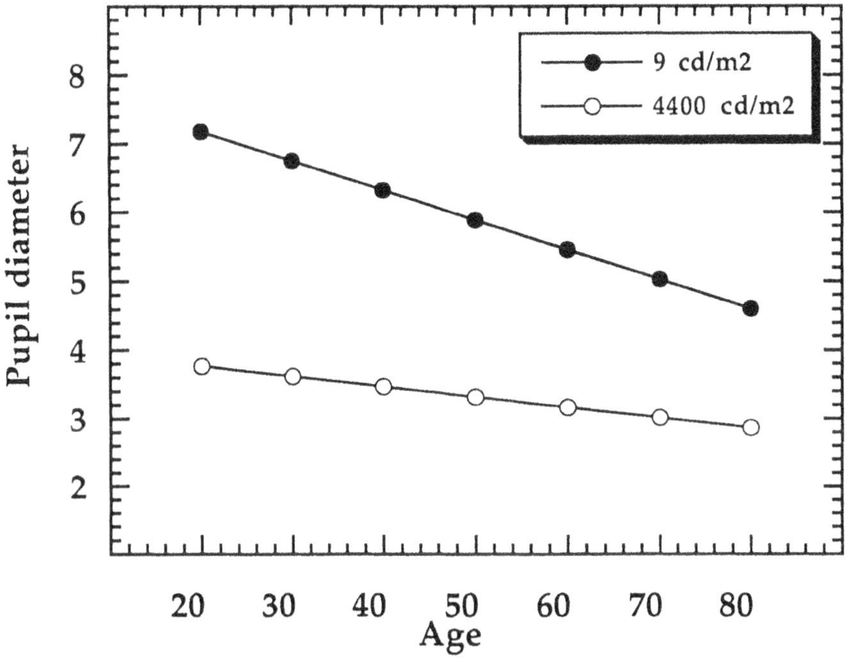

Fig. 5. The pupil diameter as a function of age. The solid circles show measurements at 9 cd/m^2; the open circles show measurements at 4400 cd/m^2. Data from Winn *et al.* (1994).

Individual differences

There are substantial differences in the pupil sizes of observers viewing the same stimulus. Figure 6 shows data for 10 observers (Crawford, 1936). The range of individual differences in pupil size is comparable to other studies in the literature. Table 3 shows published data on the range of pupil size for a

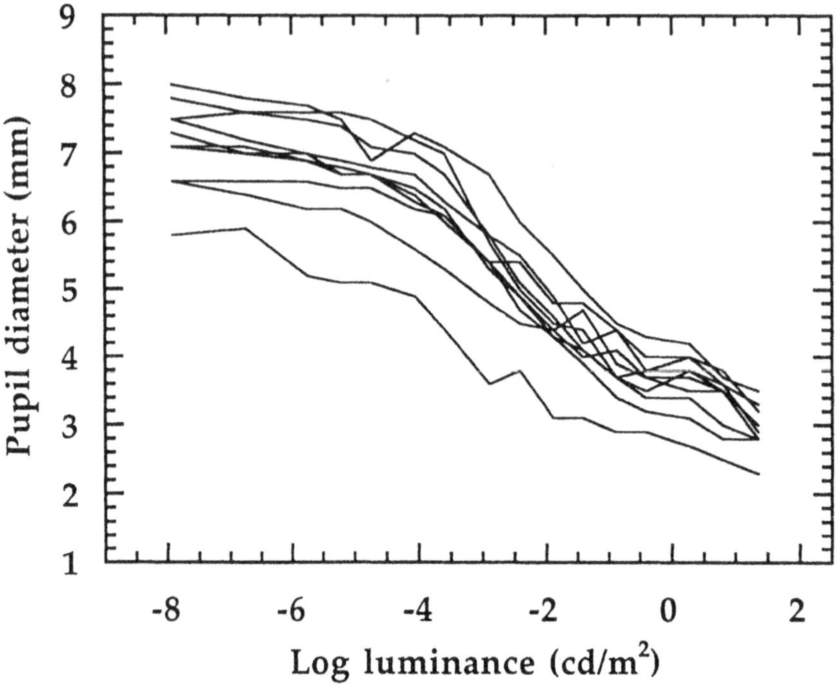

Fig. 6. Pupil diameter as a function of luminance for 10 observers. Data from Crawford (1936), Table 2.

20 cd/m^2 light level, a level typical of color CRT displays. Individual young observers may have pupil area that differ by a factor of 3 or more at this light level. The range of area variation may be even greater in older age populations due to variation in age-related miosis.

Some consequences of variation in pupil size

There are variations in the spectral sensitivity function with changes in pupil size. Pathlength through the lens is shorter near the edge. Further, the central region of the lens, the nucleus, absorbs more than the more peripheral regions (Mellerio, 1987). The pupil diameter scaling factor (1.33) in Table 1 was derived from analysis of the regularity of estimated photopigment spectra at a retinal level (Smith and Pokorny, 1975); a similar estimate (1.28) comes from Stockman *et al.* (1993) in their estimate of the lens density (van Norren and Vos lens density spectrum) of the 1964 CIE Standard Observer. Since the lens preferentially absorbs short wavelength light, the light distribution reaching the retina is

Table 3. Observer variation in pupil size for a typical CRT monitor light level (20 cd/m^2).

Study		No. of observers	Diameter		Area ratio
			Largest	Smallest	Largest/Smallest
~50° field					
Reeves (1918)		6	5.1	3.8	1.8
Crawford (1936)		10	3.6	2.4	2.3
Spring and Stiles (1948)		12	7.0	3.1	5.1
Bouma (1965)		5	7.0	4.6	2.3
Weighted average					3.2
10° field					
Winn *et al.* (1994)	<40 years	~30	8.4	4.8	3.1
	>40 years	~60	7.0	2.3	4.8

a function of pupil size. Figure 7, constructed using the Table 1 template, shows the density difference expected for lights equated at long wavelengths for dilated and constricted pupils.

Changes in pupil size cause shifts in the relative excitation of the rod and cone mechanisms. Cones are directionally sensitive (Stiles–Crawford effect): light that reaches the retina from regions near the edge of the lens is less effective than light coming through the lens center. Rods have a minimal Stiles–Crawford effect, so that rays entering the eye through the peripheral and central parts of the lens are about equally effective. If two lights, one through a constricted and one through a dilated pupil, are equated for cone vision, the light through the dilated pupil will produce about 1.4 times greater stimulation of the rods than the light through the constricted pupil. The luminance dependence of this effect is shown in Figure 8.

Table 4 shows techniques by which pupil size may be known. Artificial pupils (Troland, 1915) are widely used in vision research. An artificial pupil is an aperture smaller that the natural pupil, placed as close as possible to the natural pupil. An alternate technique sometimes used in Maxwellian view systems places a limiting aperture in a plane conjugate to the natural pupil. There are circumstances for which physical and optical artificial pupils are not optically equivalent (Applegate *et al.*, 1992; Jacobs *et al.*, 1992). Artificial pupils may be undesirable for a variety of experimental situations. Included in this group are experiments assessing visual function under 'natural' viewing conditions, experiments requiring binocular view (binocular artificial pupils are difficult to align and require rigid head stabilization) and experiments that must utilize as much of the source luminance as possible. Included in this latter category are experiments in which the stimuli are generated on a CRT monitor.

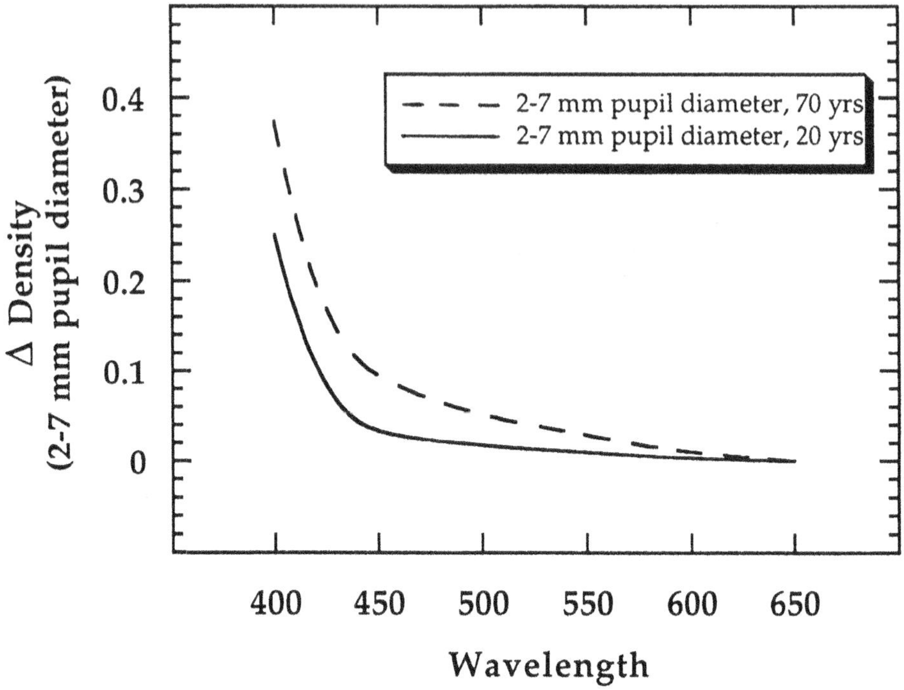

Fig. 7. The change in light level (expressed in density units) as a function of wavelength. The calculations are based on a change in pupil diameter from 2 to 7 mm. The solid line is calculated for a 20-year-old observer; the dashed line is calculated for a 70-year-old observer. Data sources as in Figure 1.

Macular pigment

The macular region of the human retina contains a non-photolabile pigment, which selectively filters light arriving at the receptors. The pigment is most dense in the Henle fiber layer but is also present in the inner plexiform layer in many retinas (Snodderly *et al.*, 1984). Current evidence indicates that two caratenoids contribute to the macular pigment spectrum, zeaxanthin and lutein (Bone *et al.*, 1988; Handelman *et al.*, 1991).

Both the spatial distribution and maximal optical density of macular pigment vary widely among observers, though the macular pigment density is closely similar in the two eyes of a given observer (Hammond and Fuld, 1992). The entoptic appearance of Maxwell's spot has been interpreted as a visualization of the spatial distribution of macular pigment (Palmer, 1978). For many observers Maxwell's spot is seen to be darkest in the centre of the fovea but for others there is considerable spatial structure (Miles, 1954). Moreland and Bhatt (1984) measured the dependence of macular pigment density on eccentricity and also

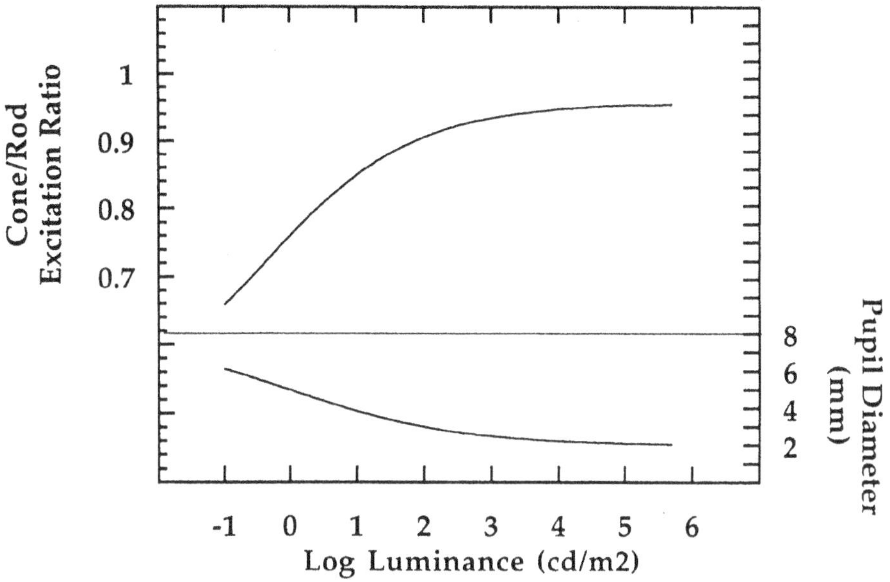

Fig. 8. The cone/rod excitation ratio (left vertical scale) and pupil diameter (right vertical scale) plotted as a function of luminance level. This plot takes into account the reduced efficiency of cone photoreceptors with pupil diameter, due to the Stiles–Crawford effect. Calculations based on LeGrand (1968, p. 106).

summarized the results of previous studies (Ruddock, 1963; Stabell and Stabell, 1980; Viénot, 1983). The range of density variation in the population has been estimated at 8-fold (Werner *et al.*, 1987; Hammond and Fuld, 1992) to 10-fold (Ruddock, 1965; Bone and Sparrock, 1971; Pease *et al.*, 1987). For a 2° field size, individuals may vary from 0.0 to >1.0 optical density at 460 nm.

Wyszecki and Stiles (1967, 1982) presented a density spectrum for macular pigment. Vos (1972) derived a macular pigment density spectrum from more extensive published data and an analysis of the differences between the 10° CIE and the Judd 2° Standard Observer color matching functions. Wyszecki and Stiles (1967, 1982) suggested a mean peak spectral density of 0.5 (for an unspecified foveal field diameter). Vos (1972) proposed a spectral density function for average macular pigment with a peak density near 0.35 for a 2° field; his analysis is based on a survey of a large number of studies all of which employed psychophysical procedures to estimate the macular pigment spectrum. Figure 9 shows the Wyszecki and Stiles and the Vos spectra scaled to have equal area. The principal differences are in the magnitude of the long wavelength hump and the rate of decrease in density at short wavelengths. Spectra from microspectrophotometry of unfixed tissue (Snodderly *et al.*, 1984b) and retinal dichroism (Bone *et al.*, 1992) are similar to the Vos spectrum in showing a shallow long wavelength shoulder.

Table 4. How can pupil size be known?

Control of pupil size

- Artificial pupil (an aperture smaller than the natural pupil placed either before the eye or in a plane conjugate to the eye pupil)
  - Advantages
    - Allows precise calculation of retinal illuminance
    - Can improve spatial resolution by a choice of diameter which minimizes ocular optical aberrations
  - Disadvantages
    - Decreases retinal illuminance relative to viewing with the natural pupil
    - Requires precise location of eye
      - Chin rest; mouth bite
- Mydriasis or miosis (drug-induced dilation or constriction)
  - Advantages
    - Allows free viewing
    - Mydriasis yields maximal retinal illumination
  - Disadvantages
    - Not the same pupillary diameter for different observers nor for the same observer on different days
    - Discomfort of the observer
    - Administration must be under medical supervision

Measurement of pupil size

- Advantages
  - Experiment may be under 'natural viewing' conditions
  - Allows maximal retinal illuminance
- Disadvantages
  - Experimenter measures rather than specifies retinal illuminance. Repeat trials not necessarily at same retinal illuminance level
- Strategies for measuring pupil size
  - Direct comparison with a scale
    - Advantages
      - Inexpensive instrumentation (plastic ruler)
      - Easy to perform
    - Disadvantages
      - Measurement interferes with steady state pupil size
      - Single trial method, continuous recording not possible
  - Entopic method (visualize a shadow of own pupil from near point source)
    - Advantages
      - Inexpensive instrumentation (flashlight bulb and pinhole)
      - Easy to perform
    - Disadvantages
      - Measurement interferes with steady state pupil size
      - Single trial method, continuous recording not possible
  - Infrared pupillometry
    - Photographic
      - No on-line information
      - Analysis cumbersome, requires frame-by-frame measurement
    - Non-imaging (photoelastic)
      - Absolute calibration difficult
      - Requires calibration for each observer
    - Video camera with VCR and/or computer
      - On-line analysis systems are expensive

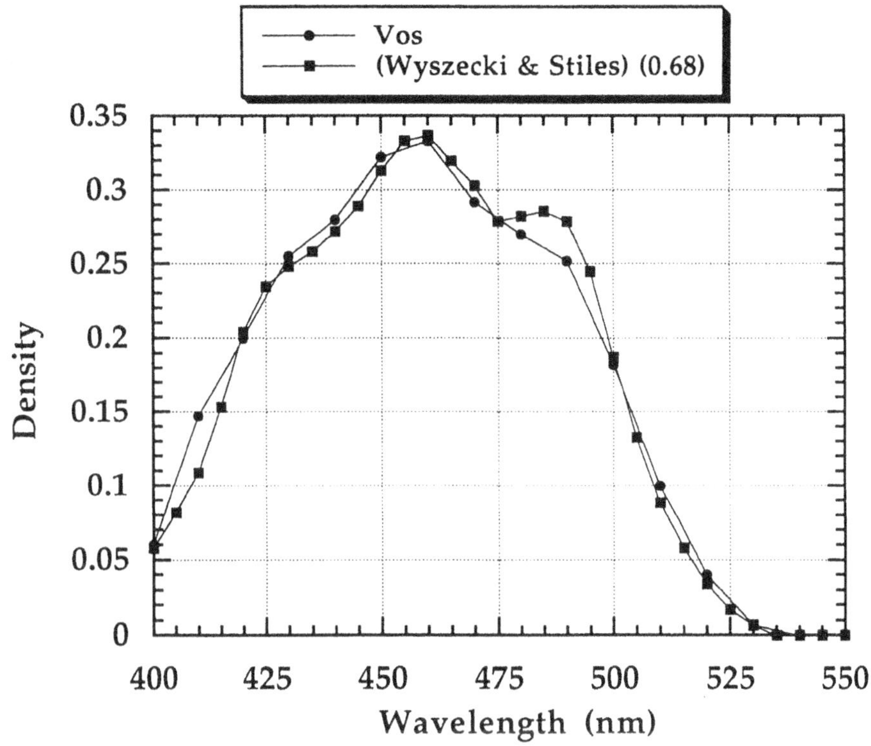

Fig. 9. The optical density of macular pigment as a function of wavelength. Data estimates of Vos (closed circles) and Wyszecki and Stiles (closed squares) are shown. The two data sets have been normalized to equal area. The functions differ in the short wavelength region (400–420 nm) and in the shoulder at 485 nm.

Discussion

As outlined above, there is considerable inter-individual variation in the attenuating characteristics of the lens, pupil and macular pigment. The spectral specificity of lens and macular pigment is largely confined to shorter wavelengths. Variation in pupil size causes large variation in retinal illuminance but also has a concomitant spectral component. How can we determine when information of prereceptoral attenuation of individual observers is needed? This is task dependent. If a measurement is confined to wavelengths longer than 530 nm, then neither lens (Fig. 2) nor macular pigment (Fig. 8) variation will impair interpretation of data from many experiments. Some frequently employed paradigms are quite insensitive to prereceptoral factors. Interpretation of data used for the determination of the presence or absence of a receptor type using either a color matching or an increment threshold technique with selective chromatic adaptation will not be affected by prereceptoral factors. However, if color matching is used to obtain precise information concerning receptoral spectral sensitivity, techniques which circumvent prereceptoral filtering are needed (He and Shevell, 1994). Likewise, prereceptoral factors will not influence measurements of increment thresholds at retinal illuminances which produce Weber behavior with homochromatic stimuli (i.e. when both adapting and test wavelength is the same). This is also true for experimental paradigms in which little change in spectral composition is required to reach threshold, for example studies of normal chromatic discrimination at moderate to high light levels (Smith and Pokorny, 1995). Alternatively, increment thresholds at dim light levels where the visual system behaves in a linear fashion will be modified by prereceptoral attenuation. Interpretation of sensitivity measurements with heterochromatic adaptation and test fields also needs to take into account prereceptoral factors.

A few guidelines may be noted. If spectral equivalence is needed but precise retinal illuminance is not, calibrate the observer for the stimuli used in the experiment with heterochromatic photometry and the establishment of tritan pairs. Knowledge of pupil size is needed to estimate retinal illuminance accurately. If lens density estimates are made, they should be for the pupil size of the experiment. Estimates of macular pigment density should be for the field size of the experiment, preferably using the methodology of the experiment.

Acknowledgements

Preparation of this paper was supported in part by USPHS NIH NEI grant EY00901 and by an unrestricted grant to the Department of Ophthalmology and Visual Sciences from Research to Prevent Blindness. We thank Jack Moreland and Steven Shevell for comments on a draft of the manuscript.

References

Alexandridis, E. (1985). The Pupil (Telger, T., transl.). Springer, New York.

Alpern, M. and Campbell, F.W. (1962). The spectral sensitivity of the consensual light reflex. J. Physiol. (Lond.) 164: 478–507.

Applegate, R.A., Bradley, A. and Thibos, L.N. (1992). Visual acuity and pupil size in Maxwellian and free view systems with and without refractive error. Adv. Color Vision Tech. Dig. 4: 170–174.

Bando, M., Nakajima, A. and Satoh, K. (1981). Spectrophotometric estimation of 3-OH L-kynurenine O-beta-glucoside in the human lens. J. Biochem. 89: 103–109.

Berman, S.M., Fein, G., Jewett, D.L., Saika, G. and Ashford, E. (1992). Spectral determinants of steady-state pupil size with full field of view. J. Illum. Eng. Soc. 21: 3–13.

Boettner, E A. and Wolter, J. (1962). Transmission of the ocular media. Invest. Ophthal. Vis. Sci. 1: 776–83.
Bone, R.A., Landrum, J.T. and Cains, A. (1992). Optical density spectra of the macular pigment in vivo and in vitro. Vision Res. 32: 105–110.
Bone, R.A., Landrum, J.T., Fernandez, L. and Tarsis, S.L. (1988). Analysis of the macular pigment by HPLC: retinal distribution and age study. Invest. Ophthal. Vis. Sci. 29: 843–849.
Bone, R.A. and Sparrock, J.M.B. (1971). Comparison of macular pigment densities in human eyes. Vision Res. 11: 1057–1064.
Bouma, H. (1965). Receptive systems mediating certain light reactions of the pupil of the human eye. Philips Research Reports Suppl. No. 5: 1–167.
Cooper, G.F. and Robson, J.G. (1969). The yellow colour of the lens of man and other primates. J. Physiol. (Lond.) 203: 411–417.
Crawford, B.H. (1936). The dependence of pupil size upon external light stimulus under static and variable conditions. Proc. Roy. Soc. B (Lond.) 121(B): 376–395.
Crawford, B.H. (1949). The scotopic visibility function. Proc. Phys. Soc. B 62: 321–334.
de Groot, S.G. and Gebhard, J.W. (1952). Pupil size as determined by adapting luminance. J. Opt. Soc. Am. 42: 492–495.
Delori, F., Staurenghi, G., Arend, O., Goger, D. and Weiter, J. (1994). Estimates of ocular media absorption from fundus reflectometry. Vision Science and Its Applications, Tech. Dig. Series 2: 220–223. Optical Society of America, Washington, D.C.
Delori, F.C. and Burns, S.A. (1996). Fundus reflectance and the measurement of crystalline lens density. J. Opt. Soc. Am. A13: 215–226.
Delori, F.C. and Pflibsen, K.P. (1989). Spectral reflectance of the human ocular fundus. Appl. Opt. 28: 1061–77.
Hammond, B.R.J.-r. and Fuld, K. (1992). Interocular differences in macular pigment density. Invest. Ophthal. Vis. Sci. 33: 350–355.
Handelman, G.J., Snodderly, D.M., Krinsky, N.I., Russett, M.D. and Adler, A. (1991). Biological control of primate macular pigment. Biochemical and densitometric studies. Invest. Ophthal. Vis. Sci. 32: 257–267.
He, J.C. and Shevell, S.K. (1994). Individual differences in cone photopigments of normal trichromats measured by dual Rayleigh-type color matches. Vision Res. 34: 367–376.
Jacobs, R.J., Bailey, I.L. and Bullimore, M.A. (1992). Artificial pupils and Maxwellian view. Appl. Opt. 31: 3668–3677.
Johnson, C.A., Adams, A.J. and Lewis, R.A. (1989). Evidence for a neural basis of age-related visual field loss in normal observers. Invest. Ophthal. Vis. Sci. 30: 2056–2064.
Johnson, C.A., Howard, D.L., Marshall, D. and Shu, H. (1993). A noninvasive video-based method for measuring lens transmission properties of the human eye. Optom. Vis. Sci. 70: 944–955.
Laurens, H. (1923). The relative physiological value of spectral lights. III. The pupilomotor effects of wavelengths of equal energy content. Am. J. Physiol. 64: 97–119.
LeGrand, Y. (1948). Optique Physiologique. Tome 2. Lumiére et Couleurs. Editions de la Rev. d'opt., Paris.
LeGrand, Y. (1968). Light, Colour and Vision. Second Edition. Chapman and Hall, London.
Lerman, S. and Borkman, R. (1976). Spectroscopic evaluation and classification of the normal, aging, and cataractous lens. Ophthal. Res. 8: 335–353.
Loewenfeld, I.E. (1993). The Pupil: Anatomy, Physiology, and Clinical Applications. (Vol. 1 & 2). Iowa State University Press, Ames, IA.
Lutze, M. and Bresnick, G.H. (1991). Lenses of diabetic patients 'yellow' at an accelerated rate similar to older normals. Invest. Ophthal. Vis. Sci. 32: 194–199.
Mellerio, J. (1987). Yellowing of the human lens: nuclear and cortical contributions. Vision Res. 27: 1581–1587.
Miles, W.R. (1954). Comparison of the functional and structural areas in human fovea. I. Method of entoptic plotting. J. Neurophysiol. 17: 22–38.

Moon, P. and Spenser, D.E. (1944). Visual data applied to lighting design. J. Opt. Soc. Am., 34: 605–617.
Moreland, J.D. and Bhatt, P. (1984). Retinal distribution of macular pigment. In: Verriest, G. (ed.), Colour Vision Deficiencies VII, Doc. Ophthal. Proc. Ser., 39: 127–132. Kluwer, Dordrecht.
Moreland, J.D., Torczynski, E. and Tripathi, R. (1991). Rayleigh and Moreland matches in the ageing eye. In: Drum, B., Moreland, J. and Serra, A. (eds.), Color Vision Deficiencies X, Doc. Ophthal. Proc. Ser., 54: 347–352. Kluwer, Dordrecht
Palmer, D.A. (1978). Maxwell spot and additivity in tetrachromatic matches. J. Opt. Soc. Am. 68: 1501–1505.
Pease, P.L., Adams, A.J. and Nuccio, E. (1987). Optical density of human macular pigment. Vision Res. 27: 705–710.
Pokorny, J., Smith, V.C., and Lutze, M. (1987). Aging of the human lens. Appl. Opt., 26: 1437–1440.
Reeves, P. (1918). The visibility of radiation. Tran. Illum. Eng. Soc. 13: 101–109.
Ruddock, K.H. (1963). Evidence for macular pigmentation from colour matching data. Vision Res. 3: 417–429.
Ruddock, K.H. (1965). The effect of age upon colour vision. II. Changes with age in light transmission of the ocular media. Vision Res. 5: 47–58.
Said, F.S. and Weale, R.A. (1959). The variation with age of the spectral transmissivity of the living human crystalline lens. Gerontologia, 3: 213–231.
Sample, P.A., Esterson, F.D., Weintraub, R.N. and Boynton, R.M. (1988). The aging lens: In vivo assessment of light absorption in 84 human eyes. Invest. Ophthal. Vis. Sci. 29: 1306–1311.
Savage, G.L., Haegerstrom-Portnoy, G., Adams, A.J. and Hewlet, S.E. (1993). Age changes in the optical density of human ocular media. Clin. Vis. Sci. 8: 97–108.
Smith, V.C. and Pokorny, J. (1975). Spectral sensitivity of the foveal cone photopigments between 400 and 500 nm. Vision Res. 15: 161–171.
Smith, V.C. and Pokorny, J. (1995). Chromatic discrimination axes, CRT phosphor spectra and individual variation in color vision. J. Opt. Soc. Am. A12: 27–35.
Snodderly, D.M., Auran, J.D. and Delori, F.C. (1984). The macular pigment. II. Spatial distribution in primate retinas. Invest. Ophthal. Vis. Sci. 25: 674–685.
Snodderly, D.M., Brown, P.K., Delori, F.C. and Auran, J.D. (1984). The macular pigment. I. Absorbance spectra, localization, and discrimination from other yellow pigments in primate retinas. Invest. Ophthal. Vis. Sci. 25: 660–673.
Spector, A., Roy, D. and Stauffer, J. (1975). Isolation and characterization of an age-dependent polypeptide from human lens with non-tryptophan fluorescence. Exp. Eye Res. 21: 9–24.
Spring, K.H. and Stiles, W.S. (1948). Variation of pupil size with change in the angle at which the light stimulus strikes the retina. Br. J. Ophthalmol. 32: 340–346.
Stabell, U. and Stabell, B. (1980). Variation in density of macular pigmentation and in short-wave cone sensitivity with eccentricity. J. Opt. Soc. Am. 70: 706–711.
Stockman, A., MacLeod, D.M. and Johnson, N.E. (1993). Spectral sensitivities of the human cones. J. Opt. Soc. Am. A10: 2491–2521.
ten Doesschate, J. and Alpern, M. (1967). Effect of photoexcitation of the two retinas on pupil size. J. Neurophysiol. 30: 562–576.
Trezona, P.W. (1983). Luminance level conversions to assist lighting engineers to use fundamental visual data. Lighting Res. Technol. 15: 83–88.
Troland, L.T. (1915). The theory and practise of the artificial pupil. Psych. Rev. 22: 167–176.
van Assendelft, O.W. (1970). Spectroscopy of Hemoglobin Derivatives. C.C. Thomas, Springfield, IL.
van den Berg, T.J. and Tan, K.E. (1994). Light transmittance of the human cornea from 320 to 700 nm for different ages. Vision Res. 34: 1453–1456.
van den Berg, T.J.T.P. (1993). Quantal and visual efficiency of fluorescence in the lens of the human eye. Invest. Ophthal. Vis. Sci. 34: 3566–3573.
van den Berg, T.J.T.P. and Felius, J. (1995). Relationship between spectral transmittance and slit lamp color of human lenses. Invest. Ophthal. Vis. Sci. 36: 322–329.

van Heyningen, R. (1971). Fluorescent glucoside in the human lens. Nature 230: 393–394.
van Norren, D. and van der Kraats, J. (1989). Retinal densitometer with the size of a fundus camera. Vision Res. 29: 369–374.
van Norren, D. and Vos, J.J. (1974). Spectral transmission of the human ocular media. Vision Res. 14: 1237–1244.
Viénot, F. (1983). Can variation in macular pigment account for the variation of colour matches with retinal position? In Mollon, J.D. and Sharpe, L.T. (eds.), Colour Vision: 107–116. Academic Press, New York.
Vos, J.J. (1972). Literature review of human macular absorption in the visible and its consequences for the cone receptor primaries (2F1972-17). TNO Report, Institute for Perception, Soesterberg, The Netherlands.
Weale, R.A. (1988). Age and the transmittance of the human crystalline lens. J. Physiol. (Lond.) 395: 577–587.
Werner, J.S., Donnelly, S.K. and Kliegl, R. (1987). Aging and human macular pigment density. Vision Res. 27: 257–268.
Winn, B., Whitaker, D., Elliott, D.B. and Phillips, N.J. (1994). Factors affecting light-adapted pupil size in normal human subjects. Invest. Ophthal. Vis. Sci. 35: 1132–1137.
Wyszecki, G. and Stiles, W. (1967). Color Science – Concepts and Methods, Quantitative Data and Formulas. Wiley, New York.
Wyszecki, G. and Stiles, W.S. (1982). Color Science – Concepts and Methods, Quantitative Data and Formulae. 2nd Edition. Wiley, New York.
Xu, J., Pokorny, J. and Smith, V.C. (in press). Optical density of the human lens. J. Opt. Soc. Am. A.

Visual Sciences Section
University of Chicago
939 East 57th Street
Chicago, IL 60637, U.S.A.

Author Index

C.R. Cavonius (ed.), Colour Vision Deficiencies XIII, pp. 513–514.

Documenta Ophthalmologica Proceedings Series

1. J. François (eds.): *Symposium on Light-Coagulation.* Argon Laser and Xenon Arc (Ghent, Belgium, 1972). 1973 ISBN 90-6193-141-X
2. J.T. Pearlman (ed.): *10th ISCERG Symposium* (Los Angeles, Calif., USA, 1972). 1973 ISBN 90-6193-142-8
3. H.E. Henkes (ed.): *Photography, Electro-Ophthalmology and Echo-Ophthalmology in Ophthalmic Practice.* 1973 ISBN 90-6193-143-6
4. E. Dodt & J.T. Pearlman (eds.): *11th ISCERG Symposium* (Bad Nauheim, Germany, 1973). 1974 ISBN 90-6193-144-4
5. W.J. Holmes (ed.): *Public Health Ophthalmology.* Papers Presented at the Conference on the Prevention of Impaired Vision and Blindness (Paris, France, 1974). 1975 ISBN 90-6193-145-2
6. A. Th. M. van Balen (ed.): *First International Symposium on Artificial Lensimplantation* (Utrecht, The Netherlands, 1974). 1975 ISBN 90-6193-146-0
7. A.F. Deutman (ed.): [Symposium on] *New Developments in Ophthalmology.* (Nijmegen, The Netherlands, 1975) 1976 ISBN 90-6193-147-9
8. O. Hockwin (ed.): *Progress of Lens Biochemistry Research.* In Honour of Prof. Dr. med. J. Nordmann. 1976 ISBN 90-6193-148-7
9. J.J. de Laey (ed.): *International Symposium on Fluorescein Angiography* (Ghent, Belgium, 1976). 1976 ISBN 90-6193-149-5
10. R. Alfieri & P. Solé (eds.): *12th ISCERG Symposium* (Clermont-Ferrand, France, 1974). 1976 ISBN 90-6193-150-9
11. E. Auerbach (ed.): *Experimental and Clinical Amblyopia. 13th ISCERG Symposium* (Kibbutz Ginossar, Israel, 1975). 1977 ISBN 90-6193-151-7
12. E.L. Greve (ed.): *Symposium on Medical Therapy in Glaucoma* (Amsterdam, The Netherlands, 1976) 1977 ISBN 90-6193-152-5
13. T. Lawwill (ed.): *ERG, VER and Psychophysics. 14th ISCERG Symposium* (Louisville, USA, 1976). 1977 ISBN 90-6193-153-3
14. E.L. Greve (ed.): *2nd International Visual Field Symposium* (Tübingen, Germany, 1976). 1977 ISBN 90-6391-154-1
15. J. François & A. De Rouck (eds.), J.T. Pearlman and J. Kelsey (co-eds.): *Electrodiagnosis, Toxic Agents and Vision. 15th ISCEV Symposium* (Ghent, Belgium, 1977). 1978 ISBN 90-6193-155-X
16. H.-J Merté (ed.): *Genesis of Glaucoma.* Contributions of the Wessely Symposium in Munich (October 1974) with final Considerations by H. Goldmann. 1978 ISBN 90-6193-156-8
17. A.F. Deutmann & J.R.M. Cruysberg (eds.): *5th International Congress on Neurogenetics and Neuro- Ophthalmology* (Nijmegen, The Netherlands, 1977). 1978 ISBN 90-6193-159-2
18. O. Hockwin & W.B. Rathbun (eds.): *Progress in Anterior Eye Segment. Research and Practice.* In Honour of Prof. J.E. Harris. 1979 ISBN 90-6193-158-4
19. E.L. Greve (ed.): *3rd International Visual Field Symposium* (Tokyo, Japan, 1978). 1979 ISBN 90-6193-160-6
20. J. François, S.I. Brown & M. Itoi (eds.): *Proceedings of the Symposium of the International Society for Corneal Research* (Kyoto, Japan, 1978). 1979 ISBN 90-6193-157-6

Documenta Ophthalmologica Proceedings Series

21. J. François, E. Maumenee & I. Esente (eds.): *First International Congress on Cataract Surgery* (Florence, Italy, 1978). 1979 ISBN 90-6193-162-2
22. E.L. Greve (ed.): *Glaucoma Symposium.* Diagnosis and Therapy (Amsterdam, The Netherlands, 1979). 1980 ISBN 90-6193-164-9
23. E. Schmöger & J.H. Kelsey (eds.): *Visual Electrodiagnosis in Systemic Diseases.* Proceedings of the 17th ISCEV Symposium (Erfurt, GDR, 1979) 1980 ISBN 90-6193-163-0
24. A. Hamburg (ed.): *Symposium on Uveal Melanomas.* On the Occasion of the Snellen Medal Presentation to Dr W.A. Manschot (Utrecht, The Netherlands, 1979). 1980 ISBN 90-6193-722-1
25. H. Zauberman (ed.): *Proceedings of the Conference on Subretinal Space* (Jerusalem, Israel, 1979). 1981 ISBN 90-6193-721-3
26. E.L. Greve & G. Verriest (eds.): *4th International Visual Field Symposium* (Bristol, UK, 1980). 1981 ISBN 90-6193-165-7
27. H. Spekreijse & P.A. Apkarian: *Visual Pathways: Electrophysiology and Pathology.* 18th ISCEV Symposium (Amsterdam, The Netherlands, 1980). 1981 ISBN 90-6193-723-X
28. H.C. Fledelius, P.H. Alsbirk & E. Goldschmidt (eds.): *3rd International Conference on Myopia* (Copenhagen, Denmark, 1980). 1981 ISBN 90-6193-725-6
29. J.M. Thijssen & A.M. Verbeek (eds.): *Ultrasonography in Ophthalmology.* Proceedings of the 8th SIDUO Congress (Nijmegen, The Netherlands). 1981 ISBN 90-6193-724-8
30. L. Maffei (ed.): *Pathophysiology of the Visual System.* Proceedings of a Workshop (Pisa, Italy 1980). 1981 ISBN 90-6193-726-4
31. G. Niemeyer & Ch. Huber (eds.): *Techniques in Clinical Electrophysiology of Vision.* 19th ISCEV Symposium (Horgen-Zürich, Switzerland, 1981). 1982 ISBN 90-6193-727-7
32. A.Th.M. van Balen & W.A. Houtman (eds.): *Strabismus Symposium* (Amsterdam, The Netherlands, 1981). 1982 ISBN 90-6193-728-0
33. G. Verriest (ed.): *Colour Vision Deficiencies VI.* Proceedings of the 6th Symposium of the International Research Group on Colour Vision Deficiencies (Berlin- Steglitz, Germany, 1981). 1982 ISBN 90-6193-729-9
34. A. Roucoux & M. Crommelinck (eds.): *Physiological and Pathological Aspects of Eye Movements.* Proceedings of a Workshop (Pont d'Oye Castle, Habay-la-Neuve, Belgium, 1982). 1982 ISBN 90-6193-730-2
35. E.L. Greve & A. Heijl (eds.): *5th International Visual Field Symposium* (Sacramento, Calif., USA, 1982). 1983 ISBN 90-6193-731-0
36. R. Birngruber & V.-P. Gabel (eds.): *Laser Treatment and Photocoagulation of the Eye.* Proceedings of an International Symposium (Munich, Germany, 1982). 1984 ISBN 90-6193-732-9
37. H.E.J.W. Kolder (ed.): *Slow Potentials and Microprocessor Applications.* 20th ISCEV Symposium (Iowa City, USA, 1982). 1983 ISBN 90-6193-733-7

GPSR Compliance

The European Union's (EU) General Product Safety Regulation (GPSR) is a set of rules that requires consumer products to be safe and our obligations to ensure this.

If you have any concerns about our products, you can contact us on ProductSafety@springernature.com

In case Publisher is established outside the EU, the EU authorized representative is:

Springer Nature Customer Service Center GmbH
Europaplatz 3
69115 Heidelberg, Germany

Batch number: 10755461

Printed by Printforce, the Netherlands